# Mathematics for Physical Science

# Mathematics for Physical Science

Editor: Lucy Flynn

NY RESEARCH PRESS
New York

Published by NY Research Press
118-35 Queens Blvd., Suite 400,
Forest Hills, NY 11375, USA
www.nyresearchpress.com

Mathematics for Physical Science
Edited by Lucy Flynn



International Standard Book Number: 978-1-63238-646-5 (Hardback)



**Cataloging-in-Publication Data**

Mathematics for physical science / edited by Lucy Flynn.
p. cm.
Includes bibliographical references and index.
ISBN 978-1-63238-646-5
1. Mathematical physics. 2. Physics. 3. Mathematics. I. Flynn, Lucy.
QC23.2 .M38 2019
530.15--dc23

# Contents

# Preface

Mathematical physics is an interdisciplinary science of physics and mathematics that is concerned with the formulation of mathematical models and their application to solve problems in physics. Mathematical models such as Lagrangian and Hamiltonian formulations have applications in classical mechanics. The field also uses partial differential equations, quantum theory, statistical mechanics and relativity to solve complex problems. This book is a compilation of topics that discuss the most relevant methods and principles of mathematics that are applied to the solution of physical problems. It further comprises of researches done by experts across the globe to advance the theoretical and applied aspects of mathematical physics. This book is aimed at physicists, mathematicians and students in this domain.

Significant researches are present in this book. Intensive efforts have been employed by authors to make this book an outstanding discourse. This book contains the enlightening chapters which have been written on the basis of significant researches done by the experts.

Finally, I would also like to thank all the members involved in this book for being a team and meeting all the deadlines for the submission of their respective works. I would also like to thank my friends and family for being supportive in my efforts.

**Editor**

# A Modified $N$=2 Extended Supersymmetry

**Djeghloul N* and Tahiri M**

*Laboratory of Theoretical Physics of Oran (LPTO), University of Oran, Algeria*

**Abstract**

A modification of the usual extended $N$=2 super symmetry algebra implementing the two dimensional permutation group is performed. It is shown that one can found a multiplet that forms an off-shell realization of this alternative extension of standard super symmetry.

**Keywords:** Extended $N$=2 Super symmetry; Non-linear algebra; Off-shell super multiplets.

## Introduction

The present paper deals with the possibility of modifying the usual extended $N$=2super Poincare algebra via a suitable implementation of the symmetric group $S_2$. This construction has the interesting advantage that one can find a multiplet that realizes the modified extended super symmetry algebra off-hell. This multiplet involves the same fields as those of the standard double tensor multiplet [1] (see also [2] for explicit construction). It is worth mentioning that this construction is not a standard extension of super symmetry in the sense that it relies on a non-local invariance represented by the symmetric group $S_2$.This is what the term *modified* underlies. The obtained transformations still transform bosons in fermions and vice versa.

We will first show that a suitable modification of the $N$=2 super symmetry algebra is possible in the context of nonlinear extension of standard Lie algebras [3]. In this context, we introduce the symmetric group $S_2$within the standard extended $N$=2 super Poincare algebra [4]. This is what is explicitly performed in section two.

In section three, we show that the multiplet containing two Weyl fermions, two real scalar fields and two 2-form gauge potentials is an off-shell multiplet of the modified super algebra so that no auxiliary fields are needed. Finally we show that the construction of a nilpotent Becci-Rouet-Stora-Tyutin BRST operator can be considered.

## Modified $N$=2 Super Symmetry Algebra

The possibility of modifying the extended $N$=2 super Poincare algebra is based on the observation that the free Lagrangian density and thus the free action of two scalar fields $\varphi_i$($i$=1, 2) or two spinor fields $\psi^a$ ($a$=1, 2) (here after a repeated index means a summation, indices a are never lowered and indices $i$ are never raised)

$$L\varphi = -\partial_\mu \varphi_i^* \partial^\mu \varphi_i \qquad (1)$$

$$L\psi = -i\overline{\psi}^a \overline{\sigma}^\mu \partial_\mu \psi^a, \qquad (2)$$

It is manifestly invariant under a permutation operation 1⇔ 2. So the symmetric group $S_2$ [5] defines a discrete symmetry of these models. The action of the identity operator $s^1$ and the transposition operator $S_2$ can be written as

$$\begin{aligned} s^1\varphi_i &= \delta_{ik}\varphi_k, s^1\psi^a = \delta^{ab}\psi^b, \\ s^2\varphi_i &= \eta_{ik}\varphi_k, s^2\psi^a = \eta^{ab}\psi^b, \end{aligned} \qquad (3)$$

Where $\delta^{ab}$={1 for $a$=$b$ and 0 for $a$≠$b$}, $\delta^{ik}$={1 for $i$=$k$ and 0 for $i$≠$k$}, $\eta^{ab}$={0 for $a$=$b$ and 1 for $a$≠$b$} and $\eta_{ik}$={1 for $i$=$k$ and 0 for $i$≠$k$}.

Furthermore, we define the modified translation operator $P_\mu^a$ as a successive application of a permutation operator $s^a$ defined by (3) and the four dimensional translation operator $P_\mu$

$$P_\mu^a = s^a P_\mu, \ a = 1,2 \text{ and } \mu = 0,1,2,3 \qquad (4)$$

$P_\mu^1$ is the usual translation (since $s^1$ is just the identity) while $P_\mu^2$ is the combination of a translation and the transposition operator. The action of $P_\mu^2$ is then given by

$$\delta'_\kappa \varphi_1 = \kappa^\mu \partial_\mu \varphi_2, \delta'_\kappa \varphi_2 = \kappa^\mu \partial_\mu \varphi_1, \qquad (5)$$

$$\delta'_\kappa \psi^1 = \kappa^\mu \partial_\mu \psi^2, \delta'_\kappa \psi^2 = \kappa^\mu \partial_\mu \psi^1, \qquad (6)$$

Where $\kappa^\mu$ is an infinitesimal real constant four-vector parameter. One can easily see that, as it is the case for the usual translation, this transformation leads also to an invariance of the Lagrangian densities (1) and (2). One explicitly finds that $\delta'_\kappa L_\varphi$ and $\delta'_\kappa L_\psi$ are total derivatives, i.e., $\delta'_k L_\varphi = -\partial_\nu(\kappa^\nu(\partial_\mu\varphi_2^*\partial^\mu\varphi_1 + \partial_\mu\varphi_1^*\partial^\mu\varphi_2))$ and $\delta'_k L_\psi = -i\partial_\nu(\kappa^\nu(\overline{\psi}^2\overline{\sigma}^\mu\partial_\mu\psi^1 + \overline{\psi}^1\overline{\sigma}^\mu\partial_\mu\psi^2))$.

Moreover, the infinitesimal transformations $\delta'$ defined by (5) and (6) Forman abelian algebra. For two successive transformation $\delta'$ of parameters κ and ζ weget $\delta'_\varsigma \delta'_\kappa X = \varsigma^\nu \kappa^\mu \partial_\nu \partial_\mu X$, where $X$ stands for all the fields. This leads obviously to

$$(\delta'_\varsigma \delta'_\kappa = \delta'_\kappa \delta'_\varsigma)X = 0 \qquad (7)$$

One can also remark that these transformations commute with usual translations, i.e.,

$$(\delta_a \delta'_k - \delta'_k \delta_a)\psi = 0 \qquad (8)$$

Where, as usual, a translation $\delta$ of parameter $a$ is defined by $\delta_a X = a^\mu \partial_\mu X$.

Finally, it is straight forward to check that the commutator of $\delta'$ with rotations $R$ (i.e., transformations of the Lorentz group) closes on $\delta'$. We have

***Corresponding author:** Djeghloul N, Laboratory of Theoretical Physics of Oran (LPTO), University of Oran, Algeria, E-mail: ndjeghloul@gmail.com

$$(R'_{\omega}\delta'_{\kappa}-\delta'_{\kappa}R_{\omega})X=\delta'_{\omega.\kappa}X, \tag{9}$$

Where $\omega\cdot k=-\omega^{\mu}{}_{\nu}k^{\nu}$ is the infinitesimal parameter of the resulting $\delta'$ transformation In deriving (9), we used the fact that a rotation $R$ of infinitesimal parameter ω acts on any four- vector as $R_{\omega}V^{\mu}=-\ {}^{\mu}{}_{\nu}V^{\nu}$, on any spinor $\psi$ as $R_{\omega}\psi=-\frac{1}{2}\omega_{\mu\nu}\sigma^{\mu\nu}\psi$ with $\sigma^{\mu\nu}=\frac{1}{4}(\sigma^{\mu}\overline{\sigma}^{\nu}-\sigma^{\nu}\overline{\sigma}^{\mu})$ and leaves any scalar fields invariant.

Therefore, we can define a modified construction for the extended $N$=2 supersymmetry algebra relying on the nonlinear extension of a Lie algebra. In this context [3], the defining commutator contains, in addition to linear terms, terms that are multilinear in generators, i.e., $[T_a,T_b]=f^{c}_{ab}T_c+V^{cd}_{ab}T_cT_d$ for quadratically nonlinear algebras. As it was pointed out in [6], such nonlinear generalization has also to satisfy Jacobi identities. As extension of the standard supersymmetry construction where the anti-commutator of two extended supersymmetry transformations closes on translation, we postulate that it closes also on the composition of a translation $P_{\mu}$ and a transposition $s^2$, such that

$$\{Q_{i\alpha},\overline{Q}_{j\dot{\alpha}}\}=2\sigma^{\mu}_{\alpha\dot{\alpha}}\tau^{a}_{ij}p^{a}_{\mu}, \tag{10}$$

Where $\tau^{a}=(\tau^{a}_{ij})$ are the two 2×2 matrices given by

$$\tau^{1}=\begin{pmatrix}1&0\\0&1\end{pmatrix},\ \tau^{2}=\begin{pmatrix}1&0\\0&1\end{pmatrix} \tag{11}$$

These two matrices form a representation of $S_2$ and satisfy the following relations

$$\tau^{a}_{ij}\tau^{b}_{jk}=\delta^{ab}\delta_{ik}+\eta^{ab}\eta_{ik}, \tag{12}$$

$$\tau^{a}_{ij}\tau^{a}_{kl}=\delta_{il}\delta_{jk}+\eta_{il}\eta_{jk}, \tag{13}$$

In view of what precedes on the commutation relations of this modified translation and the other generators of the Poincare algebra (translations and rotations), the other commutators of the as modified $N$=2super Poincare algebra read

$$[P^{a}_{\mu},P^{b}_{\nu}]=0 \text{ and } [M_{\alpha\beta},P^{a}_{\mu}]=-\eta_{\alpha\mu}P^{a}_{\beta}+\eta_{\beta\mu}P^{a}_{\alpha}, \tag{14}$$

Where $M_{\alpha\beta}$ are the generators of the rotations and all other commutators are identical to those of usual extended $N$=2 super Poincare algebra. Moreover, it is straight forward to check that the modified super symmetry algebra (10) is consistent with all possible Jacobi identities of the whole algebra.

It is worth noting that $P^{2}_{\mu}$ satisfies, just as the usual translation, $P^{2}_{\mu}P^{2\mu}=m^{2}$ since permutation operators satisfy $(s^{a})^{2}=1$ ($a$=1,2). The Casmir invariant operator $P^2$ expressed thus as $\frac{1}{2}\sum_{a}P^{a}_{\mu}P^{a\mu}=m^{2}$ shows as usual, that all members of the same multiplet representation are of same masses.

One can also see that the reduction to the simple $N$=1 case leads obviously to standard results since the permutation operations on a set of one object are trivial.

We will now show that one can find an off-shell representation of this algebra, i.e., a multiplet that realizes this modified $N$=2super symmetry algebra *off shell.*

## An off-shell Representation

We first start with the same field contents that of the double tensor multiplet (as a generalization of the N=1 super symmetric multiplet of the gauge spinor super field [7]. We will show that such a multiplet forms a representation of the above introduced modified $N$=2 supersymmetric algebra (10) and more over, a consistent off-shell construction can be performed. This multiplet contains two Weyl fermions ψ and χ, two real scalar fields $\varphi_i i=1,2$ and two real 2-formgauge potentials $B_{i\mu\nu}$, $\mu(\nu)$= 0, 1, 2, 3 and i=1,2. All conventions and notations are the same as in the previous section. In what follows we work in the two-component formalism and adopt the standard conventions of Wess and Bagger [8]. The Lagrangian density of this multiplet reads

$$L=-i\overline{\psi\sigma}^{\mu}\partial_{\mu}\psi-i\overline{\chi\sigma}^{\mu}\partial_{\mu}\chi-\frac{1}{2}\partial_{\mu}\varphi_{i}\partial^{\mu}\varphi_{i}+\frac{1}{2}H_{i\mu}H^{\mu}_{i}, \tag{15}$$

Where $H_{i\mu}$ are the Hodge-duals of the field strengths of the 2-form gauge potentials, i.e.

$$H^{\mu}_{i}=\frac{1}{2}\varepsilon^{\mu\nu\rho\sigma}\partial_{\nu}B_{i\rho\sigma}, \tag{16}$$

With $\varepsilon^{\mu\nu\rho T}(\varepsilon^{0123}=+1)$ being the four-dimensional Levi-Civita tensor.

To see that this is indeed a representation of the modified $N$=2 super symmetric algebra defined by (10), we first check that (15) is invariant, up to total derivatives, under the following modified extended $N$=2 super symmetric transformations

$$\delta\psi=i\sigma^{\mu}\overline{\xi}_{1}\partial_{\mu}\varphi_{1}+i\sigma^{\mu}\overline{\xi}_{2}\partial_{\mu}\varphi_{2}+\sigma^{\mu}\overline{\xi}_{1}H_{1\mu}+\sigma^{\mu}\overline{\xi}_{2}H_{2\mu}, \tag{17}$$

$$\delta\chi=i\sigma^{\mu}\overline{\xi}_{1}\partial_{\mu}\varphi_{2}+i\sigma^{\mu}\overline{\xi}_{2}\partial_{\mu}\varphi_{1}+\sigma^{\mu}\overline{\xi}_{1}H_{2\mu}+\sigma^{\mu}\overline{\xi}_{2}H_{1\mu}, \tag{18}$$

$$\delta\varphi_{1}=\xi_{1}\psi+\xi_{2}\chi+h.c., \tag{19}$$

$$\delta\varphi_{2}=\xi_{1}\chi+\xi_{2}\psi+h.c., \tag{20}$$

$$\delta H^{\mu}_{1}=2i\xi_{1}\sigma^{\mu\nu}\partial_{\nu}\psi+2i\xi_{2}\sigma^{\mu\nu}\partial_{\nu}\chi+h.c., \tag{21}$$

$$\delta H^{\mu}_{2}=2i\xi_{1}\sigma^{\mu\nu}\partial_{\nu}\chi+2i\xi_{2}\sigma^{\mu\nu}\partial_{\nu}\psi+h.c., \tag{22}$$

Then recasting the spinor fields $\psi$ and $\chi$ as defined in previous section, such that $\psi^1=\psi$ and $\psi^2=\chi$ one can easily write the above transformations as

$$\delta\psi^{a}=i\sigma^{\mu}\overline{\xi}_{i}\tau^{a}_{ij}\partial_{\mu}\varphi_{j}+\sigma^{\mu}\overline{\xi}_{i}\tau^{a}_{ij}H_{j\mu,} \tag{23}$$

$$\delta\varphi_{i}=\tau^{a}_{ij}\xi_{j}\psi^{a}+\tau^{a}_{ij}\overline{\xi_{j}\psi}^{a}, \tag{24}$$

$$\partial H^{\mu}_{i}=2i\tau^{a}_{ij}\xi_{j}\sigma^{\mu\nu}\partial_{\nu}\psi^{a}-2i\tau^{a}_{ij}\overline{\xi_{j}\sigma}^{\mu\nu}\partial_{\nu}\overline{\psi}^{a}, \tag{25}$$

A direct computation leads explicitly

$$\delta L=-\partial_{\mu}(\psi^{a}\sigma^{\mu}\overline{\sigma}^{\nu}\xi_{i}\tau^{a}_{ij}\partial_{\nu}\varphi_{j}+i\overline{\psi}^{a}\overline{\sigma}^{\mu}\sigma^{\nu}\overline{\xi}_{i}\tau^{a}_{ij}H_{j\nu}-\partial_{\mu}(\tau^{a}_{ij}\xi_{j}\psi^{a}\partial^{\mu}\varphi_{i}-iH^{\mu}_{i}\tau^{a}_{ij}\xi_{j}\psi^{a}h.c.)$$

We are now able to compute the action of the commutator of two successive modified $N$=2 super symmetric transformations of parameters $\xi(\xi_1,\xi_2)$ and $\zeta(\zeta_1,\zeta_2)$ on each field of the multiplet. Starting with the scalar fields $\varphi_i$, we first get

$$\delta_{\zeta}\delta_{\xi}\varphi_{i}=i(\xi_{k}\sigma^{\mu}\overline{\varsigma}_{k}+\overline{\xi}_{k}\overline{\sigma}^{\mu}\varsigma_{k})\partial_{\mu}\varphi_{i}+i(\xi_{k}\sigma^{\mu}\eta_{kl}\overline{\varsigma}_{l}+\overline{\xi}_{k}\overline{\sigma}^{\mu}\eta_{kl}\zeta_{l})\eta_{ij}\partial_{\mu}\varphi_{j},$$
$$+(\xi_{k}\sigma^{\mu}\overline{\zeta}_{k}-\overline{\xi}_{k}\overline{\sigma}^{\mu}\varsigma_{k})H_{i\mu}+i(\xi_{k}\sigma^{\mu}\eta_{kl}\overline{\varsigma}_{l}-\overline{\xi}_{k}\overline{\sigma}^{\mu}\eta_{kl}\zeta_{l})\eta_{ij}H_{j\mu} \tag{26}$$

Using $\overline{\xi}_{i}\overline{\sigma}^{\mu}\varsigma_{j}=-\varsigma_{j}\sigma^{\mu}\overline{\xi}_{i}$, we see that the terms proportional to $\partial_{\mu}\varphi$ are anti symmetric under the substitution $\xi$ □ □□ $\varsigma$ such that they are doubled in the commutator $(\delta_{\varsigma}\delta_{\xi}-\delta_{\xi}\delta_{\varsigma})\varphi_{i}$, while the terms proportional to $H$ are symmetric under the same substitution, thus they disappear when computing this commutator. Explicitly, we get

$$(\delta_{\zeta}\delta_{\xi}-\delta_{\xi}\delta_{\zeta})\varphi_{i}=-2i(\varsigma_{k}\sigma^{\mu}\overline{\xi}_{k}-\xi_{k}\sigma^{\mu}\overline{\varsigma}_{k})\partial_{\mu}\varphi_{i}-2i(\varsigma_{k}\sigma^{\mu}\eta_{kl}\overline{\xi}_{l}-\xi_{k}\sigma^{\mu}\eta_{kl}\overline{\varsigma}_{l})\eta_{ij}\partial_{\mu}\varphi_{j}, \tag{27}$$

Which in regard to (10), shows that $(\delta_{\zeta}\delta_{\xi}-\delta_{\xi}\delta_{\zeta})$ on the scalar fields $\varphi_i$ closes off shell. We turn now to compute the commutator on the spinor fields. A direct evaluation of $\delta_\zeta\,\delta_\xi\,\psi^a$ shows that the terms in equations of motion of $\psi^a$ cancel due to the contribution of the variation of $H_{i\mu}$, we then obtain

$$\delta_\varsigma\delta_\xi\psi^a_\alpha = -2i\varsigma_k\sigma^\mu\bar{\xi}_k\partial_\mu\psi^a_\alpha + -2i\varsigma_k\sigma^\mu\eta_{kl}\bar{\xi}_l\eta^{ab}\partial_\mu\psi^b_\alpha,$$
$$-2i\bar{\varsigma}_k\bar{\xi}_k\sigma^\mu_{\alpha\dot\alpha}\partial_\mu\bar{\psi}^{a\dot\alpha} - 2i\bar{\varsigma}_k\eta_{kl}\bar{\xi}_l\sigma^\mu_{\alpha\dot\alpha}\eta^{ab}\partial_\mu\bar{\psi}^{b\dot\alpha} \tag{28}$$

Whereweusedtheidentities $(\sigma^\mu\bar{\sigma}^\nu + \sigma^\nu\bar{\sigma}^\mu)^\beta_\alpha = -2\eta^{\mu\nu}\delta^\beta_\alpha, \sigma^\mu_{\alpha\dot\alpha}\bar{\sigma}^{\beta\dot\beta}_\nu = -2\delta^\beta_\alpha\delta^{\dot\beta}_{\dot\alpha}$ and the definition $\bar{\sigma}^{\mu\dot\alpha\alpha} = \varepsilon^{\dot\alpha\dot\beta}\varepsilon^{\alpha\beta}\sigma^\mu_{\beta\dot\beta}$ with $\varepsilon_{12} = \varepsilon^{21} = -1$ is the two-dimensional Levi-Civita tensor. Noticing that the factors of the terms involving the equations of motion of $\bar{\psi}^a$ are symmetric under the substitution $\bar{\xi} \leftrightarrow \bar{\varsigma}$ we end up with the following commutator

$$(\delta_\varsigma\delta_\xi - \delta_\xi\delta_\varsigma)\psi^a = -2i(\varsigma_k\sigma^\mu\bar{\xi}_k - \xi_k\sigma^\mu\bar{\varsigma}_k)\partial_\mu\psi^a$$
$$-2i(\varsigma_k\sigma^\mu\eta_{kl}\bar{\xi}_l - \xi_k\sigma^\mu\eta_{kl}\bar{\varsigma}_l)\eta^{ab}\partial_\mu\psi^b_\alpha, \tag{29}$$

Which close off-shell.

Finally, we check the closure on the fields $H^\mu_i$. Using the identity $\bar{\xi}_i\bar{\sigma}^{\mu\nu}\bar{\sigma}^\rho\varsigma_j = -\varsigma_j\sigma^\rho\bar{\sigma}^{\mu\nu}\bar{\xi}_i$ and rearranging terms, we first find

$$\delta_\varsigma\delta_\xi H^\mu_i = -2(\xi_k\sigma^{\mu\nu}\sigma^\rho\bar{\varsigma}_k - \varsigma_k\sigma^\rho\bar{\sigma}^{\mu\nu}\bar{\xi}_k)\partial_\nu\partial_p\varphi_i - 2(\xi_k\sigma^{\mu\nu}\sigma^\rho\eta_{kl}\bar{\varsigma}_l - \varsigma_k\sigma^\rho\bar{\sigma}^{\mu\nu}\eta_{kl}\bar{\xi}_l)\eta_{ij}\partial_\nu\partial_\rho\varphi_j$$
$$+2i(\xi_k\sigma^{\mu\nu}\sigma^\rho\bar{\varsigma}_k + \varsigma_k\sigma^\rho\bar{\sigma}^{\mu\nu}\bar{\xi}_k)\partial_\nu H_{ip} + 2i(\xi_k\sigma^{\mu\nu}\sigma^\rho\eta_{kl}\bar{\varsigma}_l - \varsigma_k\sigma^\rho\bar{\sigma}^{\mu\nu}\eta_{kl}\bar{\xi}_l)\eta_{ij}\partial_\nu H_{jp}. \tag{30}$$

When evaluating the commutator $(\delta_\eta\delta_\xi - \delta_\xi\delta_\eta)H^\mu_i$, we can see that all terms proportional to $\partial_\nu\partial_\rho\varphi$ are of type $\xi_k(\sigma^{\mu\nu}\sigma^\rho + \sigma^\rho\bar{\sigma}^{\mu\nu})\bar{\varsigma}_l - (\bar{\xi} \leftrightarrow \bar{\varsigma})$ which is identical to $i\varepsilon^{\mu\nu\rho T}\xi_k\sigma_T\bar{\varsigma}_l$ so that all $\varphi$ contributions in the resulting commutator vanish. At the same time $\partial_\nu H_{i\rho}$ contributions involve terms of type $\xi_k(\sigma^{\mu\nu}\sigma^\rho - \sigma^\rho\bar{\sigma}^{\mu\nu})\bar{\varsigma}_l - (\xi \leftrightarrow \varsigma)$ which is identical to $(\eta^{\mu\rho}\xi_k\sigma^\nu\bar{\varsigma}_l - \eta^{\nu\rho}\xi_k\sigma^\mu\bar{\varsigma}_l) - (\bar{\xi} \leftrightarrow \bar{\varsigma})$ so that we end up with the following commutator.

$$(\delta_\varsigma\delta_\xi - \delta_\xi\delta_\varsigma)H^\mu_i = -2i(\varsigma_k\sigma^\nu\bar{\xi}_k - \xi_k\sigma^\nu\bar{\varsigma}_k)\partial_\nu H^\mu_i - 2i(\varsigma_k\sigma^\nu\eta_{kl}\bar{\xi}_l - \xi_k\sigma^\nu\eta_{kl}\bar{\varsigma}_l)\eta_{ij}\partial_\nu H^\mu_j, \tag{31}$$

Where the identities $\sigma^\mu\bar{\sigma}^\nu\sigma^\rho - \sigma^\rho\bar{\sigma}^\nu\sigma^\mu = 2i\varepsilon^{\mu\nu\rho\tau}\sigma_\tau$, $\sigma^\mu\bar{\sigma}^\nu\sigma^\rho + \sigma^\rho\bar{\sigma}^\nu\sigma^\mu = 2(\eta^{\mu\rho}\sigma^\nu - \eta^{\nu\rho}\sigma^\mu - \eta^{\mu\nu}\sigma^\rho)$ and $\bar{\varsigma}_k\bar{\sigma}^\mu\sigma^\nu\bar{\sigma}^\rho\bar{\xi}_l = -\bar{\xi}_l\sigma^\rho\bar{\sigma}^\nu\sigma^\mu\bar{\varsigma}_k$ are used as well as the identity $\partial_\mu H^\mu_i = 0$ which follows from the definition(16). This ends the proof that the modified *N*=2 super symmetric transformations (23)-(25) form a super symmetric algebra that closes off shell. The *N*=2 multiplet $(\psi^a, \phi_i, H^\mu_i)$ is then an off-shell multiplet of the modified *N*=2 super symmetric algebra defined by (10). It is well known [2] that the double tensor multiplet model has also special gauge invariance. Similarly for the above-introduced multiplet, it is easy to check that the Lagrangian density (15) is also invariant upon the gauge transformation

$$\delta_\wedge B^{\mu\nu}_i = \partial^\mu \wedge^\nu_i - \partial^\nu \wedge^\mu_i \tag{32}$$

Where $\wedge^\mu_i$ are the space time dependent gauge parameters. Since, by construction (16), the Hodge-duals $H^\mu_i$ are obviously invariants upon such a transformation, we have, to make this gauge transformation appear, to replace $H^\mu_i$ by the corresponding gauge potentials $B^{\mu\nu}_i$ within the transformations (23)-(25). We find

$$\delta\psi^a = i\sigma^\mu\bar{\xi}_i\tau^a_{ij}\partial_\mu\varphi_j + \frac{1}{2}\varepsilon^{\mu\nu\rho\sigma}\sigma_\mu\bar{\xi}_i\tau^a_{ij}B_{j\rho\sigma}, \tag{33}$$

$$\delta B^{\mu\nu}_i = -2\tau^a_{ij}\xi_j\sigma^{\mu\nu}\psi^a - 2\tau^a_{ij}\bar{\xi}_j\bar{\sigma}^{\mu\nu}\bar{\psi}^a, \tag{34}$$

While the transformations of the scalar fields (24) remains the same. We now show that the commutator $(\delta_\zeta\,\delta_\xi - \delta_\xi\,\delta_\zeta)$ on the gauge potentials $B^{\mu\nu}_i$ closes, as previously, off shell on the combination of translation and permutations but also on the above defined gauge transformation. After a similar computation to (31), we find

$$(\delta_\varsigma\delta_\xi - \delta_\xi\delta_\varsigma)B^{\mu\nu}_i = -2i(\varsigma_k\sigma^\lambda\bar{\xi}_k - \xi_k\sigma^\lambda\bar{\varsigma}_k)\partial_\lambda B^{\mu\nu}_i$$
$$-2i(\varsigma_k\sigma^\lambda\eta_{kl}\bar{\xi}_l - \xi_k\sigma^\lambda\eta_{kl}\bar{\varsigma}_l)\eta_{ij}\partial_\lambda B^{\mu\nu}_i + \partial^\mu \wedge^\nu_i - \partial^\nu \wedge^\mu_i, \tag{35}$$

Where the gauge parameters $\wedge$ are given by

$$\wedge^\mu_i = 2i \wedge^{\mu\nu}_{ijk}(\varsigma_j\sigma_\nu\bar{\xi}_k - \xi_j\sigma_\nu\bar{\varsigma}_k),\ \wedge^{\mu\nu}_{ijk} = \tau^a_{ij}\tau^a_{kl}(\eta^{\mu\nu}\varphi_l - B^{\mu\nu}_l). \tag{36}$$

In deriving (35) the identity $\varepsilon_{k\tau\rho\sigma}\varepsilon^{k\mu\nu\lambda} = -[\delta^\mu_\tau(\delta^\nu_\rho\delta^\lambda_\sigma - \delta^\nu_\sigma\delta^\lambda_\rho) - \delta^\nu_\tau(\delta^\mu_\rho\delta^\lambda_\sigma - \delta^\mu_\sigma\delta^\lambda_\rho) + \delta^\lambda_\tau(\delta^\mu_\rho\delta^\nu_\sigma - \delta^\mu_\sigma\delta^\nu_\rho)]$ issued. As it is generally the case in super symmetric gauge theories, these gauge parameters are field dependent. It is worth noting that comparatively to the standard approach [2] of the double tensor multiplet, in addition to the fact that in the context presented here the off shell construction is possible. The obtained gauge parameters (36) do noting volve explicitly the space time coordinates.

Even if the structure of the modified algebra (10) differs from the usual one, we can ,in view of the off-shell closure obtained above, consider the construction of a nilpotent BRST operator. Indeed, starting from the modified *N*=2 super symmetry transformations (23) - (25), and upon the usual replacement of the symmetry parameters by the corresponding host fields of opposite statistics, the corresponding BRST construction follows naturally.

Defining the BRST operator$\Delta$ on the fields $\psi^a, \varphi_i, B^{\mu\nu}_i$ as

$$\Delta\psi^a = i\sigma^\mu\bar{\xi}_i\tau^a_{ij}\partial_\mu\varphi_j + \frac{1}{2}\varepsilon^{\mu\nu\rho\sigma}\sigma_\mu\bar{\xi}_i\tau^a_{ij}B_{j\rho\sigma} + c^\rho\partial_\rho\psi^a + \kappa^\rho\eta^{ab}\partial_\rho\psi^b, \tag{37}$$

$$\Delta\varphi_i = \tau^a_{ij}\xi_j\psi^a + \tau^a_{ij}\bar{\xi}_j\bar{\psi}^a + c^\rho\partial_\rho\varphi_i + \kappa^\rho\eta_{ij}\partial_\rho\varphi_j, \tag{38}$$

$$\Delta B^{\mu\nu}_i = -2\tau^a_{ij}\xi_j\sigma^{\mu\nu}\psi^a - 2\tau^a_{ij}\bar{\xi}_j\bar{\sigma}^{\mu\nu}\bar{\psi}^a + c^\rho\partial_\rho B^{\mu\nu}_i + \kappa^\rho\eta_{ij}\partial_\rho B^{\mu\nu}_i + \partial^\mu \wedge^\nu_i - \partial^\nu \wedge^\mu_i \tag{39}$$

And on the ghosts fields $\xi_i(\bar{\xi}_i), c^\mu, \kappa^\mu$ and $\wedge^\mu_i$ as

$$\Delta\xi_i = 0, \tag{40}$$

$$\Delta c^\mu = -2i(\xi_k\sigma^\mu\bar{\xi}_k), \tag{41}$$

$$\Delta\kappa^\mu = -2i(\xi_k\sigma^\mu\eta_{kl}\bar{\xi}_l) \tag{42}$$

$$\Delta\kappa^\mu = -2i(\xi_k\sigma^\mu\eta_{kl}\bar{\xi}_l), \tag{43}$$

$$\Delta \wedge^\mu_i = 2iT^a_{ij}T^a_{kl}(\eta^{\mu\nu}\varphi_j - B^{\mu\nu}_j)(\xi_k\sigma_\nu\bar{\xi}_l), \tag{44}$$

It is straight forward to show its off-shell nil potency, i. e , $\Delta^2 X = 0, \forall X$.

## Conclusion

The main result of this work is that an alternative *N*=2 extension of standard supersymmetry is possible. This is done in the context of nonlinear extensions of standard Lie algebra by a suitable introduction of the symmetric group $S_2$.The additional nonlinear term being a composition of translation and transposition. The obtained algebra being a quadratically nonlinear extension of the standard *N*=2 super symmetric algebra is however a non usual construction. Indeed, this latter contains structurally the permutation transformations that are obviously non-local, while, in deriving the general realization of supersymmetry algebra, only continuous groups (in particular Lie groups) are usually considered (see e.g. [8]).This kind of construction will be analyzed in detail elsewhere.

The presented result is different from the standard extended $N$=2 super symmetry, i.e., the anti commutator of two modified super symmetric transformations must close (at least on shell) on a mix of translation and permutations, but leads to a consistent algebraic construction. The reduction to the N=1 case leads to usual super symmetry due to the triviality of the group $S_1$ which contains only the identity. Such a modified extended $N$=2 super symmetric algebra (10) admits as representation a multiplet that contains the same fields as the double tensor multiplet (which is in particular, relevant to type II B super string vacua [9]). We have shown that an off-shell construction is possible, i.e., without relying on field equation. This result has to be compared with the usual double tensor multiplet for which, inspite of the fact that the bosonic and fermionic degrees of freedom balance at both on-shell and off-shell levels, the off-shell construction fails.

Moreover, if a systematic procedure can be considered in order to give the off-shell version of any given open gauge (local) theory [10], no such systematic approach is available in the context of global (rigid) symmetries such as extended matter super symmetry, even if specific models exist where the construction of off-shell realization is possible, i.e., the so-called 0(2n) super multiplets [11] (see also [12] for a modern review). We believe that the approach developed here in which the symmetric group $S_2$ (or equivalently the group $Z_2$) shows up within the nonlinear extension of the usual $N$=2 super symmetry algebra can offer a new perspective for investigating the off-shell structure of extended super symmetric models (e.g $N$=4 super symmetric models).

## References


1. De Wit B, Kaplunovsky V, Louis J and Lust D (1995) Quantum Gravity Mathematical Models and Experimental Bounds. Nucl Phys B 451-453.
2. Brandt F (2000) Supersymmetry Algebra Cohomology IV Primitive Elements in All Dimensions from D=4 to D=11. Nucl Phys B 543-587.
3. Kent A (1988) Normal ordered Lie algebras. Princeton preprint IASSNS HEP 88 04.
4. Frappat L, Sciarrino A and Sorba P (2000) Dictionary on Lie Algebras and Superalgebras. Academic Press.
5. Jones HF (1998) Groups, Representations and Physics. IOP Publishing Ltd.
6. Schoutens K, Sevrin A, Van Nieuwenhuizen P (1989) Quantum BRST Charge for Quadratically Non linear Lie Algebra. Commun and Math Phys 124: 87-103.
7. Siegel W (1979) Gauge Spinor Superfield as Scalar Multiplet. Phys Lett B 85: 333-334.
8. Wess J, Bagger J (1983) Supersymmetry and Supergravity. NJ Princeton University Press, Princeton.
9. Louis J, Förger K (1997) Holomorphic Couplings in String Theory. Nucl Phys Proc Suppl 55: 33-64.
10. Djeghloul N, Tahiri M (2002) From On-Shell to Off-Shell Open Gauge Theories. Phys Rev D 66: 065010
11. Lindstrom U, Rocek M (1988) New Hyperkahler Metrics and New Supermultiplets. Commun Math Phys 115: 21-29.
12. Kuzenko SM (2010) Lectures on Nonlinear Sigma-Models in Projective Superspace. J Phys. A Math Theor 43: 443001.

# 2

# Confluent Hypergeometric Equation via Fractional Calculus Approach

**Rodrigues FG* and Capelas de Oliveira E**

*Department of Applied Mathematics, IMECC – UNICAMP, Brazil*

**Abstract**

In this paper, using the theory of the so-called fractional calculus we show that it is possible to easily obtain the solutions for the confluent hypergeometric equation. Our approach is to be compared with the standard one (Frobenius) which is based on the ordinary calculus of integer order.

**Keywords:** Fractional calculus; Confluent hypergeometric equation; Mathematical-physics

## Introduction

Investigations in the so-called fractional calculus, that is, the theory of integration and differentiation of arbitrary order, have increased considerably in the last fifty years [1-7]. Nowadays, the topic is broadly dispersed in several distinct applications in science such as fractional viscoelasticity [8,9], fractional harmonic oscillators [9,10], fractional frictional forces [10], fractional fluid dynamics [11] and fractional signal processing [12,13], just to mention a few examples. Such investigations indicates that the descriptions of previously established theories by means of ordinary calculus (of integer order) are now being reviewed by means of fractional calculus with promising results such as the increase in accuracy between theoretical and experimental data.

Following this trend of approaching "old problems" by means of "new tools", recently we have approached Bessel's differential equation of order p [14] by a methodology of the fractional calculus as proposed by the authors [15]. Although we were successful in obtaining Bessel's solutions of the first kind, the methodology itself was not rigorously accurate (from a mathematical point of view) to justify the second linear independent solution, specifically $J_p$ for $0 < p \neq -1, -2, \ldots$ inspiring us to continue with other investigations about the methodology by applying it to a more general setting, the confluent hypergeometric equation. Also, as conjectured in the final considerations [14], we have pointed out that there might be some "minor patches" needed to be done to the definitions of the fractionalized versions of the integral and differential operators we have used in that paper (which was the Riemann- Liouville formulation and that we will present it in the next section of this paper). Nevertheless, the fractional methodology is still very promising and therefore we continue its investigations and, as mentioned above, we now apply it to obtain solutions of the confluent hypergeometric equation.

The paper is organized as follows: In section 2 we review the Riemann-Liouville integrodifferential operator particularly presenting an alternative definition by means of the Hadamard's finite part integral. In section 3 we discuss the confluent hypergeometric equation be means of the fractional methodology. In section 4 we present our conclusions and some remarks.

## Riemann-Liouville Integrodifferential Operator

In this section we describe the Riemann-Liouville formulations; hence we will present two main definitions of the fractional integral operator and fractional differential operator [16-20]. Some properties of those operators are also discussed.

**Definition 1:** *Let* $\Omega = [a, b] \subset \mathbb{R}$ *be a finite real interval. Suppose* $f \in L_1$ *[a, b] is a Lebesgue integrable function in [a, b]. Then the expressions* $\mathcal{I}^{\nu}_{a+}f$ *and* $\mathcal{I}^{\nu}_{b-}f$ *established by the equalities below*

$$\left(\mathcal{I}^{\nu}_{a+}f\right)(x) \equiv \frac{1}{\Gamma(\nu)}\int_a^x (x-t)^{\nu-1} f(t)\,\mathrm{d}t, \tag{1}$$

*with* $x > a$, $Re(\nu) > 0$ *and*

$$\left(\mathcal{I}^{\nu}_{b-}f\right)(x) \equiv \frac{1}{\Gamma(\nu)}\int_x^b (t-x)^{\nu-1} f(t)\,\mathrm{d}t, \tag{2}$$

with $x < b$, $Re(\nu) > 0$ and where $r(\nu)$ is the gamma function, defines the Riemann-Liouville fractional integral operator (RLFI) of order $\nu \in \mathbb{C}$. The integrals in Equation (1) and Equation (2) are usually called the left and right versions, respectively.

**Definition 2:** Let $\Omega = [a, b] \subset \mathbb{R}$ be a finite real interval. Suppose $f \in AC^n$ *[a, b]* is an absolute continuous function until order **n** - 1. Then the eepressions $\mathcal{I}^{\nu}_{a+}f$ and $\mathcal{I}^{\nu}_{b-}f$ established by the equalities below

$$\left(\mathcal{D}^{\nu}_{a+}f\right)(x) \equiv \mathcal{D}^{\mathbf{n}}_{a+}\left[\left(\mathcal{I}^{\mathbf{n}-\nu}_{a+}f\right)(x)\right] = \left(\frac{\mathbf{d}}{\mathbf{d}x}\right)^{\mathbf{n}}\left(\mathcal{I}^{\mathbf{n}-\nu}_{a+}f\right)(x)$$

$$= \frac{1}{\Gamma(\mathbf{n}-\nu)}\left(\frac{\mathbf{d}}{\mathbf{d}x}\right)^{\mathbf{n}}\int_a^x (x-t)^{\mathbf{n}-\nu-1} f(t)\,\mathbf{d}t, \tag{3}$$

with x > a and

$$\left(\mathcal{D}^{\nu}_{b-}f\right)(x) \equiv \mathcal{D}^{\mathbf{n}}_{a+}\left[\left(\mathcal{I}^{\mathbf{n}-\nu}_{b-}f\right)(x)\right] = \left(-\frac{\mathbf{d}}{\mathbf{d}x}\right)^{\mathbf{n}}\left(\mathcal{I}^{\mathbf{n}-\nu}_{b-}f\right)(x)$$

$$= \frac{1}{\Gamma(\mathbf{n}-\nu)}\left(-\frac{\mathbf{d}}{\mathbf{d}x}\right)^{\mathbf{n}}\int_x^b (t-x)^{\mathbf{n}-\nu-1} f(t)\,\mathbf{d}t, \tag{4}$$

with $x < b$ and $[Re(\nu)]$ being the integer part function of $Re(\nu)$ and $\mathbf{n} = [Re(\nu)] + 1$, defines the Riemann-Liouville fractional differential operators (RLFD) of order $\nu \in \mathbb{C}$ $(Re(\nu) \geq 0)$. The derivatives in Equation (1) and Equation (2) are usually called the left and right versions, respectively.

***Corresponding author:** Rodrigues FG, Department of Applied Mathematics, IMECC – UNICAMP, Brazil, E-mail: fabior@mpcnet.com.br

Before continuing with the presentation of new definitions and results, we inform the reader that in this work we will restrict ourselves to the left versions (Equation (1) and Equation (3)) only, and consider the order v of the operators to be a real number.

Mind that the two definitions are reduced to the classical integer cases whenever we choose $v = n \in \mathbb{N}$. In particular, it can be shown [16,20] that if $v = n = 0$, then we have the identity operator **I**:

$$\mathcal{D}_{a+}^{0} = \mathbf{I} = \mathcal{I}_{a+}^{0} := \lim_{v \to 0} \mathcal{I}_{a+}^{v}.$$

We also point out that if $f(x)$ is of the form $(x - a)^{\beta-1}$ for $\beta > 0$, then [17]

$$\left(\mathcal{I}_{a+}^{v}(t-a)^{\beta-1}\right)(x) = \frac{\Gamma(\beta)}{\Gamma(\beta+v)}(x-a)^{\beta+v-1}, \tag{5a}$$

$$\left(\mathcal{D}_{a+}^{v}(t-a)^{\beta-1}\right)(x) = \frac{\Gamma(\beta)}{\Gamma(\beta-v)}(x-a)^{\beta-v-1}. \tag{5b}$$

As a consequence of these rules and the properties of $\mathcal{I}_{a+}^{v}$ and $\mathcal{D}_{a+}^{v}$ are continuous operators relative to the index $v$ [16], then if $f(x) \in C^{\infty}[a, \epsilon]$ is analytic in the interval $[a, \epsilon]$ for some $\epsilon > a$, then $f$ is representable as a power series

$$f(x) = \sum_{k=0}^{\infty} c_k (x-a)^k, \tag{6}$$

where $c_k$ are constants and, in this case, applying the rules as in Equation (5a) and Equation (5b) we may integro differentiate functions as per Equation (6) according to the formulas [15,16,20]:

$$\left[\mathcal{I}_{a+}^{v}\sum_{k=0}^{\infty} c_k (t-a)^k\right](x) = \sum_{k=0}^{\infty} c_k \mathcal{I}_{a+}^{v}(x-a)^k = \sum_{k=0}^{\infty} c_k \frac{\Gamma(k+1)(x-a)^{k+v}}{\Gamma(k+1+v)} \tag{7}$$

and

$$\left[\mathcal{D}_{a+}^{v}\sum_{k=0}^{\infty} c_k (t-a)^k\right](x) = \sum_{k=0}^{\infty} c_k \mathcal{D}_{a+}^{v}(x-a)^k = \sum_{k=0}^{\infty} c_k \frac{\Gamma(k+1)(x-a)^{k-v}}{\Gamma(k+1-v)}. \tag{8}$$

We now present a "unified" version of the RLFI and RLFD under a single operator symbol $\mathfrak{D}_{a+}^{v}$. We shall call it the Riemann-Liouville fractional integro differential operator or simply integro differential operator. That is, for $v \in \mathbb{R}$ and assuming that $f$ is such that $\mathcal{D}_{a+}^{v} f$ and $\mathcal{I}_{a+}^{v} f$ are well defined, then

$$\mathfrak{D}_{a+}^{v} = \begin{cases} \mathcal{D}_{a+}^{v}, & v > 0 \text{ (RLFD)}, \\ \mathbf{I}, & v = 0, \\ \mathcal{I}_{a+}^{-v}, & v < 0 \text{ (RLFI)}, \end{cases} \tag{9}$$

so that we have the following situation for $v > 0$

$$\left(\mathcal{D}_{a+}^{-v} f\right)(x) = \left(\mathcal{I}_{a+}^{v} f\right)(x), \tag{10}$$

and

$$\left(\mathcal{D}_{a+}^{v} f\right)(x) = \left(\mathcal{I}_{a+}^{-v} f\right)(x). \tag{11}$$

This suggests that the RLFD can be interpreted as an analytical extension of the operator RLFI. However we must take some caution here. As we have pointed out [14], the integral

$$\left(\mathcal{I}_{a+}^{-v} f\right)(x) = \frac{1}{\Gamma(-v)}\int_a^x (x-t)^{-v-1} f(t)\,\mathbf{d}t,$$

usually diverges even if $f(x) \in L_1\,[a, b]$. To avoid this "pitfall" we need to consider the so-called Hadamard's finite part integral [16,21]:

$$\mathcal{H}\int_a^b (x-a)^{-\beta} f(x)\,\mathbf{d}x := \sum_{k=0}^{\mathbf{n}} \frac{f^{(k)}(a)(b-a)^{k+1-\beta}}{(k+1-\beta)k!} + \int_a^b (x-a)^{-\beta} R_{\mathbf{n}}(x,a)\,\mathbf{d}x, \tag{12}$$

where $\mathbf{n} = [\beta] + 1$, $\beta \notin \mathbb{N}$ and

$$R_{\mathbf{n}}(x,a) = \frac{1}{\mathbf{n}!}\int_a^x (x-t)^{\mathbf{n}} f^{(\mathbf{n}+1)}(t)\,\mathbf{d}t$$

is the remainder of the nth degree Taylor polynomial of $f$ expanded at $x = a$.

An alternative definition for the Hadamard's finite part integral can be stated in terms of the following theorem [16].

**Theorem 3:** Let $1 < \beta \notin \mathbb{N}$ and $\mathbf{m} = [\beta] + 1$. For $f \in C^m\,[a, b]$ is **m** times continuously differentiable, then

$$\frac{1}{\Gamma(1-\beta)}\mathcal{H}\int_a^b (x-a)^{-\beta} f(x)\,\mathbf{d}x = \sum_{k=0}^{\mathbf{m}-1} \frac{f^{(k)}(a)(b-a)^{k+1-\beta}}{\Gamma(k+2-\beta)} + \mathcal{I}_{a+}{}^{\mathbf{m}-\beta+1} f^{(\mathbf{m})}(b).$$

In view of this result, we can interpret the RLFD as an integral as stated in the next theorem [16].

**Theorem 4:** Let $0 < v \notin \mathbb{N}$ and $n = [v] + 1$. If $f \in C^n\,[a, b]$ and $x \in [a, b]$, then

$$\left(\mathcal{D}\quad f\right)(x) = \frac{}{\Gamma(-\ )}\mathcal{H}\int\ (x-t)^{-\ -}\ f(t)\ \ t$$

So in fact, the formal interpretation of Equation (10) and Equation (11) must be

$$\left(\mathcal{D}_{a+}^{v} f\right)(x) = \mathcal{H}\left(\mathcal{I}_{a+}^{-v} f\right)(x), \tag{13}$$

$$\left(\mathcal{D}_{a+}^{-v} f\right)(x) = \mathcal{H}\left(\mathcal{I}_{a+}^{v} f\right)(x), \tag{14}$$

for $v \neq 1, 2, \ldots$. Nevertheless, for simplicity in notation, we shall use the ones established in Equation (10) and Equation (11), but having the interpretations as in Equation (13) and Equation (14).

We now summarize, without presenting the proofs[1], some composition laws for the $\mathfrak{D}_{a+}^{v}$ operator that will be needed for the next section.

The first of them is the following:

$$\mathfrak{D}_{a+}^{-p}\left[\mathfrak{D}_{a+}^{q} f(x)\right] = \mathfrak{D}_{a+}^{q-p} f(x) - \sum_{k=0}^{\mathbf{n}-1} \frac{(\mathfrak{D}_{a+}{}^{q-k-1} f)(a)}{\Gamma(p-k)}(x-a)^{p-k-1}, \tag{15}$$

where $p, q \in \mathbb{R}^+$ and $\mathbf{n} = [q] + 1$.

Then for $v > 0$ and $m \in \mathbb{N}$, we have

$$\left(\mathfrak{D}_{a+}^{m}\mathfrak{D}_{a+}^{v} f\right)(x) = \left(\mathfrak{D}_{a+}^{m+v} f\right)(x), \tag{16}$$

$$\left(\mathfrak{D}_{a+}^{v}\mathfrak{D}_{a+}^{m} f\right)(x) = \left(\mathfrak{D}_{a+}^{m+v} f\right)(x) - \sum_{k=0}^{m-1} \frac{f^{(k)}(a)(x-a)^{k-v-m}}{\Gamma(1+k-v-m)}. \tag{17}$$

Also, assuming that the derivatives exists, we have the equalities

$$\mathfrak{D}_{a+}^{p}\left[\mathfrak{D}_{a+}^{q} f(x)\right] = \mathfrak{D}_{a+}^{p+q} f(x) - \sum_{k=0}^{\mathbf{m}-1} \frac{(\mathfrak{D}_{a+}^{q-k-1} f)(a)}{\Gamma(-p-k)}(x-a)^{-p-k-1}, \tag{18}$$

[1]The proofs can be found in several distinct sources. We mention the main literature where these results can be found, [16,17,19,20].

$$\mathfrak{D}_{a+}^{q}\left[\mathfrak{D}_{a+}^{p}f(x)\right]=\mathfrak{D}_{a+}^{p+q}f(x)-\sum_{k=0}^{\mathbf{n}-1}\frac{(\mathfrak{D}_{a+}^{p-k-1}f)(a)}{\Gamma(-q-k)}(x-a)^{-q-k-1}, \quad (19)$$

where p, q > 0, **n** = [p] + 1 and **m** = [q] + 1.

Note that the equalities in Equation (18) and Equation (19) occurs if, and only if, the summands are nulls and this is guaranteed if

$$(\mathfrak{D}_{a+}^{\beta-k-1}f)(a)=0,\ \left(k=0,1,\ldots,\mathbf{n}-1\right), \quad (20)$$

$$(\mathfrak{D}_{a+}^{\alpha-k-1}f)(a)=0,\ \left(k=0,1,\ldots,\mathbf{m}-1\right). \quad (21)$$

The author [19] shows in section 2.3.7 of his book that if *f(x)* have enough continuous derivatives, then the conditions in Equation (20) and Equation (21) are equivalent, respectively, to

$$f^{(k)}(a)=0,\ \left(k=0,1,\ldots,\mathbf{n}-1\right), \quad (22)$$

$$f^{(k)}(a)=0,\ \left(k=0,1,\ldots,\mathbf{m}-1\right). \quad (23)$$

Hence, if $f^{(k)}(a) = 0$, $(k = 0, 1, \ldots, \mathbf{r} - 1)$, where $\mathbf{r} = \max\{\mathbf{m}, \mathbf{n}\}$, then

$$\mathfrak{D}_{a+}^{\alpha}\left[\mathfrak{D}_{a+}^{\beta}f(x)\right]=\mathfrak{D}_{a+}^{\alpha+\beta}f(x)=\mathfrak{D}_{a+}^{\beta}\left[\mathfrak{D}_{a+}^{\alpha}f(x)\right].$$

Finally, we also mention the following result

$$\mathfrak{D}_{a+}^{p}\left[\mathfrak{D}_{a+}^{-q}f(x)\right]=\mathfrak{D}_{a+}^{p-q}f(x),\ p,q\in\mathbb{R}^{+} \quad (24)$$

valid whenever *f(x)* is at least continuous and when $p \geq q \geq 0$ that the derivative $\mathcal{D}_{a+}^{p-q}f(x)$ exists. Finalizing this section, we present a generalization of the Leibniz rule for integrodifferentiating the product of functions [14,16].

**Theorem 5 (Fractional Leibniz Rule):** Let *f* and *g* be analytical in *[a, b]*, then

$$\mathfrak{D}_{a+}^{\nu}[fg]=\sum_{k=0}^{\infty}\binom{\nu}{k}f^{(k)}\left(\mathfrak{D}_{a+}^{\nu-k}g\right),\ \nu\in\mathbb{R} \quad (25)$$

where $\binom{\nu}{k}$ are the generaliied binomial coeJlcients written in terms of the gamma function.

## Confluent Hypergeometric Equation

Inspired by a similar technique was developed [14] where we investigated how to obtain a solution of the Bessel's equation, in this section we verify the possibility of obtaining solutions of the confluent hypergeometric equation by means of fractional calculus methodology.

We start by considering the standard form of the confluent hypergeometric equation

$$x\frac{\mathbf{d}^2u}{\mathbf{d}x^2}+(c-x)\frac{\mathbf{d}u}{\mathbf{d}x}-au=0, \quad (26)$$

defined for x > 0 with *a, c* ∈ℝ. We now assume that for each *u* = *u(x)* satisfying Equation (26), there exists an *f* = *f(x)* differ integrable such that

$$u=\mathfrak{D}_{0+}^{-1-\alpha}f, \quad (27)$$

where α is an unknown parameter yet to be determined.

So rewriting Equation (26), it follows that

$$x\frac{\mathbf{d}^2}{\mathbf{d}x^2}\mathfrak{D}_{0+}^{-1-\alpha}f+(c-x)\frac{\mathbf{d}}{\mathbf{d}x}\mathfrak{D}_{0+}^{-1-\alpha}f-a\mathfrak{D}_{0+}^{-1-\alpha}f=0 \quad (28)$$

which by means of Eq.(24), assume the form

$$x\mathfrak{D}_{0+}^{1-\alpha}f+(c-x)\mathfrak{D}_{0+}^{-\alpha}f-a\mathfrak{D}_{0+}^{-1-\alpha}f=0, \quad (29)$$

and yet, by the action of $\mathfrak{D}_{0+}^{\alpha}$ applied to the left of each term in Eq.(29) can be rewritten as

$$\mathfrak{D}_{0+}^{\alpha}\left[x\mathfrak{D}_{0+}^{1-\alpha}f\right]+\mathfrak{D}_{0+}^{\alpha}\left[(c-x)\mathfrak{D}\%_{0+}^{-\alpha}f\right]-\mathfrak{D}_{0+}^{\alpha}\left[a\mathfrak{D}\%_{0+}^{-1-\alpha}f\right]=0. \quad (30)$$

Now, if we make use of the fractionalized version of the Leibniz rule as established by Theorem 5, it can be verified the following equalities

$$\mathfrak{D}_{0+}^{\alpha}\left[x\mathfrak{D}_{0+}^{1-\alpha}f\right]=x\frac{\mathbf{d}f}{\mathbf{d}x}+\alpha f, \quad (31)$$

$$\mathfrak{D}_{0+}^{\alpha}\left[(c-x)\mathfrak{D}_{0+}^{-\alpha}f\right]=(c-x)f-\alpha\mathfrak{D}_{0+}^{-1}f, \quad (32)$$

whenever *f* satisfies the following conditions

$$\left(\mathfrak{D}_{0+}^{-\alpha-k}f\right)(0)=0,\mathrm{k}=0,1,\ldots,[1-\alpha].$$

Hence, when Equation (31) and Equation (32) are substituted back in Equation (30) it returns

$$x\frac{\mathbf{d}f}{\mathbf{d}x}+(\alpha+c-x)f+(-a-\alpha)\mathfrak{D}_{0+}^{-1}f=0. \quad (33)$$

Imposing that *α* = - *a*, the last term of Eq.(33) vanishes and the equation reduces to the following equivalences

$$x\frac{\mathbf{d}f}{\mathbf{d}x}+(\alpha+c-x)f=0\Leftrightarrow\frac{1}{\mathrm{f}}\frac{\mathbf{df}}{\mathbf{d}x}=1+\frac{(\mathrm{a\text{-}c})}{\mathrm{x}} \quad (34)$$

which is a separable differential equation whose solution is given by

$$f(x)=Ke^{x}x^{a-c}, \quad (35)$$

with *K*∈ℝ. an arbitrary constant.

Finally, remembering the results of Equation (27) and again applying the fractional Leibniz rule we have that

$$u(x)=\mathfrak{D}_{0+}^{-1+a}\left[Ke^{x}x^{a-c}\right]$$

$$=K\sum_{n=0}^{\infty}\binom{-1+a}{n}\frac{\mathbf{d}^ne^x}{\mathbf{d}x^n}\mathfrak{D}_{0+}^{-1+a-n}\left[x^{a-c}\right]$$

$$=Ke^{x}\sum_{n=0}^{\infty}\binom{-1+a}{n}\frac{\Gamma(a-c+1)}{\Gamma(n+2-c)}x^{n-c+1}$$

$$=Kx^{1-c}e^{x}\sum_{n=0}^{\infty}\binom{-1+a}{n}\frac{\Gamma(a-c+1)}{\Gamma(n+2-c)}x^{n}. \quad (36)$$

The solution *u(x)* just obtained does not resembles, at least a priori, to the canonical form ${}_1F_1(a;c;x)$ that we commonly encounter when solving the confluent hypergeometric equation by means of the Frobenius method, for example. However, it can be promptly modified by means of a sequence of algebraic manipulations as it follows. We begin by rewriting the binomial coefficients in term of the gamma function

$$\binom{-1+a}{n}=\frac{\Gamma(a)}{\Gamma(a-n)n!}. \quad (37)$$

Also, by the famous reflection formula

$$\Gamma(z)\Gamma(1-z)=\frac{\pi}{\sin\pi z}, \quad (38)$$

we have the identities

$$\Gamma(a-n)\Gamma(1-a+n)=\pi\sin\left[\pi(a-n)\right]=\pi(-1)^n\sin(\pi a). \quad (39)$$

Then, substituting Equation (37) and Equation (39) into the expression of Equation (36), we obtain

$$u(x)=K\frac{\Gamma(a)\Gamma(a-c+1)}{\pi\sin(\pi a)}x^{1-c}e^x\sum_{n=0}^{\infty}\frac{\Gamma(1-a+n)}{(-1)^n\Gamma(n+2-c)}\frac{x^n}{n!}. \quad (40)$$

Now if we make use of the Pochhammer symbols $(z)_n=\frac{\Gamma(z+n)}{\Gamma(z)}$ to write down

$$(1-a)_n=\frac{\Gamma(1-a+n)}{\Gamma(1-a)} \text{ and } (2\text{-c})_n=\frac{\Gamma(\text{n+2-c})}{\Gamma(2\text{-c})},$$

then

$$u(x)=K\frac{\Gamma(a)\Gamma(1-a)}{\pi\sin(\pi a)}\frac{\Gamma(a-c+1)}{\Gamma(2-c)}x^{1-c}e^x\sum_{n=0}^{\infty}\frac{(1-a)_n}{(2-c)_n}\frac{(-x)^n}{n!}, \quad (41)$$

nd yet, noticing that if n = 0 in Equation (39), then

$$\frac{\Gamma(a)\Gamma(1-a)}{\pi\sin(\pi a)}=1,$$

and it follows that

$$u(x)=K\frac{\Gamma(a-c+1)}{\Gamma(2-c)}x^{1-c}e^x\sum_{n=0}^{\infty}\frac{(1-a)_n}{(2-c)_n}\frac{(-x)^n}{n!}. \quad (42)$$

With a trivial proper choice of the constant $K$ we may have the equality

$$K\frac{\Gamma(a-c+1)}{\Gamma(2-c)}=1$$

such that our solution for the equation Equation (26) obtained by means of fractional calculus reduces to

$$\begin{aligned} u(x) &= x^{1-c}e^x\sum_{n=0}^{\infty}\frac{(1-a)_n\%(-x)^n}{(2-c)_n\quad n!} \\ &= x^{1-c}e^x{}_1F_1(1-a;2-c;-x). \end{aligned} \quad (43)$$

Finally, using the following identity

$${}_1F_1(a;c;x)=e^x{}_1F_1(c-a;c;-x), \quad (44)$$

it is possible to recognize that

$${}_1F_1(a-c+1;2-c;x)=e^x{}_1F_1(1-a;2-c;-x)$$

so that solution u(x) is of the form

$$u(x)=x^{1-c}{}_1F_1(a-c+1;2-c;x), \quad (45)$$

which we know to be a linear independent solution with ${}_1F_1(a;c;x)$ as long as $c \notin \mathbb{Z}$

Although this methodology does not explicit a second linear independent solution, once we have one of them the other can be found using, for example, the reduction of order method.

## Concluding Remarks

This paper discussed the solution of the confluent hypergeometric equation by means of a methodology associated with fractional calculus. Using the Riemann-Liouville formulation we have obtained one such solution in a very simple way. Comparing this proposed fractional methodology with the standard Frobenius one, we can argue in favor of the former by saying that although the classical Frobenius approach seems to be a more "natural approach", the computational effort of the fractional method seems to be much simpler. Also, the fractional method expose the possibility of highlighting (possibly new) non-trivial relations between the special functions of the mathematical physics, suggesting a fertile ground of study and research in this area. As previously mentioned in the end of the last section, for the second linearly independent solution one can, for example, use the reduction of order method. Although, at this moment, we do not envision a clear approach to get a second linearly independent solution using such fractional method, it is passive of further investigations. Nevertheless, the most important conclusion is that fractional calculus can be used to discuss an ordinary differential equation as an interesting alternative to obtain solutions and such mathematical tool should be encouraged to be investigated and used more frequently in theoretical and applied problems.

### References


1. Tenreiro Machado JA, Galhano AMSF, Trujillo JJ (2014) On Development of Fractional Calculus During the Last Fifty Years. Scientometrics 98: 577-582.
2. Capelas de Oliveira E, Tenreiro Machado JA (2014) A review of Definitions for Fractional Derivatives and Integral. Math Prob Eng.
3. Tenreiro Machado JA, Galhano AMSF, Trujillo JJ (2013) Science Metrics on Fractional Calculus Development since 1966. Fract Calc Appl Anal 16: 479-500.
4. Garra R (2014) Hilfer-Prabhakar derivatives and some applications. Appl Math Comp 242: 576-589.
5. Katugampola UN (2014) A new fractional derivative with classical properties.
6. Khalil R (2014) A new definition of fractional derivative. J Comput Appl Math 264: 65-70.
7. Ortigueira MD, Tenreiro Machado JA (2014) What is a fractional derivative? J Comp Phys 293: 4-13.
8. Mainardi F (2010) Fractional Calculus and Waves in Linear Viscoelasticity. Imperial College Press, London.
9. Rodrigues FG, Oliveira EC (2015) Introduction to the Techniques of the Fractional Calculus to Investigate some Models of Mathematical Physics (in portuguese).
10. Herrmann R (2011) Fractional Calculus, An Introduction for Physicists. Word Scientific Publishling Singapore.
11. Tarasov VE (2010) Fractional Dynamics: Applications of Fractional Calculus to Dynamics of Particles. Fields and Media. Springer-Verlag.
12. Das S (2008) Functional Fractional Calculus for System Identification and Controls. Springer-Verlag, Berlin.
13. Solteiro Pires EJ , Tenreiro Machado JA, de Moura Oliveira PB (2003) Fractional Order Dynamics in a GA Planner. Signal Processing 83: 2377-2386.
14. Rodrigues FG, Oliveira EC (2015) Solution of the Bessel equation via fractional calculus, (in portuguese), at Press Rev Bras Ens Fis.
15. Oldham KB, Spanier J (2002) The Fractional Calculus, Theory and Applications of Differentiation and Integration to Arbitrary Order. Dover Publications, Inc., New York.
16. Diethelm K (2010) The Analysis of Fractional Differential Equations. Springer-Verlag, Berlin.
17. Kilbas AA, Srivastava HM, Trujillo JJ (2006) Theory and Applications of Fractional Differential Equations. Elsevier BV, Amsterdam.
18. Miller KS, Ross B (1993) An Introduction to the Fractional Calculus and Fractional Differential Equations. John Wiley and Sons, Inc., New York.
19. Podlubny I (1998) Fractional Differential Equations. Academic Press, San Diego.
20. Samko SG, Kilbas AA , Marichev OI (1993) Fractional Integrals and Derivatives, Theory and Applications. Gordon and Breach Science Publishers, Amsterdam.
21. Hadamard J (1923) Lectures on Cauchy's Problem in Linear Partial Differential Equations. Yale Univ. Press, New Haven.

# New Analytical Formulae to Calibrate HPGe Well-type Detectors Efficiency and to Calculate Summing Corrections

**Abbas MI[1]* and Ibrahim OA[2]**

[1]*Physics Department, Faculty of Science, Alexandria University, Alexandria, Egypt*
[2]*Physics Department, Faculty of Science, Beirut Arab University, Beirut, Lebanon*

**Abstract**

Direct mathematical method is used in this paper to calculate the photo peak efficiency of a system containing a radioactive source placed in the well of a well type detector. Attenuation of the source and detector's walls are included. In addition summing corrections are also included. The comparison between published experimental values and present calculated values are given. This work proved quite success in predicting the photo peak efficiency of certain detectors even for a radionuclide emitting rays with different energies.

**Keywords:** Photo peak efficiency; Detectors; Photons; Azimuth angle

## Introduction

The direct mathematical method is an analytical method created by Younis S. Selim and Mahmoud I. Abbas to calculate the total or photo peak efficiency of radioactive detectors [1-4].Then it was used by several authors. The comparison between the analytical values and experimental ones shows that the error is very small. Besides, the direct mathematical method is built on simple physical concept and mathematical integrals and it does not need long computer programs like other analytical methods. This analytical method is used to calculate the photo peak efficiency of a radioactive source placed in the well of a well type HPGe detector.

In this paper, the direct mathematical method is explained, then a detailed mathematical perspective is described and a comparison between analytical, experimental and those obtained by Monte Carlo method is given. In addition, new method to correct for summing corrections is discussed.

## Description of our system

Our system is made of a well type semiconductor detector made by Canberra. Its active material (Ge) is mounted in aluminum end cap of thickness 0.5 mm and having an external shielding of lead. The source should be put in the detector well; it can be a "point source" with very small size or a volumetric source that can fit in the well.

The radioactive source contains radioactive nuclide emitting gamma rays. Each nuclide can emit gamma rays with specific energy; the rays emitted can be attenuated by the source material and container walls as well as the detector's walls. Other rays will be absorbed by the detector material. The rays absorbed with in the detector active medium will deposit their energy in it. A ray can deposit all its energy or part of it depending on the way of interaction. The energy deposited will be turned into an electric signal. This electric signal is sent to an amplifier, then to a multichannel analyzer that will record the pulse. Finally a computer will display all the signals depending on their energy Figure 1.

The total efficiency of the system is defined to be the total number of rays detected divided by the total number of rays emitted while the photo peak efficiency is defined to be the number of rays detected within certain energy divided by the number of rays emitted at this energy. In this work we will focus on photo peak efficiency [5].

## Direct mathematical method

This method uses simple mathematical idea to calculate total efficiency as well as photo peak efficiency. It starts from this perspective, when a photon enters the detector active medium, the probability for it to not interact with the detector is given by $e^{-\mu . d}$ where μ is the linear attenuation coefficient expressed in $cm^{-1}$ and d is the distance covered in the detector active medium. So the probability for a photon striking a detector to interact with its material is done by:

$$P = 1 - e^{-\mu_t . d} \quad (1)$$

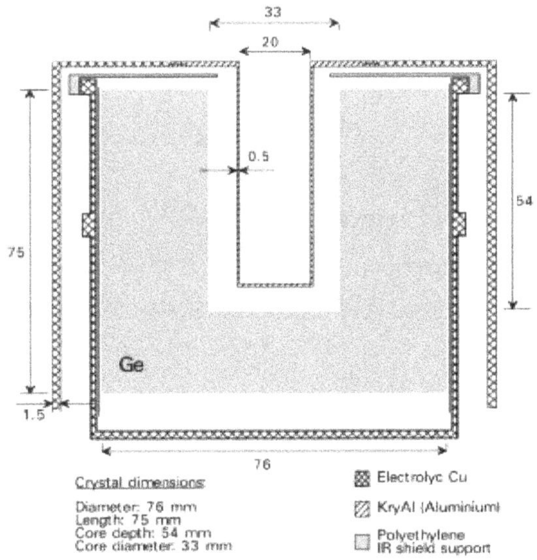


**Figure 1:** Demonstrates the dimensions of the detector.

***Corresponding author:** Abbas MI, Physics Department, Faculty of Science, Alexandria University, Alexandria, Egypt
E-mail: mabbas@physicist.net

Where, is the total attenuation coefficient and it depends on the detector material type and photon's energy. Also the photon might be absorbed by the walls of the detector or by the source material itself. The probability to be absorbed by certain non-detecting material is $e^{-\mu . 1}$. The "attenuation factor" caused by several walls or barriers can be done by [6-10]:

$$a_f = \Pi_i e^{(-\mu_i' . l_i)} i \in [0; n]. \tag{2}$$

Where, $\mu_i'$ is the attenuation coefficient of the *ith* absorber and $l_i$ is the distance covered in that absorber. So in general, the probability for a single photon, to cross n barriers then enter the detector material, and interact with it depositing certain energy will be:

$$P = a_f . \left(1 - e^{-\mu_t . d}\right); \tag{3}$$

While the probability of the same photon to be interacted with the detector depositing all its energy will be:

$$P = a_f . \left(1 - e^{-\mu_p . d}\right); \tag{4}$$

Where, $\mu_p$ is called the photo peak attenuation coefficient. The interactions contributing to determine $\mu_p$ are: photoelectric effect and pair production. While Compton scattering does not contribute unless followed by photoelectric effect. Measuring efficiency for certain system requires two main things, the first one is full awareness of the geometry of the system to calculate the distance a photon emitted at certain polar and azimuth angle will cover in detecting and attenuating materials. Besides it requires understanding the different kinds of interactions between the photon and these materials. In addition, detector dead time and summing out effects must be known also.

So for a point source the efficiency in general will be the sum 'integration" of all rays detected at any polar or azimuth angle divided by the total number of rays emitted at certain energy or the integration-sum of the probability function (equation.3) over the entire solid angle divided by the total solid angle: 4π. The general rule is:

$$\varepsilon = \frac{\int_0^{\pi} \int_0^{2\pi} a_f \left(1 - e^{-\mu_t . d}\right) \sin\theta d\varphi d\theta}{4\pi}; \tag{5}$$

To calculate the photo peak efficiency of the same point source, we replace $\mu_t$ by $\mu_p$.

## Mathematical Perspective

### Axial point source

To derive an equation for this efficiency, we will start by studying the different path lengths allowed for a photon starting from a point source. Then we will write a mathematical equation covering the entire solid angle, which is a developed form of equation 5 to deal with different functions of the distance covered in the detector "d".

**Allowed path lengths in detecting and attenuating materials:**

For a photon created by an axial point source, five path lengths are allowed for a photon in the detecting material. All the path lengths functions are illustrated in Figures 2 and 3.

1. If the photon enters the detector from the well's bottom and leaves from the bottom the path length "$d_1$" followed will be:

$$d_1 = h/\cos(\theta) \tag{6}$$

2. If the photon enters from the well's bottom and leaves from the outer side, the path length function will be:

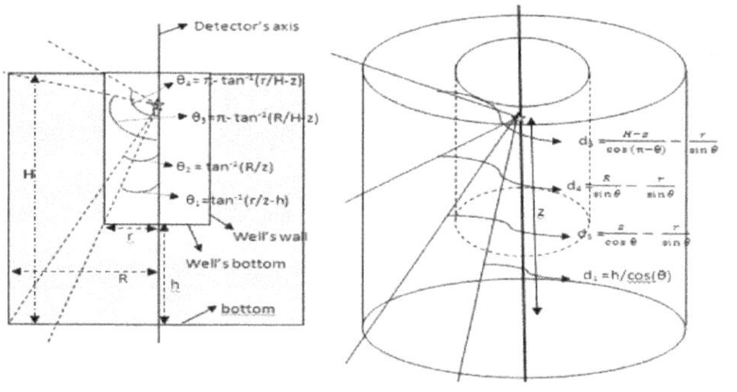

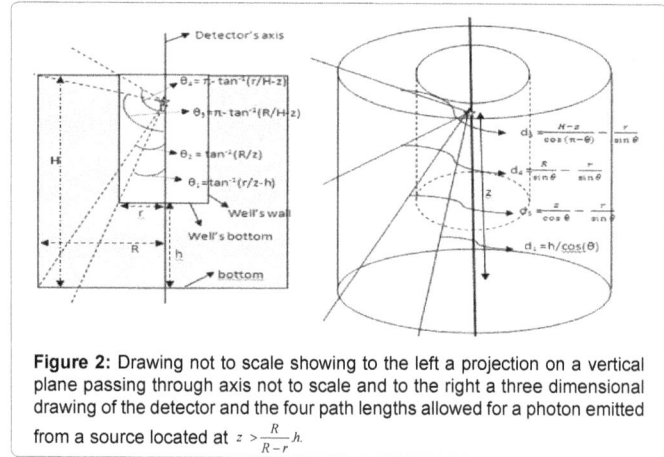

**Figure 2:** Drawing not to scale showing to the left a projection on a vertical plane passing through axis not to scale and to the right a three dimensional drawing of the detector and the four path lengths allowed for a photon emitted from a source located at $z > \frac{R}{R-r} h$.

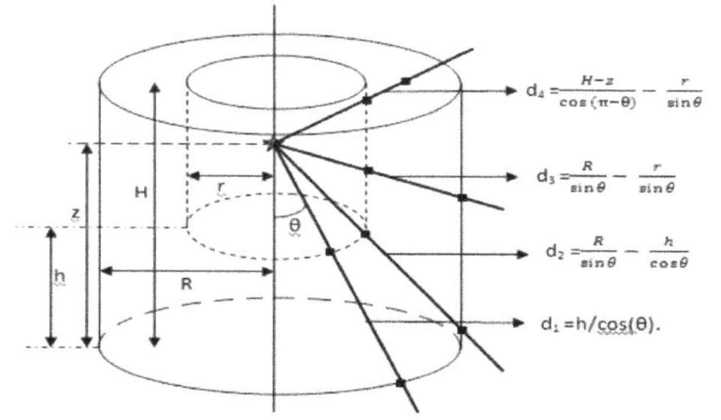


**Figure 3:** Three dimensional drawing not to scale of our detector showing the four path length allowed for a photon emitted from a point source located at height "z" less than $\frac{R}{R-r} h$.

$$d_2 = \frac{R}{\sin\theta} - \frac{h}{\cos\theta} \tag{7}$$

3. If the photon enters the detector material from the well's side and leaves from the bottom:

$$d_3 = \frac{z}{\cos\theta} - \frac{r}{\sin\theta} \tag{8}$$

4. If the photon enters the detector from the well's side and leaves from the outer side. The path length "$d_3$" followed will be:

$$d_4 = \frac{R}{\sin\theta} - \frac{r}{\sin\theta} \tag{9}$$

If the photon enters from the well's side and leaves from the face, the path length function will be:

$$d_5 = \frac{H - z}{\cos(\pi - \theta)} - \frac{r}{\sin\theta} \tag{10}$$

While the distance traveled by the photon in the non-detecting materials usually source container or detector shielding is simply, $\frac{t}{\sin\theta}$ if the photon crosses the wall of source or shielding; and $\frac{t}{\cos\theta}$ if the photon is crossing the bottom of the container or the well's bottom; t is the thickness of the absorber.

**Photo peak efficiency:** Since we are dealing with an axial point source, the azimuth angle does not affect the path length of a photon, the later depends only on the polar angle. We will start our study by defining four polar angles that are important to our study.

Which are : $\theta_1 = \tan^{-1}\frac{r}{z-h}$; $\theta_2 = \tan^{-1}\frac{R}{z}$; $\theta_3 = \pi - \tan^{-1}\frac{R}{H-z}$; and $\theta_4 = \pi - \tan^{-1}\frac{r}{H-z}$. These polar angles are shown in Figure 2.

We have to discuss two cases:

A.If the point is located at a height more than $\frac{R}{R-r}.h$. (Figure 2):

For $0<\theta<\theta_1$; the photon has to follow path length: $d_1$.

For $\theta_1<\theta<\theta_2$; the photon has to follow path length: $d_3$

For $\theta_2<\theta<\theta_3$; the photon has to follow path length: $d_4$.

For $\theta_3<\theta<\theta_4$; the photon has to follow path length: $d_5$.

For $\theta_4<\theta<\pi$; the photon will never strike the detector.

So the photo peak efficiency is given by:

$$\varepsilon = \frac{\int_0^{\theta_1}\int_0^{2\pi} p_1 \sin\theta d\varphi d\theta + \int_{\theta_1}^{\theta_2}\int_0^{2\pi} p_3 \sin\theta d\varphi d\theta + \int_{\theta_2}^{\theta_3}\int_0^{2\pi} p_4 \sin\theta d\varphi d\theta + \int_{\theta_3}^{\theta_4}\int_0^{2\pi} p_5 \sin\theta d\varphi d\theta}{4\pi} \quad (11)$$

Where $p_i = a_f \cdot \left(1 - e^{-\mu d_i}\right)$

B. If the point source located at a height less than $\frac{R}{R-r}.h$. (Figure 3):

For $0<\theta<\theta_2$; the photon has to follow the path length: $d_1$.

For $\theta_2<\theta<\theta_1$; the photon has to follow the path length: $d_2$.

For $\theta_1<\theta<\theta_3$; the photon has to follow the path length: $d_4$.

For $\theta_3<\theta<\theta_4$; the photon has to follow the path length: $d_5$.

For $\theta_3<\theta<\pi$; the photon will never strike the detector.

In this case the photo peak efficiency will be:

$$\varepsilon = \frac{\int_0^{\theta_2}\int_0^{2\pi} p_1 \sin\theta d\varphi d\theta + \int_{\theta_2}^{\theta_1}\int_0^{2\pi} p_2 \sin\theta d\varphi d\theta + \int_{\theta_1}^{\theta_3}\int_0^{2\pi} p_4 \sin\theta d\varphi d\theta + \int_{\theta_3}^{\theta_4}\int_0^{2\pi} p_5 \sin\theta d\varphi d\theta}{4\pi} \quad (12)$$

## Non Axial point Source

For a point source not located on the axis of detector the path length functions allowed depend on two things only where the photon enters the detecting material and where it leaves. These things in turn depend on the location of the source (its height "z" and the axial distance "ρ") as well as the polar and azimuth angle (Figures 4-6).

**Allowed path Lengths:** In general for a photon generated at a point source S (ρ; z), the path length it can follow can be:

- If it enters from the well's bottom and leaves from the bottom. The path length it can follow is $d_1$ (as identified before in equation 6).
- If it enters from the well's bottom and leaves from the side:

$$d_2 = \frac{\pm\rho.\cos(\varphi) + \sqrt{R^2 - (c.\sin\varphi)^2}}{\sin\theta} - \frac{z-h}{\cos\theta} \quad (13)$$

If it enters from the side and leaves from the bottom:

$$d_3 = \frac{z}{\cos\theta} - \frac{\pm\rho.\cos(\varphi) + \sqrt{r^2 - (\rho.\sin\varphi)^2}}{\sin\theta} \quad (14)$$

4. If it enters from the side and leaves from the side:

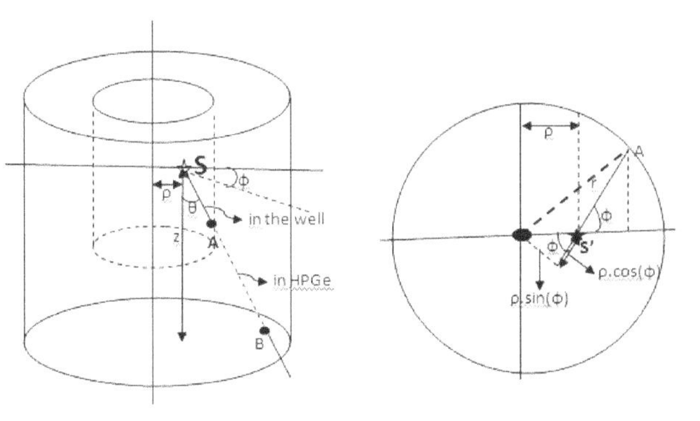


**Figure 4:** Clarifies the path length function $d_2'$; the distance covered in the detector is AB, shown as dashed line to the left. It is clear the total distance SB is z. cosθ but AB=SB- SA. To get SA we made a projection on anhorizontalplane containing the point A. the projection of SA: "S'A" is simply $S'A = \sqrt{r^2 - (\rho.\sin\varphi)^2} - \rho.\cos(\varphi)$ as it is obvious by simple geometry of triangle AHO. $SA = \frac{S'A}{\sin(\theta)}$.

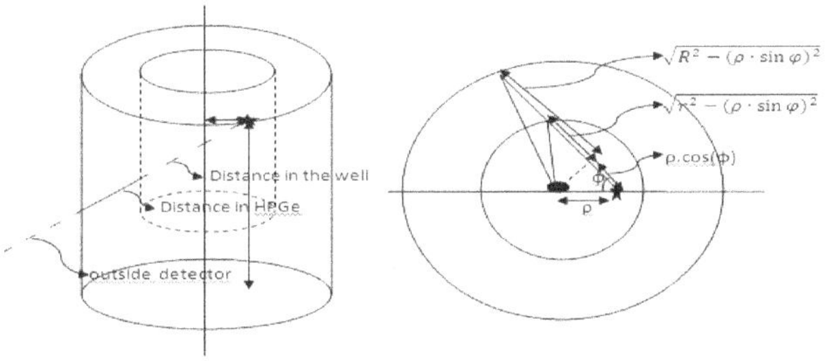


**Figure 5:** Clarifies the path length $d_4'$. The solid segment AB "to the left" represents the distance covered in HPGe Material. By projecting on a horizontal plane "to the right"we obtain A'B' the projection of AB.

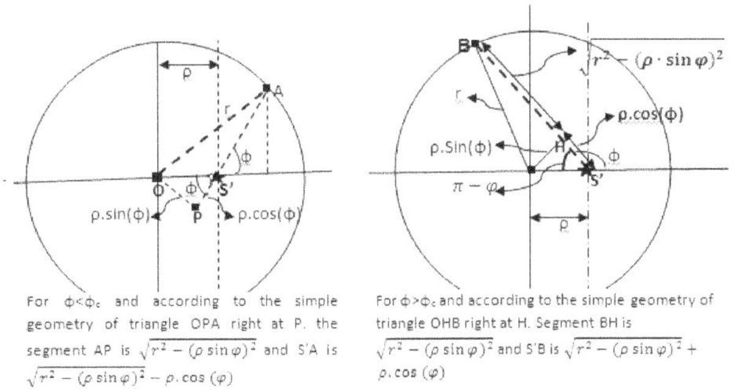


**Figure 6:** Shows a projection on a horizontal plane containing the well bottom to discuss the ± in the equation of $d_2'$ according to this figure. This "±" turns – for $\varphi<\varphi_c$ and + for $\varphi>\varphi_c$.

$$d_4 = \frac{+\sqrt{R^2 - (c.\sin\varphi)^2}}{\sin\theta} - \frac{\sqrt{r^2 - (c.\sin\varphi)^2}}{\sin\theta} \quad (15)$$

A'B' is $\sqrt{R^2 - (\rho.\sin\varphi)^2} - \sqrt{R^2 - (\rho.\sin\varphi)^2}$

5. If it enters from the side and leaves from the upper face:

$$d_5 = \frac{H-z}{\cos(\pi-\theta)} - \frac{\pm c.\cos(\varphi) + \sqrt{R^2 - (c.\sin\varphi)^2}}{\sin(\pi-\theta)} \quad (16)$$

**Photo peak Efficiency:** Studying the photo peak efficiency of a non-axial point source is more complicated than the axial source case, because we will study the path length functions allowed according to polar and azimuth angle both range from 0 to π. We define four new polar angles illustrated in Figure 7:

$$\theta_1 = \tan^{-1}\left(\frac{r-\rho}{z-h}\right);\theta_2 = \tan^{-1}\left(\frac{r+\rho}{z-h}\right);\theta_3 = \tan^{-1}\left(\frac{R-\rho}{z}\right);\theta_4 = \tan^{-1}\left(\frac{R+\rho}{z}\right)$$

**In the lower part of the detector (located below the point source)** $\left(0<\theta<\frac{\pi}{2}\right)$

A projection is shown in Figure 7 on a horizontal plane containing the well's bottom to understand the different places the photon can enter from.

According to this projection:

For $0<\theta_1$, the photon has to enter from the well's bottom.

For $\theta_1<\theta<\theta_2$, the photon can enter either from the well's bottom or well's side depending on the azimuth angle: φ.

If $\varphi<\varphi'_c$; "illustrated in figure 7" the photon enters from the well's bottom.

If $\varphi>\varphi'_c$; the photon enters from the well's wall.

$$\varphi_c' = \cos^{-1}\frac{\left(\frac{z-h}{\tan\theta}\right)^2+\rho^2-r^2}{2.\rho.\left(\frac{z-h}{\tan\theta}\right)} \quad (17)$$

For $\theta>\theta_2$; the photon must enter from the wall.

A similar projection on a horizontal plane containing the detector's bottom can show where the photon can show the detector:

For $0<\theta_3$, the photon has to leave from the bottom

For $\theta_3<\theta<\theta_4$ the photon can leave either from the bottom or from the wall depending on the azimuth angle: φ.

If $\varphi<\varphi_c$; "indicated in figure" the photon leaves from the wall

If $\varphi>\varphi_c$; the photon leaves from the wall.

$$\varphi_c = \cos^{-1}\frac{\left(\frac{z}{\tan\theta}\right)^2+\rho^2-R^2}{2.\rho.\left(\frac{z}{\tan\theta}\right)} \quad (18)$$

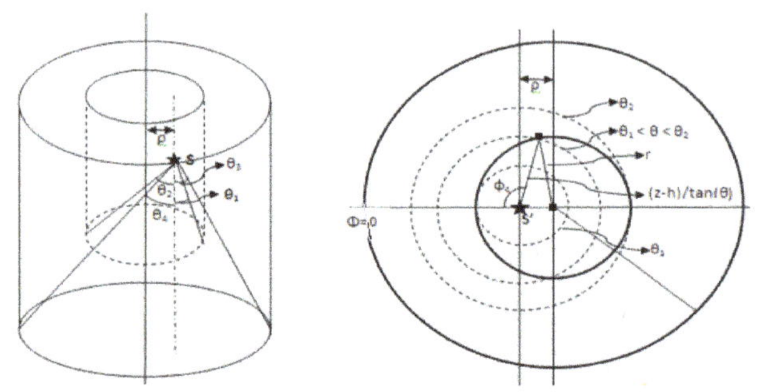


**Figure 7:** Shows a three dimensional drawings of our system with the four angles $\theta_1$; $\theta_2$; $\theta_3$ and $\theta_4$ to the left and a projection on a horizontal plane containing the well's bottom. Equation of $\varphi_c$'can be applying the cosine law in the triangle S'AO:

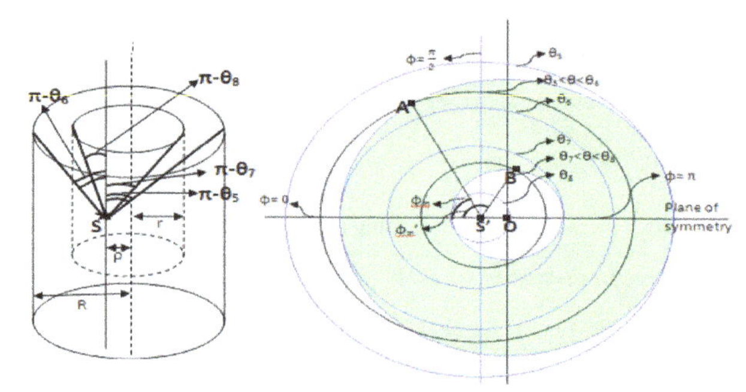


**Figure 8:** Shows to the left a 3 Dimensionalillustration the polar angles: $\theta_5$; $\theta_6$; $\theta_7$ and $\theta_8$.Projection of these polar angles $\theta_5$; $\theta_6$; $\theta_7$ and $\theta_8$ on the upper surface of the well and detector is shown to the right. Also $\varphi_m$ and $\varphi_m$' are shown for two random angles. The simple geometry of triangles S'OA and S'OB allow us to derive the equations of $\varphi_m$ and $\varphi_m$'.

| | $0<\theta_3$ | $\theta_3<\theta<\theta_4$ $\varphi<\varphi'_c$ | $\theta_3<\theta<\theta_4$ $\varphi>\varphi'_c$ | $\theta_4<\theta<\frac{\pi}{2}$ |
|---|---|---|---|---|
| $0<\theta_1$ | $d_1$ | $d_2'$ | $d_1$ | $d_2'$ |
| $\theta_1<\theta<\theta_2$ $\varphi<\varphi_c$ | $d_3'$ | $d_4'$ | $d_3'$ | $d_4'$ |
| $\theta_1<\theta<\theta_2$ $\varphi>\varphi_c$ | $d_1$ | $d_2'$ | $d_1$ | $d_2'$ |
| $\theta_2<\theta<$— | $d_3'$ | $d_4'$ | $d_3'$ | $d_4'$ |

**Table 1:** shows the path length followed by the photon depending on its polar and azimuth angle for $\theta<\frac{\pi}{2}$.

For $\theta>\theta_4$ the photon must leave from the wall.

Note that: the equation of $\varphi_c$ can be obtained similar way as $\varphi'_c$ by just replacing "r" by "R" and "z-h" by "z" shown in Table 1.

$$\overline{S'A}^2+\overline{S'O}^2-2\overline{S'O}.\overline{S'A}.\cos(\varphi'_c)=\overline{AO}^2.\ \text{So}\ \cos(\varphi'_c)=\frac{\overline{S'A}^2+\overline{S'O}^2-\overline{AO}^2}{2\overline{S'O}.\overline{S'A}}$$

**In the upper part of the detector (located above the point source)** $\theta>\frac{\pi}{2}$**:**

To study the path length function followed by a photon striking the upper part of the detector we will start by defining five new angles:

$$\theta_5=\frac{\pi}{2}+\tan^{-1}\tan^{-1}\frac{H-z}{R+\rho};\ \theta_6=\frac{\pi}{2}+\tan^{-1}\frac{H-z}{R-\rho};\theta_7=\frac{\pi}{2}+\tan^{-1}\frac{H-z}{r+\rho};\theta_8=\frac{\pi}{2}+\tan^{-1}\frac{H-z}{r-\rho}$$

For $\frac{\pi}{2}<\theta<\theta_5$; the photon must enter from the well's wall and leave from the detector's wall so it must follow path length $d_4$'.

For $\theta_5<\theta<\theta_6$; the photon must enter from the wall but it can leave either from upper side or wall depending on the azimuth angle φ. It leaves from upper face for $\varphi<\varphi_m$ and from the wall for $\varphi>\varphi_m$.

$$\varphi_m=\varphi\cos^{-1}\frac{(\frac{H-z}{\tan\theta})^2+\rho^2-R^2}{2.\rho.(\frac{H-z}{\tan\theta})} \quad (19)$$

For $\theta_6<\theta<\theta_7$; the photon must enter from the well's wall and leave from the detector's upper surface following path length $d_5$'.

For $\theta_7<\theta<\theta_8$; the photon can enter the detector from the wall and leave from upper face following path length $d_5$' and not strike the

| | $\frac{\pi}{2}<\theta<\theta_5$ | $\theta_5<\theta<\theta_6$ | $\theta_6<\theta<\theta_7$ | $\theta_7<\theta<\theta_8$ |
|---|---|---|---|---|
| $\varphi<\varphi_m'$ | $d_4'$ | $d_4'$ | $d_5'$ | $d_5'$ |
| $\Phi_m'<\varphi$ | $d_4'$ | $d_5'$ | $d_5'$ | $d_5'$ |
| $\varphi<\varphi_m$ | $d_4'$ | $d_5'$ | $d_5'$ | $d_5'$ |
| $\varphi>\varphi_m$ | $d_4'$ | $d_5'$ | $d_5'$ | 0 |

**Table 2:** Shows the different path length functions allowed for a photon crossing the upper part of the detector $(\theta>\frac{\pi}{2})$.

| E (keV) | Expt. [5] | Present work | M.C. [5] |
|---|---|---|---|
| 46.5 | 67.68 | 61.2 | 72.77 |
| 59.5 | 72.3 | 73.3 | 77.1 |
| 88 | 76.75 | 75.3 | 80.87 |
| 122 | 73 | 74.7 | 77.8 |
| 279 | 40.5 | 41.4 | 41.99 |
| 392 | 28.26 | 28.5 | 30.09 |
| 514 | 22.1 | 21.6 | 22.79 |
| 661 | 17.76 | 16.9 | 18.56 |
| 898 | 7.64 | 7.7 | 8.34 |
| 1173 | 5.6 | 5.6 | 6.25 |
| 1332 | 4.77 | 4.63 | 5.46 |
| 1863 | 3.45 | 2.9 | 3.82 |

**Table 3:** Comparison between different values calculated using our program with experimental ones for a source constituting of a plastic vial of diameter 1.5 cm containing 3 cm³ of water in which the following radioactive nuclides are dissolved.

detector for $\varphi>\varphi_m'$ shown in Table 2.

$$\varphi_m' = \cos^{-1}\frac{(\frac{H-z}{\tan\theta})^2+\rho^2-r^2}{2.\rho.(\frac{H-z}{\tan\theta})} \tag{20}$$

For $\theta>\theta_8$; the photons leave without touching the detecting material.

So the point source efficiency will be:

$$\begin{aligned}\varepsilon = \frac{1}{2\pi}\{&\int_0^{\theta_1}\int_0^{\pi} p_1.i_1 \sin\theta d\varphi d\theta, + \int_0^{\theta_1}\int_0^{\pi} p_2.i_2 \sin\theta d\varphi d\theta + \int_0^{\theta_1}\int_0^{\pi} p_1.i_3 \sin\theta d\varphi d\theta + \int_0^{\theta_1}\int_0^{\pi} p_2.i_4 \sin\theta d\varphi d\theta \\ &+ \int_{\theta_1}^{\theta_2}\int_0^{\delta c} p_3.i_1 \sin\theta d\varphi d\theta + \int_{\theta_1}^{\theta_2}\int_0^{\delta c} p_4.i_2 \sin\theta d\varphi d\theta + \int_{\theta_1}^{\theta_2}\int_0^{\delta c} p_3.i_3 \sin\theta d\varphi d\theta + \int_{\theta_1}^{\theta_2}\int_0^{\delta c} p_4.i_4 \sin\theta d\varphi d\theta + \\ &\int_{\theta_1}^{\theta_2}\int_{\delta c}^{\pi} p_1.i_1 \sin\theta d\varphi d\theta + \int_{\theta_1}^{\theta_2}\int_{\delta c}^{\pi} p_2.i_2 \sin\theta d\varphi d\theta + \int_{\theta_1}^{\theta_2}\int_{\delta c}^{\pi} p_1.i_3. \sin\sin\theta d\varphi d\theta + \int_{\theta_1}^{\theta_2}\int_{\delta c}^{\pi} p_2.i_4 \sin\theta d\varphi d\theta + \\ &\int_{\theta_2}^{\frac{\pi}{2}}\int_0^{\pi} p_3.i_1 \sin\theta d\varphi d\theta + \int_{\theta_2}^{\frac{\pi}{2}}\int_0^{\pi} p_4.i_2 \sin\theta d\varphi d\theta + \int_{\theta_2}^{\frac{\pi}{2}}\int_0^{\pi} p_3.i_3 \sin\theta d\varphi d\theta + \int_{\theta_2}^{\frac{\pi}{2}}\int_0^{\pi} p_4.i_4 \sin\theta d\varphi d\theta + \\ &\int_{\frac{\pi}{2}}^{\theta_5}\int_0^{\pi} p_4 \sin\theta d\varphi d\theta + \int_{\theta_5}^{\theta_6}\int_0^{\varphi_m'} p_5 \sin\theta d\varphi d\theta + \int_{\theta_5}^{\theta_6}\int_{\varphi_m'}^{\pi} p_4 \sin\theta d\varphi d\theta + \int_{\theta_6}^{\theta_7}\int_0^{\pi} p_5 \sin\theta d\varphi d\theta + \int_{\theta_7}^{\theta_8}\int_{\varphi_m}^{\pi} p_5 \sin\theta d\varphi d\theta\end{aligned} \tag{21}$$

Where, $i_1$ is a function that turns to 1 when $\theta<\theta_3$ and turns to zero if else. $i_2$ is a function that turns to 1 when $\theta_3<\theta<\theta_4$ and simultaneously $\varphi<\varphi_c'$; but turns to zero if else., $i_3$ is a function that turns to 1 when $\theta_3<\theta<\theta_4$ and $\varphi>\varphi_c'$; but turns to zero if else., $i_4$ is a function that turns to 1 if $\theta>\theta_4$ and turns to zero if else.

## Efficiency of a cylindrical source

The efficiency of a coaxial disc source of radius S (S<r) placed inside the well can be done by integrating over dρ and ρ ranges from 0 to S. the integration over dφ is useless because of the symmetry:

$$\varepsilon = \frac{2}{S^2}\times\int_0^S \varepsilon_P\times\rho\times d\rho \tag{22}$$

| E (keV) | Expt. [5] | Present work | M.C. [5] |
|---|---|---|---|
| 59.5 | 59.9 | 58.3 | 64.4 |
| 88 | 60.5 | 64.3 | 66.06 |
| 122 | 56.35 | 63.4 | 64.3 |
| 279 | 31.08 | 31.1 | 33.98 |
| 392 | 21.85 | 23.6 | 24.25 |
| 514 | 16.86 | 18.06 | 18.57 |
| 661 | 13.78 | 14.2 | 14.84 |
| 834 | 11.02 | 11.2 | 12.21 |
| 898 | 6.43 | 6.4 | 7.5 |
| 1173 | 4.72 | 5.1 | 5.57 |
| 1333 | 4.1 | 4.41 | 4.62 |
| 1863 | 3.03 | 2.63 | 3.32 |

**Table 4:** Comparison between different values calculated using our program, Monte Carlo method and experimental ones for a source with 8 cm³ volume of water that contains same radionuclides.

Where $\varepsilon_P$ is the efficiency of a point source as identified before in equation 21.In case of cylindrical sources of radius S and height H, we have to integrate over dρ and dh:

$$\varepsilon = \frac{2}{S^2\times H}\times\int_0^H\int_0^S \varepsilon_P\times\rho\times d\rho\times dh \tag{23}$$

This equation in its developed form was used to calculate the theoretical values represented in Tables 3 and 4.

## Summing out Correction

Since we are dealing with photo peak efficiency; it is crucial to be aware of the signals resulting from the simultaneous detection of two rays. The result will be a pulse recorded within a channel of height corresponding to the sum of energies of both rays. This is frequent when a radioactive isotope emitting two or more gamma lines in its decay. There is a possibility that one ray will interact with the detector material depositing all of its energy in it, so it is expected to be recorded in the full-energy peak (photo peak), but the other might deposit part of its energy or all of it. Thus the total energy deposited will be higher than the one corresponding to the full energy peak of the first and the first ray will never be recorded in the full-energy peak.

In our program we added the effect of the summing correction. For example Cobalt-60 emits simultaneously two gamma lines with energies 1173 and 1332keV. According to equation5, the probability of a 1173keVgamma line to be recorded in the full-energy peak should be $a_f\ (1-e^{-\mu_{p1}d})$ , where $\mu_{p1}$ is the photo peak attenuation coefficient of the photons having 1173keV energy. This probability will decrease because of the 1332keV gamma line is depositing its energy. The probability of 1332-ray to escape the detector will be $e^{-\mu_{t2}d}$, where $\mu_{t2}$ here is the total attenuation coefficient of the photons having 1332 keV energy. So equation 5 turns to:

$$P = a_f.(1-e^{-\mu_{p1}d}).e^{-\mu_{t2}d} \tag{24}$$

## Results and Discussion

Equations 11, 12, 23 and 24were used to get the efficiency of point sources and cylindrical sources. Comparison between the calculated values in our work, those calculated using Monte Carlo method and experimental values is tabulated in the following table.

$^{210}$Pb (46 keV), $^{241}$Am (59 keV), $^{109}$Cd (88 keV), $^{57}$Co (122 keV), $^{203}$Hg (279 keV), $^{113}$Sn (391 keV), $^{85}$Sr (514 keV), $^{137}$Cs (661 keV), $^{54}$Mn (834 keV), $^{88}$Y (898 and 1836 keV) and $^{60}$Co (1173 and 1332 keV).

## Conclusion

The present work offers straightforward mathematical expressions to calibrate HPGe well type detectors over a large energy range without the need for standard sources, norlong computer programs as for other simulation methods. In addition, the summing corrections are also presented in a simple mathematical expression. Tables 3 and 4 shows the differences between analytical and experimental values are very small. Direct mathematical method proves quite success even when some radionuclides placed in the well of the detector emit several gamma lines.

### References

1. Selim YS, Abbas M I (1996) Direct Calculation of the total efficiency of cylindrical scintillation detectors for extended circular sources. Radiation Physics and Chemistry 48: 23-27.
2. Selim YS, Abbas MI, Fawzy MA (1998) Analytical calculation of the efficiencies of gamma scintillators efficiencies. I: total efficiency for coaxial disk sources. Radiant Physics and Chemistry 53: 589-592.
3. Selim YS, Abbas MI (2000) Analytical calculations of gamma scintillators efficiencies. II: total efficiency for wide co-axial disk sources. Radiant Physics and Chemistry 53: 15-19.
4. Abbas MI (2001) HPG detector photo peak efficiency calculation including self-absorption and coincidence corrections for Marinilli beaker sources using compact analytical expressions. Appl Radiant Isotope 54: 761-768.
5. Laborie JM, Petit GL, Girard DMA (2002) Monte Carlo calculation of the efficiency response of a low-background well-type HPG detector. Nuclear Instruments and Methods in Physics Research 479: 618-630.
6. Badawi MS, El-Khatib AM, Krar ME (2013) New Numerical Simulation Approach to Calibrate the NaI (Tl) Detectors Array Using Non-Axial Extended Spherical Sources. Journal of Instrumentation 8: 1-11.
7. Badawi MS, Elzaher MA, Thabet AA, El-khatib AM (2013) An empirical formula to calculate the full energy peak efficiency of scintillation detectors. Appl Radiant Isotopes 74: 46-49.
8. Jaderstrom H (2010) Coincidence summing correction in Genie 2000 Canberra Solutions.
9. Abbas MI (2010) Analytical formulae for borehole scintillation detectors efficiency calibration. Nuclear Instruments and Methods in Physics Research 622: 171-175.
10. Abbas MI (2010) A new analytical method to calibrate cylindrical phoswich and $LaBr_3$ (Ce) scintillation detectors. Nuclear Instruments and Methods in Physics Research 621: 413-418.

# 4

# An Existence Result for Impulsive Stochastic Functional Differential Equations with Multiple Delays

**Anguraj A* and Banupriya K**

*P.S.G. College of Arts and Science, Coimbatore, Tamil Nadu, India*

**Abstract**

In this paper we consider Impulsive stochastic neutral functional differential equations with multiple delays. By using Schaefer's fixed point theorem, we prove the existence of solutions for stochastic differential equations with impulses.

**Keywords:** Impulsive differential equations; Stochastic differential equations; Multiple delay; Fixed point theorem

**AMS Mathematical Subject Classification:** 34A37, 34K50, 34K45, 60H10.

## Introduction

The theory of impulsive differential equations is an important area of scientific activity. Many evolution processes are characterized by the fact that at certain moments of time they experience a change of state abruptly. These short term perturbations act instantaneously, that is in the form of impulses. For example, that many biological phenomena involving thresholds, optimal control models in economics and frequency modulated systems, do exhibit impulsive effects. So the impulsive differential equations appear as a natural description of observed evolution phenomena of several real world problems. Existence of solutions of impulsive differential equations has been studied by many authors. If the impulses are random the solution becomes a stochastic process. Existence of solutions of differential equations with random impulses have been studied by many authors [1-3].

Furthermore, besides impulsive effects, stochastic effects likewise exist in real systems. There is a wide range of interesting process in robotics, economics and biology that can be described as differential equations with non-deterministic dynamics such phenomena are described by stochastic differential equations. The solution of stochastic differential equation is a stochastic process. However the solution of differential equation with random impulses is different from the solution of stochastic differential equations. Existence, Uniqueness and qualitative analysis of solutions of stochastic differential equations have discussed by several authors [4,5].

Since both impulsive and stochastic effects exist it is very difficult to investigate the existence of solution of impulsive stochastic differential equations. In [6] Anguraj and Vinodkumar discussed the existence, uniqueness and stability of impulsive stochastic semi linear neutral functional differential with infinite delays. Lakrib [7] discussed about the existence results for impulsive neutral functional differential equations with multiple delays. Based on the existing literature, stochastic impulsive differential equations involved mainly on controlability and stability. To the best of our knowledge, there is no work reported on impulsive stochastic differential equations with multiple delays. The purpose of this paper is to discuss about the existence results of impulsive stochastic neutral functional differential equations with multiple delays. Our approach is based on Schaefer's fixed point theorem.

In this paper we study the existence results for stochastic impulsive differential equations with multiple delays

$$d[x(t)-f(t,x_t)]=[g(t,x_t)+\sum_{i=1}^{p}x(t-\tau_i)]dt+a(t,x_t)dB_t \qquad (1.1)$$

$$a.e.,\ t\in J=[0,1], t\neq t_k, k=1,2,...m.$$

$$\Delta_x(t_k)=I_k(x(t_k)),\ k=1, 2, ...m$$

$$x_0=\varphi$$

where $f: J\times R^n\to R^n$, $g: J\times R^n\to R^n$ and $a: J\times R^n\to R^n$ are Borel Measurable functions, $J_0=[-r, 0]$, r=max $\{\tau i : i=1, 2, ...p\}$ and $\varphi$: $[-r, 0]\to R_n$. Further-more the fixed moments of time $t_k$ satisfy $0=t_0<t_1<.....<t_m<t_{m+1}=1$, where $x(t_k^+)$ and $x(t_k^-)$ represent the right and left limits of x(t) at $t=t_k$, respectively. And $\Delta x(t=t_k)=x(t_k^+)-x(t_k^-), k=1,2,...,m$ represent the jump in the state at time $t_k$ with $I_k$ determining the size of the jump.

## Preliminaries

Let $(\bar{\Omega}, F\{F_t\}_{t\geq 0}, P)$ be a complete probability space with a filtration $\{Ft\}_{t\geq 0}$ satisfying the conditions that it is right continuous and $F_0$ contains all P- null sets and $w(t)=(w_1(t), ....w_m(t))^T$ is an m-dimensional Brownian motion defined on $(\bar{\Omega}, F\{F_t\}_{t\geq 0}, P)$. Let C=C $[[-r, 0], R^n]$ denote the family of all continuous $R_n$–valued function $\varphi$ on [−r, 0] with the norm $\|\phi\|=sup_{-r\leq\theta\leq 0}|\phi(\theta)|$ where |.| is Euclidean norm of $R_n$. Denote by $C^b_{F_0}[[-r,0],R^n]$ the family of all bounded F0- measurable, $C[[-r,0],R^n]$ -valued random variables $\varphi$, satisfying $\|\phi\|^2_{L^2}=sup_{-r\leq\theta\leq 0}E|\phi(\theta)|^2<\infty$, where E denotes the expectation of stochastic process [8-10]. The initial condition $\phi\in C^b_{F_0}[[-r,0],R^n]$.

Let PC(J, $R^n$) the space of piecewise continuous functions x: J → $R^n$ such that x is continuous everywhere except for $t=t_k$ at which $x(t_k^-)$ and $x(t_k^+)$ exist and $x(t_k^-)=x(t_k), k=1,...m$. If we set Ω={x: $J_1\to R^n$, x ∈ $R^n$ ∩ PC (J, $R^n$)} where $J_1$=[−r, 1] then Ω is a Banach space normed by $\|x\|=sup\{|x(t)|:t\in J_1\},\ x\in\Omega.$

***Corresponding author:** Anguraj A, P.S.G. College of Arts and Science, Coimbatore-641 014, Tamil Nadu, India
E-mail: angurajpsg@yahoo.com

Obviously, for any $x \in \Omega$ and $t \in J$, the function $x_t$ defined by $x_t(\theta)=x(t+\theta)$, for $\theta \in J_0$, belongs to $R_n$.

By $L^1(J, R^n)$ we denote the Banach space of measurable functions $x: J \to R^n$ which are Lebesgue integrable, normed by

$$\|x\| = \int_0^1 x(t)dt$$

**Definition 2.1:** A stochastic process $\{x(t) \in C^b_{F_0}[[-r,0],R^n], t \in [-r,1]\}$, is called a solution of equations (1.1) if

(i) $x(t)$ is $F_t$ adapted;

(ii) $x(t)$ satisfies the integral equation

$$x(t) = \begin{cases} \phi(t) & for\ t \in J_0, \\ \phi(0) - f(0,\phi(0)) + f(t,x_t) + \int_0^t g(s,x_s)ds + \sum_{i=1}^{p} \int_{-\tau_i}^{0} \phi(s)ds \\ +\sum_{i=1}^{p} \int_0^{t-\tau_i} x(s)ds + \int_0^t a(s,x_s)dB(s) + \sum_{0<t_k<t} I_k(x(t_k^-)) & for\ t \in J \end{cases} \quad (2.1)$$

**Definition 2.2:** (Schaefer fixed point theorem). Let X be a normed linear space and let $\Gamma: X \to X$ be a completely continuous map, that is, it is a continuous mapping which is compact on each bounded subset of X. If the set $\zeta=\{x \in\in X: \lambda x=\Gamma x \text{ for some } \lambda>1\}$ is bounded, then $\Gamma$ has a fixed point [11].

## Hypotheses

$H_1$: The function f, g: $J \times R^n \to R^n$ is such that

$$\|f(t,x_t\|^2 \vee \|g(t,x_t\|^2 \le c_1 \|x_t\|^2 + c_2$$

for each $t \in [0, 1]$ and $c1, c2>0$ are constants.

$H_2$: The function a: $J \times R^n \to R^n$ is Caratheodory, that is,

(i) $t \mapsto a(t,x)$ is measurable for each $x \in R$,

(ii) $x \mapsto a(t,x)$ is continuous for a.e $t \in J$.

H3: There exists a function $q \in L^1(J, R^n)$ with $q(t)>0$ for a.e $t \in J$ and a continuous non decreasing function $\psi: [0, \infty) \to [0, \infty)$ such that

$$|a(t,x)| \le q(t)\psi(\|x\|)$$

for a.e $t \in J$ and each $x \in R^n$ with

$$\int_C^\infty \frac{ds}{s+\psi(s)} = \infty \quad (3.1)$$

Where

$$C = \frac{1}{1-2c_1}\left[\|\phi\|\left(1+c_1+\sum_{i=1}^{p}\tau_i\right)+3c_2\right]$$

$H_4$: The function $I_k$: $R \to R$ and there exist positive constant $c_k$, k=1, 2, ....m, such that $\|I_k(x)\| \le c_k,\ x \in R$

**Theorem 2.1:** Suppose that the conditions $(H_1)$ – $(H_4)$ are satisfied then there exists a solution of the problem (1.1) on $J_1$ [12-14].

**Proof:** Transform the problem (1.1) – (1.3) into a fixed point problem. Consider the operator $\Gamma: \Omega \to \Omega$ defined by

$$\Gamma(x(t)) = \begin{cases} \phi(t) for\ t \in J_0, \\ \phi(0) - f(0,\phi(0)) + f(t,x_t) + \int_0^t g(s,x_s)ds + \sum_{i=1}^{p} \int_{-\tau_i}^{0} \phi(s)ds \\ +\sum_{i=1}^{p} \int_0^{t-\tau_i} x(s)ds + \int_0^t a(s,x_s)dB(s) + \sum_{0<t_k<t} I_k(x(t_k^-))\ for\ t \in J \end{cases}$$

**Step 1:** $\Gamma$ has bounded values for bounded sets in $\Omega$.

Let B be a bounded set in $\Omega$. Then there exists a real number $\rho>0$ such that $E\|x\|^2 \le \rho$, for all $x \in B$.

Let $x \in B$ and $t \in J$, we have

$$\Gamma x(t) = \phi(0) - f(0,\phi(0)) + f(t,x_t) + \int_0^t g(s,x_s)ds + \sum_{i=1}^{p} \int_{-\tau_i}^{0} \phi(s)ds$$

$$+\sum_{i=1}^{p} \int_0^{t-\tau_i} x(s)ds + \int_0^t a(s,x_s)dB(s) + \sum_{0<t_k<t} I_k(x(t_k^-))$$

$$E\|\Gamma x(t)\|^2 \le E\|\phi\|^2 + c_1E\|\phi\|^2 + c_2 + c_1E\|x_t\|^2 + c_2 + E\int_0^1 \|g(s,x_s)\|^2 ds + \sum_{i=1}^{p}\tau_i E\|\phi\|^2$$

$$+p(E\|x_t\|^2) + E\int_0^1 \|a(s,x_s)\|^2 ds + \sum_{0<t_k<t} EI_k(x(t_k^-))$$

$$\le E\|\phi\|^2(1+c_1+\sum_{i=1}^{p}\tau_i) + 2c_2 + (c_1+p)E\|x_t\|^2 + \int_0^1 (c_1E\|x_s\|^2 + c_2)ds$$

$$+\int_0^1 q(s)\psi(E\|x_s\|^2)ds + m\sum_{0<t_k<t} c_k^2$$

$$\le \|\phi\|^2(1+c_1+\sum_{i=1}^{p}\tau_i) + 2c_2 + (c_1+p)\rho + (c_1\rho + c_2) + \psi(\rho)\int_0^1 q(s)ds$$

$$+m\sum_{0<t_k<t} c_k^2$$

$$\le E\|\phi\|^2(1+c_1+\sum_{i=1}^{p}\tau_i) + 3c_2 + 2c_1\rho + p\rho + \psi(\rho)\|q\| + m\sum_{0<t_k<t} c_k^2$$

$$\le \eta\ (say)$$

If $t \in J_0$, then $E\|\Gamma x\|^2 \le \|\phi\|^2$ and hence $E\|\Gamma x\|^2 \le \eta$ for all $x \in B$, that is $\Gamma$ is bounded on bounded subsets of $\Omega$.

**Step 2:** $\Gamma$ maps bounded sets into equicontinuous sets.

Let B be as in Step 1 and $x \in B$. Let t and $h \ne 0$ be such that $t, t+h \in J \setminus \{t_1, t_2, ....t_m\}$

Now

$$E\|\Gamma x(t+h) - \Gamma x(t)\|^2 \le E\|f(t+h,x_{t+h}) - f(t,x_t)\|^2 + E\int_t^{(t+h)} \|g(s,x_s)\|^2 ds$$

$$+phE\|x\|^2 + E\int_t^{(t+h)} \|a(s,x_s)\|^2 ds + \sum_{t<t_k<t+h} E\|I_k(x(t_k^-))\|^2$$

$$\le E\|f(t+h,x_{t+h}) - f(t,x_t)\|^2 + E\int_t^{(t+h)} \|g(s,x_s)\|^2 ds$$

$$+php + E\int_t^{(t+h)} \|a(s,x_s)\|^2 ds + \sum_{t<t_k<t+h} E\|I_k(x(t_k^-))\|^2$$

as $h \to 0$, the right hand side of the above inequality tends to zero. This implies the equicontinuity on $J \setminus \{t_1, t_2, ....t_m\}$.

It remains to examine at $t=t_i$, i=1, 2, ....m. Let $h \ne 0$ be such that $\{t_k : k \ne i\} \cap [t_i - |h|, t_i - |h|] = \varnothing$ Thus we have

$$\|\Gamma x(t_i+h) - \Gamma x(t_i)\| \le E\|f(t+h,x_{t+h} - f(t,x_t)\|^2$$

$$+E\int_{t_i}^{t_i+h}(c_1\|x_s\|^2+c_2)ds+p\rho h+\psi(\rho)\int_{t_i}^{t_i-h}q(s)ds$$

The right hand side of the above inequality tends to zero as h → 0. The equicontinuity on $J_0$ follows from the uniform continuity of φ on this interval.

**Step 3:** Now we show that Γ is continuous

Let $\{x_n\}\subset\Omega$ be a sequence such that $x_n\to x$. We will show that $\Gamma x_n\to\Gamma x$.

For t ∈ J

$$E\|\Gamma x_n-\Gamma x\|^2=E\left\|f(t,x_{nt})-f(t,x_t)+\int_0^1(g(s,x_{ns})-g(s,x_s))ds+p\int_0^1(x_n(s)-x(s))ds\right.$$

$$\left.+\int_0^1(a(s,x_n)-a(s,x))dB(s)+\sum_{k=1}^m(I_k(x_n(t_k^-))-I_k(x(t_k^-))\right\|$$

$$\le 3E\|f(t,x_{nt})-f(t,x_t)\|^2+3E\int_0^1\|(g(s,x_{ns})-g(s,x_s))\|^2ds$$

$$+3pE\int_0^1\|x_n(s)-x(s)\|^2ds+3E\int_0^1\|(a(s,x_n)-a(s,x))\|^2ds$$

$$+3m\sum_{k=1}^m E\|(I_k(x_n(t_k^-))-I_k(x(t_k^-))\|^2 \quad (3.2)$$

Using $H_3$ it can be easily shown that the function $t\mapsto g(t,x_{nt})-g(t,x_t)$ is Lebesgue integrable. By the continuity of f and $I_k$, k=1, 2, ...m and the dominated convergence theorem, the right hand side of inequality (3.2) tends to zero as n → ∞, which completes the proof that Γ is continuous [15-18].

As a sequence of steps 1 to 3, together with the Arzela-Ascoli theorem, we conclude that Γ is completely continuous.

To complete the proof of the theorem, it suffices to prove the following step.

**Step 4:**

There exists a priori bound of the set

ς={x ∈ Ω: λx=Γx for some λ>1}

Let x ∈ ς and λ>1 be such that λx=Γx. Then x|[−r, $t_1$] satisfies for each t ∈ [0, $t_1$],

$$x(t)=\lambda^{-1}[\phi(0)-f(0,\phi(0))+f(t,x_t)+\int_0^t g(s,x_s)ds+\int_0^t a(s,x_s)dB(s)$$

$$+\sum_{i=1}^p\int_{-\tau_i}^0\phi(s)ds+\sum_{i=1}^p\int_0^{t-\tau_i}x(s)ds]$$

It is easy to verify that

$$E\|x(t)\|^2\le E\|\phi\|^2\left(1+c_1+\sum_{i=1}^p\tau_i\right)+2c_1\|x_t\|^2$$

$$+3c_2+\int_0^t\left(q(s)\psi(E\|x_s\|^2)+pE\|x_s\|^2\right)ds \quad (3.3)$$

Consider the function $v_1(t)=sup\{E\|x(s)\|^2:s\in[-r,t]\}$, for t ∈ [0, $t_1$]. We have $\|x_t\|^2\le v_1(t)$ for all t ∈ [0, $t_1$] and there is a point t* ∈ [−r, t] such that $v_1(t)=\|x_t^*\|^2$ If t* < 0, we have $v_1(t)<\|\phi\|^2$ for all t ∈ [0, $t_1$]. Now, if t* ≥ 0 from (3.3) it follows that,

for t ∈ [0, $t_1$],

$$v_1(t)\le E\|\phi\|^2\left(1+c_1+\sum_{i=1}^p\tau_i\right)+2c_1v_1(t)+3c_2+\int_0^t\left(q(s)\psi(v_1(s))+pv_1(s)\right)ds$$

And hence

$$v_1(t)\le C_1^1+C_1^2\int_0^t Q(s)[\psi(v_1(s))+v_1(s)]ds$$

Where

$$C_1^1=C_1^2\left[\|\phi\|^2\left(1+c_1+\sum_{i=1}^p\tau_i\right)+3c_2\right]$$

$C_1^2=\dfrac{1}{1-2c_1}$ And Q(t)=max {q(t), p} , f or t ∈ [0, $t_1$]

Set,

$$w_1(t)=C_1^1+C_1^2\int_0^t Q(s)[\psi(v_1(s))+v_1(s)]ds \quad for\ t\in[0,t_1]$$

Then we have $v_1(t)\le w_1(t)$ for all t ∈ [0, $t_1$]

A direct differentiation of $w_1$ yields

$$\begin{cases}w_1'(t)\le Q(t)[\psi(w_1(t))+w_1(t))], & a.e,\ t\in[0,t_1]\\ w_1(0)=C_1^{(1)}\end{cases}$$

By integration, this gives

$$\int_0^t\frac{w_{1'}(s)}{\psi[(w_1(s))+w_1(s)]}ds\le\int_0^t Q(s)ds\le\|Q\| \quad t\in[0,t_1] \quad (3.4)$$

By a change of variables, inequality (3.4) becomes,

$$\int_{C_1^1}^{w_1(t)}\frac{ds}{\psi(s)+s}\le\|Q\|,\quad t\in[0,t_1]$$

By (3.1) and the mean value theorem, there is a constant $M_1=M_1(t_1)>0$ such that $w_1(t)\le M_1$ for all t ∈ [0, $t_1$]

That is $v_1(t)\le M_1$ for all t ∈ [0, $t_1$]

At last we choose $M_1$ such that $\|\phi\|\le M_1$ to get

$$sup\{E\|x(t)\|^2:t\in[-r,t_1]\}=v_1(t_1)\le M_1$$

Now, consider $x|_{[-r,t_2]}$, satisfies for each t ∈ [0, $t_2$]

$$x(t)=\lambda^{-1}[\phi(0)-f(0,\phi(0))+f(t,x_t)+\int_0^t g(s,x_s)ds+\int_0^t a(s,x_s)dB(s)$$

$$+\sum_{i=1}^p\int_{-\tau_i}^0\phi(s)ds+\sum_{i=1}^p\int_0^{t-\tau_i}x(s)ds+I_1(x(t_1))]$$

$$E\|x(t)\|^2\le E\|\phi\|^2\left(1+c_1+\sum_{i=1}^p\tau_i\right)+2c_1\|x_t\|^2+3c_2$$

$$+\int_0^t[q(s)\psi(E\|x\|_s^2+pE\|x\|_s^2]ds+supE\|I_1(u)\|^2:\|u\|\le M_1$$

Denote $v_2(t)=sup\{E\|x(s)\|^2:s\in[-r,t]\}$, for t ∈ [0, $t_2$], Then for each t ∈ [0, $t_2$], we have $\|x(t)\|$, such that $\|x_t\|^2\le v_2(t)$.

Let t* ∈ [−r, t] be such that $v_2(t)=\|x(t^*)\|^2$.

If t* < 0, we have $v_2(t)\le\|\phi\|^2$ for all t ∈ [0, $t_2$].

Now if t* ≥ 0, then by (3.5) we have for t ∈ [0, $t_2$]

$$v_2(t)\le E\|\phi\|^2\left(1+c_1+\sum_{i=1}^p\tau_i\right)+2c_1v_2(t)+3c_2+\int_0^t[q(s)\psi(v_2(s))+pv_2(s)]$$

$$+sup\{E\|I_1(u)\|^2:\|u\|\le M_1\}$$

$$v_2(t)\le C_2^1+C_2^2\int_0^t[Q(s)\psi(v_2(s))+pv_2(s)]$$

Where

$$C_2^1C_2^2E\|\phi\|^2\left(1+c_1+\sum_{i=1}^p\tau_i\right)+3c_2+supE\|I_1(u)\|^2:\|u\|\le M_1$$

$$C_2^2 = \frac{1}{1-2c_1}$$

And Q(t)=Max {q(t), p} , for t ∈ [0, $t_2$].

If we set,

$$w_2(t) = C_2^1 + C_2^2 \int_0^t Q(s)[\psi(v_2(s)) + v_2(s)]ds \ \ for \ \ t \in [0, t_2]$$

Then we get $v_2(t) \leq w_2(t)$ for all t ∈ [0, $t_2$] and

$$\begin{cases} w_{2'}(t) \leq Q(t)[\psi(w_2(t)) + w_2(t)]., \ a.e, \ t \in [0, t_2] \\ w_2(0) = C_2^1 \end{cases}$$

These yields,

$$\int_0^t \frac{w_{2'}(t)}{\psi(w_2(s)) + pw_2(s)} ds \leq \int_0^t Q(s)ds \leq \|Q\|, \ \ t \in [0, t_2]$$

$$\int_{C_2^1}^{w_2(t)} \frac{ds}{\psi(s) + s} \leq \|Q\|, t \in [0, t_2]$$

Again by (3.1) and the mean value theorem, there is a constant $M_2=M_2$ $(t_1, t_2)>0$ such that $w_2(t) \leq M_2$ for all t ∈ [0, $t_2$ ], and then $v_2(t) \leq M_2$ for all t ∈ [0, $t_2$]. Finally, if we

Choose $M_2$ such that φ ≤ $M_2$, we get,

$$sup\left\{E\|x(t)\|^2 : t \in [-r, t_2]\right\} = v_2(t_2) \leq M_2$$

Continue this process for $x|_{[-r,t3]}$, .... $x|J_1$, we obtain that there exists a constant M=M ($t_1$, ...$t_m$)>0 such that x ≤ M.

This finish to show that the ζ is bounded in Ω.

As a result the conclusion of theorem holds and consequently the problem (1.1) has a solution x on $J_1$. This completes the proof [19].

## References


1. Anguraj A, Shujin Wu, Vinodkumar A (2011) The existence and exponential stability of semilinear functional differential equations with random impulses under non-uniqueness. Nonlinear Analysis 74: 331-342.
2. Wu S (2007) The Euler scheme for random impulsive differential equations. Applied Mathematics and Computation 191: 164-175.
3. Wu SJ, Han D (2005) exponential stability of functional differential systems with impulsive effect on random moments. Computer and Mathematics with Applications 50: 321-328.
4. Ren Y, Xia N (2009) A note on the existence and uniqueness of the solution to neutral stochastic functional differential equations with infinite delay.applied mathematics and computation 214: 457-461.
5. Sakthivel R, Luo J (2009) asymptotic stability of nonlinear impulsive stochastic differential equations. Statistics and Probability Letters 9: 1219-1223.
6. Anguraj A, Vinodkumar A (2009) Existence, Uniqueness and stability Results of Impulsive Stochastic Semi linear Neutral Functional Differential Equations with Infinite delays. Electronic Journal of Qualitative Theory of Differential Equations 67: 1-13.
7. Lakrib M (2008) An existence result for impulsive neutral functional differentia equations with multiple delay. Electronic Journal of Differential equations 36: 1-7.
8. Oksendal B (2002) Stochastic Differential Equations An Introduction with Applications: (5thedn), springer, Berlin, Germany.
9. Mao X (1997) stochastic differential equations and their Applications. Horwood Publishing, Chichester, UK.
10. Balachandran K, Kiruthika S (2010) Existence of solutions of Abstract fractional Impulsive Semi linear evolution Equations, Electronic. Journal of Qualitative Theory of Differential Equations 4: 1-12.
11. Sakthivel R, Luo J (2009) asymptotic stability of impulsive stochastic partial differential equations with infinite delays. J Math Anal Appl 356: 1-6.
12. Pan L (2012) Existence of mild solution for impulsive stochastic differential equations with non-local conditions. Differential Equations and Applications 4: 485-494
13. Guendouzi T, Mehdi K (2013) existence of mild solutions for impulsive fractional stochastic equations with infinite delay. Malaya Journal of Mathematik 4: 30-43.
14. Anguraj A, Karthikeyan K (2009) Existence of solutions for impulsive neutral functional differential equations with nonlocal conditions. Nonlinear Anal 70: 2717-2721.
15. Yang Z, Xu D, Xiang L (2006) Exponential p-stability of impulsive stochastic differential equations with delays. Physics Letters 359: 129-137.
16. Benchora M, Henderson J, Ntouyas SK (2006) Impulsive Differential equations and Inclusions. Hindawi Publishing Corporation, New York, USA.
17. Bao H, Cao J (2009) Existence and Uniqueness of solutions to neutral stochastic functional differential equations with infinite delay. Applied mathematics and Computation 215: 1732-1743.
18. Sakthivel R, Luo J (2009) asymptotic stability of nonlinear impulsive stochastic differential equations. Statistics and Probability Letters 9: 1219-1223.
19. Hernandez EM, Rabello M, Hernan R, Henriquez HR (2007) existence of solutions for impulsive partial neutral functional differential equations. J Math Anal Appl 331: 1135-1158.

# 5

# Control of Laminar Fluid Flow and Heat Transfer in a Planar T-Channel with Rotating Obstacle

**Palraj Jothiappan***

*Department of Mathematics, PSG College of Arts and Science, Coimbatore, Tamil Nadu, India*

**Abstract**

In the present study, a laminar flow in a planar 2D right angled T-channel in the presence of a rotating heated cylindrical obstacle placed in the junction area is numerically studied to control the heat transfer and fluid flow. The effect of Reynolds number ($20 \leq Re \leq 300$) and cylinder rotation angle ($-5 \leq \omega \leq 5$) on the fluid flow and heat transfer characteristics are studied numerically. It is observed that the flow field and heat transfer rate are influenced by the variation of these parameters.

**Keywords:** Finite element method; Navier-Stokes equation; Rotating obstacle; Rotating cylinder; T-channel; Recirculation length; Heat transfer

## Nomenclature

c: Temperature, K

$C_p$: Specific heat, J/kg K

D: Diameter of the cylinder, m

H: Width of the branch, m

h: Local heat transfer coefficient, $Wm^{-2}K^{-1}$

k: Thermal conductivity, $Wm^{-1}K^{-1}$

$L_1$: Upstream length of main branch, m

$L_2$: Downstream length of main branch, m

$L_3$: Side branch length, m

n: Unit normal vector

Nu: Local Nusselt number, hH/k

$\overline{Nu}$: Spatial-averaged Nusselt number

P: Pressure, Pa

Pr Prandtl number, $\mu C_p/k$

Re: Reynolds number, UH/v

t: Time, s

u: X-component velocity, $ms^{-1}$

v: Y-component velocity, $ms^{-1}$

x,y: Cartesian coordinates, m

Greek Symbols

$\rho$: Density, $kgm^{-3}$

: Kinematic viscosity, $m^2 s^{-1}$

$\omega$: Cylinder rotation angle.

## Introduction

The incompressible Navier-Stokes system is one of the main equations studied in mathematical physics and fluid mechanics fields [1]. The Navier Stokes equations describe the flow of incompressible, newtonian fluids. The equations are transient, nonlinear and velocity is non-trivially coupled with pressure. A lot of research has been devoted to finding efficient ways of linearzing, coupling and solving these equations. Therefore, to utilize the computational power of modern high-performance computers, much effort is thrown into the development of efficient computing methods for the Navier-Stokes equations [2-5]. FEniCS is a generic open source software framework that aims at automating the discretization of differential equations through the finite element method [6,7]. The three most common flow configurations are branching T-channel, impacting T-channel and combining T-channel. The present study is concerned with the flow in branching T-channel. Branching fluid flow and heat transfer in a 90 degree T-channel is of considerable importance to a wide variety of applications in the biomedical and engineering fields. Flow through a T-channel has a wide range of applications, such as biomechanical applications, phase separation, oil and gas pipelines, polymer and pharmaceutical industries, irrigation systems, wastewater treatment, ventilation systems and in many other areas. Boundary-layer development, flow separation, and secondary flows that occur in these complex flows give rise to significant modification of heat and mass transport, as reported by various experimental groups see [8,9]. Heat transfer and fluid flow characteristics over a backward facing [10,11] and forward facing [12] step in a channel with the insertion of rotating obstacles has received some attention in the literature [13,14].

The mechanics of such flow are complex and not well understood exhibiting nontrivial flow patterns which include zones of recirculation and stream wise vortices. The distribution of the flow into various branches depends on the flow resistances of these branches and in general, it is even impossible to predict the direction of flow through branches under given pressure drops. It has been a geometrical model of choice because in addition to its simplicity, its flow features demonstrate the most common flow behaviour at arterial bifurcations. Pollard has experimentally investigated the laminar and turbulent fluid

***Corresponding author:** Palraj Jothiappan, Department of Mathematics, PSG College of Arts and Science Coimbatore, India
E-mail: palrajpsg@gmail.com

flow and heat transfer in T channel prior to 1978 [15]. Kawashima et al. experimentally studied the turbulent heat transfer in a two-dimensional right-angled confluence [16]. They observed that the axial variation of the local Nusselt number was affected by the cross-sectional area and flow rate ratios, but was affected insignificantly by the Reynolds number. Hayes et al. [17] studied the steady laminar flow in a 90 degree planar branch; they found that the fractional flow in the main duct increases with increasing Reynolds number for the case of constant exit pressures at the outlet of each branch. Neary and Sotiropoulos presented numerical solutions for the steady 3-D laminar flows through a 90 rectangular cross section [18]. They compared solutions with experimental measurements to elucidate the flow topology patterns and showed that both length and width of the separation zone decrease with increasing discharge ratio. Dhiman investigated the flow characteristics of non-Newtonian power-law fluids in a right-angled horizontal T-channel in the laminar regime [19]. They observed that for a particular power-law index, the length of recirculation zone increases in the side branch with increasing Reynolds number. Hassan and Kim studied numerically the mixing of a high-pressure coolant injected into the leg of a pressurized water reactor during the loss of coolant accident [20]. Sparrow investigated the Effect of a Mixing Tee on turbulent heat transfer in a tube [21]. The mixing of the two streams of air gave rise to a remarkable augmentation of the heat transfer coefficient compared with those in a conventional thermal entrance region.

Thus, based on the above discussion, flow in a T-channel for Newtonian and non-Newtonian fluids has been investigated extensively both experimentally and numerically to obtain the basic information of flow separation and reattachment phenomena in the laminar flow regime. The two-dimensional (2D) laminar flow for Newtonian fluids in a T-channel in the presence of a rotating heated circular cylinder placed in the junction is not investigated yet and the lack of such results motivated the current study. The 2-D simulations are deemed adequate to represent actual three-dimensional situations when the aspect ratios of the ducts forming the T-channel are large, as in the experiments of Liepsch [22] and Khodadadi [23].The main objective of this study is to investigate the characteristics of two-dimensional (2D) laminar flow for Newtonian fluids in a T-channel in the presence of a rotating heated circular cylinder placed in the junction area over a range of Reynolds numbers ($20 \leq Re \leq 300$) and cylinder rotation angle ($-5 \leq \omega \leq 5$). To the best of author's knowledge and based upon the above literature survey such a study has not been seen in the literature.

## Numerical Simulations

### Geometry of the computational domain

In this section, the computational domain and configurations of the obstacle are presented. A schematic description of the physical problem is shown in Figure 1a. All branches have the same width H. The inlet condition was set as fully developed with the corresponding velocity pro le having a parabolic shape. The minimum lengths of the bifurcation channels that guaranty a fully developed flow at the outlets depend on the flow characteristics. According to Shah and London [24] the pipe length required to develop a Newtonian flow is given by

$$\frac{L}{H}=\frac{0.315}{0.068\mathrm{Re}+1}+0.044\mathrm{Re}$$

It helps in estimating the required geometry dimensions. The Reynolds number is defined as: $Re=UH/v$ where $v$ is the viscosity of the fluid. The distance between the inlet plane and the junction of the channel $L_1$ is taken as 5H, and the downstream distance between the

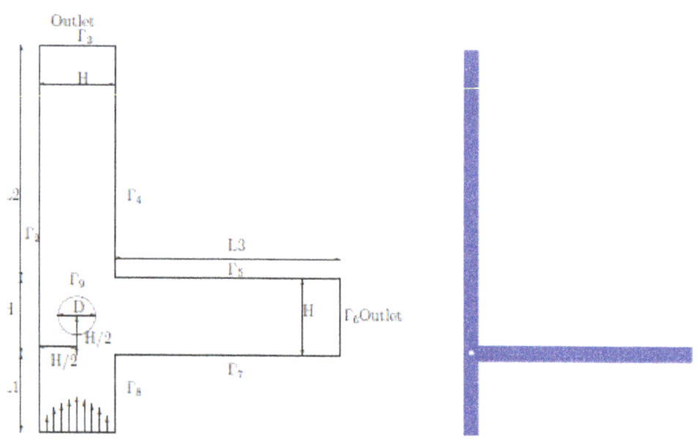


**Figure 1:** (a) Schematic description of the problem Inlet (b) Mesh Distribution.

junction and the exit plane $L_2$ is taken as 20H. The length of side branch $L_3$ is taken as 15H. The circular cylinder with diameter D=0.2 is kept at channel junction. The diameter of the cylinder is so chosen as to keep the H/D ratio in the appropriate range so as to ensure the steady nature of the flow in the downstream channel as has been previously described by Singha and Sinhamahapatra [13] for the channel flow. The distance from the centre of the cylinder to the inlet plane of the channel is 6H. The circular heated cylinder $\Gamma_9$ is maintained at constant temperature higher than the inlet temperature. The wall boundaries $\Gamma_2,\Gamma_4,\Gamma_5,\Gamma_7$ and $\Gamma_8$ of the channel are treated as no slip boundary with adiabatic condition. At the inlet of the channel $\Gamma_1$ a parabolic velocity (U) and a uniform temperature are imposed and pressure is set to zero at the out flow boundaries $\Gamma_3$ and $\Gamma_6$ Working fluid is air with a Prandtl number $P_r$=0.71. The flow is assumed to be two dimensional, Newtonian, incompressible and in the laminar flow regime.

### Governing equations and boundary conditions

The governing equations for two dimensional, incompressible, laminar and unsteady cases can be written as follows,

Continuity equation

$$\mathrm{Div}\ u=0 \tag{2.1}$$

Momentum equation

$$\frac{\partial u}{\partial t}+u.\nabla u=-\nabla p+\frac{1}{\mathrm{Re}}\nabla u \tag{2.2}$$

Energy equation

$$\frac{\partial c}{\partial t}+u.\nabla c=\frac{1}{\mathrm{Re\,Pr}}\Delta c \tag{2.3}$$

Where c is temperature.

The boundary conditions are given as follows,

At the channel inlet $\Gamma_1$, velocity is unidirectional parabolic and temperature is uniform

$$u=U,\ v=0,\ c=0 \tag{2.4}$$

On the channel walls $\Gamma_2,\Gamma_4,\Gamma_5,\Gamma_7$ and $\Gamma_8$ adiabatic wall with no-slip boundary conditions are assumed,

$$u=0,\ v=0,\ \frac{\partial c}{\partial n}=0 \tag{2.5}$$

Where n denote the surface normal directions

On the cylinder surface specified velocity components with constant temperature boundary condition is used

$U=-\omega(y-y_0)$, $v=\omega(x-x_0)$, and c=1

At the out flow boundary main branch $\Gamma_3$

$$\frac{\partial u}{\partial y}=0, \frac{\partial v}{\partial y}=0, \frac{\partial c}{\partial y}=0 \quad (2.6)$$

At the out flow boundary side branches $\Gamma_6$

$$\frac{\partial u}{\partial x}=0, \frac{\partial v}{\partial x}=0, \frac{\partial c}{\partial x}=0 \quad (2.7)$$

Once the velocity and temperature fields are obtained, the local Nusselt number is defined as

$$Nu=\frac{hH}{k}=\frac{\partial c}{\partial n} \quad (2.8)$$

Where h represents the local heat transfer coefficient and k denotes the thermal conductivity of air.

The spatial-averaged Nusselt number along the surface of a cylinder is calculated as

Follows:

$$\overline{Nu}=\frac{1}{S}\int_0^S Nu dS \quad (2.9)$$

Where S represents the surface area of the cylinder, n denote the surface normal directions of the cylinder.

The governing equations (2.1-2.3) along with the above boundary conditions (2.4-2.7) are solved for the fluid flow and heat transfer in a T-channel over a Rotating heated circular cylinder obstacle to obtain velocity, pressure and temperature Fields.

## Finite Element Scheme and Implementation

To construct a time discretization scheme for the Navier-Stokes equations (2.1-2.3). Our choice is a Fully-implicit Crank-Nicolson scheme as described in [24] for retaining the basic non-linearity in the Navier-Stokes equations. The time centred Crank-Nicolson scheme gives nonlinear equations to be solved at each time level and redefining u be the velocity at the new time level and $u_1$ be the velocity at the previous time level, we arrive at these spatial problems:

$$U=0$$

$$\frac{\text{u-u}_1}{\Delta t}+u_1.\nabla u_1=-\nabla p+\frac{1}{\text{Re}}\Delta U$$

$$\frac{c-c_1}{\Delta t}+\text{U}.\nabla C=\frac{1}{\text{Re}\,\text{Pr}}\Delta C$$

With

$$U=\frac{1}{2}(\text{u}+\text{u}_1), \text{C}=\frac{1}{2}(\text{c}+\text{c}_1)$$

Denoting the arithmetic averages needed in Crank-Nicolson time integration. The corresponding variation formulation involves the integrals

$$\int_\Omega ((\frac{u-u_1}{\Delta t})v_u+(u_1.\nabla u_1).v_u-p\nabla U+\frac{1}{\text{Re}}\nabla U:\nabla v_u+v_p\nabla.\text{U})\,\text{dx}=0$$

$$\int_\Omega ((\frac{c-c_1}{\Delta t})v_c+(U.\nabla C).v_c+\frac{1}{\text{RePr}}\nabla C.\nabla v_c)\,\text{dx}=0 \quad (3.1)$$

Where $v_u$; $v_p$ and $v_c$ are test functions for the test spaces of u; p; and c respectively.

Finite element software basically has to solve three sub problems: Mesh generation, system-matrix assembly and solution of the resulting linear systems of equations. We use Mshr for mesh generation and FEniCS [6] for matrix assembly and the solution of linear systems. CBC PDE Sys is built upon the finite element package FEniCS [6]. FEniCS is a powerful development environment for performing finite element modelling, including strong support for symbolic automatic differentiation, native parallel support and parallel interface with linear algebra solvers such as PET Sc and Trillions, and automatic code generation and compilation for compiled performance from an interpreted language interface. The Python scripting environment makes the generation and linking of new code straightforward. A finite element method consisting of 12673 vertices and 23860 triangular elements as shown in Figure 1b, is used in this study. The mesh is finer near the walls of the channel and cylinder to resolve the higher gradients in the thermal and velocity boundary layer and in the vicinity of the junction. The sensitivity analysis of the simulation results with the number of elements, mesh size are also assured to obtain an optimal grid distribution with accurate results and minimal computational time. Three different grid sizes are tested and the convergence in the recirculation zone along the bottom wall downstream of the side branch is checked for Reynolds numbers 100 and w=0, tabulated in Table 1. Grid sensitivity analysis for Re=100 and ω=0.

## Validation study

The numerical solution procedure used here has been benchmarked with standard results for the incompressible flow of Newtonian fluids in a T-channel reported in the literature [15,17]. The Figure 2 shows the validation of reattachment length for Newtonian fluids on varying Reynolds numbers. The minimum deviation for the percentage in the error is obtained for the results of [15] is found to be about 0.31, whereas the maximum deviation is around 2.74. This validates the present numerical solution, the comparison results shows good overall

| Grid Size | $L_r$=D | Relative difference (%) |
|---|---|---|
| 17979 | 2.8510 | 0.0947 |
| 23680 | 2.8532 | 0.0175 |
| 34719 | 2.8537 | 0.0000 |

**Table 1:** Grid sensitivity analysis for Re=100 and ω=0.

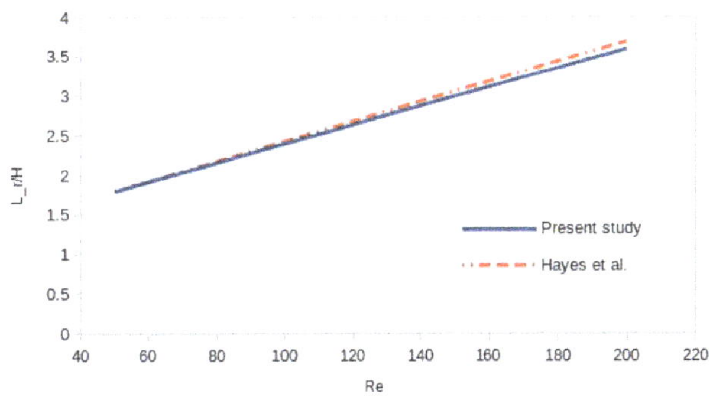


**Figure 2**: Recirculation length (Lr=H) Vs Reynolds numbers (Re).

agreement and slight differences for higher Reynolds number. This difference is believed to be due to the differences in the domain and grid sizes used by others as opposed to the present study.

## Results and Discussion

The main parameters that affect the fluid flow and thermal characteristics are Reynolds number, Prandtl number, and distance between the inlet to the junction, cylinder position and diameter, expansion ratio and cylinder rotation angle. In the current study, the numerical simulations are performed for the Reynolds number ($20 \leq Re \leq 300$) and cylinder rotation angle ($-5 \leq \omega \leq 5$) re-examined for fluid flow structures and convective heat transfer enhancement over a T-channel.

### Effects of cylinder rotation angle

Figures 3 and 4 indicates the effect of varying cylinder rotation angle (ω) on the streamlines and isotherms at blockage ratio of 20 percentages for fixed values of Reynolds number 300 and Prandtl number 0.71. The case ω=0 corresponds to a stationary heated cylinder which is shown in Figures 3b, 3e and 3h. In this case, streamlines are slightly.

Defected towards the side branch in the presence of the motionless cylinder and several vortices appear behind the obstacle and in the vicinity of the cylinder close to the bottom wall of side branch. The case ω=0, no recirculation zone is occur in the side branch till the Re=18. The flow separation is delayed with cylinder rotation angle (ω). When the cylinder rotates in the clockwise direction (negative value of ω) (Figures 3a, 3d and 3g) flow is accelerated in the region between the left channel wall and the cylinder due to the contraction effect. Some portion of the flow is directed towards the right of the cylinder and related to this effect, the vortex appear on the upper right of the cylinder for motionless cylinder case disappears. The size and extent of the recirculation bubble appearing on the side branch bottom wall increase compared to motionless cylinder case. As it can be seen in the Figures 3a, 3d and 3g the extent and strength of the recirculation bubble can be controlled with cylinder rotation angle.

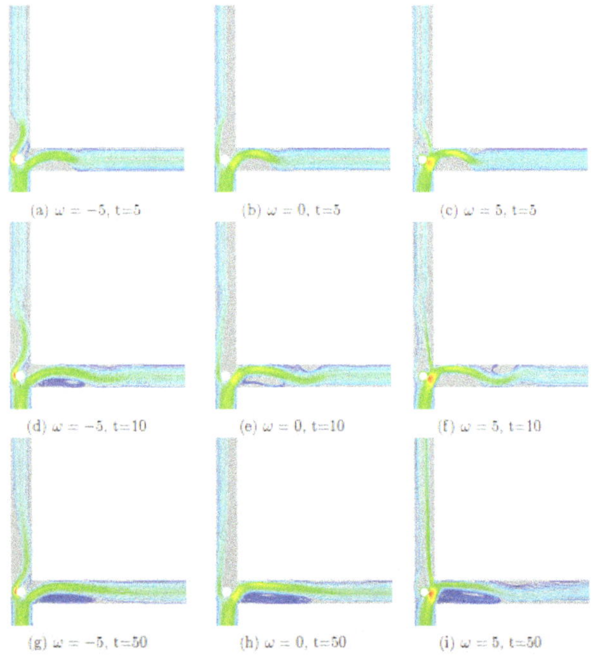


**Figure 3:** Effect of cylinder rotation angle on the streamlines for fixed value of Re=300.

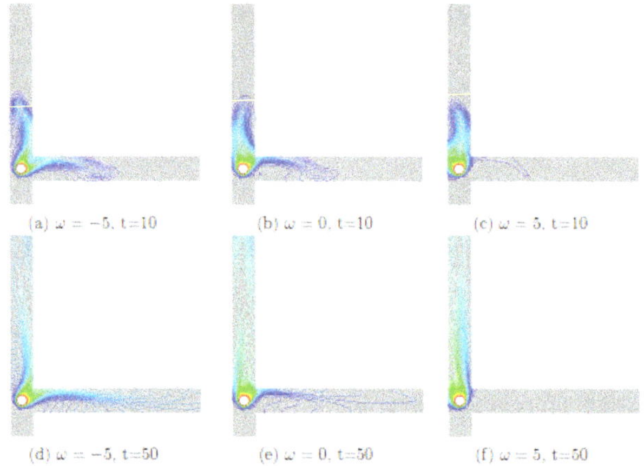


**Figure 4:** Effect of cylinder rotation angle on the isotherms for fixed value of Re=300.

A positive value of the rotation (ω) indicates counter clockwise rotation of the cylinder (Figures 3c, 3f and 3i). When the cylinder rotates in this direction, more flow is accelerated towards the main branch due to the combined effect of contraction area and rotation. The motion of the fluid owing through the right of the cylinder and the size of the recirculation zone on the side branch upper wall are affected compared to motionless cylinder case. The flow structure near the upper and right of the cylinder are affected to some extent. The size and extent of the recirculation bubble appearing on the side branch bottom wall decreases compared to motionless cylinder case. Figure 4 shows the effect of cylinder rotation on the isotherms for Reynolds number Re=300. For counter clockwise rotation direction, the isotherms fluctuates more less on the side branch (Figures 4a and 4d) and right and bottom parts of the cylinder due to the formation of the vortices compared to motionless cylinder case. When the cylinder rotates in clockwise direction, the isotherms fluctuates more on the side branch and less clustering of the isotherms for the side branch due to the recirculation region and indicates poor heat transfer characteristic for this region since more flow is directed towards the spacing between the main branch left channel wall and cylinder. The case ω=5, the isotherms fluctuates both main and side branches, while comparing to the other cases. Local Nusselt number distributions along the surfaces of the circular cylinder are demonstrated in Figure 5b. Introducing a heated cylinder (motionless or rotating) enhances the thermal transport from the main branch and the side branch due to the flow acceleration towards the main and side branches. For the counter clockwise rotation of the cylinder, heat transfer is less effective on side branch. Since more flow is entrained into the side branch and defection of the flow patterns upwards on the top of the side branch. For clockwise rotation of the cylinder, better heat transfer characteristic is observed at surface of the side branch due to the flow acceleration towards the side branch.

### Wake length

The variation of the recirculation wake) length ($L_r$=H) in the side branch (defined as the distance from the junction of T- channel to the point of attachment of the fluid with the bottom wall) as a function of Reynolds number and angle of rotation ω is shown in Figure 5a.

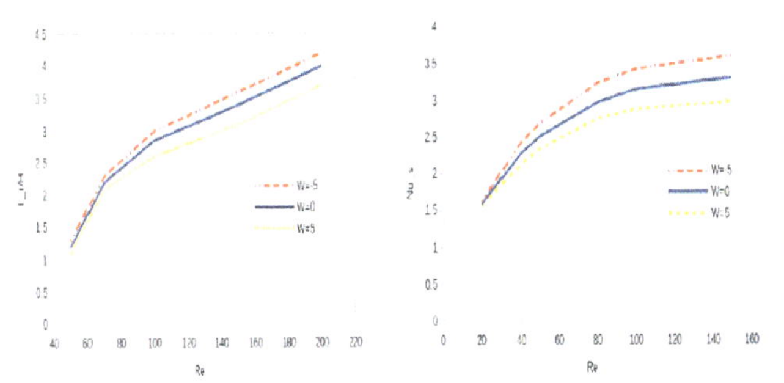

**Figure 5:** (a) Variation of recirculation length Lr=H and (b) averaged Nusselt number with Reynolds number at Different values of cylinder rotation angle.

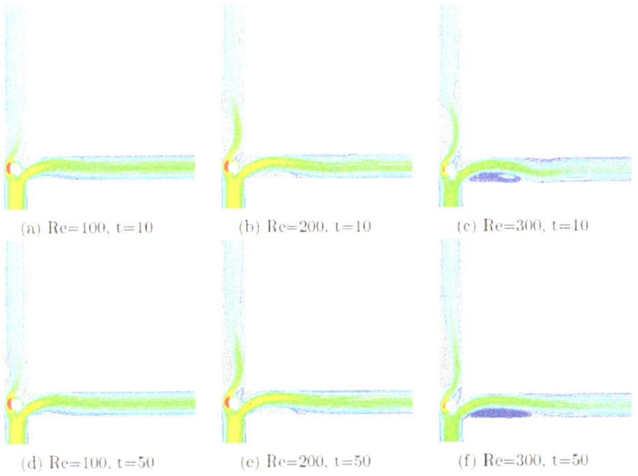


**Figure 6:** Effect of Reynolds number on the streamlines for ω=5.

The length of the recirculation zone is seen to increase in a non-linear fashion with an increase in Reynolds number for a particular angle of rotation. The recirculation length is also seen to increase with decreasing angle of rotation for a fixed Reynolds number. Thus, the dimensional considerations suggest that the recirculation length is a function of Reynolds number and angle of rotation.

## Effects of Reynolds number

Figures 6 shows the effect of varying Reynolds numbers on the flow patterns at fixed values of !=5 and different time level. The fluid while travelling from main branch to side branch maintains contact with the wall of the side branch. As the Reynolds number is increased, beyond a critical point the fluid gets separated from the lower wall of the side branch and a closed recirculation region is developed.

Further downstream of the reattachment points, the flow regains its fully developed flow behaviour. With a gradual increase in the value of the Reynolds number (Re>18), the size of recirculation region increases. The flow patterns for the case of Newtonian fluids in a T-channel are found to be in close agreement with those of Hayes et al. [17] and Neary and Sotiropoulos [18]. As Reynolds number increases, the flow begins to separates behind the cylinder causing vortex shedding which is an unsteady phenomenon and recirculation region is seen and increases in size and strength. At Reynolds number of 200, some portion of the fluid separated from the bottom wall and separates from the main branch side wall and again separates behind the cylindrical obstacle as seen in Figure 6. When Reynolds number increases, the flow separated from the bottom wall is entrained into the wake of the cylindrical obstacle and formation of the vortices is seen behind the cylindrical obstacle shown in Figures 6e and 6f. Figure 7 indicates the effect of varying Reynolds numbers on the isotherms for fixed value of rotation angle ω=5. When the cylinder rotates in this direction, the isotherms fluctuates more on the side branch and less clustering of the isotherms for the side branch due to the recirculation region and indicates better heat transfer characteristic for this region since more flow is directed towards the spacing between the main branch left channel wall and cylinder.

## Conclusion

In this study, laminar flow in a planar right angled T-channel in the presence of a rotating heated circular cylinder placed in the junction area is numerically studied. The effect of Reynolds number (20 ≤ Re ≤ 300) and cylinder rotation angle (-5 ≤ ω ≤ 5) on the fluid flow and heat transfer characteristics are numerically investigated.

The numerical methodology has been extensively validated against previous numerical and experimental studies. The grid and computational domain were chosen after extensive testing by varying various grid densities. Detailed observations of flow pattern, recirculation length, local and averaged Nusselt number for the onset of flow separation and viscosity variation along the bottom wall of the side branch have been presented.

The result shows that the length of the recirculation zone increases on increasing Reynolds number for a particular angle of cylinder rotation. It also increases on decreasing the angle of cylinder rotation for a fixed value of Reynolds number. As the Reynolds number increases local Nusselt number also increases. When the cylinder rotates in the clockwise direction, more flow is entrained into the wake of the cylinder and some portion of the flow is directed towards the right of the cylinder and related to this effects the size and extent of the recirculation bubble appearing on the side branch lower wall and main branch side wall. Length and intensity of the recirculation zone behind the obstacle are considerably affected with the installation of the cylindrical obstacle and rotation angle. Adding the rotating cylindrical obstacle alters the isotherm plots. It is seen that there is significant change on the clustering of the isotherm patterns and the location where this steep temperature gradient occurs in the flow with

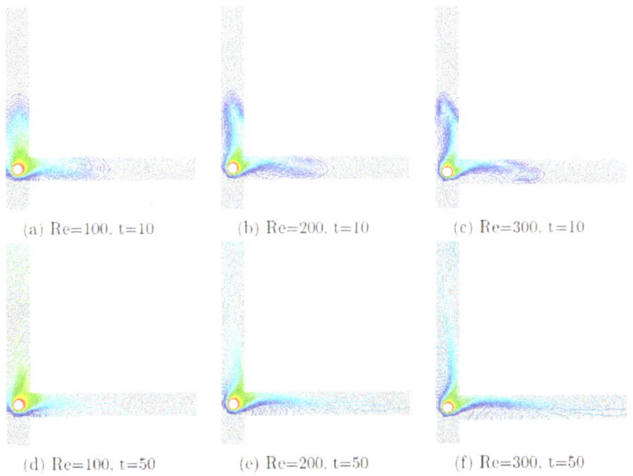


**Figure 7:** Effect of Reynolds number on the isotherms for ω=5.

the installation of the rotating cylindrical obstacle. The result shows the effect of rotation on the local Nusselt number distribution is more pronounced at low Reynolds number.

## References


1. Girault V, Raviart PA (1986) Finite element method for Navier-Stokes equations. Theory and algorithms. Springer-Verlag New York.
2. Anteparaa O, Lehmkuhla O, Borrellb R, Chivaa J, Olivaa A (2015) Parallel adaptive mesh refinement for large-eddy simulations of turbulent flows. Journal of Computers and Fluids 110: 48-61.
3. Bianchia E, Groppib G, Schwiegera W (2015) Numerical simulation of heat transfer in the near-wall region of tubular reactors packed with metal open-cell foams. Che Engg Journal 264: 268-279.
4. Lemmer A, Hilfer R (2015) Parallel domain decomposition method with non-blocking communication for flow through porous media. J Comput Physics 281: 970-981.
5. Burlutskiy E, Turangan CK (2015) A computational fluid dynamics study on oil-in-water dispersion in vertical pipe flows. Che Engg Research and Design 93: 48-54.
6. Logg A, Mardal KA, Wells GN (2012) Automated solution of differential equations by the finite element method. Springer.
7. Selim K, Logg A, MG Larson (2012) An adaptive finite element splitting method for the incompressible Navier-Stokes equations. Methods Appl Mech Eng 209: 54-65.
8. Sherry M (2010) An experimental investigation of the recirculation zone formed down-stream of a forward facing step. J Wind Eng Ind Aerodyn 98: 888-894.
9. Selimefendigil F, OztopHF (2014) Numerical study and identification of cooling of heated blocks in ulsating channel flow with a rotating cylinder. Int J Th Science 79: 132-145.
10. Selim K, Logg A, MG Larson (2012) An adaptive finite element splitting method for the incompressible Navier-Stokes equations. Methods Appl Mech Eng 209: 54-65.
11. Eymard R, Gallouet T, Herbin R (2003) Finite volume methods. Handbook of Numerical Analysis 7: 713-1020.
12. Selime fendigil F, Oztop HF (2013) Numerical analysis of laminar pulsating flow at a backward facing step with an upper wall mounted adiabatic thin flim. Comp Fluids 88: 93-107.
13. Singha S, Sinhamahapatra KP (2010) Flow past a circular cylinder between parallel walls at low Reynolds numbers.Ocean Eng 37: 757-769.
14. Hussain SH, Hussein AK (2011) Mixed convection heat transfer in a differentially heated square enclosure with a conductive rotating circular cylinder at different vertical locations. Int Commun Heat Mass Transf 38: 263-274.
15. Pollard A (1981) Computer modeling of flow in tee-junctions. PCH Phys Chem Hydro dyn 2: 203-227.
16. Kawashima Y (1983) The heat transfer characteristics of a two-dimensional, right angled, T-shaped flow junction. Int Chem Eng 23: 510-516.
17. Hayes RE, Kumar NK, Nasr-El-Din H (1989) Steady laminar flow in a 90 degree planar branch. Journal of Computers and Fluids 17: 537-553.
18. Neary VS, Sotiropoulos F (1995) Numerical investigation of laminar flows through 90-degree diversions of rectangular cross-section. Journal of Computers and Fluids 25: 95-118.
19. Khandelwal V, Dhiman A, Baranyi L (2015) Laminar flow of non-newtonian shear-thinning fluids in a T-channel. Comput Fluids 108: 79-91.
20. Hassan YA, Kim JH (1984) Three-dimensional analysis of mixing of negatively buoyant jet injected into a confined cross flow.
21. Sparrow EM, Kemink RG (1979) The effect of a mixing tee on turbulent heat transfer in a tube. Int J Heat Mass Transfer 22: 909-917.
22. Liepsch D, Moravec S, Rastogi AK ,Vlachos NS (1982) Measurements and calculation of laminar flow in a ninety degree bifurcation. J Biomech 15: 473-485.
23. Khodadadi JM, Nguyen TM, Vlachos NS (1986) Laminar forced convective heat transfer in a two-dimensional 90-degree bifurcation. Numer Heat Transfer 9: 677-695.
24. Shah RK, London AL (1978) Laminar flow forced convection in ducts. Academic Press, New York.

6

# $L^P$ Donoho-Stark Uncertainty Principles for the Dunkl Transform on $\mathbb{R}^d$

**Fethi Soltani***

*Department of Mathematics, Faculty of Science, Jazan University, Saudi Arabia*

**Abstract**

In the Dunkl setting, we establish three continuous uncertainty principles of concentration type, where the sets of concentration are not intervals. The first and the second uncertainty principles are $L^p$ versions and depend on the sets of concentration $T$ and $W$, and on the time function $f$. The time-limiting operators and the Dunkl integral operators play an important role to prove the main results presented in this paper. However, the third uncertainty principle is also $L^p$ version depends on the sets of concentration and he is independent on the band limited function $f$. These uncertainty principles generalize the results obtained for the Fourier transform and the Dunkl transform in the case p=2.

**Keywords:** Dunkl transform; Dunkl integral operators; Concentration uncertainty principles

## Introduction

According to the classical uncertainty principle a function $f(t)$ is essentially zero outside an interval of length $\Delta t$ and its Fourier transform $\hat{f}(w)$ is essentially zero outside an interval of length $\Delta w$, then $\Delta t\ \Delta w \geq 1$; a function and its Fourier transform cannot both be highly concentrated [1,2]. The uncertainty principle is widely known for its"philosophical" applications: in quantum mechanics, it shows that a particle's position and momentum cannot be determined simultaneously [3]; in signal processing, it establishes limits on the extent to which the "instantaneous frequency" of a signal can be measured [4]. However, it has also technical applications, such as in the theory of partial differential equations [5,6]. In this paper, we consider $\mathbb{R}^d$ with the Euclidean inner product $\langle .,. \rangle$ and norm $|y| := \sqrt{\langle y, y\rangle}$. For $\alpha \in \mathbb{R}^d \setminus \{0\}$, let $\sigma_\alpha$ be the reflection in the hyperplane $H_\alpha \subset \mathbb{R}^d$ orthogonal to α:

$$\sigma_\alpha y := y - \frac{2\langle \alpha, y\rangle}{|\alpha|^2}\alpha$$

A finite set $\Re \subset \mathbb{R}^d \setminus \{0\}$ is called a root system, if $\Re \cap \mathbb{R}.\alpha = \{-\alpha, \alpha\}$ and $\sigma_\alpha \Re = \Re$, for all $\alpha \in \Re$. We assume that it is normalized by $|\alpha|^2 = 2$, for all $\alpha \in \Re$. For a root system $\Re$, the reflections $\sigma_\alpha, \alpha \in \Re$, generate a finite group $G \subset O(d)$, the reflection group associated with $\Re$. All reflections in $G$ correspond to suitable pairs of roots. For a given $\beta \in \mathbb{R}^d \setminus \bigcup_{\alpha\in\Re} H_\alpha$, we fix the positive subsystem $\Re_+ := \{\alpha \in \Re : \langle \alpha, \beta\rangle > 0\}$. Then for each $\alpha \in \Re$ either $\alpha \in \Re_+ or -\alpha \in \Re_+$. Let $k$: $\Re \to C$ be a multiplicity function on $\Re$ (a function which is constant on the orbits under the action of $G$). As an abbreviation, we introduce the index

$$\gamma = \gamma_k := \sum_{\alpha\in\Re_+} k(\alpha)$$

Throughout this paper, we will assume that the multiplicity is nonnegative, that is, $k(\alpha) \geq 0$, for all $\alpha \in \Re$. Moreover, let $W_k$ denote the weight function

$$w_k(y) := \prod_{\alpha\in\Re_+} |\langle \alpha, y\rangle|^{2k(\alpha)}, y \in \mathbb{R}^d$$

which is $G$-invariant and homogeneous of degree 2γ. Let $c_k$ be the Mehta-type constant given by

$$c_k := \left(\int_{\mathbb{R}^d} e^{-|y|^2/2} w_k(y)\,dy\right)^{-1}$$

We denote by $\mu_k$ the measure on $\mathbb{R}^d$ given by $d\mu_k(y) := c_k w_k(y)dy$; and by $L_k^p, 1 \leq p \leq \infty$, the space of measurable functions $f$ on $\mathbb{R}^d$, such that

$$\|f\|_{L_k^p} := \left(\int_{\mathbb{R}^d} |f(y)|^p d\mu_k(y)\right)^{1/p} < \infty,\ 1 \leq p < \infty$$

$$\|f\|_{L_k^\infty} := ess\ \sup_{y\in\mathbb{R}^d} |f(y) < \infty|$$

For $f \in L_k^1$ the Dunkl transform is defined [6] by

$$F_k(f)(x) := \int_{\mathbb{R}^d} E_k(-ix, y) f(y)\, d\mu_k(y), x \in \mathbb{R}^d,$$

where $E_k(-ix, y)$ denotes the Dunkl kernel. (For more details see the next section). Many uncertainty principles have already been proved for the Dunkl transform, namely by Rösler [7] and Shimeno [8] who established (by two different methods) the Heisenberg-Pauli-Weyl inequality. Kawazoe and Mejjaoli gave some related versions of the uncertainty principle (Cowling-Price's theorem, Miyachi's theorem, Beurling's theorem and Donoho-Stark's theorem). Recently, the author [9,10] proved a general forms of the Heisenberg-Pauli-Weyl inequality and he also established a logarithmic uncertainty principle [11].

Let $T$ and $W$ be a measurable subsets of $\mathbb{R}^d$. We say that a function $f \in L_k^p, 1 \leq p \leq 2$, is $\varepsilon$-concentrated to $T$ in $L_k^p$, is concentrated to T in $L_k^p$-norm, if there is a measurable function $g(t)$ vanishing outside $T$ such that $\|f - g\|_{L_k^p} \leq \varepsilon \|f\|_{L_k^p}$. Similarly, we say that $F_k(f)$ is $\varepsilon$-concentrated to $W$ in $L_k^p$-norm, $q = p/(p-1)$, if there is a function $h(w)$ vanishing outside $W$ with $\|F_k(f) - h\|_{L_k^q} \leq \varepsilon \|F(f)\|_{L_k^q}$.

Based on the ideas of Donoho and Stark , we show a continuous-time uncertainty principle of concentration type for the $L_k^p$ theory: If $f$ is $\varepsilon_T$-concentrated to $T$ in $L_k^p$ norm, $1 < p \leq 2$, and $F_k(f)$ is $\varepsilon_W$-concentrated to $W$ in $L_k^q$ norm, $q = p/(p-1)$, then

$$\|F_k(f)\|_{L_k^q} \leq \frac{(\mu_k(T))^{1/q}(\mu_k(W))^{1/q} + \varepsilon_T}{1 - \varepsilon_w}\|f\|_{L_k^p}$$

***Corresponding author:** Fethi Soltani, Department of Mathematics, Faculty of Science, Jazan University, P.O. Box 114, Jazan, Kingdom of Saudi Arabia
E-mail: fethisoltani10@yahoo.com

Next, we prove another version of continuous-time uncertainty principle of concentration type for the $L_k^1 \cap L_k^p$ theory: If $f \in L_k^1 \cap L_k^p, 1 < p \le 2$, is $\varepsilon_T$ -concentrated to $T$ in $L_k^1$ -norm and $F_k(f)$ is $\varepsilon_W$ -concentrated to $W$ *in* $L_k^q$ -norm, *q=p/(p-1)*, then

$$\|F_k(f)\|_{L_k^q} \le \frac{(\mu_k(T))^{1/p}(\mu_k(W))^{1/q}}{(1-\varepsilon_T)(1-\varepsilon_w)}\|f\|_{L_k^p}$$

Let $B_k^p(W)$, $1 \le p \le 2$, be the set of functions $g \in L_k^p$ that are bandlimited to $W$ (i.e. $g \in B_k^p(W)$ implies $S_W g=g$). Here $S_W$ is the Dunkl integral operator given by

$$F_k(S_W f) = F_k(f)1_W$$

where $1_W$ is the indicator function of the set $W$. We say that $f$ is ε-bandlimited to $W$ in $L_k^p$ norm if there is a $g \in B_k^p(W)$ with

$$\|f-g\|_{L_k^p} \le \varepsilon \|f\|_{L_k^p}$$

The space $B_k^p(W)$ leads to establish the following version of continuous-bandlimited uncertainty principle for $L_k^p$ theory: If $f$ is $\varepsilon_T$ -concentrated to $T$ and $\varepsilon_W$ -bandlimited to $W$ in $L_k^p$ norm, $1 \le p \le 2$, then

$$\frac{1-\varepsilon_T-\varepsilon_W}{1+\varepsilon_W} \le (\mu_k(T))^{1/p}(\mu_k(W))^{1/p}$$

This paper is organized as follows. The Section 2 is devoted to recalling some basic properties of the Dunkl transform $F_k$: Plancherel theorem, inversion formula and Hausdorff-Young inequality, which are tools to prove the main results presented in this paper. In Section 3, we introduce some properties of the time-limiting operators and the Dunkl integral operators. These operators play an important role to establish the concentration uncertainty principles in the next sections. In Section 4, we present two continuous-time uncertainty principles of concentration type. These principles depend on the sets of concentration $T$ and $W$, and on the time function $f$. In the last section, we establish continuous-bandlimited uncertainty principle of concentration. This principle depends also on the sets of concentration $T$ and $W$, but he is independent on the bandlimited function $f$.

## The Dunkl transform on $\mathbb{R}^d$

The Dunkl operators $D_j$; $j$=1,..., $d$, on $\mathbb{R}^d$ associated with the finite reflection group $G$ and multiplicity function $k$ are given, for a function $f$ of class $C^1$ on $\mathbb{R}^d$, by

$$D_j f(y) := \frac{\partial}{\partial y_j} f(y) + \sum_{\alpha \in \Re_+} k(\alpha)\alpha_j \frac{f(y)-f(\sigma_\alpha y)}{\langle \alpha, y\rangle}$$

For $y \in \mathbb{R}^d$, the initial problem $D_j u(.,y)(x) = y_j u(x,y), j=1,...,d,$ with μ(0, $y$) = 1 admits a unique analytic solution on $\mathbb{R}^d$, which will be denoted by $E_h(x, y)$ and called Dunkl kernel [12,13]. This kernel has a unique analytic extension to $\mathbb{C}^d \times \mathbb{C}^d$.

The Dunkl kernel has the Laplace-type representation [14]

$$E_k(x,y) = \int_{\mathbb{R}^d} e^{\langle y,z\rangle} d\Gamma_x(z), x \in \mathbb{R}^d, y \in \mathbb{C}^d$$

where $\langle y,z\rangle := \sum_{i=1}^d y_i z_i$ *and* $\Gamma_x$ is a probability measure on $\mathbb{R}^d$ such that

$$supp(\Gamma_x) \subset \{z \in \mathbb{R}^d : |z| \le |x|\}.$$

In our case,

$$|E_k(ix,y)| \le 1, x, y \in \mathbb{R}^d. \quad (2.1)$$

The Dunkl kernel gives rise to an integral transform, which is called Dunkl transform on $\mathbb{R}^d$, and was introduced by Dunkl in, where already many basic properties are established. Dunkl's results have been completed and extended later by De Jeu. The Dunkl transform of a function $f$ in $L_k^1$, is defined by

$$F_k(f)(x) := \int_{\mathbb{R}^d} E_k(-ix,y) f(y) d\mu_k(y)$$

We notice that $F_0$ agrees with the Fourier transform $F$, that is given by

$$F_k(f)(x) := (2\pi)^{-d/2} \int_{\mathbb{R}^d} e^{-i\langle x,y\rangle} f(y) dy, x \in \mathbb{R}^d$$

Some of the properties of Dunkl transform $F_k$ are collected bellow.

**(a)** $L$ $L^\infty$ ***-boundedness:*** For all $f \in L_k^1, F_k(f) \in L_k^\infty$ and

$$\|F_k(f)\|_{L_k^\infty} \le \|f\|_{L_k^1} \quad (2.2)$$

**(b)** ***Inversion theorem:*** Let $f \in L_k^1$, such that $F_k(f) \in L_k^1$. Then

$$f(x) = F_k(F_k(f))(-x), a.e. x \in \mathbb{R}^d \quad (2.3)$$

**(c)** ***Plancherel theorem:*** The Dunkl transform $F_k$ extends uniquely to an isometric isomorphism of $L_k^2$ onto itself. In particular,

$$\|f\|_{L_k^2} = \|F_k(f)\|_{L_k^2} \quad (2.4)$$

**(d)** ***Hausdorff-Young inequality:*** Using relations (2.2) and (2.4) with Marcinkiewicz's interpolation theorem [15,16], we deduce that for every $1 \le p \le 2$, and for every $f \in L_k^p$ the function $F_k(f)$ belongs to the space $L_k^q$, *q=p/(p-1), and* $\|F_k(f)\|_{L_k^q} \le \|f\|_{L_k^p}$ (2.5)

## The Dunkl integral operators

Let T and W be a measurable subsets of $\mathbb{R}$. We introduce the time-limiting operator $P_T$ [1] by

$$P_t f := f 1_T \quad (3.1)$$

And, we introduce the Dunkl integral operator $S_W$ by

$$F_k(S_W f) = F_k(f) 1_W \quad (3.2)$$

In the case k=0, the operator $S_W$ is the frequency-limiting operator given in [1].

**Theorem 3.1:** If $\mu_k(W) < \infty$ *and* $f \in L_k^p, 1 \le p \le 2,$

$$S_W f(x) = \int_W E_k(ix,y) F_k(f)(y) d\mu_k(y)$$

Proof. Let f $\in L_k^p$, 1 ≤ p ≤ 2 and let q=p/(p−1). Then by (2.1), H¨older's inequality and (2.5),

$$\|F_k(f)1_W\|_{L_k^1} = \int_W |F_k(f)(w)| d\mu_k(w)$$

$$\le (\mu_k(W))^{1/p} \|F_k(f)\|_{L_k^q}$$

$$\le (\mu_k(W))^{1/p} \|f\|_{L_k^q}$$

And

$$\|F_k(f)1_W\|_{L_k^2} = \left(\int_W |F_k(f)(w)|^2 d\mu_k(w)\right)^{1/2}$$

$$\le (\mu_k(W))^{\frac{q-2}{2q}} \|F_k(f)\|_{L_k^q} \le (\mu_k(W))^{\frac{q-2}{2q}} \|f\|_{L_k^p}$$

Thus $F_k(f)1_W \in L_k^1 \cap L_k^2$ and by (3.2)

$$S_W f = F_k^{-1}(F_k(f)1_W)$$

This combined with (2.3) gives the result.

**Lemma 3.2:** If 1 ≤ p ≤ 2, q=p/(p−1) and f ∈ $L_k^p$, then

$$\|F_k(S_W f)\|_{L_k^q} \le \|f\|_{L_k^p}$$

**Proof:** Let f∈ $L_k^p$, 1 ≤ p ≤ 2 and let q=p/(p−1). From (2.5) and (3.2),

$$\|F_k(S_W f)\|_{L_k^q} = \left(\int_W |F_k(f)(w)|^q d\mu_k(w)\right)^{1/q} \le \|F_k(f)\|_{L_k^q} \le \|f\|_{L_k^p}$$

This yields the desired result.

**Lemma 3.3:** Let T and W be measurable subsets of $\mathbb{R}^d$. If 1 < p ≤ 2, q = p/(p−1) and f∈ $L_k^p$, then

$$\|F_k(S_W P_T f)\| \le (\mu_k(T))^{1/q}(\mu_k(W))^{1/q}\|f\|.$$

**Proof:** Assume that $\mu_k(T) < \infty$ and $\mu_k(W) < \infty$.

Let f∈ $L_k^p$, 1 < p ≤ 2 and let q=p/(p−1). From (3.2),

$$\|F_k(S_W P_T f)\| = 1_W F_k(P_T f)$$

Thus

$$\|F_k(S_W P_T f)\|_{L_k^q} = \left(\int_W |F_k(P_T f)(w)|^q d\mu_k(w)\right)^{1/q} \tag{3.3}$$

So

$$F_k(P_T f)(w) = \int_T E_k(-iw,t) f(t) d\mu_k(t)$$

and by Holder's inequality and (2.1),

$$|F_k(P_T f)(w)| \le \left(\int_T |E_k(-iw,t)|^q d\mu_k(t)\right)^{1/q}\left(\int_T |f(t)|^p d\mu_k(t)^{1/p}\right)$$

$$\le (\mu_k(T))^{1/q}\|f\|_{L_k^p}$$

Then by (3.3),

$$\|F_k(S_W P_T f)\|_{L_k^q} \le (\mu_k(T))^{1/q}(\mu_k(W))^{1/q}\|f\|_{L_k^p}$$

Thus, the proof is complete.

## Concentration uncertainty principle

Let T and W be a measurable subsets of $\mathbb{R}^d$. We say that a function f ∈$L_k^p$, 1 ≤ p ≤ 2, is ε -concentrated to T in $L_k^p$-norm, if there is a measurable function g(t) vanishing outside T such that $\|f-g\|_{L_k^p} \le \varepsilon\|f\|_{L_k^p}$. Similarly, we say that $F_k(f)$ is ε-concentrated to W in $L_k^q$-norm, q=p/(p−1), if there is a function h(w) vanishing outside W with $\|F_k(f)-h\|_{L_k^q} \le \varepsilon\|F_k(f)\|_{L_k^q}$.

If f is $\varepsilon_T$ -concentrated to T in $L_k^p$-norm (g being the vanishing function) then by (3.1),

$$\|f-P_T f\|_{L_k^p} = \left(\int_{\mathbb{R}^d \setminus T} |f(t)|^p d\mu_k(t)\right)^{1/p} \le \|f-g\|_{L_k^p} \le \varepsilon_T\|f\|_{L_k^p} \tag{4.1}$$

and therefore f is $\varepsilon_T$ -concentrated to T in $L_k^p$-norm if and only if $\|f-P_T f\|_{L_k^p} \le \varepsilon_T\|f\|_{L_k^p}$.

From (3.2) it follows as for $P_T$ that $F_k(f)$ is $\varepsilon_W$-concentrated to *W* in $L_k^q$-norm, q=p/(p – 1), if and only if

$$\|F_k(f)-F_k(S_W f)\|_{L_k^q} \le \varepsilon_W\|F_k(f)\|_{L_k^q} \tag{4.2}$$

The following theorem, states the first continuous-time uncertainty principle of concentration type for the theory.

**Theorem 4.1:** Let T and W be a measurable subsets of $\mathbb{R}^d$ and f∈ $L_k^p$, 1 < p ≤ 2. If f is $\varepsilon_T$ -concentrated to T in $L_k^p$-norm and $F_k(f)$ is $\varepsilon_W$-concentrated to W in $L_k^q$-norm, q=p/(p−1), then

$$\|F_k(f)\|_{L_k^q} \le \frac{(\mu_k(T))^{1/q}(\mu_k(W))^{1/q}+\varepsilon_T}{1-\varepsilon_W}\|f\|_{L_k^p}.$$

**Proof:** Let f∈ $L_k^p$, 1 < p ≤ 2 and let q=p/(p−1). From (4.1), (4.2) and Lemma 3.2 it follows that

$$\|F_k(f)-F_k(S_W P_T f)\|_{L_k^q} \le \|F_k(f)-F_k(S_W f)\|_{L_k^q}$$

$$+\|F_k(S_W f)-F_k(S_W P_T f)\|_{L_k^q}$$

$$\le \varepsilon_W\|F_k(f)\|_{L_k^q} + \|f-P_T f\|_{L_k^p}$$

$$\le \varepsilon_W\|F_k(f)\|_{L_k^q} + \varepsilon_T\|f\|_{L_k^p}$$

The triangle inequality and the Lemma 3.3 show that

$$\|F_k(f)\|_{L_k^q} \le \|F_k(S_W P_T f)\|_{L_k^q} + \|F_k(f)-F_k(S_W P_T f)\|_{L_k^q}$$

$$\le \left[(\mu_k(T))^{1/q}(\mu_k(W))^{1/q}+\varepsilon_T\right]\|f\|_{L_k^p} + \varepsilon_W\|F_k(f)\|_{L_k^q}$$

which gives the desired result.

Next, the second continuous-time uncertainty principle of concentration type for the $L_k^1 \cap L_k^p$ theory is given by the following theorem.

**Theorem 4.2:** Let T and W be a measurable subsets of $\mathbb{R}^d$ and $f \in L_k^1 \cap L_k^p, 1 < p \le 2.$. If f is $\varepsilon_T$ -concentrated to T in $L_k^1$-norm and $F_k(f)$ is $\varepsilon_W$-concentrated to W in $L_k^q$-norm, q=p/(p−1), then

$$\|F_k(f)\|_{L_k^q} \le \frac{(\mu_k(T))^{1/p}(\mu_k(W))^{1/q}}{(1-\varepsilon_T)(1-\varepsilon_W)}\|f\|_{L_k^p}$$

**Proof:** Assume that $\mu_k(T) < \infty$ and $\mu_k(W) < \infty$.

Let $f \in L_k^1 \cap L_k^p, 1 < p \le 2.$ Since $F_k(f)$ is $\varepsilon_W$ -concentrated to W in $L_k^q$-norm, q=p/(p−1), then

$$\|F_k(f)\|_{L_k^q} \le \varepsilon_W\|F_k(f)\|_{L_k^q} + \left(\int_W |F_k(f)(w)|^q d\mu_k(w)\right)^{1/q}$$

$$\le \varepsilon_W\|F_k(f)\|_{L_k^q} + (\mu_k(W))^{1/q}\|F_k(f)\|_{L_k^\infty}$$

Thus by (2.2),

$$\|F_k(f)\|_{L_k^q} \le \frac{(\mu_k(W))^{1/q}}{1-\varepsilon_W}\|f\|_{L_k^1} \tag{4.3}$$

On the other hand, since f is $\varepsilon_T$ -concentrated to T in $L_k^1$-norm,

$$\|f\|_{L_k^1} \le \varepsilon_T\|f\|_{L_k^1} + \int_T |f(t)| d\mu_k(t)$$

$$\le \varepsilon_T\|f\|_{L_k^1} + (\mu_k(T))^{1/p}\|f\|_{L_k^p}$$

Thus

$$\|f\|_{L_k^1} \le \frac{(\mu_k(T))^{1/p}}{1-\varepsilon_T}\|f\|_{L_k^p} \tag{4.4}$$

Combining (4.3) and (4.4) we obtain the result of this theorem.

**Conclusion 4.3:** The first uncertainty principle (Theorem 4.1) depends on the time function f. However, for p=q=2, we obtain

$1-\varepsilon_T-\varepsilon_W \leq (\mu_k(T))^{1/2}(\mu_k(W))^{1/2}$ and the inequality is independent on the time function f. Also, the second uncertainty principle (Theorem 4.2) depends on the time function f. In a particular case when p=q=2, we obtain $(1-\varepsilon_T)(1-\varepsilon_W) \leq (\mu_k(T))^{1/2}(\mu_k(W))^{1/2}$ and the inequality is independent on the time function f.

These uncertainty principles generalize the results obtained for the Fourier transform and the Dunkl transform in the case p=q=2.

## Another uncertainty principle

Let $B_k^p(W), 1 \leq p \leq 2$, be the set of functions $g \in L_k^p$ that are bandlimited to W (i.e. $g \in B_k^p(W)$ implies $S_W g = g$).

We say that f is ε-bandlimited to W in $L_k^p$-norm if there is a $g \in B_k^p(W)$ with $\|f-g\|_{L_k^p} \leq \varepsilon \|f\|_{L_k^p}$

Then, the space $B_k^p(W)$ satisfies the following property.

**Lemma 5.1.** Let T and W be a measurable subsets of $\mathbb{R}^d$. For $g \in B_k^p(W)$, $1 \leq p \leq 2$,

$$\|P_T g\|_{L_k^p} \leq (\mu_k(T))^{1/p} (\mu_k(E))^{1/p} \|g\|_{L_k^p}$$

**Proof.** If $\mu_k(T)=\infty$ or $\mu_k(W) = \infty$, the inequality is clear.

Assume that $\mu_k(T) < \infty$ and $\mu_k(W) < \infty$.

For $g \in B_k^p(W)$, $1 \leq p \leq 2$, from Theorem 3.1,

$$g(t) = \int_W E_k(iw,t) F_k(g)(w)\, d\mu_k(w)$$

and by (2.1) and H¨older's inequality,

$$g(t) \leq (\mu_k(W))^{1/p} \|F_k(g)\|_{L_k^q} \leq (\mu_k(w))^{1/p} \|g\|_{L_k^p}, q = p/(p-1)$$

Hence,

$$\|P_T g\|_{L_k^p} = \left(\int_T |g(t)|^p d\mu_k(t)\right)^{1/p} \leq (\mu_k(T))^{1/p} (\mu_k(W))^{1/p} \|g\|_{L_k^p}$$

which yields the result.

**Theorem 5.2:** Let T and W be a measurable subsets of $\mathbb{R}^d$ and f ∈ $L_k^p$, $1 \leq p \leq 2$. If f is $\varepsilon_W$-bandlimited to W in $L_k^p$-norm, then

$$\|P_T g\|_{L_k^p} \leq \left[(1+\varepsilon_W)(\mu_k(T))^{1/p} (\mu_k(W))^{1/p} + \varepsilon_W\right] \|f\|_{L_k^p}$$

**Proof:** Let f ∈ $L_k^p$, $1 \leq p \leq 2$. Since f is $\varepsilon_W$-bandlimited in $L_k^p$-norm, by definition there is a g in $B_k^p(W)$ with $\|f-g\|_{L_k^p} \leq \varepsilon_W \|f\|_{L_k^p}$. For this g, we have

$$\|P_T f\|_{L_k^p} \leq \|P_T g\|_{L_k^p} + \|P_T(f-g)\|_{L_k^p} \leq \|P_T g\|_{L_k^p} + \varepsilon_W \|f\|_{L_k^p}.$$

Then by Lemma 5.1 and the fact that $\|g\|_{L_k^p} \leq (1+\varepsilon_W)\|f\|_{L_k^p}$ we get the result.

Next, the third continuous bandlimited uncertainty principle of concentration type for the $L_k^p$-norm is given by the following.

**Corollary 5.3:** Let T and W be measurable subsets of $\mathbb{R}^d$ and f ∈ $L_k^p$, $1 \leq p \leq 2$. If f is $\varepsilon_T$-concentrated to T and $\varepsilon_W$-bandlimited to W in $L_k^p$-norm, then

$$\frac{1-\varepsilon_T-\varepsilon_W}{1+\varepsilon_W} \leq (\mu_k(T))^{1/p} (\mu_k(W))^{1/p}$$

**Proof:** Let f ∈ $L_k^p$, $1 \leq p \leq 2$. Since f is $\varepsilon_T$-concentrated to T in $L_k^p$-norm then by (4.1),

$$\|f\|_{L_k^p} \leq \varepsilon_T \|f\|_{L_k^p} + \|P_T f\|_{L_k^p}$$

Thus,

$$\|f\|_{L_k^p} \leq \frac{1}{1-\varepsilon_T} \|P_T f\|_{L_k^p}$$

By (5.1) and Theorem 5.2 we deduce the desired inequality of Corollary 5.3.

**Conclusion 5.4:** The third uncertainty principle (Corollary 5.3) is independent on the bandlimited function f for every $1 \leq p \leq 2$. This uncertainty principle generalizes the result obtained in when p=q=2.

**Acknowledgement**

The author is very grateful to the Reviewers of the Journal for their important comments on this work.

**References**

1. Donoho DL, Stark PB (1989) Uncertainty principles and signal recovery. SIAM J Appl Math 49: 906-931.
2. Kawazoe T, Mejjaoli H (2010) Uncertainty principles for the Dunkl transform. Hiroshima Math J 40: 241-268.
3. Heisenberg W (1949) The physical principles of the quantum theory: Dover, The University of Chicago Press in1930, NewYork.
4. Gabor D (1946) Theory of communication. J Inst Elec Engrg 93: 429-457.
5. Fefferman CL (1983) The uncertainty principle. Bull Amer Math Soc 9: 129-206.
6. Dunkl CF (1992) Hankel transforms associated to finite reflection groups. Contemp Math 138: 123-138.
7. Rosler M (1999) An Uncertainty principle for the Dunkl transform. Bull Austral Math Soc 59: 353-360.
8. Shimeno N (2001) A note on the uncertainty principle for the Dunkl transform. J Math Sci Univ Tokyo 8: 33-42.
9. Soltani F (2013) Heisenberg-Pauli-Weyl uncertainty inequality for the Dunkl transform on Rd. Bull Austral Math Soc 87: 316-325.
10. Soltani F (2013) A general form of Heisenberg-Pauli-Weyl uncertainty inequality for the Dunkl transform. Int Trans Spec Funct 24: 401-409.
11. Soltani F (2014) Pitt's inequality and logarithmic uncertainty principle for the Dunkl transform on R. Acta Math Hungar 143: 480-490.
12. Dunkl DF (1991) Integral kernels with reflection group invariance. Canad J Math 43: 1213-1227.
13. De Jeu MFE (1993) The Dunkl transform. Invent Math 113: 147-162.
14. Rosler M (1999) Positivity of Dunkl's intertwining operator. Duke Math J 98: 445-463.
15. Stein EM (1956) Interpolation of linear operators. Trans Amer Math Soc 83: 482-492.
16. Stein EM, Weiss G (1971) Introduction to Fourier analysis on Euclidean spaces: Princeton Univ Press Princeton NJ.

# Alternative Interpretation of the Lorentz-transformation

**Deyssenroth H***

*Senior Researcher, Germany*

**Abstract**

The Lorentz-Transformation (LT) is the basis of the Theories of Relativity, which are capable of describing the experimentally manifold confirmed relativistic phenomena that deviate from classical physics. Here I present a proof that results in an alternative interpretation of the LT. In particular, the LT cannot be applied to high relative velocities and related space-time modeling – one of the most important tools in physics and astronomy – and will lead to a dead end. Two experiments are proposed to test this idea.

**Keywords:** Lorentz-transformation; Doppler effect; Frequency; Galilean transformation

## Introduction

### The relativistic doppler effect

Suppose two reference frames A and B with identical emitters and detectors move with constant velocities against each other, but their velocities against a fixed point are not known. Due to the measured change of frequency, observers in those systems could calculate the relative velocity between them. However, in classical physics a formula for this model does not exist.

If an observer in A can assume that he is at rest and B moves with velocity –v in his direction, then the classical Doppler-formula [1] is valid (with $\beta = v/c$).

$$f_{AB} = f_0 \frac{1}{1-\beta} \quad (1)$$

*This formula is not based on a transmission medium like air for sound. It is sufficient to assume a constant velocity in relation to a reference point outside of this test system.*

Alternatively, if system B is at rest, and the observer in A moves to B with the velocity +v then a different Doppler-formula is valid.

$$f_{AA} = f_0 (1+\beta) \quad (2)$$

This is observed outside of frames A and B where the information is transmitted with constant velocity c, independent of the movements of A and B.

*Take for example standing on a hill and consider frequency changes observed by drivers (A and B) in moving cars emitting sound at a given frequency in a valley.*

Does the observer in B arrive at the same conclusion as the observer in A?

$$f_{BA} = f_{AB}? \quad (3)$$

$$f_{BB} = f_{AA}? \quad (4)$$

No. If A and B have different velocities $v_A$ and $v_B$ in reference to a fixed point, the observers will measure different frequencies despite the relative velocity v being the same for both. This is a consequence of (1) and (2) being scaled differently.

If the velocities $v_A$ and $v_B$ are not known, one has to estimate the relative velocity v. The best approximation is the mean value of the frequencies $f_{AB}$ and $f_{AA}$

$$f = \sqrt{f_{AB} f_{AA}} = f_0 \sqrt{(1+\beta)/(1-\beta)} = k f_0 \quad (5)$$

This is the formula of the Relativistic Doppler effect [2]. k is the Bondi-k-Calculus factor from which all formulas of the special theory of relativity (STR) can be derived [3-4]

According to this formula, the observed frequencies are symmetric and identical for A and B. This is a mathematical method to compensate the unknown absolute velocities of A and B. It is valid for all information-transmissions, including water waves and light, given the correct constant transmission velocity c. There is no physical mechanism included in this formula.

The extension of formula (5) with $\sqrt{(1+\beta)}$ yields for the boost in x-direction (y'=y, z'=z)

$$f = f_0\, \gamma (1+\beta) \quad (6)$$

and extended with $\sqrt{1-\beta}$

$$f = f0/[\gamma(1-\beta)] \quad (7)$$

These are the classical Doppler formulas but now with a correction factor $\gamma$ that compensates the lack of knowledge of the absolute velocities of A and B. It is based on the geometric mean of the observations in A and B.

$$\gamma = 1/\sqrt{1-\beta^2} \quad (8)$$

We regard the binomial expression $(1-\beta^2) = (1+\beta)(1-\beta)$:

The opposite signs of $\beta$ do not belong to the system B that moves back and forth towards A. This results in the consideration of the simultaneous movement of A and B to or from each other with the relative velocity v. The factor $\gamma$ is the geometric mean of these velocity proportions.

If A and B stipulate to send mutual N pulses within their local time $T_0$ then the equation (7) becomes

***Corresponding author:** Deyssenroth H, Senior Researcher, Germany
E-mail: deyssenroth@t-online.de

$f=N/T=(N/T_0)/\gamma(1-\beta)$ (9)

$T=T_0\gamma(1-\beta)=\gamma(T_0-x_0 v/c^2)$ (10)

With

$x_0=cT_0$ (11)

The distance that information covers within the duration of time $T_0$ to get to an 'event-point' $(x_0,t_0 | x,t)$.

This is the well-known Lorentz transformation [5] for time, which is also valid for sound. With (11) we get the LT for the x-coordinate observed by the observer at 'rest' in A

$x=\gamma(x_0-vT_0)$ (12)

Now it is evident that the maximum speed c is reasonable for these formulas only. In practice, the speed of objects can be greater than c, such as in air or in water.

**The derivation of $\gamma$**

The well-known Galilean transformation provides two formulas [6-7]:

$X'=X(1-\beta)$ (13)

which describes the reference frame B=F' moving away from A=F, and

$X=X'(1+\beta)$ (14)

which describes the reference frame A=F moving away from B=F'

If we assume that the speed of light is the same in both reference frames c=X/T=X'/T' then the time in F' must be transformed as well

$$T'=\frac{X'}{c}=T-T\left(\frac{v}{c}\right)$$

The second term is a movement term, which can also be expressed by X

$$T'=T-X\left(\frac{v}{c^2}\right) \quad (13a)$$

This - together with (13) - is the Voigt transformation, the predecessor of the Lorentz transformation [2].

How would a transformation look like if we assume, that both frames move in direction x towards or away from each other - similar to the Doppler Effect? We are searching for a common factor for both transformations that allows the simultaneous opposite movements of A and B in direction x.

$X'=\gamma X(1-\beta)$ (13b)

$X=\gamma X'(1+\beta)$ (14b)

We multiply and get

$XX'=\gamma^2 XX'(1-\beta^2)$ (15)

and

$\gamma=1/\sqrt{1-\beta^2}$ (16)

Here the $\gamma$-factor is also a geometric mean of the velocity parts $\pm$ v/c of A and B. In particular, it is reasonable for the case when the absolute velocities of A and B are not known.

From (13b) follows the Lorentz transformation

$X'=\gamma(X-vT)$ (17)

And with $c=\frac{X'}{T'}=\frac{X}{T}$

$T'=X'/c=\gamma[T-X(v/c^2)]$

**The invariance of the space-time interval**

Is the interpretation of the LT via the geometric mean actually justified? We regard the simplified equations of a spherical wave in the reference systems F and F' for a boost in direction x which follows from Einstein's second postulate of relativity [3].

$$s^2=c^2t^2-x^2=s'^2=c^2t'^2-x'^2=0,\ (y=y'=z=z'=0)$$

and an object that moves with a speed < c in space-time

$$s^2=c^2t^2-x^2=s'^2=c^2t'^2-x'^2>0$$

we set x = vt and x' = vt'

It is now apparent that the geometric means s and s' are built by the sides of the rectangles of the space-time intervals a and b | a' and b'

$$s^2=\underset{a}{(ct+vt)}\underset{b}{(ct-vt)}=s'^2=\underset{a'}{(ct'+vt')}\underset{b'}{(ct'-vt')} \quad (18)$$

Shaping a rectangle with sides a and b into a square by retaining the same size of area, a side of this square is the geometric mean of sides a and b.

In this case one side of the square $c^2t^2$ resp. $c^2t'^2$ is linearly elongated to side a resp. to a' and the other side is linearly shortened to b resp. to b' but such that the areas $s^2$ and $s'^2$ remain the same. For that purpose one side of the square $s^2$ must be multiplied by a stretching factor k and the other side by a compression factor 1/k:

$$s^2=ks\left(\frac{1}{k}\right)s \quad (19)$$

Therefore,

$a=ct+vt=ks$ (20)

$b=ct-vt=(1/k)s \Rightarrow k/s=1/(ct-vt)$ (21)

As the invariance of areas $s^2$ and $s'^2$ is the basic condition for the LT we can also shrink side a and extend side b:

$$a'=ct-vt=\left(\frac{1}{k}\right)s \quad (22)$$

$$b'=ct+vt=ks \quad (23)$$

As ab=a'b' and therefore $a/a'=b'/b=k^2$ we get (besides the trivial result v=0)

$$k^2=\frac{ct+vt}{ct-vt} \quad (25)$$

Again, this is the formula of the Relativistic Doppler effect.

Geometrically the same areas $s^2$ and $s'^2$ are illustrated by the following graphic:

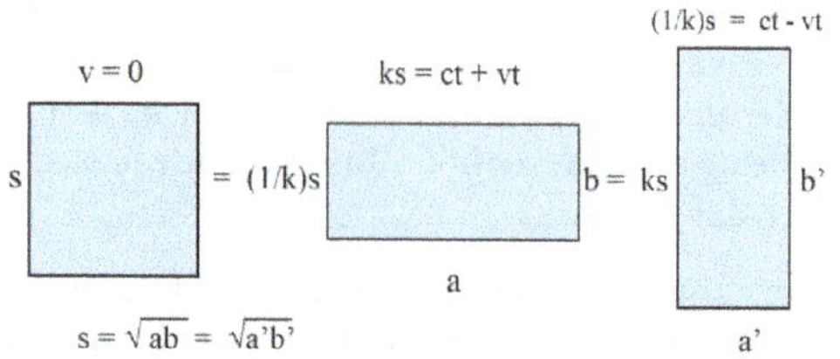


reference system F | F' moves away from F | two possible interpretations |

The straightened up rectangle a'b' allows two interpretations

1) F moves away from F' with velocity –v or

2) F' approaches to F with velocity –v

Again, the common factor k is the formula of the Relativistic Doppler effect but in this case a geometric mean due to the invariance of $s^2=s'^2$ . The second interpretation is invalid because F' cannot move back and forth at the same time. The first interpretation supports the results of the other two derivations of the LT: The frames A=F and B=F' must move in opposite directions simultaneously.

The above three derivations of the LT [8] were carried out by different methods. But the outcome is the same in each case and allows the following interpretation:

***The Lorentz transformation is – like the Doppler effect - based on a frame outside of reference frames A and B, in which light propagates isotropically. The Lorentz-transformed physical values are geometric means*** [4] ***as a result of the simultaneous movement of frames A and B in opposite directions.***

The advantage of this interpretation is that we can now understand the STR intuitively, but the consequences of this interpretation are significant.

## Consequences

a) The geometric mean only makes sense if frames A and B can move against each other. This is not the case for example with experiments at CERN or with the gedankenexperiment in textbooks explaining the STR, where an observer on an embankment observes a train. It is also invalid to apply the LT to two fixed points on a rotating system, since they cannot move together.

b) The mean value is an estimate with an error that grows with the relative velocity v between A and B. However, at particle accelerators this velocity is very high. Therefore, the application of the LT cannot yield a reliable result.

c) The LT cannot be applied to distance X between frames A and B, as is done in various textbooks to explain the slower decay of muons at high velocities when they descend from a height of 10 km to Earth's surface. The invalidity of this approach is demonstrated in the following derivation of the length contraction of X:

B moves away from A (Galilean transformation):

$$X'=X\,(1-\beta) \quad (13b)$$

B approaches A

$$X'=X\,(1+\beta) \quad (13c)$$

$$\Rightarrow X'^2=X^2(1-\beta^2)$$

$$X'=(1/\gamma)\,X \quad (14c)$$

This is the formula of the length contraction of X=cT where the distance x=vT between A and B is included. This is the result of the illogical assumption that B moves to A and moves away from A at the same time.

d) The application of the LT to sound should demonstrate the known relativistic effects near the speed of sound as well, but this is not the case.

e) The application of the LT to the Maxwell-equations results in the statement that light is transmitted isotropically in all reference frames:

$$c^2t^2 - \mathbf{r}^2=c^2t'^2 - \mathbf{r}'^2=0 \quad (18a)$$

The above considerations show that this is an incorrect mathematical construct based on a geometric mean. An isotropic light transmission is possible only if the light will be carried along with the light source. But this contradicts the second STR postulate which is experimentally verified. How could photons 'know' that their reference system was declared by a scientist as being at rest or as moving?

f) The Doppler Effect is based on the limited speed of sound or light. The observed changes of frequency are not real in the observed frame. In the Relativistic Doppler effect, the factor γ describes the situation that only the relative velocity v between A and B is known. In this case the relativistic change of frequency f is not real either in the observed system. As t=Nf, the time dilation is also not real. As the STR is based on the Relativistic Doppler effect k (Bondi) all relativistic effects are not real in the observed frame F'. The experimental findings however show that the mass increase and the time dilation (e.g. decay of muon) are real. Therefore, these findings must have another reason and cannot be caused by the LT.

## Two Symmetry-Experiments

It has now become possible to test the above concept regarding the γ correction with the geometric mean.

*1. Does an observed clock run faster?*

If a car with clock A moves to a fixed clock B, what time difference would an observer in A measure in B? According to the rules of LT the observer in A can regard his frame at rest and B as a moving frame. The Hafele-Keating experiments [11] however show a slower running clock in its own system. In this case, A would measure a faster running clock in B, which is inconsistent with the STR. This experiment can now be demonstrated with new optical clocks.

*2. Does the light clock really work?*

Because of isotropic light emission, a laser pulse (in north-south direction) should hit a detector positioned exactly opposite in its own reference frame A [9-11]. Is this valid? Regarding isotropy as the result of a geometric mean real physics should show that the laser pulse would arrive behind the detector with respect to the movement direction of A. This experiment can now be demonstrated with a squeezed laser.

**Such fundamental experiments are necessary to verify or disprove the LT as basis for the STR** [12].

## Conclusion

The STR experiments produce results that are in accordance with theory. This is akin to the Epicycle theory of the middle Ages, which accurately predicted the orbits of the planets. This was also a coordinate transformation without physical mechanisms. The STR shows that the classical physical formulas (containing space-, time-coordinates or mass) must be multiplied by γ, but as discussed above this cannot be accomplished with the LT. Therefore, scientist should be open-minded and consider an unknown interaction between matter and an unknown medium. This should also solve the problem of the STR being incompatible with Quantum theory. Space-time modeling is useful for the time being but only a vague description of relativistic phenomena. However, lack of physical substantiation leads to a dead end in the long run. There must be a different explanation for gravity. Another option is to continue in position: Nobody can understand the STR/GTR and all seem to be happy.

### References

1. Andrade C (1959) Doppler and the Doppler Effect. Einstein Online.
2. D'Inverno R Introducing Einsteins's Relativity. Physics pages.
3. Einstein A (1905) Zur Electro dynamic bewegter Bodies. Wiley Online Library.
4. Crawley, Michael J (2005) Statistics: An Introduction using R. John Wiley and Sons.
5. Lorentz HA (1909) The Theory of Electrons and the Propagation of Light. Nobel Lecture.
6. Goenner H (1996) Einfuehrung in die spezielle und allgemeine Relativitaetstheorie. Berlin, Heidelberg: Spektrum Akademischer Verlag.
7. http://oyc.yale.edu/sites/default/files/notes_relativity_3.pdf
8. Voigt W (1887) Ueber das Doppler'sche Princip Springer.
9. Rindler W (1977) Essential Relativity. Springer.
10. Bondi H (1980) Relativity and Common Sense. Philosophical research online.
11. Hafele JC, Keating RE (1972) Around the world atomic clocks: Predicted relativistic time gains. Science 177:166-168.
12. Mermin ND (1968) Space and Time in Special Relativity. McGraw-Hill.

# 8

# Guarcs in the Inside Hadronic Four-Dimensional Euclidean Space with Real Time

**Eugene Kreymer***

*Institute for Physics and Engineering, Donetsk, 83114, Ukraine*

**Abstract**

The paper represents the results of the study of the four-dimensional Euclidean space with real time (E-space), where $0 \leq ||VE|| \leq \infty$, in sub-hadronic physics. This closed space has a metric that distinguished from the Minkowski space and the results obtained in the model are different from physical law in the Minkowski space. As it follows from the model of Lagrangian Mechanics, quarks in the central-symmetric attractive potential, kinetic energy of quark diminishes while the speed grows as the quarks exchange their energy-mass with gluons possessing a zero rest mass, so that to ensure the permanent proton mass. This dependence describes the dynamical relation of constituent and current quarks masses.

In the quantified motion model it has been stated, that the oscillations of the particles are cyclic, including alternating localization and translation phases, the action per cycle for a free particle equals $\bar{h}$. The calculation of charge distribution density in proton, carried out on the basis of this model, conforms to the results of the experimental research. All relations between physical values in the E-space, mapped in the Minkowski space, correspond to the principles of SR and are Lorentz-covariant and the infinite velocity is equal to the velocity of light in the Minkowski space. These models have a transparent physical sense.

**Keywords:** Dynamics of quarks in the proton; Euclidean invariants; Motion of quarks and gluons; Quantum cyclic motion; Charge distribution in the proton

## Introduction

Non-perturbative effects are of great importance for the theory of space inside hadron. Supposing a sequence of QCD problems are concentrated in the branch of occurrences that can be described through the transition from the Minkowski space $M(x_{M0}, x_{M1}, x_{M2}, x_{M3})$ (M-space) into the Euclidean space inside hadron via the analytical extension of the time axis onto the lower semi plane $x_{Ei0}=ix_{M0}$. In this case we get the Euclidean space with the imaginary time $\mathbf{E}_{im}(x_{Ei0,} x_{E1} x_{E2}, x_{E3},)$ ($\mathbf{E}_{im}$ is space), and $\mathbf{X}_{Ei=}\mathbf{X}_{M}$ is automatically $\mathbf{V}_{Ei}=i\mathbf{V}_{M}$ and $\boldsymbol{0} \leq ||V_{Ei}|| \leq \boldsymbol{1}$. The use of such a space has brought to great results: the QCD valuum models, lattice calculations, string theory and so on. However, e.g. QCD in lattice can now be used only for the description of a limited class of hadronic elements of the matrix. There is no common and self-congruent description of the QCD vacuum heretofore, as well as confinement occurrence and a spontaneous disturbance of the chiral invariance. In the common case the rotation group of the Euclidean space in the plane $(x_{E0,} \mathbf{X}_{E})$ presupposes that $\boldsymbol{0} \leq ||V_{E}|| \leq \infty$, while $\mathbf{E}_{im} \boldsymbol{0} \leq ||V_{Ei}|| \leq \boldsymbol{1}$. At the same time $\mathbf{E}_{im}$ is not even a subspace of the Euclidean space, because it is not closed in respect of the operation of composition of vectors. Thus, an infinite velocity causing non-local (instantaneous) interactions and contained in some NQCD models lies outside the frames of $\mathbf{E}_{im}$ - space. Non-local quark non-perturbative vacuum condensate plays a crucial role while creating realistic hadrons models [1]. At the same time the space correlation functions look like the curve of decreasing exponent [2] whose negative parameters include the distance of $z=x-y$ while $x_{Ei0=}$const.

In correspondence with [3,4] physics of non-locality starts to be seen at the distance of $\lambda \approx 0,2\text{fm}$. The correlation length λ determines the spatial declining of bound gauge-invariant bilocal correlator of field gradient.

In other studies, a minimal Gauss model, offered in [5], is used for condensates in a non-perturbative vacuum. The parameter of non-locality λ characterizes an average square of quarks' impulse in the QCD vacuum. Its estimations by means of QCD in lattice have shown the following range of probable values: $\lambda_q^2 = 0.45 \pm 0.1 \text{ГэВ}^2$ [6,7]. $\mathbf{E}_{im}$- is homomorphic in respect of the M-space and non-local, in other words, the instantaneous interactions even at some low λ value contradict with S principles.

It gives a reason to consider that the use of merely a part of four-dimensional Euclidean space volume in the models with $\mathbf{E}_{im}$ does not allow using its potential to the full extent. The article expounds the first steps in the research of the inside hadronic four-dimensional Euclidean space with real time model $E(x_{E0,} x_{E1,} x_{E2,} x_{E3})$ (E-space), where $\boldsymbol{0} \leq ||V_{E}|| \leq \infty$, and its aim is to show the expedience of the studies in the **E**-space as s probable prospective direction of sub-hadronic physics development. The article contains researches of the E-space properties in protons and it is presupposed that the obtained correlations have a common nature and can cover all the hadrons. Moreover it has been considered been considered that the models in the **E**-space will not be an alternative for the theoretical developments in $\mathbf{E}_{im}$, but will extend their possibilities. The following requirement is the basic condition enabling this model to exist:

**Requirement 1:** Space-time relations and regularities in the **E**-space model mapped into the **M** -space must correspond to the principles of SR and be Lorentz-covariant.

***Corresponding author:** Eugene Kreymer, Institute for Physics and Engineering, Donetsk, 83114, Ukraine, E-mail: elkreymer@gmail.com

## Inside Hadronic Euclidean Frames of Reference

In the **E**-space no frames of reference, which are microscopic in reality, can be physically implemented. To determine the spatial coordinates the laboratory frame of reference LFR with the coordinates ($x_{M0,}$ $x_{M1,}$ $x_{M2,}$ $x_{M3}$) has been used, where hadron rests, and $d\mathbf{x}_{E=}$ $d\mathbf{x}_M$. The own time of particles in the LFR is admitted to be the temporal coordinate $x_{E0,}$ Thus the **E**-space is "subsidiary" towards the **M**-space.

**Definition 1**: Inner hadronic four-dimensional Euclidean Frame of Reference ($x_{E0,}x_{E1},x_{E2},x_{E3}$) EFR, is a system, where the space coordinates are indexed by the coordinates of ($x_{M1,}$ $x_{M2,}$ $x_{M3}$) LFR and the own time of the particles is equal to the own time of the particles in the LFR

$$dx_{E0i} = ds_{Mi} = dx_{M0i}\sqrt{1-\mathbf{v}_{Mi}^2}, \tag{2.1}$$

Where $V_{Mi}$ is the velocity of the i-number particle in the LFR. The transition to the other IFR is carried out by means of Lorenz transformation. The **E** -space of the real particles corresponds to the **M**-space upper closed cone $\bar{V}_+ := \{x_M \in M / x_M^2 \geq 0, +x_{M0} \geq 0\}$ and $x_{E0}$=[0;∞), that ensures the execution of the causality principle. The EFR has an invariant which taking into consideration the Definition 1 is equal to

$$dx_{E0}^2 + d\mathbf{x}_E^2 = dx_{M0}^2 \tag{2.2}$$

Then there is symmetry between EFR and LFR: the time of one space is the invariant of the other.

From (2.1) and (2.2) it follows that

$$\mathbf{v}_M = \frac{\mathbf{v}_E}{\sqrt{1+v_E^2}}, \tag{2.3}$$

Where $\mathbf{v}_E$ - is the velocity of the particle in the EFR. And, correspondingly

$$\mathbf{v}_E = \frac{\mathbf{v}_M}{\sqrt{1-v_M^2}}. \tag{2.4}$$

If $\mathbf{v}_E \to \infty$, then $\mathbf{v}_M \to 1$. There is also 4-vector of velocity in the EFR

$$u_E = \left(\frac{1}{\sqrt{1+v_E^2}};\frac{\mathbf{v}_E}{\sqrt{1+v_E^2}}\right)$$

And its invariant is equal to the invariant of the corresponding relativistic 4-vector.

*SO*(2) Group of the rotation of plane ($x_{E0,}\mathbf{x}_E$), cannot be applied in the EFR, because the existence of the infinite velocity makes the time absolute, and $x_{E0}$ can take no negative values. In accordance with (2.2), E-group of position-vector rotations E which describes the particles moving with different velocity is valid in the EFR. This group does not mix the temporal $x_{E0}$ and the spatial coordinates $R_{E(}(x_{E1,}x_{E2,}x_{E3})$. Mapping kinetic parameters of the particle in ERF observed in into LRF putting the fundamental quadratic forms

$[E] = (dx_{E\mu}, g_{E\mu\nu}dx_{E\nu}) = ds_E^2$, where $g_{E\mu\nu}$-Kronecker symbol, μ, ν=0,1,2,3 and

$[M] = (dx_{M\mu}, g_{M\mu\nu}dx_{M\nu}) = ds_M^2$, where $g_{M\mu\nu}$-metric tensor. The translation matrix

$$\varphi_{EM} : [E] \mapsto [M] = (dx_E^O, g_E dx_E) \mapsto (K_{EM}x_E^O, G_{EM}g_E K_{EM}x_E) = (dx_M^O, g_M dx_M)$$

must involve kinematic $K_{EM}$ and metric $G_{EM}$ transformations. With (2.2) we obtain the kinematic transformation matrix $\|K_E\| = diag(\sqrt{1+\mathbf{v}_E^2}, 1, 1, 1)$. the metric transformation matrix $\|G_{EI}\|$= diag (1, -1, -1, -1). There is a distinction of properties of the studied space from the Minkowski space that emerges because of different metric: **E**-group of radius-vector rotations $x_{M0}$ does not mix the temporal $x_{E0}$ and spatial coordinates $R_E(x_{E1},x_{E2},x_{E3})$.

## The Model of E-invariant Lagrange Mechanics Particle

### 4-vector energy- momentum

Lagrange function of the free particle

$$L_E = m\sqrt{1+(\mathbf{v}_E)^2} \cdot \tag{3.1}$$

The momentum of the particle

$$\mathbf{p}_E = \frac{m\mathbf{v}_E}{\sqrt{1+\mathbf{v}_E^2}}, \tag{3.2}$$

And the kinetic energy

$$E_E = L_E - \mathbf{v}_E\frac{\partial L_E}{\partial \mathbf{v}_E} = \frac{m}{\sqrt{1+\mathbf{v}_E^2}}. \tag{3.3}$$

This equation is valid under condition that $E_E \geq 0$.. At the same time

$$E_{E+}^2P_E^2=m^2 \tag{3.4}$$

From (3.4) we can make a conclusion that there is a 4-vector of energy-momentum in the EFR, and its invariant is equal to the invariant of the corresponding relativistic 4-vector and it is one more symmetry between the LFR and EFR. Translating the 4-vector of the particle in LFR through (2.3), we obtain $E_{EM} = m\sqrt{1-\mathbf{v}_M^2}$ и $\mathbf{p}_{EM} = m\mathbf{v}_M$. These values stay $\mathbb{E}$ -invariant.

Formula (3.3) testifies to an unusual behavior in the **E**-space of the kinetic energy: it diminishes when the speed grows. The next unit will demonstrate that it is so because of the energy-mass exchange between quarks and gluons.

### Mechanics of quark in the proton

Here we use the model where quarks are considered electrically neutral particles, and we admit that in the center of a proton there is a hypothetical source creating central-symmetrical attractive potential V(*r*) of strong interactions. It is considered that this simplified model will provide the possibility to determine some peculiarities of quarks motion in the proton.

The E-invariant Lagrange function of the quark in the potential V(*r*)

$$L_{EVq} = m_q\sqrt{1+\mathbf{v}_{Eq}^2} - V(r), \tag{3.5}$$

where $m_q$ - constituent mass of the quark. **E**-invariance of this function is ensured by $d\mathbf{x}_E = d\mathbf{x}_M$ potential V(*r*) will be identical for each proton in LFR.

On the analogy with (3.3) the energy of the system "quark – potential V(*r*) »

$$E_V = E_{Eq} - V(r) = \text{const}. \tag{3.6}$$

If a particle is under the influence of power $\mathbf{F} = -\nabla V(r)$ parallel to the velocity, that it will change the momentum as follows:

$$\frac{d\mathbf{p}_{Eq}}{dx_{E0}} = \frac{m_q}{\left(1+\mathbf{v}_E^2\right)^{\frac{3}{2}}}\frac{d\mathbf{v}_{Eq}}{dx_{E0}} = \mathbf{F}. \tag{3.7}$$

The alteration of the energy

$$\frac{dE_{Eq}}{dx_{E0}} = -\frac{\mathbf{v}_{Eq}m_q}{\left(1+\mathbf{v}_{Eq}^2\right)^{\frac{3}{2}}}\frac{d\mathbf{v}_{Eq}}{dx_{E0}} = -\mathbf{F}\mathbf{v}_{Eq}. \tag{3.8}$$

From which

$$d\mathbf{p}_{Eq} = \mathbf{F}dx_{E0}\ (a);\quad dE_{Eq} = -\mathbf{F}\mathbf{v}_{Eq}dx_{E0} = \nabla V(\mathbf{x})\ (b). \tag{3.9}$$

From the (3.9b) and (3.6) it follows that $E_V$=0. The zero-value of $E_V$ is a result of the fact that gluons have not been taken into account. To ensure the constant proton mass, the alteration of the quark kinetic motion must be compensated by the relevant alteration of gluons energy - mass. Taking gluons into account.

$L_{EV}=L_{Eq}(\mathbf{v}_q)+L_{EG}(\mathbf{v}_G)-V(r)$, where $L_{EG}(\mathbf{v}_G)$ is the Lagrange function for gluons.

The preserved energy of "quark - gluon - potential V(*r*) system makes

$$E_{VqG} = L_{Eq}(\mathbf{v}_q) - \mathbf{v}_q\frac{\partial L_{Eq}(\mathbf{v}_q)}{\partial(\mathbf{v}_q)} + \mathbf{v}_G\frac{\partial L_{EG}(\mathbf{v}_G)}{\partial(\mathbf{v}_G)} - L_{EG}(\mathbf{v}_G) - V(r) = m_q. \tag{3.10}$$

This equation has a solution, if $\mathbf{v}_G=\mathbf{v}_q$. Then in the potential V(*r*)

$$L_{EGV} = m_q\sqrt{1+(\mathbf{v}_q)^2} - V(r) - m_q \tag{3.11}$$

Gluon momentum is

$$\mathbf{p}_{EG} = \frac{m_q\mathbf{v}_q}{\sqrt{1+\mathbf{v}_q^2}}, \tag{3.12}$$

And the energy

$$E_{EG} = m_q - \frac{m_q}{\sqrt{1+\mathbf{v}_q^2}}. \tag{3.13}$$

From Esq. (3.3) and (3.13) we can draw a conclusion, that the energy - mass of the quark translates into the energy - mass of the gluon, and their sum makes equals $m_q$. At the same time $\mathbf{P}_{EG}=\mathbf{P}_{Eq}$ and gluons are moving along with quarks creating valon. As a result, the constituent mass of quarks includes zero rest mass. This determines the dynamical relation of constituent and current quarks' masses. The quark mass diminishes as it approaches to the centre of a proton. This corresponds to the existing idea that quark has a minimum mass under a big transferred to it $q^2$ momentum. There are some scientific studies devoted to the NQCD, in which gluons are described as possessors of dynamical energy - mass [8]. Contains an approximate solution of Dyson-Schwinger equation, where a propagator of non-perturbative gluon is regulated by the dynamical generated mass of a gluon. The usage of this propagator gives an opportunity to calculate sections of pp- scattering and achieve a good concord of calculations with experimental data for an effective gluon mass of 370 MeV [9], this value corresponds to $m_q$ in the nucleon. The fact that gluon has peculiarities of a massive particle is confirmed by calculations in lattice [10,11]. In the papers [12,13] different non-zero masses of gluons have also been studied. Let us examine the quark motion in the linearly increasing potential V(*r*) =*cr*. The zero orbital moment of a proton along with experimental studies of the charge distribution in proton means that the quark is vibrating along the diameter towards the center of a proton. Let us presuppose that the quark vibrates under the power of $|F_z|$=constant along the ***z*** axis which has a null in of the center a proton. Basing n the eq. (3.2), (3.3) and (3.9a) we obtain

$$\frac{dz}{dx_{E0}} = \frac{p_{Ezq}}{E_{Eq}} = \frac{F_z x_{E0}}{\sqrt{m_q^2 - F_z^2 x_{E0}^2}} = \frac{a_{clz}x_{E0}}{\sqrt{1-(a_{clz}x_{E0})^2}}, \tag{3.14}$$

Where $a_{clz} = F_z / m_q$ is a "classic" acceleration? Then

$$z = -\sqrt{a_{clz}^{-2} - x_{E0}^2} \tag{3.15}$$

and

$$v_{Ezq} = \frac{\sqrt{a_{clz}^{-2} - z^2}}{z}. \tag{3.16}$$

The dependence of the quark energy on the radius is $E_{Eq} = mza_{clz}^{-1}$. From the (3.6) we can draw a conclusion that $V(r_{max}=r_p)=m_q$, where $r_p$ is the radius of a proton and $a_{clz}^{-1} = r_p$. Under the condition that $-a_{clz}^{-1} \le x_{E0} \le a_{clz}^{-1}$ in the coordinates ($x_{E0}$, *z*) quark makes a circumference with a radius $a_{clz}^{-1}$. But the allowable values are $x_{E0}$=[0;∞) and this formula must be specified. The half period of quark vibration is $\theta = 2a_{clz}^{-1}$ and to preserve $x_{E0}$ in the given range of values we need to put (3.15) it in the following way:

$$z = \pm\sqrt{a_{clz}^{-2} - (t_{E0} - a_{clz}^{-1})^2}, \tag{3.17}$$

Where $t_{E0} = x_{E0} - [n]2a_{clz}^{-1}$, [*n*] is the biggest whole number in $x_{E0} / 2a_{clz}^{-1}$. The digits before the root take turns depending on the alteration of [*n*].

Thus, a vibrating quark makes two half circumferences with *z*>0 and *z*<0, moved at $2a_{clz}^{-1}$. Figure 1 shows the graph of the quark oscillations. The calculation involves the rms radius of the proton $r_p$=0.84fm.

Here we can show how the formula (3.6) is functioning. Under *z*=0 and V=0 the speed makes $v_{Ezq}$=∞ and $E_{Eq}$=0 (points A, C, E). Under $z = \pm a_{clz}^{-1}$ and $v_{Ezq}=0$ as well as $E_{Eq}=m_q$, as well as $V=m_q$ (points B, D). And therefore $E_{Eq}$-V=0.

This brings up a question: how do the oscillations of quarks provide total zero momentum in the motionless proton while they are oscillations? Under multi-particle interactions, a symmetric disposition of particles corresponds to the minimum of energy and therefore a proton possesses a spherical symmetry and that means that 3 quarks make diametric oscillations creating a space angle π and their impulses are getting balanced. This supposition correlates with analytical studies described in [14]; according to them effective fields in baryons has a Y-shaped configuration of quarks' plane making an equilateral triangle. This conclusion has also been confirmed by calculations in lattice [15].

## Models of E-Invariant Quantized Motion of Massive Particles

A peculiarity of inside hadronic E-space is that its size in the three-

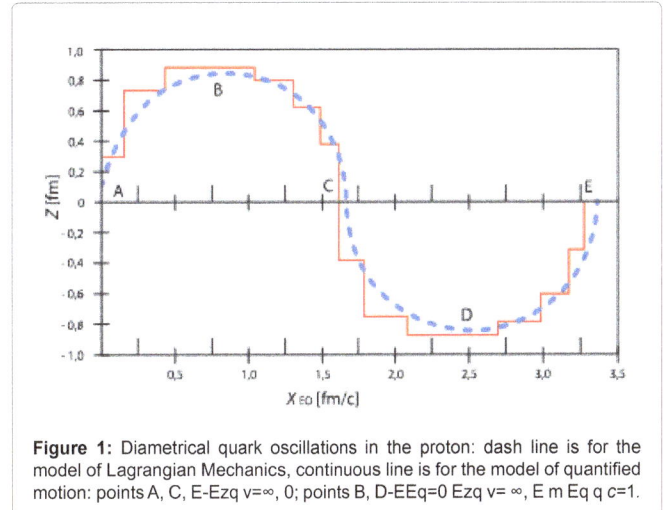

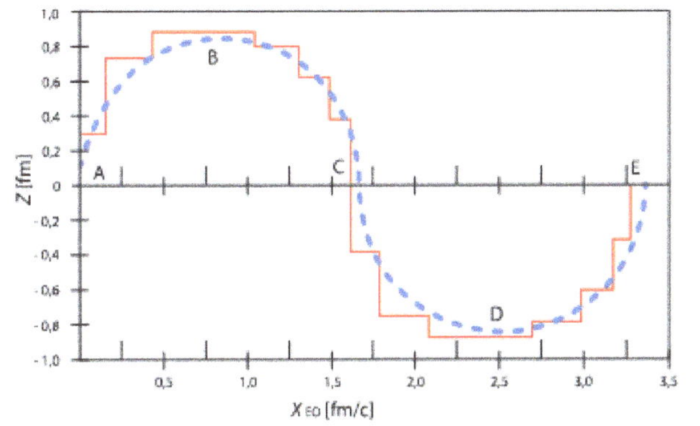


**Figure 1:** Diametrical quark oscillations in the proton: dash line is for the model of Lagrangian Mechanics, continuous line is for the model of quantified motion: points A, C, E-Ezq v=∞, 0; points B, D-EEq=0 Ezq v= ∞, E m Eq q c=1.

dimensional space is comparable to the Compton quark wave length and the maximum value of quark kinetic energy makes $m_q$. According to the quantum mechanics the minimum quark energy in the limited space must excess its mass. This is also applicable for oscillators' energy in the quantum field theory. Thus the wave equations cannot be applied in our case, including probability interpretation. Though the quarks' behavior in hadrons has a casual nature and the definite metric of the E-space enables to precede straight forward to the probability characteristics.

## Free scalar particle

The model is oriented towards the inside hadronic space, in which a particle cannot be free, so this part is of a methodic character.

Let us introduce the probabilistic space indexed by **E**-elements and defined by three quantities (Ω,Σμ) where Ω is a multitude of eve, Σ σ is algebra of Ω subsets and $\mu$ is a positive measure normalized, and μ(Ω) ≤1. If $X_E$ is the real random variable and $X_A \in \Omega$, then the distribution of $X_E$ is the probabilistic measure on Ω $\mu$ =P($X_{E0}$<$\mathbf{X}_E$<$X_E$)

**Definition 2:** The state of the particle is described by the function $\hat{O}(x_{E0},\mathbf{x}_E)=1-\mu=\bar{\mu}$, belonging to **E** and selected for $\Phi(x_{E0}=\infty,\mathbf{x}_E=\infty)=0$.

If the functions $\varphi(x_{E0,}\mathbf{x}_E)=\Phi(x_E)$ describe a scalar particle then its Lagrangian will equal Ò

$$L=\frac{1}{2}(\partial_\mu\varphi)^2+\frac{m^2}{2}\varphi^2, \tag{4.1}$$

From which in a usual way we can get a Klein-Gordon-Fock equation in the **E** - space

$$(\partial_0^2+\Delta)\varphi(x_{E0,}\mathbf{x}_E)=m^2\varphi(x_{E0},\mathbf{x}_E), \tag{4.2}$$

Correspondingly to the (3.4)

The obvious function $\varphi(x_{E0,}\mathbf{x}_E)=\exp[-(p_{E0}x_{E0}+\mathbf{p}_E\mathbf{x}_E)]$ may not seem to be the solution of the (4.2), as it will give the conditional expectation value $E(\mathbf{x}_E|x_{E0}=0)=\mathbf{p}_E^{-1}$. Under $\mathbf{P}_E$=0 we obtain the non-physical value $E(\mathbf{x}_E|x_{E0}=0,\mathbf{v}_E=0)=\infty$. It is also impossible to use the transition to *K*-representation through a Fourier transformation, as the frequency $k_{E0}$ and the wave vector $\mathbf{K}_E$ not satisfy the (3.4).

Despite the time coordinate, (4.2) describes the static state. However the infinite velocity in **E** - space, makes it possible to transform the equation for the description of dynamic systems. Now represent $\varphi(x_{E0},\mathbf{x}_E)$ as the product of two functions $\varphi(x_{E0},\mathbf{x}_E)=\varphi_1(x_{E0})\varphi_{tr}(\mathbf{x}_E)$, each depending on just one variable. Such separation has the following physical meaning. For $\mathbf{v}_E$=0 the function $\varphi_1(x_{E0})=\varphi(x_{E0},\mathbf{x}_E)_{\mathbf{x}_E=0}$ will describes the localization phase, and the function $\varphi_{tr}(\mathbf{x}_E)=\varphi(x_{E0},\mathbf{x}_E)_{x_{E0}=0}$ the translation phase for $\mathbf{v}_E$=∞. These phases cannot exist simultaneously, and supposing the average duration of the localization phase is $\tau_E=\langle x_{E0}\rangle$, and that of the translation phase is - $\boldsymbol{\chi}_E=\langle\mathbf{x}_E\rangle$, and the average phase change occurring with $\tau_E$ and $\boldsymbol{\chi}_E$, then after every cycle of phase change we get the motion of the particle at the average velocity of $\langle\mathbf{v}_E\rangle=\boldsymbol{\chi}_E/\tau_E$. Such a separation is due to the infinite velocity.,

Probabilistic approach in compliance with definition 2, consider $\hat{O}(x_{E0},\mathbf{x}_E)$=P($X_{E0}$>$x_{E0}$,$\mathbf{X}_E$>$\mathbf{x}_E$) being the multidimensional random vector. The random projection of this vector on, for example, axis $x_{E0}$ defines the probability of event $P(X_{E0}>x_{E0})$ and requires the condition $P(\mathbf{X}_E>\mathbf{x}_E)$=1 to be commonly met. The latter condition is met for $\mathbf{X}_E$=0, that is $\varphi_1(x_{E0})=P(X_{E0}>x_{E0})$. Accordingly, u, $\varphi_{tr}(\mathbf{x}_E)=P(\mathbf{X}_E>\mathbf{x}_E)$ for $P(X_{E0}>x_{E0})$ =1, i.e. $x_{E0}$=0. As a result, we arrive at (4.2).

Separating the variables it is necessary to take into account that $\tau_E$ and ÷$_E$ must be a 4 - vectors: $\tau_E^2+\boldsymbol{\chi}_E^2=\Theta^2$, and $\Theta^2$ -**E**-invariant and $\Theta=c(x_{E0})\tau_E=c(\mathbf{x}_E)\boldsymbol{\chi}_E$, where

$$c(x_{E0})=(1+\mathbf{v}_E^2)^{-\frac{1}{2}}\text{ (a)};\quad c(\mathbf{x}_E)=\mathbf{v}_E(1+\mathbf{v}_E^2)^{-\frac{1}{2}}\text{ (b)}. \tag{4.3}$$

From this it follows that equations for each phase of the i-cycle are to be solutions of (4.1)

$$\partial_0^2\varphi_1(x_{E0})=k_E^2\varphi_1(x_{E0})\text{ (a)};\quad \Delta\varphi_{tr}(\mathbf{x}_E)=\mathbf{k}_E^2\varphi_{tr}(\mathbf{x}_E)\text{ (b)}, \tag{4.4}$$

where

$$k_E=m\sqrt{1+\mathbf{v}_E^2}=E_E\left(1+\mathbf{v}_E^2\right)\text{ (a)};\quad \mathbf{k}_A=\frac{m\sqrt{1+\mathbf{v}_A^2}}{\mathbf{v}_E}\text{ (b)}. \tag{4.5}$$

Equation (4.4a) has the following solution $\varphi_i(x_{E0})=C_1\exp(-k_Ex_{E0})+C_2\exp(+k_Ex_{E0})$. As attractive potential V*(r)* ≥ 0 equally affects the particle as well as the antiparticle, according to (3.6) $A_E=\sqrt{m^2-\mathbf{p}_E^2}\geq 0$ and correspondingly, $K_E$ ≥ 0. Considering that $x_{E0}\geq 0$ from Definition 2 it follows that $C_1=1, C_2=0$, and there remains the decreasing exponent. The probability density $f_1(x_{E0})=-d\varphi_1(x_{E0})/dx_{E0}=k_E\exp(-k_Ex_{E0})$ possesses necessary properties: the densities are not negative and the integral of the densities over all values of $x_{E0}$ equals unity. The mathematical expectation of the localization phase duration

$$\tau_E=\frac{1}{k_E}=\frac{1}{m\sqrt{1+\mathbf{v}_E^2}}. \tag{4.7}$$

For several cycles, segments $x_{E0}$ form the simplest stream with no aftereffect. For the free particle in the translation phase the displacement vector of the particle $\mathbf{x}_E$ and vector $\mathbf{k}_A$ are co-directed and (4.4b) has the following solution

$$u_{tr}(\mathbf{x}_E)=\hat{\mathbf{x}}_E\exp(-|\mathbf{k}_E|x_E), \tag{4.8}$$

Where $x_E=|\mathbf{x}_E|$ and $\hat{\mathbf{x}}_E$ is the unit vector. Probability density $f_{tr}(\mathbf{x}_E)=\mathbf{k}_E\exp(-|\mathbf{k}_E|x_E)$, i.e. probability density is also positive and the mathematical expectation of the particle displacement in the translation phase is

$$\boldsymbol{\chi}_E=\mathbf{k}_E^{-1}=\frac{\mathbf{v}_E}{m\sqrt{1+\mathbf{v}_A^2}}. \tag{4.9}$$

As has been assumed , $\tau_E$ and $\boldsymbol{\chi}_E$ are the components of E-vector and with (2.2)

$$\tau_E^2+\boldsymbol{\chi}_E^2=m^{-2}=\tau_{MI}^2, \tag{4.10}$$

where $\tau_{MI}$ is the average cycle duration in LRF. E-invariant is a value

$$\tau_Ep_{E0}+\boldsymbol{\chi}_E\mathbf{p}_E=\hbar, \tag{4.11}$$

which equals quantum of action .

On the grounds of (4.10) we consider the cycle duration in LRF $x_{MOI}$ to be the two-dimensional random vector with random coordinates $x_{E0}$ and $\mathbf{x}_E$ , distributed by the exponential law. Then $x_{MOI}$ is also distributed by the exponential law $u_c(x_{M0l}) = \exp(-mx_{M0l})$ with the average value $\langle x_{M0l}\rangle = m^{-1}$. From equality $d\mathbf{x}_E = d\mathbf{x}\, d\mathbf{x}_E = d\mathbf{x}_{Ml}$ it follows that $\mathbf{x}_{Ml}$ is also distributed by the exponential law $u(\mathbf{x}_{Ml}) = \hat{\mathbf{x}}_{Ml}\exp(-|\mathbf{k}_M|x_{Ml})$ and $\langle \chi_{l\,l}\rangle = \langle \chi_E\rangle$. Using equations (2.3b) and (4.9) we obtain

$$\chi_{l\,l}(\mathbf{v}_l) = \mathbf{v}_l\, m^{-1}. \quad (4.12)$$

and the average velocity of $\langle \mathbf{v}_M\rangle = \chi_{Ml} / \tau_{Ml} \le 1$. The value $\tau_{Ml} p_{M0} - \chi_{Ml}\mathbf{p}_M = \hbar\sqrt{1-\mathbf{v}_M^2}$ is also relativistically covariant, and equals to relativistic Lagrangian accurate to a coefficient and changes from $\hbar$ to 0.

Thus the motion of the particle in E-space is discreet, consists of alternating translation and localization phases and the resultant action for every cycle equals a quantum of motion. The averaged graph of free - particle motion in $\mathbb{E}$ is a random step function with the average step length $\tau_E$ and the average step height $\chi_E$.

The average duration of free – particle cycle in LRF quantizes time $x_{M0l}$ into intervals with the average value $\tau_M = m^{-1}$ dependent only on particle mass. And homogeneity is not violated.

### Free spinor particle

Spinor function $\psi(x_{E0,}\mathbf{x}_E)$ also should be a solution to the Dirac equation in E and describe two phases of motion. To derive the Dirac equation model in E we need to take into account that $\mathbf{E} = \mathbb{R}_E(x_{E0} \ge 0) \oplus \mathbb{R}_E(x_{E1}, x_{E2}, x_{E3})$. The sense of such E-space partition is in the fact that in it the rotation is only possible in $\mathbf{E} = \mathbb{R}_E(x_{E0} \ge 0) \oplus \mathbb{R}_E(x_{E1}, x_{E2}, x_{E3})$ and consequently only bispinors have effect. Let us factorize (4.2)

$$\gamma_{E\mu}\partial_{E\mu}\psi(x_{E0,}\mathbf{x}_E) = -m\psi(x_{E0,}\mathbf{x}_E). \quad (4.13)$$

Matrices $\gamma_{E\mu}$ satisfy the relation $\gamma_{E\mu}\gamma_{E\nu} + \gamma_{E\nu}\gamma_{E\mu} = 2g_{E\mu\nu}$, where $g_{E\mu\nu}$ - the Kronecker symbol, and equal

$$\gamma_{E0} = \hat{a} = \mathrm{diag}(\,1,\,1),\ \gamma_{Ei} = \sigma_i. \quad (4.14)$$

Function $\psi(x_{E0,}\mathbf{x}_E)$ should describe two phases of motion

$$\psi(x_{E0,}\mathbf{x}_E) = \psi_1(x_{E0})\psi_{tr}(\mathbf{x}_E). \quad (4.15)$$

For the localization phase together with (4.5) we obtain

$$\gamma_{E0}\partial_0\psi_1(x_{E0}) = -k_E\psi_1(x_{E0}), \quad (4.16)$$

where $\psi_1(x_{E0}) = \beta\exp(-k_E x_{E0})$ - bispinor with $k_E > 0$ and $x_{E0} \ge 0$.

Equation for the translation phase is

$$\gamma_E\nabla\psi_{tr}(\mathbf{x}_E) = -\mathbf{k}_E\psi_{tr}(\mathbf{x}_E) \quad (4.17)$$

and in compliance with (4.8) the solution is $\psi_{tr}(\mathbf{x}_E) = \beta\hat{\mathbf{x}}_E\exp(-|\mathbf{k}_E|x_E)$. Then

$$\gamma_E\psi_{tr}(\mathbf{x}_E) = \frac{\mathbf{k}_E}{|\mathbf{k}_E|}\psi_{tr}(\mathbf{x}_E), \quad (4.18)$$

When movement is along axis $x_3$

$$\sigma_3\psi_{tr}(x_{E3}) = \frac{k_{E3}}{|k_{E3}|}\psi_{tr}(x_{E3}) \quad (4.19)$$

and the space of bispinor $\psi_{tr}(x_3)$ is a proper space of the diagonal matrix $\sigma_3$ with positive and negative helicity and there may be only a discrete transition between these subspaces. The duration of localization phases and the extent of translation phases are defined by formulae (4.7) and (4.11).

All the features of the quantum theory of the scalar particle are valid for spinors as well. But in the latter case we have a new detail of helicity. In **E**, the helicity of massive fermions is only observed in the translation phase, and it is a "good" quantum number, whereas in **M** the helicity of massive fermions with a nonzero mass can't be a quantum number characterizing the particle, since it can be inverted by appropriate Lorentz transformations. Nevertheless, in nature, there exist left and right fermions that are quite different particles and this is seen in **E**.

### Neutral spin or particle in the strong potential

If the particle is affected by the attractive potential which in the general case equals $V(x_0, \mathbf{x}_E)$, then (4.13) will take the form

$$\gamma_{E\mu}\partial_{E\mu}\psi(x_{E0}, \mathbf{x}_E) = -[m + V(x_{E0}, \mathbf{x}_E)]\psi(x_{E0}, \mathbf{x}_E)\partial_{E\mu}V(x_{E0}, \mathbf{x}_E). \quad (4.20)$$

If potential $V(\mathbf{x}_E)$ works then in the localization phase

$$\gamma_{E0}\partial_{E0}\psi_{IV}(x_{E0}) = -k_E\left(1 + \frac{V(\mathbf{x}_E)}{m}\right)\psi_{IV}(x_{E0}). \quad (4.21)$$

The solution to this equation is

$$\psi_{IV}(x_{E0}) = \beta\exp\left(-k_E(1 + \frac{V(\mathbf{x}_E)}{m})x_{E0}\right). \quad (4.22)$$

The average duration of the localization phase is

$$\tau_V = \frac{1}{k_E(1 + V(\mathbf{x}_E)/m)}. \quad (4.23)$$

The equation for the translation form will take the form

$$\gamma_E\nabla\psi_{trV}(\mathbf{x}_E) = -\mathbf{k}_E(1 + V(\mathbf{x}_E)/m)\psi_{trV}(\mathbf{x}_E)\nabla V(\mathbf{x}_E) \quad (4.24)$$

and the solution

$$\psi_{tr}(\mathbf{x}_E) = \beta\hat{\mathbf{x}}_E\exp\left(-(1 + V(\mathbf{x}_E)/m)|\mathbf{k}_E|x_E\right). \quad (4.25)$$

The average extent of the translation phase is

$$\chi_V = \frac{1}{\mathbf{k}_E(1 + V(\mathbf{x})/m)}. \quad (4.26)$$

With the quantized motion for $V(r)=cr$ (3.9) takes the following form

$$E(\delta\mathbf{p}_{Eqi}) = \mathbf{F}\tau_{Vi}\ (a);\quad E(\delta E_{Eqi}) = -\mathbf{F}\chi_{Vi}\ (b). \quad (4.27)$$

Equation (4.27*b*) proves that while the translation phase is on when $x_{E0}$=const, there are instant nonlocal interactions in **E**. However, when mapped in **M** they take place with speed *c*.

## Application of the Model of Quantized Motion of Quarcs to Determine Some Properties of Quarks in Protons

### Quantized motion of quarks

The calculation of the quantized motion of quarks has been done on the basis of the IVC (Figure 1) on the assumption that the quark moves along the axis *z* which passes though the centre of the proton, parameters of motion $\tau_{Vi}$ and $\chi_{Vzi}$ being of average value. The following data are used in the calculation: root-mean-square radius of the proton $r_p$=0.84 fm and $a_{cly}^{-1} = r_p = 0.84$fm, averaged constituent mass *u* and *d* of quarks 0.33Gev. This mass is included into the calculation as Compton wave-length of a quark $\lambda_q \approx 0.6$fm. The motion of a quark is divided into deceleration and acceleration portions. The initial point for the calculation (point A) is chosen at the beginning of the deceleration portion when a quark has passed through the centre of a proton and at point $x_E = 0,\ z = \lambda_q / 2 = 0,3\text{fm}$ the quark localization phase starts. The acceleration portion starts with the translation phase at point B when $v_{Ez}$=0 and the end of the translation phase coordinate $x_{E0}$ has become more than 0.84fm.

The quark deceleration in the second half-period of oscillation starts also with the localization phase at point C for $v_{Ez}=\infty$ and $z<0$ and the calculation is done in the way similar to the first half-period. Here the following peculiarity is disclosed: the coordinates of the beginning of the second oscillation (0.08fm 0.32fm) are close to the accepted coordinates of the beginning of the first oscillation (0.0, 0.3fm).

## Charge distribution in the proton

Central-symmetric motion of quarks (Section 3.1) makes it possible to confine to the calculation of the charge distribution for one quark considering that its charge equals the charge of a proton. The calculation is done on the assumption that V($r$) =$cr$ and the charge distribution is defined by the probability of the quark being at a given point of radius r=|Z| and this probability must be determined from the **M**-space "viewpoint"|

$P(r_i) = \tau_{MIVi} / \Sigma \tau_{MIVi}$ , where $\tau_{MIVi} = \sqrt{(\tau_{EVi})^2 + \chi_{Vi}^2}$ . As the calculations show that the second oscillation practically repeats the first oscillation the parameters of the first oscillation are accepted as the calculation basis. The calculation of the charge density has been done under the condition that the charge is located in the spherical layer with a unit thickness which has radius $r$. After the approximation by the exponential function the equation for the charge density calculation is obtained $\rho_c(r) = 4.0\exp(-3.9r)e/\text{fm}^3$ for validity factor $R^2$=0.85. The calculated charge distribution along the radius is $j_c(r) = 4\pi r^2 \rho_c(r)\ e/\text{fm}$ (Figure 2).

For the comparison the experimental data for the electric form-factor of the proton have been used which are usually described by dipole approximation $G = (1 + q^2/0.71)^{-2}$ [16] for the preset square of 4-momentum $q^2$. This dependence gives the experimental value of charge density $\rho_e(r) = 3.0\exp(-4.35r)\ e/\text{fm}^3$ and that of the distribution of a charge along the radius $j_e(r) = 4\pi r^2 \rho_e(r)\ e/\text{fm}$ (Figure 2).

Graph $j_c(r)$ systematically exceeds $j_e(r)$. It is connected with the fact that definitional domain $j_c(r)$ equals $0<r<0.85$ fm and the box under $j_c(r)$ equals $\approx 1$. Definitional domain $j_c(r)$ equals $0<r<\infty$ and the box under this curve on the section $0<r<0.9$fm equals 0.6.

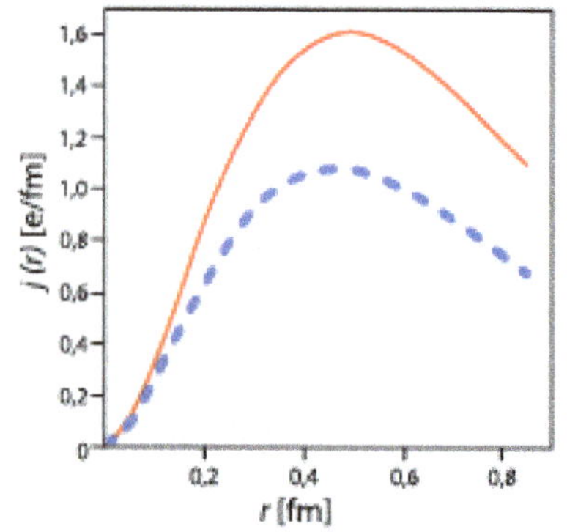


**Figure 2:** Electric charge distribution in the proton: continuous line is for calculation data, dashlineis for experimental data.

## Conclusion

It is stated that in the E-space model, Radius-vector rotations group does not mix temporal and spatial coordinates;- kinetic energy diminishes when the speed grows. This determines the existence of constituent and current quarks and describes the dynamic relation of their masses; - to describe quantum movement in the E-space, wave equations cannot be applied. The application of the random function theory has shown that the quarks' movement consists of localization and translation phases;- helicity of massive fermions can be observed only during translation phase and is a "good" quantum number;- an infinite velocity and non-local interactions connected with it while mapping in the **M**-space does not upset the RS-principles: the maximum interaction transmission velocity and the maintenance of causality principle; -the proton charge calculation result plausibly agrees with the experimental data; the four-dimensional values in the E-space are the 4-vector with scalar invariants which have analogies in the **M**-space;- the **E**-invariant models have a transparent physical content and are no alternative for the existing QCD methods, but expand their possibilities.

### References

1. Dorokhov A, Tomio L (2000) Pion structure function within the instanton model. Phys Rev.
2. Dorokhov A, Esaibegyan S, Mikhailov S (2000) Eur J Phys C 13: 331.
3. Dosch H (1987) condensate and effective linear potential. Phys Lett B 190: 177.
4. Dosch H, Simonov YU (1988) Gluon The area law of the Wilson loop and vacuum field correlators. Phys Lett B 205:339.
5. Bakulev B, Mikhailov S, Stefanis N (2001) Erratum to QCD-based pion distribution amplitudes confronting experimental data. Phys Lett B 508: 279.
6. DElia M, DiGiacomo A, Meggiolaro E (1999) phys Rev D 59.
7. Bakulev A, Mikhailov S (2002) Lattice measurements of nonlocal quark condensates, vacuum correlation length, and pion distribution amplitude in QCD. Phys Rev D 65: 114511.
8. Cornwall J (1982) Dynamical mass generation in continuum quantum chromo dynamics. Phys Rev D 26: 1453.
9. Halzen G, Krein A, Natale A (1993) Relating the QCD Pomeron to an effective gluon mass. Phys Rev D 47: 295.
10. Bernard C (1982) Monte Carlo evaluation of the effective gluon mass. Phys Lett B 108: 431.
11. Mandula J, Ogilvie O (1987) The gluon is massive: A lattice calculation of the gluon propagator in the Landau gauge. Phys Lett B 185: 127.
12. Djordjevic M, Gyulassy M (2003) Phys Lett B 560.
13. Djordjevic M (2006) Transition radiation in QCD matter. Phy Rev C 73: 044912.
14. Kuzmenko D, Simonov YU.
15. Alexandrou C, de Forcrand PH, Jahn O (2003) Nucl Phys Proc Suppl 119: 667- 669.
16. Belkov A, Kovalenko S (1987) *PEPAN* 18: 110.

# Comparison of Macrodosimetric Efficacy of Transarterial Radioembolization (TARE) by Using 90Y Microspheres of Different Density of Activity

**Traino AC[1*], Piccinno M[1], Boni G[2], Bargellini I[3] and Bozzi E[3]**

[1]*Unit of Medical Physics, University Hospital Pisana, Italy*
[2]*Unit of Nuclear Medicine, University Hospital Pisana, Italy*
[3]*S.D Radiologia Vascular and Interventional, University Hospital Pisana, Italy*

## Abstract

**Purpose:** Transarterial 90Y microspheres radioembolization is emerging as a multidisciplinary promising therapeutic modality for primary and metastatic cancer in the liver. Actually two different type of microspheres are used, whose main characteristic is the different density of activity (activity per microsphere). In this paper the effect due to the possible different distribution of the microspheres in a target is presented and discussed from a macrodosimetric point of view.

**Material and methods:** A 100 g virtual soft-tissue target region has been builded. The administered activity was chosen to have a target average absorbed dose of 100 Gy and the number of 90Y microspheres needed was calculated for two different activity-per-microsphere values (2500) Bq/microsphere and 50 Bq/microsphere, respectively). The spheres were randomly distributed in the target and the Dose Volume Histograms were obtained for both. The cells surviving fractions (SF) for four different values of the radiobiological parameter $\alpha$ were calculated from the Linear - Quadratic model.

**Results:** The DVH obtained are very similar and the SF is almost equal for both the activity-per- microsphere values.

**Conclusions:** This macrodosimetric approach shows no radiobiological difference between the glass and resin microspheres. Thus the different number of microspheres seems to have no effect when the number of spheres is big enough that the distance between the spheres in the target can be considered small compared to the range of the β-particles of 90Y.

**Keywords:** Microspheres; β-particles; Linear-quadratic model; TheraSphere

## Introduction

Radioembolization with $^{90}$Y microspheres via hepatic arterial administration has been shown to be effective in the treatment of primary and metastatic liver cancer (HCC), as well as in unresectable colon carcinoma metastases [1-5] Radioembolization is a loco-regional liver directed therapy that involves transcatheter delivery of microspheres embedded with $^{90}$Y. $^{90}$Y microspheres are injected into the arterial supply of the liver, where they preferentially flow into hyper vascularized tumor zones, with a higher irradiation of tumour tissue compared to the normal liver parenchyma, with a consequent tumour-tissue necrosis. Actually $^{90}$Y can be delivered to the hepatic tumor as either a constituent of a glass microsphere, TheraSphere®, or as a biocompatible resin-based microsphere, SIR-Spheres®. TheraSphere® was approved by the USA Food and Drug Administration for unresectable HCC in December 1999 under a Humanitarian Device Exemption. SIR-Spheres® was approved in March 2002 for colorectal cancer metastatic to the liver in conjunction with continuous infusion of intrahepatic floxuridine (FUDR) [6].

The characteristics of these two different kinds of 90Y microspheres are summarized in Table 1 [7]. From a dosimetric point of view, the main difference between SIR-Spheres® and TheraSphere® is the density of activity that in one case (SIR-Spheres® whose activity per microsphere is ~50 Bq) is much lower than in the other (TheraSphere® whose activity per microsphere is 2500 Bq).

In principle this difference could have an impact on the radiobiological effectiveness of the treatment [8], due to the much higher number of microspheres needed to have the same activity in the target tissue in one case (SIR-Spheres®) compared to the other (TheraSphere®). This because an higher number of microspheres could mean a more homogeneous distribution of activity (and consequently of target absorbed dose).

| $^{90}$Y-Microspheres | | |
|---|---|---|
| Material | Resin | Glass |
| Sphere size (mm) | 20-60 | 20-30 |
| Activity per sphere (Bq) | 40-70 | 2500 |
| Specific gravity | Low | High |
| Handling for dispensing | Required | Not required |
| Splitting one vial for more patients | Possible | Not possible |

**Table 1:** Main characteristics of the two available kinds of 90Y microsphere: Glass spheres (TheraSphere®) and resin spheres (SIR-Spheres®).

***Corresponding author:** Traino AC, Professor, Unit of Medical Physics, University Hospital Pisana, Italy, E-mail: c.traino@ao-pisa.toscana.it

In this paper a comparison of these two different densities of activity is presented and discussed, showing that, neglecting other differences between the two types of 90Y microspheres (different material and consequent different specific gravity of SIR-Spheres® compared to TheraSphere®, for example) the therapeutic effectiveness of the two radioembolization tools is almost the same from a macrodosimetric point of view.

## Materials and Methods

A cubical target was simulated to test the expected difference between the two different activities-per- sphere tools. The mass of the target was 100 g, its density was 1.04 g/cm$^3$, that is the density of the soft tissues [9]. The activity needed for an average target absorbed dose of 100 Gy was calculated by using the MIRD formalism:

$$D(r_T,\infty) = S(r_T \leftarrow r_T)\int_0^{\infty} A_0 e^{-\lambda t} dt = S(r_T \leftarrow r_T)\frac{A_0}{\lambda}$$

where $D(r_T,\infty)$=100 Gy is the target absorbed dose; $S(r_T,r_T)$=5.08 mGy/MBq*h is the S-value for a self-irradiating 100 g spherical target treated with 90Y; $A_0$=administered activity and λ=0.011 h-1 is the physical decay constant of 90Y. From Equation 1 it follows $A_0$=213 MBq.

The target volume was divided into $N$=21x21x21 square voxels of 2.21 mm size.

A number $n_{sph}$ of 90Y embedded microspheres of the same size were randomly distributed into the target, according with the equation:

$$n_{sph} = \frac{A_0}{\delta_a}$$

Where $\delta_a$ represents two different densities of activity of 2500 Bq/sphere and 50 Bq/sphere respectively. For an administered activity of 213 MBq there will be *nsph* =8.52 × 10$^4$ 90Y microspheres corresponding to a density of activity $\delta_a$ =2500 Bq/sphere (glass spheres) and *nsph* =4.26 × 10$^6$ 90Y microspheres corresponding to $\delta_a$ =50 Bq/sphere (resin spheres). The size was considered the same, 30μm, for both the type of microspheres.

Software was built by using the open-source environment GNU-Octave to randomly distribute a number *nsph* of microspheres in the target, to perform a 3D dosimetric calculation at the voxel level by using the MIRD 17 method [10] and to show the Dose Volume Histogram (DVH) corresponding to the different distributions of the 90Y microspheres.

The absorbed dose *di* in each voxel was calculated, according to the MIRD 17 method, by using the equation:

$$d_i = \sum_{h=0}^{n} \tilde{A}_h S_{i\leftarrow h} = \sum_{h=0}^{n} S_{i\leftarrow h}\int_0^{\infty} A_0 e^{-\lambda t} dt$$

where $\tilde{A}_h$ is the cumulated activity in the ith-voxel and *Si←h* are the dose conversion factors at the voxel level. The *Si←h* values for the voxel size of 2.21 mm were taken [9] Note that *Ah* is the activity in each *h* voxel, calculated by multiplying the number of microspheres randomly placed in that voxel by the density of activity $\delta_a$.

## Results

The Dose Volume Histograms for $\delta_a$ =2500 Bq/sphere and $\delta_a$ =50 Bq/sphere are shown in Figure 1. DVHs for the two values of δa considered are very similar.

The surviving fraction SF was calculated for each of the two values of $\delta_a$ by using the Linear-

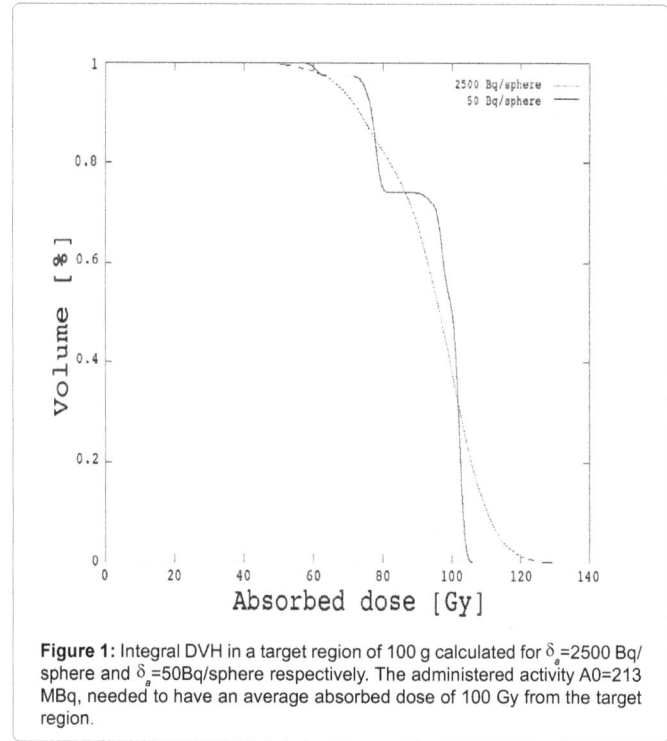

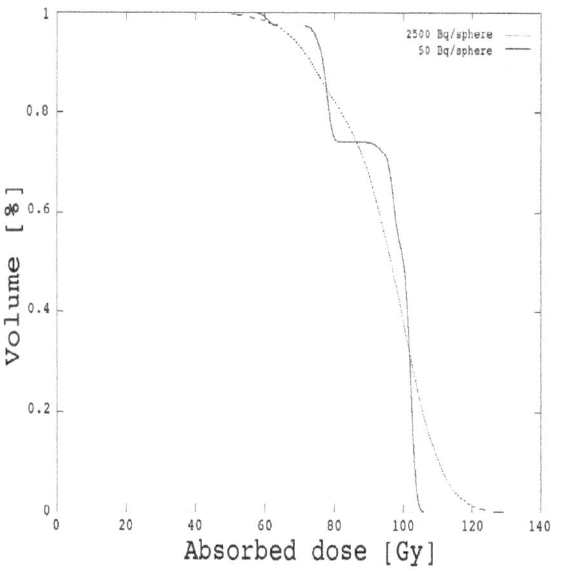


**Figure 1:** Integral DVH in a target region of 100 g calculated for $\delta_a$=2500 Bq/sphere and $\delta_a$=50Bq/sphere respectively. The administered activity A0=213 MBq, needed to have an average absorbed dose of 100 Gy from the target region.

| | Surviving Fraction | |
|---|---|---|
| a (1/Gy) | 2500 Bq/sphere | 50 Bq/sphere |
| 1 | 5.90E-21 | 4.30E-27 |
| 0.1 | 1.02E-04 | 6.00E-05 |
| 0.01 | 0.32 | 0.32 |
| 0.001 | 0.89 | 0.89 |

**Table 2:** Cells surviving fractions for different values of a and different values of $\delta_a$ (2500 Bq/sphere and 50 Bq/sphere). The calculation was based on equation 4 with a /b=10 and G=0.023.

Quadratic (LQ) model:

$$SF = \frac{1}{N}\sum_{i=1}^{N} e - (\alpha d_i + G\beta d_i^2)$$

In Equation 4 β=/10 and the Lea-Catcheside factor G=0.023. This last factor for a mono-exponential decreasing dose-rate can be written as [11].

$$G = \frac{\lambda}{\mu + \lambda}$$

Where μ=ln(2)/ $t_{rep}$ is the rate of repair of sub-lethal damages (the repair half-time constant trep=1.5 h for tumor lesions was extracted from Strigari [12] and λ=0.011h-1 is the physics decay rate of 90Y. The cells surviving fraction for four different values of α (α =1; 0.1; 0.01 and 0.001 Gy-1) is shown in Table 2.

## Discussion and Conclusions

From a macrodosimetric point of view, the different number of 90Y microspheres per unit mass could have an impact on the radiobiological effect of the transarterial radioembolization therapy [8], due to the different distribution of the microspheres in the treated target region. The microspheres tend to be distributed as homogeneously as possible via the microvascularization of the target zone. The homogeneous distribution of the activity represents the better situation from a radiobiological point of view, because it causes an uniform absorbed dose by the target. In this paper the hypothesis of a random distribution of the microspheres in the target has been done, without considering the trapping of the spheres in the blood vessels.

If the microspheres are rarefied in the target, meaning that the average distance between spheres is higher than than the double range of the β- of the radionuclide employed (90Y in this case), the dose distribution will be very unhomogeneous, leaving target zones where the absorbed dose is zero Figure 2.

In the case described in this paper, doing the hypothesis of a random distribution of the 90Y microspheres in the target, for a 100 g target treated with an activity of 213 MBq, which corresponds to a target absorbed dose of 100 Gy, the average density of 90Y microspheres is 0.852 spheres/mm$^3$ if $\delta_a$ =2500 Bq/sphere and 42.6 spheres/mm$^3$ if $\delta_a$ =50 Bq/sphere. Remembering that the range of the β- particles of 90Y is about 11 mm in the soft tissues, this means that the β- Particles of the 90Y are close also for $\delta_a$ =2500 Bq/sphere, compared to their range. Thus for the glass microspheres ($\delta_a$ =2500 Bq/sphere) there are low high-activity spheres for unit mass, compared to an higher number of low-activity resin microspheres ($\delta_a$ =50 Bq/sphere). In both cases the effect of the microspheres is almost homogeneous in the volume considered, because the average distance among the spheres is lower than the range of the β- particles of 90Y Figure 3.

In Table 3 the average density (number of spheres/mm$^3$) of 90Y glass and resin microspheres ($\delta_a$=2500 Bq/sphere and $\delta_a$ =50 Bq/sphere respectively) needed for an average target absorbed dose of 100 Gy is shown for different target mass values. The average number of spheres per mm$^3$ of target has a very low variation depending on target mass. This means that the macrodosimetric effect of the glass and resin 90Y microspheres is almost the same in the target, because the microspheres are very close respect to the range of the β- particles of 90Y. This seems true for all the possible treated masses, if the required target average absorbed dose is 100 Gy.

The random distribution of the microspheres in the whole target is only a hypothesis. The real situation is different because the microspheres are vehiculated into the target by the arterial system: this means that the distribution of the spheres in the target is unhomogeneous in principle. In the real situation the 90Y activity

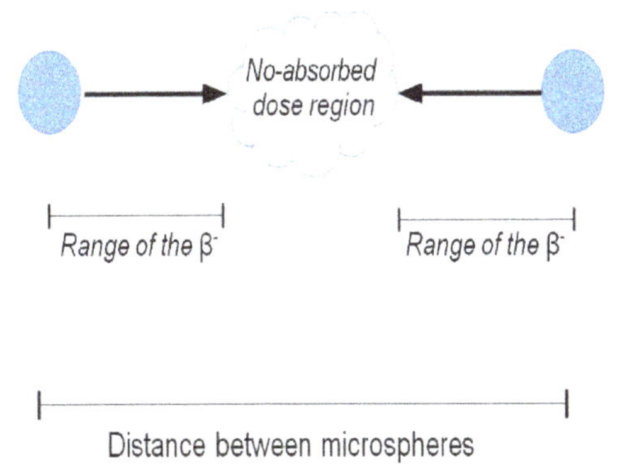

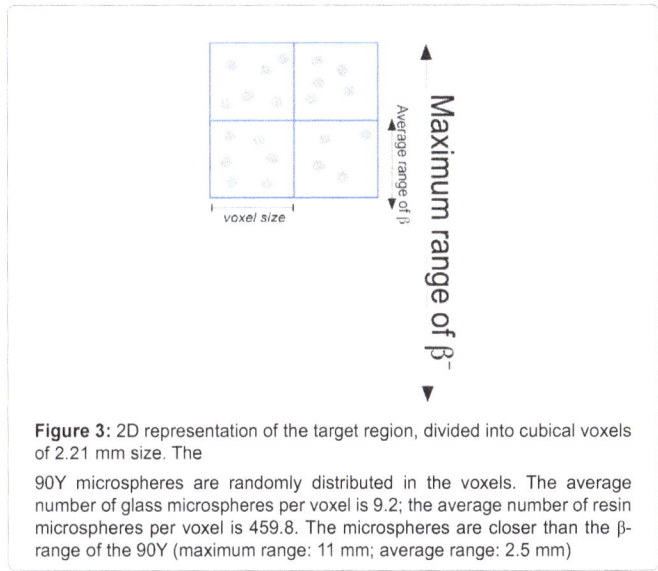

**Figure 2:** Representation of a rarefied distribution of microspheres in the target. If the distance between two microspheres is bigger than the double range of the β-particles there is a no-absorbed energy (and then no-absorbed dose) zone. This means a very unhomogeneous absorbed dose distribution in the target.

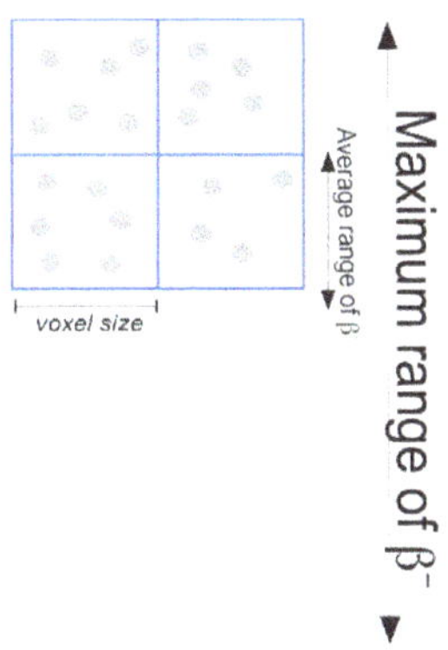


**Figure 3:** 2D representation of the target region, divided into cubical voxels of 2.21 mm size. The

90Y microspheres are randomly distributed in the voxels. The average number of glass microspheres per voxel is 9.2; the average number of resin microspheres per voxel is 459.8. The microspheres are closer than the β-range of the 90Y (maximum range: 11 mm; average range: 2.5 mm)

| Target Mass (g) | Aministered Activity (MBq) | Average Number of Glass Spheres/mm$^3$ | Average Number of Resin Spheres/mm$^3$ |
|---|---|---|---|
| 40 | 86 | 0.866 | 43.3 |
| 80 | 170 | 0.854 | 42.7 |
| 100 | 213 | 0.853 | 42.6 |
| 300 | 626 | 0.835 | 41.7 |
| 400 | 833 | 0.833 | 41.7 |
| 500 | 1041 | 0.833 | 41.7 |
| 100 | 2059 | 0.825 | 41.2 |
| 2000 | 4026 | 0.805 | 40.3 |

**Table 3:** Density (number of spheres/mm$^3$) of glass and resin microspheres ($\delta_a$ =2500 Bq/sphere and $\delta_a$ =50 Bq/sphere respectively) for different masses of soft-tissue target. The administered activity corresponds to an average target absorbed dose of 100 Gy; the density of the soft-tissue is 1.04 Kg/dm$^3$.

needed to have a certain absorbed dose (es.100 Gy) in a target volume (es. 100 g) is forced in a volume smaller than the target (arterial system in the target is smaller than the whole target). This means that there is the same number of microspheres (8.52 × 10$^4$ 90Y microspheres corresponding to a density of activity $\delta_a$ =2500 Bq/sphere and 4.26x106 corresponding to $\delta_a$ =50 Bq/sphere) randomly distributed in a volume lower than 100 g. For this reason the $^{90}$Y microspheres are much closer than the situation described in this paper, thus the macrodosimetric differences due to the different number of resin and glass microspheres in the target are probably lower.

In this paper the macrodosimetric effect due to the different number of 90Y microspheres per unit mass in a target submitted to the radioembolization procedure is treated. From this point of view, it seems that the different number of microspheres doesn't have any effect on the distribution of the target absorbed dose due to the small distance among the spheres compared to the range of the particles of 90Y. This result seems to be in agreement with those described by Gulec in their paper [13], where, starting from a microdosimetric approach, they don't find any difference in the absorbed dose due to the different number of microspheres in the target.

The microdosimetic effects due to the different density of activity (activity per sphere) between glass and resin microspheres, already treated in literature [13,14] is beyond the scope of this paper. Also the probable radiobiological effect due to the different specific gravity of the microspheres (due to the different material), which almost surely

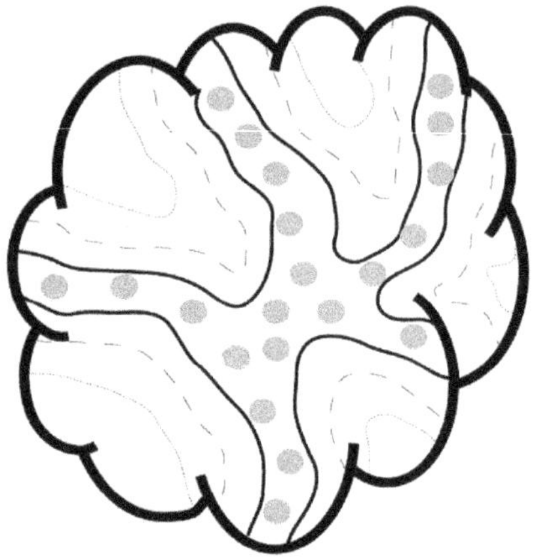

**Figure 4:** Representation of the real distribution of the 90Y microspheres in the target. Themicrospheres are vehiculated in the target by the arterial vascularization, thus their distributioncannot be homogeneous. The dotted lines represent the isodose curves in the target.

affects the distribution of the microspheres in the target, is not taken into consideration in this paper.

### References


1. Blanchard RJ, Morrow IM, Sutherland JB (1989)Treatment of liver tumors with yttrium-90 microspheres alone. CanAssoc Radiol J 40: 206-210.
2. Jacobs TF (2007) Mid-term results in otherwise treatment refractory primary or secondary liver confined tumours treated with selective internal radiation therapy (SIRT) using (90)Yttrium resin-microspheres. Eur Radiol 17: 1320-1330.
3. Dancey JE (2000) Treatment of non resectable hepatocellular carcinoma with intrahepatic 90Y-microspheres. J Nucl Med; 41: 1673-1681.
4. Geschwind JF (2004) Yttrium-90 microspheres for the treatment of hepatocellular carcinoma. Gastroenterology 127 :S194 -S205.
5. Salem R (2005) Treatment of unresectable hepatocellular carcinoma with use of 90Y microspheres (TheraSphere): safety, tumor response, and survival. J Vasc Interv Radiol16: 1627-1639.
6. Lewandowski RJ, Salem R (2006) Yttrium-90 Radioembolization of Hepatocellular Carcinoma and Metastatic Disease to the Liver. Semin Intervent Radiol; 23: 64-72.
7. Salem R, Thurston KG (2006) Radioembolization with 90Yttrium microspheres: a state-of-the-art brachytherapy treatment for primary and secondary liver malignancies. Part 1: Technical and methodologic considerations. J Vasc Interv Radiol 17: 1251-1278
8. Spreafico C, Maccauro M, Mazzaferro V, Chiesa C (2014) The dosimetric importance of the number of 90Y microspheres in liver transarterial radioembolization (TARE). Eur J Nucl Med Mol Imaging 41: 634-638.
9. Lanconelli N (2012) A free database of radionuclide voxel S values for the dosimetry of nonuniform activity distributions. Phys Med Biol 57: 517-533.
10. Bolch WE (1999) the dosimetry of nonuniformactivity distributions – radionuclide S values at the voxel level. J Nucl Med 40: 11S-36S.
11. Dale RG (1996) Dose-rate effects in targeted radiotherapy. Phys Med Biol 41: 1871-1884.
12. Strigari L, et al (2010) Efficacy and toxicity related to treatment of hepatocellular carcinoma with 90Y-SIR spheres: radiobiologic considerations. J Nucl Med 51: 1377-1385.
13. Gulec SA, Sztejnberg ML, Siegel JA, Jevremovic T, Stabin M (2010) Hepatic structural dosimetry in 90Y microspheres treatment: a Monte Carlo modeling approach based on lobular microanatomy. J Nucl Med 51: 301-310.
14. Walrand S, Hesse M, Chiesa C, Lhommel R, Jamar F (2014)The low hepatic toxicity per Gray of 90Y glass microspheres is linked to their transport in the arterial tree favoring a nonuniform trapping as observed in posttherapy PET imaging. J Nucl Med 55: 135-140.

# Foundations of a Unified Physics

**Pestov IB***
*Bogoliubov Laboratory of Theoretical Physics, Joint Institute for Nuclear Research, 141980, Dubna, Moscow Region, Russia*

**Abstract**

Since space is the primary concept of science, we put forward an idea that regularities of unified physics are concealed in a simple relation: everything in the concept of space and the concept of space in everything. With this hypothesis as a ground, a conceptual structure of a unified geometrical theory of fundamental interactions is created and deductive derivation of its main equations is produced. The formulated theory provides opportunity to understand the origin and nature of physical fields and local internal symmetry; time and energy; spin and charge; confinement and quark-lepton symmetry; dark energy and dark matter, giving by this solution of the fundamental physical problems and conforming together with this the existence of new physics in its unity.

**Keywords:** Unified physics; Unified geometrical theory; Smooth manifold

## Introduction

The creation of a unified geometrical theory of fundamental interactions is motivated by two severe constraints. It is natural to suppose that solution of the fundamental physical problems cannot be given without a more drastic revision of our fundamental concepts than any that have done before. Quite likely these innovations are so great that direct attempts to derive new ideas from experimental data will be beyond the power of human intelligence [1]. And second, we need to know the origin of the laws and, hence, to formulate a new theory which considers world as a whole (as an integral structure) and gives a possibility to understand why nature is just the way it is. To produce a single theory that provides understanding of the structure of our universe as a unique possible consequence of the only simple assumption (a first principle), we abstract away from the known concepts and laws and put forward the fundamental principle of unification which can be presented in the form of the grand relation: fundamental physical ideas, symmetries and equations are tightly connected with the idea of space and, hence, they have the geometrical origin and are formulated in the form completely independent of any outer and a priori conditions (everything in the concept of space and the concept of space in everything). Thus, the radically new idea of a unified physics is a most possible generality and the outlined first principle. The theory presented is the only one unique logically possible physical theory since it gives the solution of the most difficult and long-standing problem in the history of science - the problem of time. A new concept of time provides the formulation of the law of energy conservation (the most fundamental law of the nature) in the form which is the most general and, hence, suitable for all cases without exclusion. When we discover connections of time with other objects, the reason of the existence of time becomes clear.

The most general realization of the idea of space is the concept of physical manifold. The physical manifold is a smooth manifold [2] plus its connections with physical fields that transfer energy and momentum. The second part of this definition is especially important for understanding of the nature of the gravitational field and time [3,4]. Most physicists nowadays consider a theory be fundamental only if it does make explicit use of the concept of smooth manifold. Thus, the first principle of unification should be formulated as follows: everything in the concept of physical manifold and the concept of physical manifold in everything. With this only first principle we can get the ideas and results we need to describe and understand a realm where all intuitions derived from the life in common space and time become inapplicable.

Some words about mathematics we use. Of course, similar to the electromagnetic field all the geometrical quantities are invariant and, hence, are coordinate free and basis-free. But physical laws are formulated as systems of differential equations in partial derivatives. Hence, the using of the coordinates is desirable. Let us consider a simple example to illustrate our statement. It is clear that number is invariant but we need to remember about the interplay between an observer and computing machines which is provided by the binary number system. We also pay attention on the mathematics of some novel developments of modern physics [5,6].

The super symmetry [7,8], the superstring theory [5,6] and the M-theory [9], were analyzed by Steven Weinberg in his article [10], entitled "A Unified Physics by 2050?" We cite his general conclusion: Observations of this kind will yield valuable clues to the unified theory of all forces, but the discovery of this theory will probably not be possible without radically new ideas." Since the unified physics provides the new understanding of the nature of spin and superstring theories (with the problem of their unification) have no a particular support from the side of experiment, it is natural to reconsider the interesting idea of super symmetry in the framework of the unified physics with its new concept of time and to discover the real underlying degrees of freedom of M-theory.

## General Characteristic of Fields and Time

The geometrical structure of the physical manifold (points, curves, congruence's of curves, families of curves) determines a very restricted set of really geometrical quantities and along with that geometrical internal symmetry that makes these quantities variable and forms from them the fundamental physical fields: the positive definite Riemann metric $g_{ij}$ which is the potential of the gravitational field providing connection of the physical fields with the physical manifold

***Corresponding author:** Pestov IB, Bogoliubov Laboratory of Theoretical Physics, Joint Institute for Nuclear Research, 141980, Dubna, Moscow Region, Russia
E-mail: pestov@theor.jinr.ru

(it is characterized by the re parameterization invariance); the linear (affine) connection $P^i_{jk}$ which is the potential of the generalized electromagnetic field ( the group of geometrical internal symmetry is a general linear group); the scalar and covariant vector fields, and anti-symmetric covariant tensor fields which are connected by the geometrical internal symmetry (spin symmetry) into the real spinning field.

$$\mathbf{A} = (a, a_i, a_{ij}, a_{ijk}, a_{ijkl}), \quad i,j,k,l = 1,2,3,4.$$

The intrinsic causal structure of the physical manifold is the temporal field (the natural clock) and bilateral symmetry defined by the stream of time. The temporal field with respect to the coordinate system $u^1$, $u^2$, $u^3$, $u^4$ is denoted by $f(u)=f(u^1, u^2, u^3, u^4)$. The reading $t$ of the natural clock is defined by the formula $t=f(u)=f(u^1, u^2, u^3, u^4)$ for any point of the physical manifold. The two events at the point's $u^1$, $u^2$, $u^3$, $u^4$ and $v^1$, $v^2$, $v^3$, $v^4$ go on concurrently if $f(u^1, u^2, u^3, u^4)=f(v^1, v^2, v^3, v^4)$. The gradient of the temporal field (the stream of time) is the vector field $t$ with the components $t^i = (\nabla f)^i = g^{ij}\partial_j f = g^{ij}t_j$, where $g^{ij}$ are the contravariant components of the Riemann metric $g_{ij}$. The temporal field obeys the fundamental equation:

$$D_t f = (\nabla f)^2 = g^{ij}\frac{\partial f}{\partial u^i}\frac{\partial f}{\partial u^j} = 1 \tag{1}$$

Which means that time flows equably. This equation has general solution [11] and special solution known as the function of geodesic distance.

The stream of time defines fundamental discrete internal symmetry- bilateral symmetry. A pair of vector fields $v$ and $\bar{\mathbf{v}}$ has bilateral symmetry if the sum of these fields is collinear to the gradient of a temporal field and their difference is orthogonal to it, $\bar{\mathbf{v}} + \mathbf{v} = \lambda\mathbf{t}$, $(\bar{\mathbf{v}},\mathbf{t}) = (\mathbf{v},\mathbf{t})$, where $(\mathbf{v},\mathbf{w}) = g_{ij}v^i w^j = v^i w_i$ is a scalar product. The bilateral symmetry may be represented as a linear transformation (reflection) $\bar{v}^i = R^i_j v^j$, where $R^i_j = 2t^i t_j - \delta^i_j$.

The bilateral symmetry defines the causal structure of the physical manifold and the auxiliary metric

$$\bar{g}_{ij} = 2t_i t_j - g_{ij} = g_{ik}R^k_j, \quad \bar{g}^{ij} = 2t^i t^j - g^{ij}$$

Giving the straightforward method of introduction of a temporal field (and together with this the dynamics) into the Lagrangians (and the equations) of the fundamental physical fields [3,4,11].

From the consideration of the bilateral symmetry it follows that in the geometrical (coordinate independent) form the time reversal invariance means that a theory is invariant with respect to the transformations $T: \ t^i \to -t^i$.

It is clear that a theory will be time reversal invariant if the gradient of the temporal field appears in all formulae only as an even number of times, like $t^i t^j$.

## Equations of Gravidynamics

The difference between general covariance and reparametrization invariance is established and a new mathematical operation of constructing of the new fields by the reparametrization is introduced. The dynamical equations of the gravitational field and the law of energy conservation are represented. The comparison of these equations with the Hamilton equation (which is the cornerstone of the solution of the Poincare conjecture given by Perelman) is discussed.

The manifold $M$ is called topologically nontrivial if it cannot to be maintained by one nonsingular coordinate system. If we insist on one coordinate system, we get into singularities. If we relax that, we have no singularity. In general, the physical manifold is topologically nontrivial. To establish exact mathematical and physical meaning of general covariance and reparametrization invariance, we suppose, for simplicity, that two systems of coordinates are sufficient. Thus, there are two open neighborhoods $U$ and $\bar{U}$ such that $M = U \cup \bar{U}$ and two homeomorphisms of $U$ and $\bar{U}$ onto an open sets of $R^n$. In the domain of overlapping $U$ and $\bar{U}$, $W = U \cap \bar{U}$ two systems of coordinates $u^1$, $u^2$, $u^3$, $u^4$ and $\bar{u}^1, \bar{u}^2, \cdots, \bar{u}^n$ compete with each other and this takes the form of the functional relations

$$\bar{u}^i = \bar{u}^i(u^1, u^2, \cdots, u^n), \quad u^i = u^i(\bar{u}^1, \bar{u}^2, \cdots, \bar{u}^n), \quad i = 1,2,\cdots n,$$

Called coordinate transformations. Let the sets of functions

$$g_{ij}(u^1, u^2, \cdots, u^n), \quad \bar{g}_{ij}(\bar{u}^1, \bar{u}^2, \cdots, \bar{u}^n)$$

Represent correspondingly the gravitational field in the neighborhoods $U$ and $\bar{U}$. Then, in the simplest case the following relations should be valid in the domain of overlapping $W$

$$\bar{g}_{ij} = \frac{\partial u^{-k}}{\partial u^i}\frac{\partial u^{-l}}{\partial u^j} gkl \tag{2}$$

Thus, under the consideration of the physical fields on the topologically nontrivial manifolds the principle of general covariance holds valid.

In general, the connection between $g_{ij}(u^1, u^2, \cdots, u^n)$, $\bar{g}_{ij}(\bar{u}^1, \bar{u}^2, \cdots, \bar{u}^n)$ can be more refined. To show this, let us consider diffeomorphisms $\varphi$ of $U$ onto itself, which is defined as follows

$$\varphi : u^i \Rightarrow \varphi^i(u^1, u^2, \cdots, u^n), \quad \varphi^{-1} : u^i \Rightarrow \theta^i(u^1, u^2, \cdots, u^n),$$

$$\varphi^i(\theta^1(u), \theta^2(u), \cdots, \theta^n(u)) = u^i, \quad \theta^i(\varphi^1(u), \varphi^2(u), \cdots, \varphi^n(u)) = u^i.$$

The diffeomorphism $\varphi$ induces the mapping $\tilde{g} = \varphi g$ of the form

$$\tilde{g}_{ij}(u) = g_{kl}(\theta(u))\theta^k_i(u)\theta^l_j(u),$$

Where $\theta^l_j(u) = \partial_j \theta^l(u)$. The last formula describes the process of getting of the new fields by the reparametrization. Hence, in the most general case we instead of equation (2) have the following equation of the analitic continuation

$$\bar{g}_{ij} = \frac{\partial \bar{u}^k}{\partial u^i}\frac{\partial \bar{u}^l}{\partial u^j}\tilde{g}_{kl}.$$

If $g_{ij}(u)$ and $\tilde{g}_{ij}(u)$ are solutions of some equation, this equation is called invariant with respect to the reparametrization defined by the diffeomorphisms $\varphi$. Inversely, if equation is invariant with respect to reparametrization, it is possible to construct a set of new solutions of the same equation having a given solution. Thus, in the general case to establish connection between $g$ and $\bar{g}$, the intermediate factor $g$ should be taken into account. Reparametrization invariance expresses in the exact mathematical form that in general we do not insist on one gravitational potential $g_{ij}(u)$ in the given coordinate patch. It is very important for the understanding the nature of possible singularities (true or not).

From our consideration it follows that equations of the physical fields on the topologically nontrivial manifolds should be general covariant and invariant with respect to the reparametrization. The simultaneous consideration of these two things is very important since, for example, the Maxwell equations alone are generally covariant but they are not invariant with respect to the reparametrization and this is not the case if we consider the genuine Riemann metric alone. By this the central role of the gravitational field in a unified physics is defined.

We denote

$$\Gamma^i_{jk} = \frac{1}{2} g^{il}(\partial_j g_{kl} + \partial_k g_{jl} - \partial_l g_{jk}), \quad R_{ijk}{}^l = \partial_i \Gamma^l_{jk} - \partial_j \Gamma^l_{ik} + \Gamma^l_{im}\Gamma^m_{jk} - \Gamma^l_{jm}\Gamma^m_{ik}$$

And get the Ricci tensor $R_{jk}=R_{ijk}{}^i$ and covariant derivative $\nabla_i$ with respect to $\Gamma^i_{jk}$. With this we introduce momentum of the gravitational field

$$P^i_j = \frac{1}{2} g^{ik} D_t g_{jk} = \frac{1}{2} g^{ik}(\nabla_j t_k + \nabla_k t_j) = g^{ik}\nabla_j t_k = \nabla_j t^i,$$

Where $D_t$ is the Lie derivative along a stream of time. The stress tensor of the gravitational field can be written in the following form:

$$S^i_j = h^i_k R^k_l h^l_j + D_t P^i_j + \varphi P^i_j,$$

Where $\varphi = \nabla_i t^i$ and $h^i_j = \delta^i_j - t_j t^i$. for a density of kinetic and potential energy of the gravitational field we have the following expressions:

$$T = \frac{1}{2}(P^i_i)^2 - \frac{1}{2} P^i_j P^j_i = \frac{1}{2}(TrP)^2 - \frac{1}{2} Tr(P^2),$$

$$U = \frac{1}{2} S = \frac{1}{2} S^l_l.$$

The first group of equations of Gravidynamics reads

$$D_t P^i{}_j + \varphi P^i{}_j + S^i{}_j + N^i{}_j = \frac{1}{2}\varepsilon h^i{}_j, \quad (3)$$

Where $N^i_j$ is the stress tensor of the generalized electromagnetic field and spinning field, $\varepsilon$ is the energy density of the gravitational field and other fields $\varepsilon=\varepsilon_g+\varepsilon_m$ The second group of equations of Gravidynamics reads

$$G_j = \Pi_i, \quad (4)$$

Where

$$G_j = h^i_j(\nabla_l P^l_i - \partial_i P^l_l)$$

is the energy flow vector of the gravitational field and $\Pi_i$ is the energy flow vector of the other fields. We pay attention to that the heat equation of Hamilton (tightly connected with the solution of the Poincare conjecture) can be written in our notation as follows:

$$P^i_j + S^i_j = 0$$

Equation (3) can be derived from the energy functional $L=T-U$ and according to the Hamilton paper; this equation is more natural than his heat equation. Thus, we have there very interesting physical and mathematical problem to consider connection between gravity and topology in the new framework.

The law of energy conservation reads.

$$D_t(\sqrt{g}\varepsilon) = 0, \quad \varepsilon = \varepsilon_g + \varepsilon_m = T + U + \varepsilon_m.$$

Since the scalar temporal field enters into the Lagrangians of the physical fields in the form of the gradient of the scalar field

$$t_i = \partial_i f(u),$$

The laws of a unified physics are invariant with respect to transformations of the form

$$f(u) \Rightarrow f(u) + a,$$

Where $a$ is a constant, this is internal symmetry that defines the law of energy conservation as a fundamental physical law of the universe which is true in all cases.

## Equations of Generalized Electromagnetic Field

Here we establish dynamical equations of the generalized electromagnetic field (shortly gef) which are defined by gef symmetry and describe a new form of matter which interacts only gravitationally (dark matter) [12-14]. The problem of invisible mass is acknowledged to be among the greatest puzzles of modern cosmology and field theory. The most direct evidence for the existence of large quantities of dark matter in the Universe comes from the astronomical observation of the motion of visible matter in galaxies. One neither knows the identity of dark matter nor whether there are one or more types of its structure elements. The most commonly discussed theoretical elementary particle candidates are a massive neutrino, a syper symmetric neutralino, and the axion. So, at the present time there is a good probability that the set of known fields is by no means limited to those fields. Moreover, we are free to look for deeper reasons for the existence of a new form of energy unusual in many respects.

The parallel displacement of vector fields $\bar{\mathbf{v}}$ and $V$ with bilateral symmetry $\bar{v}^i = R^i_j v^j$ can be produced only by a pair of connections $\bar{P}^i_{jk}$ and $P^i_{jk}$ with bilateral symmetry. From the law of parallel displacement we have

$$\bar{P}^i_{jk} = R^i_m P^m_{jn} R^n_k + R^i_m \partial_j R^m_k.$$

Being generalized the bilateral symmetry takes status of gef symmetry *GL (n, R)* with the law of transformation

$$\bar{P}^i_{jk} = U^i_m P^m_{jn} V^n_k + U^i_m \partial_j V^m_k,$$

Where $V^i_j$ is the component of the operator $U^{-1}$ inverse to the operator $U$, $U^i_k V^k_j = \delta^i_j$. For brevity, we use the matrix notation

$$\mathbf{U} = (U^k_l), \quad \mathbf{P}_i = (P^k_{il}), \quad \mathbf{E} = (\delta^k_l), \quad \mathbf{H}_{ij} = (H_{ijl}{}^k), \quad \mathrm{Tr}\mathbf{U} = U^k_k.$$

The transformations of gef symmetry take the form

$$\bar{\mathbf{P}}_i = \mathbf{U}\mathbf{P}_i\mathbf{U}^{-1} + \mathbf{U}\partial_i\mathbf{U}^{-1} = \mathbf{P}_i + \mathbf{U}\mathrm{D}_i\mathbf{U}^{-1},$$

Where $\mathrm{D}_i$ is the natural differential operator associated with gef symmetry only

$$\mathrm{D}_i\mathbf{U} = \partial_i\mathbf{U} + \mathbf{P}_i\mathbf{U} - \mathbf{U}\mathbf{P}_i = \partial_i\mathbf{U} + [\mathbf{P}_i, \mathbf{U}].$$

The Riemann tensor of $P^i_{jk}$

$$\mathbf{B}_{ij} = \partial_i\mathbf{P}_j - \partial_j\mathbf{P}_i + [\mathbf{P}_i, \mathbf{P}_j]$$

is reducible with respect to the transformations

$$\bar{\mathbf{B}}_{ij} = \mathbf{U}\mathbf{B}_{ij}\mathbf{U}^{-1},$$

Since

$$\mathbf{B}_{ij} = (\mathbf{B}_{ij} - \frac{1}{4}\mathrm{Tr}(\mathbf{B}_{ij})\mathbf{E}) + \frac{1}{4}\mathrm{Tr}(\mathbf{B}_{ij})\mathbf{E}.$$

Hence, the strength tensor of the generalized electromagnetic field is given by the formula

$$\mathbf{H}_{ij} = \mathbf{B}_{ij} - \frac{1}{4}\mathrm{Tr}(\mathbf{B}_{ij})\mathbf{E}, \quad \mathrm{Tr}(\mathbf{H}_{ij}) = 0$$

and the singlet state of the gef defines the strength tensor of the electromagnetic field $F_{ij} = \mathrm{Tr}(\mathbf{B}_{ij})$.

The ground state of gef is defined by the equation $B_{ij} = 0$ which means that this state transfers a new form of energy. We give general solution of this equation. Let four linear independent vector fields $E^\mu_i$ be given and one can construct purely algebraically components of the four covector fields $E^\mu_i$, so that $E^i_\mu E^\mu_j = \delta^i_j$ holds valid. Setting $P^i_{jk} = L^i_{jk}$, where $L^i_{jk} = E^i_\mu \partial_j E^\mu_k$, get general solution of the equation in question. Let us introduce a tensor field $Q^i_{jk} = P^i_{jk} - L^i_{jk}$ but in what follows we shall consider the irreducible deviation tensor

$$T^i_{jk} = Q^i_{jk} - \frac{1}{4} Q^l_{jl} \delta^i_k, \quad \mathbf{T}_j = \mathbf{Q}_j - \frac{1}{4} \mathrm{Tr}(\mathbf{Q}_j)\mathbf{E}.$$

Let us introduce electric and magnetic strength of the generalized electromagnetic field by the direct and inverse mapping

$$\mathbf{J}_i = t^k \mathbf{H}_{ik}, \quad \mathbf{M}_i = \frac{1}{2} e_{ikjl} t^k \mathbf{H}^{jl} = t^k \overset{*}{\mathbf{H}}$$

$$\mathbf{H}_{ik} = -t_i \mathbf{J}_k + t_k \mathbf{J}_i - e_{ikjl} t^j \mathbf{M}^l,$$

The Lagrangian of gefdynamics takes the form

$$\mathcal{L}_p = -\frac{1}{4} \mathrm{Tr}(\mathrm{H}_{ij} \tilde{\mathrm{H}}^{ij}) - \frac{\mu^2}{2} \mathrm{Tr}(\mathrm{T}_i \tilde{\mathrm{T}}^i) \quad (5)$$

Where μ is a constant of dimension of $cm^{-1}$

$$\tilde{\mathbf{H}}^{ij} = \overline{g}^{ik} \overline{g}^{jl} \mathbf{H}_{kl} = 2t^i \mathbf{J}^j - 2t^j \mathbf{J}^i + \mathbf{H}^{ij}, \tilde{\mathbf{T}}^i = \overline{g}^{ik} \mathbf{T}_k.$$

The first pair of equations of gefdynamics can be represented in the following form:

$$t^k \mathrm{D}_k \mathbf{M}^i - \mathbf{M}^k \partial_k t^i + \varphi \mathbf{M}^i = -e^{ijkl} t_j \mathrm{D}_k \mathbf{J}_l,$$

$$\frac{1}{\sqrt{g}} \mathrm{D}_i (\sqrt{g} \mathbf{M}^i) = 0,$$

where $\varphi = \nabla_i t^i$.

The second pair of equations of gefdynamics reads

$$t^k D_k \mathbf{J}^i - \mathbf{J}^k \partial_k t^i + \varphi \mathbf{J}^i = e^{ijkl} t_j \mathrm{D}_k \mathbf{M}_l + \mu^2 \mathbf{S}^i,$$

$$\frac{1}{\sqrt{g}} \mathrm{D}_i (\sqrt{g} \mathbf{J}^i) = \mu^2 \mathbf{S},$$

Where

$$\mathbf{S}_i = h^j_i \mathbf{T}_j, \quad \mathbf{S} = t^k \mathbf{T}_k.$$

Since

$$\mathrm{Tr}(\mathbf{J}_i) = \mathrm{Tr}(\mathbf{M}_i) = \mathrm{Tr}(\mathbf{T}_i) = 0,$$

The system of equations in question is simultaneous. This system of equations describes a new form of matter which is the structure element of the unified physics and has the characteristics of an entity known from the observations as Dark Matter.

## Equations of Spin Dynamics

Spin is already a great unifying principle in the theoretical physics, but its potential is far from being exhausted. In a Unified Physics the Spin dynamics includes the region of physical phenomena which are now investigated in the Standard Model. Thus, spin symmetry is geometrical internal symmetry which provides new and complete understanding of spin and, hence, theory of elementary particles as well. Here we derive equations of Spin dynamics and with this process introduce operators of electrical charge and neutrino charge and discuss some questions connected with this innovation [16].

The spin symmetry group is a general linear group *GL (16, R)* that acts in the space of the spinning fields. We constructed a natural general covariant basis in the Lie algebra *gl* $(2^n, R)$ of *GL* $(2^n, R)$ and uncover spin as the bipolar structure on the group of spin symmetry *GL* $(2^n, R)$ [11]. The bipolar structure having two dual sets of commuting operators exists which define the Lie algebra of two dual groups *S* and $\tilde{S}$. This result expresses in the mathematical form that the spinning field has internal angular momentum and magnetic momentum and spinning field is the space of the two-valued representation of the dual groups S and $\tilde{S}$.

Let us define the positive definite scalar product and the auxiliary scalar product in the linear space in question as follows:

$$\langle \mathbf{A} | \mathbf{B} \rangle = \sum_{p=0}^{4} \frac{1}{p!} a_{i_1 \cdots i_p} b_{j_1 \cdots j_p} g^{i_1 j_1} \cdots g^{i_p j_p}, \quad (6)$$

$$\langle \mathbf{A} | \mathbf{B} \rangle = \sum_{p=0}^{4} \frac{1}{p!} a_{i_1 \cdots i_p} b_{j_1 \cdots j_p} \overline{g}^{i_1 j_1} \cdots \overline{g}^{i_p j_p}.$$

The following relation between these two scalar products holds valid:

$$\langle \mathbf{A} | \mathbf{B} \rangle = (\mathbf{A} | \mathrm{R}\mathbf{B}),$$

Where

$$\mathrm{R} = -Q_t \tilde{Q}_t = Q_{-t} \tilde{Q}_t = Q_t \tilde{Q}_{-t}$$

And

$$Q_t : \overline{a}_{i_1 \cdots i_p} = p t_{[i_1} a_{i_2 \cdots i_p]} - t^k a_{k i_1 \cdots i_p},$$

$$Q : \overline{a}_{i_1 \cdots i_p} = (-1)^p \; (p t_{[i_1} a_{i_2 \cdots i_p]} + t^k a_{k i_1 \cdots i_p}), \quad (p = 0,1,2,3,4),$$

The square brackets [...] denote the process of alternation and $t^i = g^{ij} t_j$ The operator R gives the straight forward representation of the bilateral symmetry and shows that this representation opens a new possibility to treatise the bilateral symmetry in the case of the spinning field. We can also define that a system of two spinning fields *A* and $\overline{\mathbf{A}}$ possesses the bilateral symmetry if

$$\overline{\mathbf{A}} = Q_t \mathbf{A}, \quad or \quad \overline{\mathbf{A}} = Q_t \mathbf{A}.$$

In view of this, let us consider two skew-symmetrical bilinear forms

$$[\mathbf{A}, \mathbf{B}] = (Q_t \mathbf{A} | \mathbf{B}), \quad [\tilde{\mathrm{A}}, \tilde{\mathrm{B}}] = (\tilde{Q}_t \mathbf{A} | \mathbf{B}).$$

Since $[\mathbf{A}, \mathbf{A}] = 0, \quad [\tilde{\mathbf{A}}, \tilde{\mathbf{A}}] = 0,$

It is clear that the introduction of this causal structure is adjoined with the consideration of two independent real spinning fields. To understand this new situation it is important to recognize the following results.

Symplectic scalar products are invariant with respect to the transformations of the group *SL (2, R)* called in what follows the group of pseudo charge symmetry. The generators of this group $h_1, h_2, h_3$ are represented by the real matrices $\tau_1, \tau_2, \tau_3$ which satisfy the following relations:

$$\tau_1^2 = -1, \quad \tau_2^2 = 1, \quad \tau_1 \tau_2 + \tau_2 \tau_1 = 0, \quad \tau_1 \tau_2 = \tau_3, \quad h_a = \frac{1}{2} \tau_a, \quad a = 1,2,3.$$

We consider the matrix $\tau_1 = Q_p$ as an operator of pseudo charge, since its eigenvalues are equal to $\pm i$. In what follows we should like to deal with eigenvectors of this operator and, hence, with a complex spinning field and its conjugated.

$$\Psi = \frac{\sqrt{2}}{2}(\mathbf{A} + i\mathbf{B}), \quad \overset{*}{\Psi} = \frac{\sqrt{2}}{2}(\mathbf{A} - i\mathbf{B})$$

(as a manifestation of broken pseudo charge symmetry). It is very important to recognize that this provides the existence of the probability measure in the space of solutions of the spin dynamics equations for the complex spinning field [11].

The Lagrangian of Spin dynamics associated with this new representation of bilateral symmetry

$$L_t = -\frac{i}{2}\langle \tilde{Q}_t \dot{\Psi} | \Pi \Psi \rangle + \frac{i}{2}\langle \tilde{Q} \Psi | \Pi \dot{\Psi} \rangle + im\langle \tilde{Q}_t \dot{\Psi} | \Psi \rangle, \quad (7)$$

is not invariant with respect to time reversal $T: \ t_i \to -t_i$, since $L_{-t} = -L_t$. For simplicity we do not write out the dual Lagrangian [11]. The main differential operators of Spin dynamics $\Pi$ and $\tilde{\Pi}$ are defined here by the new geometrical object, fundamental connection of Spin dynamics LOO

$$\bar{\Gamma}^i_{jk} = \Gamma^i_{jk} + t^i \nabla_j t_k - t_k \nabla_j t^i \quad (8)$$

This is characterized by the equations

$$\bar{\nabla}_i t_j = 0, \quad \bar{\nabla}_i g_{jk} = 0, \quad \bar{\nabla}_i g^{jk} = 0,$$

where $\bar{\nabla}_i$ is the covariant derivative with respect to this connection. The operator $\Pi$ is defined as follows:

$$(\Pi \mathbf{A})_{i_1 \cdots i_p} = p\bar{\nabla}_{[i_1} a_{i_2 \cdots i_p]} + \bar{\nabla}^k a_{ki_1 \cdots i_p} - 2t^k \bar{\nabla}_t a_{ki_1 \cdots i_p},$$

since $g^{kl}\bar{\nabla}_{\gamma} = 2t^k \bar{\nabla}_t - \bar{\nabla}^k$. For simplicity, we do not consider the dual operator $\tilde{\Pi}$.

Equation $\delta L_t = 0$ can be written as

$$\Pi \Psi + \frac{1}{2}\varphi Q_t \Psi = m\Psi, \quad (9)$$

and complex conjugated, where $\varphi = \nabla_i t^i$.

The Lagrangian of Spin dynamics associated with the customary representation of bilateral symmetry and invariant with respect to the time reversal takes the form

$$L_t = \frac{1}{2}\langle \mathbf{A} | \Pi \mathbf{A} \rangle - \frac{m}{2}\langle \mathbf{A} | \mathbf{A} \rangle \quad (10)$$

Equation $\delta L_t = 0$ can be written as follows:

$$\Pi A + \frac{1}{2}\varphi Q_t A = mA, \quad (11)$$

Since

$$\Pi \tilde{Q}_t - \tilde{Q}_t \Pi = 0, \quad \langle \tilde{Q}_t \mathbf{A} | \mathbf{B} \rangle = -\langle \mathbf{A} \tilde{Q}_t \mathbf{B} \rangle,$$

the Lagrangian (10) is invariant with respect to the transformations

$$\bar{\mathbf{A}} = exp(\alpha \tilde{Q}_t)\mathbf{A}.$$

For the density of the corresponding conserved current $C^k$ one can find the expression

$$\rho = t_i C^i = \langle Q_t \mathbf{A} | \tilde{Q}_t \mathbf{A} \rangle = (\mathbf{A} | \mathbf{A})$$

This shows that there is a natural probability measure in the space of solution of equation (11). This fundamental result was originally recognized in [17] and now it gives us an opportunity to identify the operator $\tilde{Q}_t$ with an operator of the electrical charge $\tilde{Q}_t = Q_e$.

Let $A_i$ be components of the potential of the electromagnetic field a. The Lagrangian of interaction of the real spinning field with the electromagnetic field is

$$L_{int} = -\frac{1}{2}qA_i c^i = -\frac{1}{2}q\phi\rho - \frac{1}{2}q\phi_i J^i = -\frac{1}{2}q\langle L_a A | \tilde{Q}_t A \rangle$$

Where $q$ is the constant of interaction and

$$\phi = t^i A_i, \quad \Phi_i = A_i - \phi t_i, \quad t^i \Phi_i = 0$$

is, respectively, scalar and vector potential of the electromagnetic field. Equation

$$(\Pi + \frac{1}{2}\varphi Q_t)A = mA, \mathrm{q}L_a \tilde{Q}_t A = mA \quad (12)$$

Describes electromagnetic interactions and can be considered as a basis for chemistry because the real spinning field carries only electrical charge. For comparison, the Lagrangian (7) is a basis for the strong interaction [11]. The operator $L_a$ is define as follows: $L_\mathbf{a} = Q_\mathbf{a} Z - (\mathbf{t}, \mathbf{a})(Q_t Z - Q_t)$, where the numerical diagonal operator is introduced $Z: \bar{a}_{i_1 \cdots i_p} = (-1)^p a_{i_1 \cdots i_p}, \quad p = 0,1,2,3,4.$

The operator of the neutrino charge $Q$ should be connected with the parity non-conservation and, hence, it is defined by the orientation of the physical manifold. The orientation is the anti-symmetrical tensor $e_{ijkl}$ ("element of volume") normalized as $e_{1234} = \sqrt{g}$, where $g = Det(g_{ij})$.

We define the operator $\tilde{H}$ in the component form

$$(\tilde{H}\mathbf{A})_{i_1 \cdots i_p} = \frac{1}{(4-p)!}(-1)^{\frac{p(p+1)}{2}} e_{i_1 \cdots i_p j_1 \cdots j_{4-p}} a^{j_1 \cdots j_{4-p}}$$

and derive the following relations:

$$\tilde{H}^2 = E, \quad \tilde{H}\tilde{Q}_t + \tilde{Q}_t \tilde{H} = 0, \quad \tilde{H}Q_t = Q_t \ \tilde{H} = 0.$$

The operator of the neutrino charge $Q_v$ takes the form

$$Q_v = \tilde{Q}_t \ \tilde{H}Q_t.$$

The operators $Q_e$ and $Q_v$ anticommute

$$Q_e Q_v + Q_v Q_e = 0, \quad Q_e^2 = Q_v^2 = -E.$$

The operators *Qe, Qv* and $Qev = Q_e Q_v$ define the representation of *SU (2)* group. The Lagrangian of Spin

Dynamics (8) is invariant with respect to the transformations

$$\bar{\Psi} = exp(\alpha Q_e)\Psi, \quad \bar{\Psi} = exp(\beta Q_v)\Psi, \quad \bar{\Psi} = exp(\gamma Q_{ev})\Psi,$$

But it is remarkable that the Lagrangian (10) is invariant only with respect to the transformation generated by the operator of the electric charge. We see that in one case the spinning field carries only the electrical charge but in the second case it carries the electrical charge, the neutrino charge and pseudo charge. In both the cases the spinning field carries the internal spin which is generated by the geometrical internal symmetry defined by the anti -symmetrical tensor field $S_{ij}$ obeying the equation $\bar{\nabla}_i S_{jk} = 0$. We have no possibility to discuss here this important symmetry [11] to be anywhere in detail.

## Conclusions

Let us mention some results of the Unification. Of course, the principal result is the dynamical equations of the gravitational field, spinning field and generalized electro-magnetic fields which form the nucleus of the unified physics as a system and this system will reveal its internal content in the process of its exploitation. In addition, we should like to formulate the following theses as the evidence that New Physics is the Unified Physics.

1. It is discovered that in the Universe there are two causal structures (the two natural clocks). One of them is defined by general solution of the equation of temporal field (1). Another solution and a new causal structure emerge as special solution of this equation. The new causal

structure gives natural explanation of the confinement and the baryon number conservation. Indeed, we cannot invoke the artificial concept of force to explain the confinement because for any force there is a more powerful one. But causality is something else again. The baryon number conservation simply expresses in the symbolic form the existence of the second causal structure and nothing else. The quark - lepton symmetry means that leptons live in the area of the usual causal structure and quarks live in the area of the new causal structure. Mutual transitions are possible.

2. The role of the gravitational field in the physics of elementary particles is recognized since the polarization of spin symmetry (spin) cannot be discovered without the gravitational field. Spin dynamics can be considered as a new purely rational realization of the Standard Model derived from the only first principle. The observed generations of quarks and leptons are different states of the complex spinning field and the number of these generations equals four. The spinning field describes the physical entity retaining the elementarity and is considered to be characterized solely by equation (12), electrical charge and the internal spin in one case and equation (9), pseudo charge, electric charge, neutrino charge and the internal spin in the other case.

3. The problem of Dark Energy is solved. Dark Energy is simply the energy of the gravitational field.

4. The problem of Dark Matter is solved. The Dark Matter is heavy light that inter-acts only gravitationally. Heavy photons are quants of the generalized electromagnetic field. Massless photons represent a singlet state of this field.

5. The most general and universal law of energy conservation is discovered which holds valid in all cases and is an opportunity for a new contributor of energy.

## References


1. Dirac PAM (1931) Quantised Singularities in the Electromagnetic Field. Proc Roy Soc 60.
2. Bernard F, Schutz (1982) Geometrical methods of mathematical physics. Cambridge University Press.
3. Pestov IB (2005) In: Horizons in World Physics, Vol. 248 by A. Reimer, Nova Science, New York.
4. Pestov IB (1996) New concept of time and gravity.
5. Green MB, Schwarz JH, Witten E (1987) Superstring Theory volume 1, Cambridge University Press.
6. Green MB, Schwarz JH, Witten E (1987) Superstring Theory volume 2, Cambridge University Press.
7. Weinberg S (2000) The Quantum Theory of Fields. Cambridge University Press.
8. Smolin L (2006) The Trouble with Physics. Houghton Mifflin.
9. Miemiec A, Schnakenburg I (2006) Basics of M-Theory. Fortsch Phys 54: 5-72.
10. Weinberg S (1999) Scientific American.
11. Pestov IB (2011) Physics of Atomic Nuclei 74: 1084
12. Pestov IB (2000) Gauge theory of oriented media. J Phys A: Math Gen 33: 3027.
13. Pestov IB (2006) In: Dark Matter: New Research. Ed. by Val Blain Nova Science, New York.
14. Pestov IB (2013) Geometrization of the Electromagnetic Field and Dark Matter. Physics of Particles and Nuclei 44: 442-449.
15. De Rham G (1955) Varietes Differentiable.
16. Hamilton RS (1982) Three-manifolds with positive Ricci curvature. Journ of Diff Geometry 17: 255-306.
17. Pestov IB (1978) Relativistic equations determined by the operators of the exterior derivative and generalized divergence. Theor and Math Phys 34: 48-58.

# A Remark on the Hopf invariant for Spherical 4-braids

**Akhmet'ev PM***
*Professor in IZMIRAN, Troitsk, Moscow region, Russia*

**Abstract**

An approach by J.Wu describes homotopy groups $\pi_n(S^2)$ of the standard 2-sphere as isotopy classes of spherical n+1--strand Brunnian braids. The case n=3 is investigated for applications.

## Introduction

An approach by Wu describes homotopy groups $\pi_n(S^2)$ of the standard 2-sphere as isotopy classes of spherical n+1-strand Brunnian braids, for more details, Theorem 1.2. This straightforward approach is not possible for n=3, i.e. for 4-strand braids the connection with $\pi_3(S^2)$ was unknown.

The homotopy group $\pi_3(S^2)$ in an infinite cyclic group, detected by the Hopf invariant

$$H:\pi_3(S^2)\mapsto \mathbb{Z}. \tag{1}$$

An element of $\pi_3(S^2)$ is represented by a mapping $h:S^3\mapsto S^2$, which is considered up to homotopy. The Hopf invariant $H(h)$ is well-defined as the integer linking number of two oriented curves $h^{-1}(a)$, $h^{-1}(b)$, where a,b$\in S^2$ be a pair of regular points of $h$. The Hopf invariant is very important for applications.

Proposition 7.1.1, sequence (17) gets an exact sequence, which algebraically describes the group $Brunn_4$ of 4-straight Brunnian braids [1]. The key point of our elementary geometrical construction is to construct an alternative epimorphism onto the group $\mathbb{Z}\times\mathbb{Z}$, see Definition. The kernel of this epimorphism is a well-defined subgroup $Brun_4 \subset \mathrm{Br}_4$ of Brunnian braids in a new sense (let us remark that $Brunn_4$ is not a subgroup of $Brun_4$. Define the Hopf invariant as a function of isotopy classes of spherical braids in $Brunn_4$. An idea of the construction was coming from Graham and Roman [2]. However, the results by Ellis and Mikhailov are not adopted for physical applications.

The Hamiltonian provides an elegant method for generating simple geometrical examples of complicated braids and links, as is presented in Mitchell A Berger [3].

The paper is motivated by the following problems:

- Derive applications of higher-order winding numbers to generate turbulent motions of vortices in two dimensions. For a special Hamiltonian motion of 3 vortices on the plane this is done in Mitchell A Berger [3]. (Problem 1).
- To unify the approach Ch.3 to $\pi^*(S^2)$ with the Wu's approach (Problem 2) [4].

Let us clarify Problem 2. Let F be the space of functions $f:\mathbb{R}^1\to\mathbb{R}^1$ with "right" boundary conditions at the infinity. The derivative of the order 1,2, and 3 of a function $f\in F$ can nowhere be vanished simultaneously. Define the mapping, $A:F\to\Omega(\mathbb{R}^3\setminus 0)$, by the formula $A(f)=\{x\mapsto(\frac{df(x)}{dx},\frac{d^2f(x)}{dx^2},\frac{d^3f(x)}{dx^3})\}$.

V.I.Arnold (1996) conjectured that the induced homomorphism $A_n$: $\pi_n(F)\to\pi_n(\Omega(S^2))\cong\pi_n+1\ (S^2)$ is an isomorphism for $n\geq 0$. This theorem was proved by V.A. Vassiliev in the special case $n$+2, and by Eliashberg and Mishachev in the general case.

The paper is organized as following. In Section 2 we recall required definitions concerning first-order stage of the construction and determine the linking numbers of spherical 4-component braids. In Section 3 the Hopf invariant for 4-component spherical braids is defined. This is a second-order particular defined invariant: to define this invariant we should assume that the all linking numbers (there are two) of components of a spherical braid are equal to zero. Results are formulated in Theorems 4, 6. The main result is the Corollary 8. In Section 4 we give proofs.

### A possible application for turbulences (Problem 1)

Assume a motion of a large collection of $n$ vortexes (or, particles) in a bounded domain $U$ on the plane is investigated. The trajectories of vortexes (or, of particles) in the configuration space, i.e. in the Cartesian product, of the domain and the time, are represented by a braid $F$, components of the braid $F$ correspond to vortexes in the collection. Assume that the windings numbers of components of the braid $F$ are distributed as in the statement of Corollary. This means that the length of the segment (a,b), which is assumed sufficiently large, is bounded from below; the upper bound depends of the number $n$ of vortexes in the collection. We may replace $F$ by a colored braid, if $b$-$a$ is sufficiently large, using the Arnol'd collection of the short paths, we have no loss of a generality.

Otherwise, assume that the bound $k$ of the distribution of full angles of windings numbers is much less then the number $n$ of partials. Consider the normalized sum of squares of Hopf invariants

$$\Upsilon=\frac{N}{(b-a)l}\sum_{i=1}^{l}H^2(g_i),\quad N=C_n^4=\frac{n^4}{24}+O(n^3) \tag{2}$$

this sum is taken over all collection of admissible quadruples of components of $F$, the number $l$ of admissible quadruples could be sufficiently large by Corollary 8. The following statements will be proved, or disproved, elsewhere:

- $\Upsilon$ is the universal constant of the motion, which depends no of the time scale and of the time interval [a, b] itself;
- The constant $\Upsilon$ is large (correspondingly, is small), if the motion of

***Corresponding author:** Dr. Akhmet'ev PM, IZMIRAN, Troitsk, Moscow region, Russia, E-mail: pmakhmet@mi.ras.ru

the system of vortexes (or, of partials) is turbulent (correspondingly, the system is closed to an integrable system);

- Assume the sum (2) is taken over all admissible quadruples, between which the distance is smaller then $L$. Then $\Upsilon(L)$ correlates with the spacial turbulent spectra of the motion up to the scale $L$.

## Linking Numbers For Spherical Braids

By a spherical (ordered) $n$-braid we mean a collection of embeddings of the standard circles

$$f: \bigcup_{i=1}^{n} S_i^1 \subset S^2 \times S^1,$$

where the composition of this embedding with the standard projection $S^2 \times S^1 \to S^1$ on the second factor in the target space, restricted to an arbitrary component $S_i^1$, $i=1,\ldots,n$ is the identity mapping.

The set of all ordered spherical n-braids up to isotopy is denoted by $Br_n$. It is well-known that $Br_n$ is a group.

For a fixed value $t \in S^1$, a braid $f \in Br_n$ intersects the level $S^2 \times t$ by an (ordered) collection of $n$ points $\{Z_1(t),\ldots Z_n(t)\}$. Let assume that $n=4$. Denote by

$$g = g(f): S_1^1 \cup S_2^1 \cup S_3^1 \subset S^2 \times S^1,$$

the 3-component braid, obtained from $f$ by eliminating of the last component $S_4^1$.

Let us identify the sphere $S^2$ with the Riemann sphere, or with the complex projective line $C$. For a braid $f$ let us consider the collection of Mobius transformations, which transforms the points $z_1$, $z_2$, $z_3$ into 0, 1, $\infty$ correspondingly:

$$F(z;t) = \frac{(z - z_1(t))(z_2(t) - z_3(t))}{(z - z_3(t))(z_2(t) - z_1(t))}.$$

The image $F(f)$ is a 4-strand braid with the constant components $\{z_1(t), z_2(t), z_3(t)\}=\{0, 1, \infty\}$. Denote this braid by

$$F(f) = f^{norm}. \tag{3}$$

The 3-strand braid g, constructed from $f^{norm}$ is the constant in the points $\{0, 1, \infty\}$. The last component $f^{norm}(S^1)$ of $F(f)$ is represented by a closed path $z_4(t) \in \hat{\mathbb{C}} \setminus \{0,1,\infty\}, t \in \mathbb{R}^1/2\pi$. Note that, generally speaking, braids $f, f^{norm}$ are not isotopic. Moreover, if $f$ is a Brunnian in the sense [2], $f^{norm}$ is, generally speaking, not a Brunnian.

For a given (ordered) 4-component braid $f$ let us define the linking number $Lk(f)$,

$$Lk: Br_4 \to \mathbb{Z}. \tag{4}$$

Consider the following 1-form

$$\omega_0 = \frac{1}{2\pi i}\frac{dz}{z}. \tag{5}$$

By definition we get

$$d\log(z) = \frac{1}{2\pi i}\frac{dz}{z},$$

where log (z) is given by the formula:

$$\log(z) = (2\pi i)^{-1}\int \frac{dz}{z},$$

assuming that log (1)=0, as a multivalued complex function.

Define $Lk(f)$ by the formula:

$$Lk(f) = \Re\int_0^{2\pi} \frac{dz_4(t)}{z_4(t)} = \int_{f_4^{norm}} \omega_0, \tag{6}$$

where $\Re$ is the real part of the integral. By construction, $Lk(f)$ is the winding number, i.e. the integer number of rotations of the path $z_4(t)$ of $f^{norm}$ with respect to the origin and the infinity in $C$.

The permutation group $\Sigma(4)$ of the order 24 acts on the space of ordered spherical braids:

$$\Sigma(4) \times Br_4 \to Br_4. \tag{7}$$

The image of an ordered braid $f$ by a transposition $\sigma:(1,2,3,4) \mapsto (\sigma_1,\sigma_2,\sigma_3,\sigma_4)$ is well-defined by the corresponding re-ordering of components of $f$. Let us investigate the orbit of the linking numbers $Lk(f)$ with respect to (7). Simply say, we investigate how many independent linking numbers of components of braids are well-defined?

Let us consider the following exact sequences of groups:

$$0 \to A_4 \to \Sigma_4 \xrightarrow{sign} \mathbb{Z}/2 \to 0, \tag{8}$$

$$0 \to \mathbb{Z}/2\times\mathbb{Z}/2 \to A_4 \to \mathbb{Z}/3 \to 0. \tag{9}$$

The subgroup $A_4 \subset \Sigma_4$ in the sequence (8) is represented by permutations, which preserve signs (equivalently, which is decomposed into an even number of elementary transpositions). The subgroup $\mathbb{Z}/2\times\mathbb{Z}/2 \subset A_4$ in the sequence (9) is generated by the permutations$\{(1,2)(3,4);(1,3)(2,4);(1,4)(2,3)\}$.

Let us consider 2-primary subgroup $K \subset \Sigma_4$ (the dihedral group of the order 8), which is defined as the extension of the subgroup $\mathbb{Z}/2\times\mathbb{Z}/2$ from the sequence (9), which is included in the sequence (8). An epimorphism

$$\theta = (\theta_1,\theta_2): K \to \mathbb{Z}/2\times\mathbb{Z}/2, \tag{10}$$

is defined as follows: $\theta_1(\sigma)=1$ (the group $Z/2$ is in the multiplicative form), if $\sigma$ preserves a (non-ordered) partition (1,3)(2,4), and $\theta_1(\sigma)$, and $\theta_1(\sigma)=-1$, otherwise. Therefore $\theta_1$ is an epimorphism with the kernel $\mathbb{Z}/2\times\mathbb{Z}/2$ from the left subgroup of the sequence (9). The epimorpism $\theta_2(\sigma)$ is determined by the sign of a permutation $\sigma$, this is the restriction of the right epimorphism in the sequence (8) to the subgroup $K \subset \Sigma_4$. The kernel $Ker(\theta) \cong \mathbb{Z}/2$ is the center of the dihedral group $K$.

### Lemma 1

1. The function (4) is invariant with respect to the action (7) by an arbitrary permutation, which in the kernel of $\theta$ in (10).

2. The function (4) is skew-invariant with respect to the action by a permutation, which is in the kernel of $\theta_1$ (the composition of $\theta$ with the projection on the first factor, but not in the kernel of $\theta_2$ (the composition of $\theta$ with the projection on the second factor).

3. Denote by $\tilde{f} \in Br_4$ the ordered braid, which is obtained from $f \in Br_4$ by the action (7) by the element (1,2) is the product of the generators of the factors). There exists an ordered braids $f \in Br_4$, for which the linking numbers $Lk(f)$, $Lk(\tilde{f})$ are arbitrary integers.

From Lemma one may deduce the following corollary.

### Corollary 2

1. For an arbitrary braid $f \subset Br_4$ the linking number $Lk(f)$, is well-defined as the differences of the winding number of the component 2 between the components 1 and 3 with the winding number of the component 4 between the components 1 and 3.

2. For a braid , where $f \in Br_4$ is an arbitrary, $\tilde{f}$ is defined in Lemma

1, the linking number $Lk(\tilde{f})$ is well-defined as the difference of the winding number of the component 2 between the components 1 and 3 with the winding number of the component 4 between the components 2 and 3.

Corollary (2) motivates the following definition.

**Definition 3**

*Let* $f \in Br_4$ *be a (ordered) spherical braid. Define the total linking number* $LK(f) \in \mathbb{Z} \oplus \mathbb{Z}$ by the following formula:

The total linking number is a well-defined homomorphism

$LK : Br_4 \mapsto \mathbb{Z} \oplus \mathbb{Z}.$

## Hopf Invariant of Braids

Let $f \in Br_4$ be a (ordered) spherical braid with the trivial total linking number: $Lk(f)=0$. Such braids generate the subgroup in the group $Br_4$, denote this subgroup by $Brunn_4 \subset Br_4$. Let us remark that this subgroup does not coincide with the subgroup of Brunnian braids $Brun_4$, defined in Berrick et al. [1], Theorem 1.2.

**Theorem 4**

*There exists a well-defined homomorphism*

$$H : Brunn_4 \to Z, \tag{11}$$

called the Hopf invariant. The homomorphism (11) is invariant with respect to the action (7) by an arbitrary permutation, which in the kernel of $\theta_2$ in (10) (this homomorphism is defined as the sign of a permutation of straights), and is skew-invariant with respect to the action by a permutation, which is not in the kernel of $\theta_2$.

### Definition of the hopf invariant

In this section we present the construction, which is closed to Theorem 3 of Mitchell A Berger [3], using differential topology instead of homology algebra. Let $f \in Brunn_4$ be an arbitrary. Consider the braid $f^{norm}$, given by Mitchell A Berger [3]. Recall, the braid $g \in Br_3$, which consists of the straits (1-3) of $f^{norm}$, is the constant braid at the points 0, 1, ∞ in correspondingly. Consider the strait (4) of the braid $f^{norm}$. This strait is represented by an oriented closed path $z_4 : S^1 \to \hat{C} \setminus \{0 \cup 1 \cup \infty\}$. This path determines a cycle, which is an oriented boundary, because of the condition $LK(f^{norm})=0$.

Let us prove that $Lk(f)=0$. Denote the group of Mobius transformations by $M$. The standard inclusion SO(3) ⊂ M is well-defined. This inclusion is a homotopy equivalence, therefore we get $\pi_1(M) = \pi_1(SO(3) = \mathbb{Z}/2$. This proves that $LK(2f^{norm})=LK(2f)$. Because $LK(2f^{norm})=2LK(2f^{norm})$, $LK(2f)=2Lk(f)$, we get $LK(f^{norm})=Lk(f)$. The equality $Lk(f)=0$ is proved.

Consider the inclusions

$I_0 : C \setminus \{0 \cup 1 \cup \infty\} \subset C \setminus \{1 \cup \infty\},$

$I_\infty : C \setminus \{0 \cup 1 \cup \infty\} \subset C \setminus \{0 \cup 1\},$

$I_1 : C \setminus \{0 \cup 1 \cup \infty\} \subset C \setminus \{0 \cup \infty\}.$

Because $H_1(\hat{\mathbb{C}} \setminus \{1 \cup \infty\}; \mathbb{Z}) = \pi_1(\hat{\mathbb{C}} \setminus \{1 \cup \infty\}) = \mathbb{Z}$, the condition $LK(f^{norm})=0$ implies $I_0,\#([i])=0$, for the homomorphism

$I_{0,\#} : \pi_1(C \setminus \{0 \cup 1 \cup \infty\}) \to \pi_1(C \setminus \{0 \cup \infty\}).$

Analogously $I_{\infty,\#}([i])=0$, $I_1,\#([i])=0$.

There exist the following 3 maps of the standard 2-disk

$$e_0 : D_0^2 \to \hat{C} \setminus \{1 \cup \infty\}, \quad e_0 |_{\partial D^2} = z_4,$$
$$e_\infty : D_\infty^2 \to \hat{C} \setminus \{0 \cup 1\}, \quad e_\infty |_{\partial D^2} = z_4,$$
$$e_1 : D_1^2 \to \hat{C} \setminus \{0 \cup \infty\}, \quad e_1 |_{\partial D^2} = z_4.$$

Consider a 2-sphere, which is represented by a gluing $D_0^2 \cup_\partial D_\infty^2$ of the disks $D_0^2 \cup D_\infty^2$ along the common boundary, which is identified with the circle $S_4^1$. Denote this sphere by $S_1^2$. Analogously define spheres $S_0^2 = D_\infty^2 \cup_\partial D_1^2$, $S_\infty^2 = D_1^2 \cup_\partial D_0^2$. Because the target spaces of the mappings $e_0$, $e_\infty$, $e_1$ are aspherical, the corresponding mapping is well-defined up to homotopy.

Consider the following commutative diagram of inclusions:

$$\begin{array}{ccc} C \setminus \{0 \cup \infty \cup 1\} & \subset & C \setminus \{0 \cup \infty\} \\ \cap & & \cap \\ C \setminus \{\infty \cup 1\} & \subset & C \setminus \{\infty\} \end{array} \tag{12}$$

Consider the mappings $e_0 : D_0^2 \to C \setminus \{1 \cup \infty\}$, $e_1 : D_1^2 \to C \setminus \{0 \cup \infty\}$ to the left bottom and to the right upper spaces of the diagram (12) correspondingly. The mapping $e_0 \cup_\partial e_1 : S_\infty^2 \to C \setminus \{\infty\}$ is well defined by gluing of the two mappings $e_0$, $e_1$ along the common mapping $i$ of the boundaries. Consider the standard 3-ball $D_\infty^3$ (with corners along the curve $S_4^1$) with the boundary $\partial D_\infty^3 = S_\infty^2$. The mapping $e_0 \cup \partial\, e_1$ can be extended to the mapping

$$d_\infty : D_\infty^3 \to \hat{C} \setminus \{\infty\}. \tag{13}$$

The target space of this mapping is the right bottom space of the diagram (12). Because the target space of the mapping $d_\infty$ is contractible, the mapping $d_\infty$ is well-defined up to homotopy. By the analogous constructions the following mappings

$$d_1 : D_1^3 \to \hat{\mathbb{C}} \setminus \{1\}, \tag{14}$$

$$d_0 : D_0^3 \to \hat{\mathbb{C}} \setminus \{0\} \tag{15}$$

are well-defined.

The mappings (13), (14), (15) determine the mapping

$$h = h(f) : S^3 \to S^2 \tag{16}$$

as follows. Take a 3-sphere $S^3$, which is catted into 3 balls $D_\infty^3, D_1^3, D_0^3$ along the common circle $S_4^1 \subset S^3$. The sphere $S^3$ is represented as the join $S_4^1 * S_a^1$ of the two standard circle. On the circle $S_a^1$ take 3 points $x_0, x_1, x_\infty \in S_a^1$. The subsets $S_4^1 * [x_0, x_1] \subset S^3, S_4^1 * [x_1, x_\infty] \subset S^3, S_4^1 * [x_\infty, x_0] \subset S^3$ are 3 copies of 3D disks, which are glued along corresponding subdomains in its boundaries.

Let us identify $D_\infty^3 \cong S_4^1 * [x_0, x_1], D_0^3 \cong S_4^1 * [x_1, x_\infty], D_1^3 \cong S_4^1 * [x_\infty, x_0]$. The boundary $\partial D_\infty^3$ is identified with the balls $S_4^1 * \{0\} \cong D_0^2$, $S_4^1 * \{1\} \cong D_1^2$, which are glued along the common boundary $S_4^1$. The boundary $\partial D_0^3$ is identified with the balls $S_4^1 * \{1\} \cong D_1^2$, $S_4^1 * \{\infty\} \cong D_\infty^2$, which are identify along the common boundary $S_4^1$. The boundary $\partial D_1^3$ is identified with the balls $S_4^1 * \{\infty\} \cong D_\infty^2$, $S_4^1 * \{0\} \cong D_0^2$, which are identified along the same boundary $S_4^1$. The mappings $d_0$, $d_1$, $d_\infty$ on the corresponded balls are well-defined by the formulas (13-15) correspondingly. This mappings define the mapping (16) on the 3-sphere.

**Definition 5:** The Hopf invarian $H(f)$ for a braid $f \in Brunn_4$ in the formula (11) is defined as the Hopf invariant of the mapping $h$ by the formula (1). The mapping $h=h(f)$ is explicitly defined from the braid $f$ by the formula (16).

## A formula to calculate the Hopf invariant

Let us introduces an explicit formula to calculate the Hopf invariant for a braid f $\in$ $Brunn_4$. Consider the complex plane C. The 4-th strain of the braid $f^{norm}$ determines a curve on the plane without two points {0.1}, which is denoted by

$$\gamma : S^1 \to \mathbb{C} \setminus \{0 \cup 1\}. \quad (16)$$

Let us consider the closed 1-form (4). Define a complex 1-form

$$\omega_1 = \frac{1}{2\pi i}\frac{dz}{z-1}. \quad (17)$$

Define a real (multivalued) function $\lambda_0$ by integration along the path γ (*t*), $t \in [0,t] \subset S^1$ of the real part of the form (18) as following:

$$\lambda_0(t) = \Re\int_0^t \omega_0. \quad (18)$$

Define a real (multivalued) function $\lambda_1$ by integration along the path of the real part of the form (17) as following:

$$\lambda_1(t) = \Re\int_0^t \omega_1. \quad (19)$$

To take the multivalued functions (18), (19) well-defined, assume that the path $\gamma$ starts at the point $2 \in \mathbb{C} : \lambda_0(0)=2, \lambda_1(0)=2$.

Define a closed 1-form $\Psi$ (*t*) along a curve $\gamma(t) \in \mathbb{C}\setminus\{0 \cup 1\}$ by the following formula:

$$\psi(t) = \lambda_0(t)\omega_1 + \lambda_1(t)\omega_0. \quad (20)$$

Let us consider a function, which is well-defined as the real part of the integral

$$\Psi(T) = \Re\int_0^T \psi(t)d\gamma, \quad t \in [0,T] \subset S^1, \Psi(0) = 0. \quad (21)$$

### Theorem 6

*The Hopf invariant of a braid f $\in$ $Brunn_4$ in the formula (11), which is defined by Definition 5, is calculated by the formula:*

$$-H(f) = \Psi(2\pi) = \frac{1}{2}\Re\int_0^{2\pi}\psi(t)d\gamma, \quad (22)$$

where $\gamma$ is the closed path, determined by the 4-th straight of the braid $f^{norm}$ by the formula (17).

From Theorem 6 we get a corollary.

### Corollary 7

1. The Hopf invariant (11) is an epimorphism.

2. Assume there is a braid $f \in Brunn_4$ for which the braid $f^{norm}$ is represented by a commutator of the straight (4) with straights (1) and (2) (such a braid is called the Borromean rings). Then $H(f)=\pm 1$, where the sign in the formula depends on the sign of the commutator.

### Proof of corollary

It is sufficient to prove --2. The right-hand side of the formula (21) coincides with the formula (28) [1], which is simplified for the considered example. The Berger's formula is applied for the 3-uple configuration space, this gives the opposite sign for the last term in the formula (21) with respect to the origin formula. For the Borromean ring the formula (22) is non-trivial. The right side of the formula gives $H(f)=1$ for the right Borromean rings. Corollary is proved.

The following Corollary is the main result of the paper. The author hope that this result is the initial step toward the solution of the first problem, mentioned in Introduction.

### Corollary 8

*Assume we have a classical -conponent colored non-ordered braid F, n>>4, for which all pairwise winding (integer) numbers of components are distributed to the segment:* $\{-2k\pi,\ldots -2\pi, 0, 2\pi, \ldots 2k\pi\}$, *0<k<<n. Let G is the spherical braid, which is defined as the image of F by the stereography projection* $\mathbb{R}^2 \times I \to S^2 \times I$. *Then there exist at least* $K = \frac{n^4}{24(2k+1)^2} + O(n^3)$ $(K(n) \to +\infty \text{ when } n \to +\infty)$ *4-component subbraids fi* $\subset$ *F, for which LK(*$g_i$*)=0,* $g_i \subset G$*,. In particular, the squares* $H^2(g_i) \in \mathbb{N}$ *are well-defined.*

### Proof of corollary

**Proof is evident:** the number *K* of subbraids $g_i \subset F$ with trivial total linking number $LK(g_i)=0$ is explicitly estimated from below using integers *k*, *n*.

## Proofs

### Proof of lemma

Proof of Statement 1. Take an oriented 3--manifold $M^3$. Take two disjoin oriented cycles $C_I \subset M^3$, $C_{II} \subset M_3$, which represent the trivial homology class

$$0 = [C_I] = [C_{II}] \in H_1(M^3; Z). \quad (23)$$

The linking number *link* ($L_I$, $L_{II}$) is a well-defined integer the algebraic intersection coefficient of the boundary $(\Gamma_I, \partial\Gamma_I) \subset M^3$, $\partial\Gamma_I = C_I$. The linking number *link* ($C_I$, $C_{II}$) is well defined, because of the condition.

Obviously, *link* ($C_I$, $C_{II}$)=*link* ($C_{II}$, $C_I$), because the collections of signed points $A_I = \Gamma_I \cap C_{II}$ and $A_{II} = C_I \cap \Gamma_{II}$ represent the same cycle [$A_I$]=[$A_{II}$] $\in H_0(M^3; Z)$. The boundary of -[$A_I$] $\cup$[$A_{II}$] is given by the oriented curve $\Gamma_I \cap \Gamma_{II}$.

Take $M^3=S^2 \times S^1$. Take an arbitrary braid $f^{norm}$. The cycle $C_I$ is represented by the images of the following two closed paths [$z_1(t)$]=0 ×- $S^1$, [$z_3(t)$]=∞ ×$S^1$, *t* ∈[0,2π], where the path $z_1(t)$ is taken with the opposite orientation along $S^1$. The cycle $C_{II}$ is represented by the two closed paths [$z_2(t)$]=0 × -$S^1$, [z4 (t)] $\subset S^2 \times S^1$, where the path $z_2(t)$ is taken with the opposite orientation along $S^1$.

Take $\sigma_\alpha$=(1,2)(3,4), $\sigma_\alpha$ is the generator of Ker(θ). It is easy to see that $Lk(f^{norm})$=*link*($C_I$, $C_{II}$), $Lk(\sigma_\alpha \times f^{norm})$=*link*($C_{II}$, $C_I$). Statement 1 is proved.

Proof of Statement 2. Assume $\sigma_b$=(1,3), the case $\sigma_b$=(2,4) is analogous. Then $Lk(f^{norm})$=link($C_I$, $C_{II}$), *Lk*(σb × $f^{norm}$)=*link*(-$C_I$, $C_{II}$). Therefore we get $Lk(f^{norm})$=-*Lk*(σb × $f^{norm}$). Statement 2 is proved.

Lemma 1 is proved.

### Proof of Corollary 2

Statements 1,2 are obvious.

Proof of Statement 3. The straights {$z_1(t)$, $z_2(t)$, $z_3(t)$, $z_4(t)$} determines 6 pairs of -cycles in $S^2 \times S^1$. A function of winding numbers of component is given by a linear combinations of linking numbers between the corresponding pairs of cycles. To prove that such a function is well-defined, we have to assume that the each cycle is a boundary. Denote the cycle, generated by the pair of paths -$z_i$ (*t*), $z_j$ (*t*) by $C_{i,j}$. We have the following identity: link ($C_{1,2}$, $C_{3,4}$)+*link* ($C_{2,3}$, $C_{1,4}$)+link ($C_{3,1}$, $C_{2,4}$)=0, and the analogous 3 identities, which are obtained by the permutation of the indexes. Therefore we get a collection of 2 independent well-defined linking numbers. Statement is proved. Corollary 2 is proved.

## Proof of theorem

Let us prove that the homomorphism (11) is skew-invariant with respect to the action (7) by an odd permutation. Assume that the permutation $\sigma$ is given by an elementary transposition of straights with number (1-3) say by the transposition $\sigma$=(1,2). Then by the formula (16), the mappings $h(f)$ is related with the mapping $h(\sigma \times f)$ by the composition with the reflection $S^3 \to S^3$, which translates the curve $S^1_4 \subset S^3$ to itself, and permutes the points $x_0$, $x_1$ on the circle $S^1_a$. The reflection changes the homotopy class of $h$ to the opposite. This proves Theorem in this case.

Assume that $\sigma$=(1, 4) (the cases $\sigma$=(2, 4), or (3,4) are analogous). Then we may calculate the Hopf invariants of the mappings $h(f)$ and $h(\sigma \times f)$, using the formula (22) (Theorem 6 is proved below). The Mobius group is locally contractible. Therefore, the ordered braid $(\sigma \times f)$ is isotope to the braid $f$ in which the components (1,4) are re-numbered. The restriction of the considered isotopy on the common straight at $\infty \in C$ is the identity.

By Statement 2 of Corollary 7, the Hopf invariant $h(\sigma \times f^{norm})$ is defined as the length of commutators of the straight (1) with straights (4) and (2).

The Hopf invariant for $h(f)$ coincides with the commutator of the straight (4) with the straights (1,4). Therefore the Hopf invariant for $h(\sigma \times f)$ is opposite to the Hopf invariant for $h(f)$, because the sign of the commutator is changed by a permutation of components.

Theorem 4 is proved.

## Proof of Theorem 6

Consider the mapping $h: S^3 \to S^2 = \hat{C}$, which is defined by the formula (16). Take two normalized volume forms $\Omega_0, \Omega_1 \in \Lambda^2(S^2)$:

$$\iint_{S^2} \Omega_0 = \iint_{S^2} \Omega_1 = 1.$$

The forms $\Omega_0$, $\Omega_1$ are defined as the standard ill-supported forms at the points 0, 1, correspondingly. The Hopf invariant (11) is calculated by the formula:

$$H(f) = \frac{1}{2} \iiint_{S^3} h^*(\Omega_0) \wedge \beta_1 + h^*(\Omega_1) \wedge \beta_0, \qquad (24)$$

where $x \in S^3$, $h^*(\Omega_0) \in \Lambda^2(S^3)$ is the pull-back of $\Omega_0 \in \Lambda^2(S^2)$ by $h$: $S^3 \to S^2$, $\beta_0 \in \Lambda^1(S^3)$ is an arbitrary 1-form, such that $d(\beta_0) = h^*(\Omega_0)$, the 1-form $\beta_1 \in \Lambda^1(S^3)$ is defined analogously to $\beta_0$.

Evidently, the 1-forms $\beta_0$ in the integral (24) is represented in its cohomology class by a cocycle, which satisfies the condition $h^*(\Omega_0) = d\beta_0 = 0$ inside the ball $D^3_0$. This follows from the fact that the curve $h^{-1}(0)$ is outside the ball $D^3_0$. In the formula (24) the first term is well-defined up to gauge transformation $\beta_0 \mapsto \beta_0 + grad\varphi_0$. We may put $\beta_0$=0 in $D^3_0$, and keep $\beta_0$ on $D^2_1 = D^3_0 \cap D^3_\infty$.

Analogously, $d\beta_1$=0 in the ball $D^3_1$. In the second term in the integral (24), using $\beta_1 \mapsto \beta_1 + grad\varphi_1$, we get $\beta_1$=0 in $D^3_1$, and keep $\beta_1$ on $D^2_0$. Then we get the following simplification of (24):

$$H(f) = \frac{1}{2} \iiint_{D^3_\infty} h^*(\Omega_0) \wedge \beta_1 + h^*(\Omega_1) \wedge \beta_0.$$

In the ball $D^3_\infty$ the 3-form $h^*(\Omega_0) \wedge \beta_1$ is exact, we get $\alpha_1 \in \Lambda^2(D^3_\infty)$, $d\alpha_1 = h^*(\Omega_0) \wedge \beta_1$. Moreover, we may put $\alpha_1 = \beta_0 \wedge \beta_1$ over $D^2_1 = D^3_\infty \cap D^3_0$, $\alpha_0 = 0$ over $D^2_0 = D^3_\infty \cap D^3_1$.

In the ball $D^3_\infty$ the 3-form $h^*(\Omega_1) \wedge \beta_0$ is exact, we get $\alpha_0 \in \Lambda^2(D^3_\infty)$, $d\alpha_0 = h^*(\Omega_1) \wedge \beta_0$. We may put $\alpha_0 = \beta_1 \wedge \beta_0 = -\beta_0 \wedge \beta_1$ over $D^2_0$, and $\alpha_0$=0 over $D^2_1$.

Apply the 3D Gauss-Ostrogradsky formula, we get

$$\iiint_{D^3_\infty} h^*(\Omega_0) \wedge \beta_1 = \iint_{D^2_1} \beta_0 \wedge \beta_1,$$

$$\iiint_{D^3_\infty} h^* \Omega_1 \wedge \beta_0 = -\iint_{D^2_0} \beta_0 \wedge \beta_1.$$

The 2-form $\beta_0 \wedge \beta_1 \in \Lambda^2(D^2_1)$ is exact. Because in the disk $D^2_1$ the 0-form $\lambda_1$ is well defined, and $d\lambda_1 = \beta_1$, we get: $d(\lambda_1 \beta_0) = -\beta_0 \wedge \beta_1$.

Analogously, the 2-form $\beta_0 \wedge \beta_1 \in \Lambda^2(D^2_0)$ is exact. Because in the disk $D^2_0$ the 0-form $\lambda_0$ is well defined, and $d\lambda_0 = \beta_0$, we get: $d(\lambda_0 \beta_1) = \beta_0 \wedge \beta_1$.

Apply the 2D Green formula (singular points of $\beta_0$, $\beta_1$ give no contribution to the integral over the boundary) we get:

$$\iint_{D^2_1} \beta_0 \wedge \beta_1 = -\int_\gamma \lambda_1 \beta_0,$$

$$\iint_{D^2_0} \beta_0 \wedge \beta_1 = \int_\gamma \lambda_0 \beta_1,$$

where $\gamma = D^2_1 \cap D^2_0$.

The integral (24) is simplified as

$$H(f) = -\frac{1}{2} \int_\gamma \lambda_0 \beta_1 + \lambda_1 \beta_0.$$

This formula coincides with the formula (22). Theorem 6 is proved.

### Acknowledgement

The author is grateful to V.P.Leksin for the explication of the results of the paper [1], and to S.A.Melikhov for discussions. The results was presented at International Conference "Nonlinear Equations and Complex Analysis" in Russia (Bashkortostan, Bannoe Lake) during the period since March 18 (arrival day) till March 22 (departure day), 2013. The author was supported in part by Russian Foundation of Basic Research Grant No. 11-01-00822.

### References

1. JA Berrick, FR Cohen, YL Wong, J Wu (2006) Configurations, braids, and homotopy groups. J Amer Math Soc 19: 265-326.
2. Graham Ellis, Roman Mikhailov (2010) A colimit of classifying spaces, Advances in Math, 223: 2097-2113; arXiv: 0804.3581.
3. Mitchell A Berger (2001) Hamiltonian dynamics generated by Vassiliev invariants, J. Phys. A: Math. Gen. 34: 1363-1374.
4. VA Vassiliev (1994) Complements of discriminants of smooth maps: topology and applications, 2-d extended edition, Translations of Math. Monographs, 98, AMS, Providence, RI: 268.

12

# Thermal Diffusive Free Convective Radiating Flow Over an Impulsively Started Vertical Porous Plate in Conducting Field

**Kumar VR[1], Raju MC[1*], Raju GSS[2] and Varma SVK[3]**

[1]Department of Mathematics, Annamacharya Institute of Technology and Sciences, India
[2]Department of Mathematics, JNTUA College of Engineering, India
[3]Department of Mathematics, S.V. University, India

**Abstract**

In this manuscript we have studied the laminar convective heat and mass transfer flow of an incompressible, viscous, electrically conducting fluid over a fluid over an impulsively started vertical plate with conduction-radiation embedded in a porous medium in the occurrence of transverse magnetic field. An exact solution is derived by solving the dimensionless main coupled partial differential equations using Laplace transform technique. The properties of important physical parameters on the velocity, temperature, concentration, Skin friction, Sherwood number and Nusselt number have been studied through graphs.

**Keywords:** MHD; Porous medium; Thermal diffusion; Thermal radiation; Shear stress; Nusselt number and Sherwood number

## Nomenclature

$C^{/}$ : Species concentration fluid

$Cp$ : Specific heat at constant pressure

$C_w^{/}$ : Concentration of the fluid for away from the plate

$C_w^{/}$ : Concentration level near the plate/wall

$D$: Chemical molecular diffusivity

$g$: Acceleration due to gravity

$q_r$: Radiative heat flux

$Gr$: Thermal Grashof number

$Gm$: Modified Grashof number

$K_r$ : Permeability parameter

$M$: Hartmann number

$Nu$: Nusselt number

$P_r$: Prandtl number

$S_0$: Soret number

$S_h$: Sherwood number

$T_w^{/}$ : Fluid temperature at the surface

$u$: Dimensional velocity components

$S_c$: Schmidt number

$T^{/}$ : Temperature

$u_0$: Plate velocity

$\beta$ : Coefficient of volume expansion for heat transfer

$\theta$: Dimensional fluid

n: Kinematic viscosity

$\sigma$ : Electrical conductivity

C: Dimensionless species concentration

$\beta_c$: Coefficient of volume expansion for mass transfer

$\kappa$: Thermal conductivity

$\rho$ : Density

$\tau$: Shearing stress

$W$ : Condition on the wall

$\infty$ : Free stream condition

## Introduction

Several transport processes exist in industries and technology where the transfer of heat and mass occurs simultaneously as an outcome of thermal diffusion and diffusion of chemical species. Natural convection induced by the simultaneous achievement of buoyancy forces resulting from thermal and mass diffusion is of considered interest in nature and in many industrial applications such as cosmic fluid dynamics, meteorology, chemical industry, cooling of nuclear reactors, magneto hydrodynamics power generators and the earth's core. Bharat et al.[1] investigated the effects of mass transfer on MHD free convective radiation flow over an impulsively started vertical plate embedded in a porous medium. Ahmed et al. [2] discussed convective laminar radiating flow over an accelerated vertical plate embedded in a porous medium with an external magnetic field. Chamka et al. [3] studied thermal radiation and buoyancy effects on hydro magnetic flow over an accelerating porous surface with heat source or sink. Ahmed et al. [4] examined Non-linear magneto hydrodynamic flow more an impulsively started vertical plate in a saturated porous regime Laplace and Numerical approach. Ravi Kumar et al. [5] examined MHD

***Corresponding author:** Raju MC, Department of Mathematics, Annamacharya Institute of Technology and Sciences, India
E-mail: mcrmaths@yahoo.co.in

double diffusive and chemically reactive flow through porous medium bounded by two vertical plates. Palani et al. [6] studied free convection MHD flow with thermal radiation from an impulsively-started vertical plate. Ravi Kumar et al. [7] discussed heat and mass transfer effects on MHD flow of viscous fluid through non-homogeneous porous medium in occurrence of temperature dependent heat source. Chen et al. [8] discussed heat and mass transfer in MHD flow by ordinary convection from a permeable, inclined surface with variable wall temperature and concentration. Ravi Kumar et al. [9] discussed combined effects of heat absorption and MHD on convective Rivlin-Ericksen flow past a semi-infinite perpendicular porous plate with variable temperature and suction. Ahmed et al. [10] examined Numerical/Laplace transform investigation for MHD radiating heat/mass transport in a Darcian porous regime bounded by an oscillating vertical surface. Kumar et al. [11] discussed thermal radiation and mass transfer effects on MHD flow past a vertical oscillating plate among variable temperature effects variable mass diffusion. Hossain et al. [12] studied radiation effect on mixed convection along a perpendicular plate with uniform surface temperature. Ibrahim et al. [13] examined similarity solution of heat and mass transfer for normal convection over a moving vertical plate with internal heat generation and a convective boundary state in the presence of thermal radiation, viscous dissipation, and chemical reaction. Pradyumna kumar et al. [14] examined analytical solution of magnetic hydro magnetic free convective flow through porous media with time dependent temperature and concentration. Das et al. [15] discussed mass transfer effects on MHD flow and heat transfer past a vertical porous plate throughout a porous medium below oscillatory suction and heat source [16]. Seth et al. [17] studied effects of thermal radiation and rotation on unsteady hydro magnetic free convection flow past an impulsively moving vertical plate with ramped temperature in a porous medium. Das et al. [18] discussed heat and mass transfer effects on unsteady MHD free convection flow near a moving vertical plate in porous medium. Kumar et al. [19] examined magnetic field effect on transient free of charge convection flow through porous medium past an impulsively started vertical plate with fluctuating temperature and mass diffusion. Mamtha et al. [20] discussed thermal diffusion effect on MHD mixed convection unsteady flow of a micro polar liquid past a semi-infinite vertical porous plate with radiation and mass transfer. Redddy et al. [21] examined unsteady MHD free convection flow of a Kuvshinski fluid past a vertical porous plate in the presence of chemical reaction and heat source/sink. Kumar et al. [22] investigated theoretical investigation of an unsteady magnetic hydro magnetic free convection heat and mass transfer flow of a non-Newtonian liquid flow past a permeable moving perpendicular plate in the presence of thermal diffusion and heat sink. Reddy at al. [23] discussed mass transfer and heat generation effects on magnetic hydro magnetic free convection flow past an incline vertical surface in a porous medium. Senapati et al. [24] examined magnetic effects on mass and heat transfer of hydrodynamics flow past an oscillating vertical plate in presence of chemical reaction. Raju et al. [25] investigated MHD convective flow through porous medium in a vertical channel with insulated and impermeable base wall in the presence of viscous dissipation and joule heating. The Effects of mass transfer on MHD free convective radiation flow over an impulsively started vertical plate embedded in a porous medium was studied by Bharat and Nityananda [1]. We have extended this work by including the thermal diffusion effect. Though it is an extension to the previous work, it will differ in several aspects like governing equations, non-dimensional parameters, figures etc. The novelty of this study is the investigation of various physical parameter on the flow quantities in the presence of thermal diffusion.

## Mathematical Formulations

The laminar convective heat as well as mass transfer flow of an incompressible, viscous, electrically conducting fluid over an impulsively started vertical plate among conduction-radiations embedded in a porous medium in presence of transverse magnetic field has been studied. The $x'$ axis is taken the length of plate in the vertical upward direction and the $y'$ axis is taken normal to the plate. A transverse magnetic field of identical strength $B_0$ is assumed to be applied normal to the plate. It is also implicit that the thermal radiation along the plate and viscous dissipation is implicit to be negligible. The induced magnetic field and viscous dissipation is understood to be negligible. Initially it is assumed that the plate and fluid are at same temperature $T'_\infty$ in the stationary situation with concentration level $C'_\infty$ at all the points. At time, $t' > 0$ the plate is specified an impulsive motion in its own plane with velocity $u_0$. The temperature of the plate and the concentration stage are also raised to $T'_w$ and $C'_w$. They are maintained at the similar level for all time $t' > 0$. Then under the above assumption the unsteady flow with usual Boussinesq's estimate is governed by the following equations.

$$\frac{\partial u'}{\partial t'} = g\beta\left(T' - T'_\infty\right) + g\beta_c\left(C' - C'_\infty\right) + \nu\frac{\partial^2 u'}{\partial y'^2} - \left(\frac{\sigma B_0^2}{\rho} + \frac{\nu}{K'}\right)u' \tag{1}$$

$$\rho C_p \frac{\partial T'}{\partial t'} = \kappa\frac{\partial^2 T'}{\partial y'^2} - \frac{\partial q_r}{\partial y'} \tag{2}$$

$$\frac{\partial C'}{\partial t'} = D\frac{\partial^2 C'}{\partial y'^2} + D_1\frac{\partial^2 T'}{\partial y'^2} \tag{3}$$

The initial and boundary conditions are

$$\begin{aligned} &t' \le 0;\ u' = 0,\ T' = T'_\infty,\ C' = C'_\infty \ \text{ for ever } y \\ &t' > 0;\ u' = u_0,\ T' = T'_w,\ C' = C'_w \ \ at\ y = 0 \\ &t' > 0;\ u' \to 0,\ T' \to T'_w,\ C' \to C'_w\ at\ y \to 0 \end{aligned} \tag{4}$$

The radiation heat flux term is simplified by making use of the Rosseland approximation [16] as

$$q_r = \frac{-4}{3}\frac{\sigma'}{a'}\frac{\partial T'^4}{\partial y'} \tag{5}$$

Where $\sigma'$ and $a'$ are the Stefan-Boltzmann steady and the mean absorption coefficient respectively. It should be noted that by using the Roseland approximation, we limit our investigation to optically thick fluids. It temperature differences within the flow are sufficiently small, such that $T'^4$ may be expressed as a linear function of the temperature, then the Taylors series for $T'^4$ and $T'_\infty$, after neglecting higher order terms is given by

$$T'^4 \cong 4T'T'^3_\infty - 3T'^4_\infty \tag{6}$$

Substitute (5) and (6) in (2) we have

$$\rho C_p \frac{\partial T'}{\partial t'} = \left[k + \frac{16}{3}\frac{\sigma'}{a'}T'_\infty\right]\frac{\partial^2 T'}{\partial y'^2} \tag{7}$$

Let us introduce the following non-dimensional terms in (1), (7) and (3)

$$y=\frac{u_0y'}{v},u=\frac{u'}{u_0},\mathrm{Pr}=\frac{\rho vC_p}{k},Sc=\frac{v}{D},t=\frac{u_0^2t'}{v},K_r=\frac{u_0^2K'}{v^2},$$
$$\theta=\frac{T'-T'_\infty}{T'_w-T'_\infty},C=\frac{C'-C'_\infty}{C'_w-C'_\infty},M=\frac{\sigma B_0^2v}{\rho u_0^2},N_a=\frac{ka'}{4\sigma'T'^3_\infty},$$
$$Gr=\frac{vg\beta\left(T'_w-T'_\infty\right)}{u_0^3},Gm=\frac{vg\beta_c\left(C'_w-C'_\infty\right)}{u_0^3},S_0=\frac{D_1}{v}\frac{\left(T'_w-T'_\infty\right)}{\left(C'_w-C'_\infty\right)}. \tag{8}$$

Hence the non-dimensional form of (1), (2) and (3) are

$$\frac{\partial u}{\partial t}=G_r\theta+G_mC+\frac{\partial^2u}{\partial y^2}-Mu-K_r^{-1}u \tag{9}$$

$$3\,\mathrm{Pr}\,N_a\frac{\partial\theta}{\partial t}=\left(3N_a+4\right)\frac{\partial^2\theta}{\partial y^2} \tag{10}$$

$$\frac{\partial C}{\partial t}=\frac{1}{Sc}\frac{\partial^2C}{\partial y^2}+S_0\frac{\partial^2\theta}{\partial y^2} \tag{11}$$

The transformed initial and boundary conditions are

$$\begin{aligned}&t\le 0:u=0,\quad \theta=0,\quad C=0 \quad \text{for every } y\\ &t>0:u=1,\quad \theta=1,\quad C=1 \quad at \quad y=0\\ &t>0:u\to 0,\theta\to 0,C\to 0 \quad as \quad y\to\infty\end{aligned} \tag{12}$$

## Method of Solution

The equations (9) to (11) are nonlinear, coupled partial differential equations, so we want to solve them by using Laplace transform technique. Taking Laplace transform, the equations (9), (10) and (11) reduce to

$$\frac{\partial^2\bar{u}}{\partial y^2}=\left(s+M+K_r^{-1}\right)\bar{u}-G_r\bar{\theta}-G_m\bar{C} \tag{13}$$

$$\frac{\partial^2\bar{\theta}}{\partial y^2}=L_1s\bar{\theta} \tag{14}$$

$$\frac{\partial^2\bar{C}}{\partial y^2}-Scs\bar{C}=-S_0Sc\frac{\partial^2\bar{\theta}}{\partial y^2} \tag{15}$$

Where's' is the Laplace transform parameter. The boundary condition (12) reduces to the following form after applying Laplace transform.

$$\begin{aligned}&\bar{u}=\frac{1}{s},\bar{\theta}=\frac{1}{s},\bar{C}=\frac{1}{s} \quad \text{when} \quad y=0\\ &\bar{u}=0,\bar{\theta}=0,\bar{C}=0 \quad \text{when} \quad y\to\infty\end{aligned} \tag{16}$$

Solving (13), (14) and (15) with boundary condition (16) we get

$$\bar{\theta}=\frac{1}{s}e^{-y\sqrt{Ls}} \tag{17}$$

$$\bar{C}=\frac{1}{s}e^{-y\sqrt{Scs}}-\frac{L_2}{s}e^{-y\sqrt{Scs}}+\frac{L_2}{s}e^{-y\sqrt{Ls}} \tag{18}$$

$$\begin{aligned}\bar{u}=&\frac{L_{12}}{s}e^{-y\sqrt{s+L_5}}-\frac{L_8}{s-L_7}e^{-y\sqrt{s+L_5}}-\frac{L_{11}}{s-L_{10}}e^{-y\sqrt{s+L_5}}-\frac{L_8}{s}e^{-y\sqrt{sL}}+\\ &\frac{L_8}{s-L_7}e^{-y\sqrt{sL}}-\frac{L_{11}}{s}e^{-y\sqrt{sSc}}+\frac{L_{11}}{s-L_{10}}e^{-y\sqrt{sSc}}\end{aligned} \tag{19}$$

Inverting the equations (17), (18) and (19) we get

$$\theta=erfc\left(\frac{y\sqrt{L}}{2\sqrt{t}}\right) \tag{20}$$

$$C=erfc\left(\frac{y\sqrt{Sc}}{2\sqrt{t}}\right)-L_2erfc\left(\frac{y\sqrt{Sc}}{2\sqrt{t}}\right)+L_2erfc\left(\frac{y\sqrt{L}}{2\sqrt{t}}\right) \tag{21}$$

$$\begin{aligned}u=&\frac{(L_{12})}{2}\left[e^{-y\sqrt{L_5}}erfc\left(\frac{y}{2\sqrt{t}}-\sqrt{L_5t}\right)+e^{y\sqrt{L_5}}erfc\left(\frac{y}{2\sqrt{t}}+\sqrt{L_5t}\right)\right]-(L_8)\left(\frac{e^{L_7t}}{2}\right)\\ &\left[e^{-y\sqrt{L_7+L_5}}erfc\left(\frac{y}{2\sqrt{t}}-\sqrt{(L_7+L_5)t}\right)+e^{y\sqrt{L_7+L_5}}erfc\left(\frac{y}{2\sqrt{t}}+\sqrt{(L_7+L_5)t}\right)\right]-\\ &(L_{11})\left(\frac{e^{L_{10}t}}{2}\right)\left[e^{-y\sqrt{L_{10}+L_5}}erfc\left(\frac{y}{2\sqrt{t}}-\sqrt{(L_{10}+L_5)t}\right)+e^{y\sqrt{L_{10}+L_5}}erfc\left(\frac{y}{2\sqrt{t}}+\sqrt{(L_{10}+L_5)t}\right)\right]-\\ &(L_8)erfc\left(\frac{y\sqrt{L}}{2\sqrt{t}}\right)+(L_8)\left(\frac{e^{L_7t}}{2}\right)\left[e^{-y\sqrt{L}\sqrt{L_7}}erfc\left(\frac{y\sqrt{L}}{2\sqrt{t}}-\sqrt{L_7t}\right)+e^{y\sqrt{L}\sqrt{L_7}}erfc\left(\frac{y\sqrt{L}}{2\sqrt{t}}+\sqrt{L_7t}\right)\right]-\\ &(L_{11})erfc\left(\frac{y\sqrt{Sc}}{2\sqrt{t}}\right)+(L_{11})\left(\frac{e^{L_{10}t}}{2}\right)\left[e^{-y\sqrt{Sc}\sqrt{L_{10}}}erfc\left(\frac{y\sqrt{Sc}}{2\sqrt{t}}-\sqrt{L_{10}t}\right)+e^{y\sqrt{Sc}\sqrt{L_{10}}}erfc\left(\frac{y\sqrt{Sc}}{2\sqrt{t}}+\sqrt{L_{10}t}\right)\right]\end{aligned} \tag{22}$$

The Skin friction at the surface of the plate is given by

$$\begin{aligned}\tau=&-\left[\frac{\partial u}{\partial y}\right]_{y=0}=-\frac{1}{2\sqrt{t}}\left[\frac{\partial u}{\partial y}\right]_{y=0}\\ =&(L_{12})\left(\frac{1}{2}\right)\left[\sqrt{L_5}\left(-2erfc\left(\sqrt{L_5t}\right)\right)-\frac{4}{\sqrt{\pi}}e^{-L_5t}\right]-(L_8)\left(\frac{e^{L_7t}}{2}\right)\\ &\left[\sqrt{L_7+L_5}\left(-2erfc\sqrt{(L_7+L_5)t}\right)-\frac{4}{\sqrt{\pi}}e^{-(L_7+L_5)t}\right]-(L_{11})\left(\frac{e^{L_{10}t}}{2}\right)\\ &\left[\sqrt{L_{10}+L_5}\left(-2erfc\sqrt{(L_{10}+L_5)t}\right)-\frac{4}{\sqrt{\pi}}e^{-(L_{10}+L_5)t}\right]-(L_8)\left(\frac{\sqrt{L}}{\sqrt{\pi t}}\right)+\\ &(L_8)\left(\frac{e^{L_7t}}{2}\right)\left[\sqrt{LL_7}\left(-2erfc\sqrt{L_7t}\right)-\frac{4}{\sqrt{\pi}}e^{-(LL_7)t}\right]-(L_{11})\left(\frac{\sqrt{Sc}}{\sqrt{\pi t}}\right)+\\ &(L_{11})\left(\frac{e^{L_{10}t}}{2}\right)\left[\sqrt{ScL_{10}}\left(-2erfc\sqrt{L_{10}t}\right)-\frac{4}{\sqrt{\pi}}e^{-(ScL_{10})t}\right]\end{aligned} \tag{23}$$

The Nusselt number and Sherwood number at the plate are respectively

$$N_u=-\left(\frac{\partial\theta}{\partial y}\right)_{y=0}=\frac{\sqrt{L}}{\sqrt{\pi t}} \tag{24}$$

$$\text{and } S_h=-\left(\frac{\partial C}{\partial y}\right)_{y=0}=\frac{\sqrt{Sc}}{\sqrt{\pi t}}-L_2\frac{\sqrt{Sc}}{\sqrt{\pi t}}+L_2\frac{\sqrt{L}}{\sqrt{\pi t}} \tag{25}$$

## Result and Discussion

To discuss the physical implication of various parameters involved in the results (20) - (25), the numerical calculation has been carried out for the distributions of velocity, temperature, concentration, Skin friction, Nusselt number and Sherwood number. The effects of various physical parameters on these flow quantities such as Hartmann number M, Prandtl number Pr, Soret number $S_0$, Schmidt number Sc, Permeability parameter $K_r$, Grashof number Gr, modified Grashof number Gm and Radiation Parameter Na are studied though graphs. The concentration profiles are plotted in Figure 1 for various values of Schmidt number Sc. From this figure, it is noticed that the concentration decreases with an increase in the values of the Schmidt number Sc. A comparison of curves in the figure shows a decrease in concentration with an increase in Schmidt number Sc. Actually it is true, since the increase of Sc means decrease of molecular diffusivity and therefore decreases in concentration boundary layer. The effects of increasing the Soret number $S_0$ on the species concentration profiles have been shown in Figure 2. From this figure, it is noticed that an increase in Soret number $S_0$ results an increase in the concentration

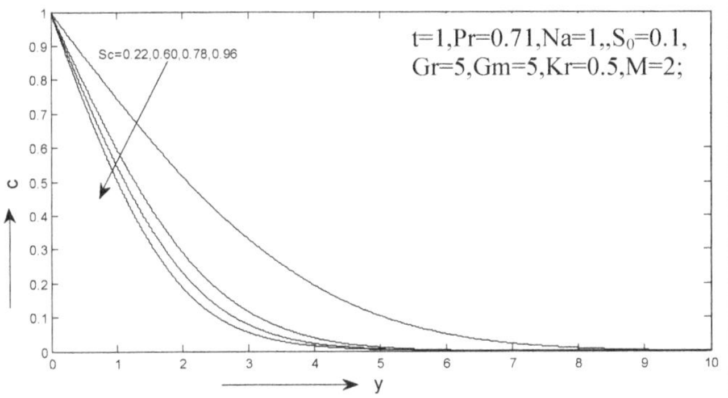


**Figure 1:** Effect of Schmidt number on Concentration.

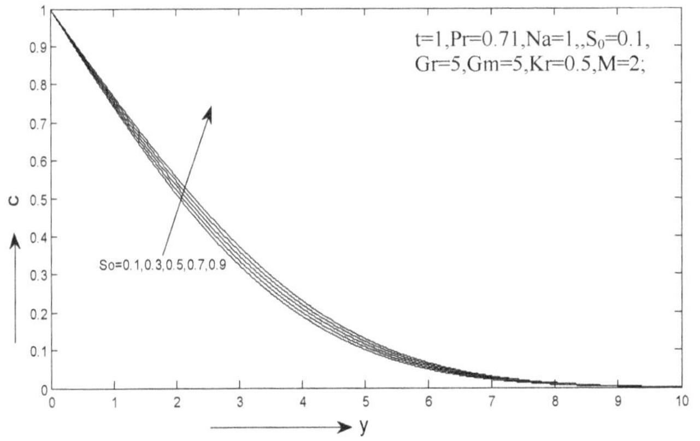


**Figure 2:** Effect of Soret number on Concentration.

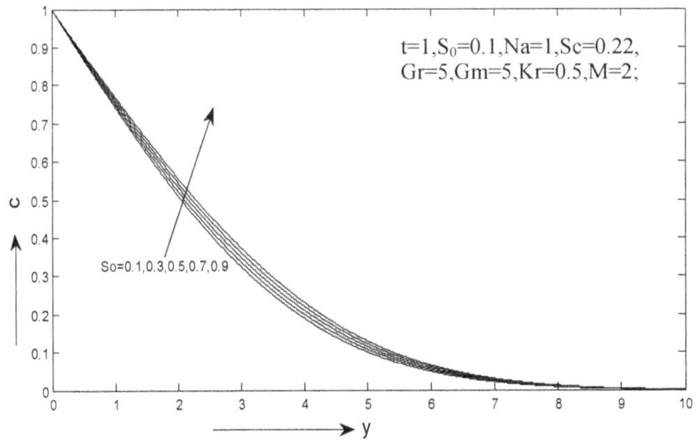


**Figure 3:** Effect of Prandtl number on Temperature.

profiles. Figure 3 revels the temperature profiles for different values of Prandtl number Pr. It is observed that the temperature decrease as an increase in the values of Prandtl number Pr. The reason is that smaller values of Prandtl number are equivalent to increase in the thermal conductivity of the fluid and therefore heat is able to diffuse away from the heated surface extra rapidly for higher values of Pr (Appendix). Hence, in the case of larger Prandtl number the thermal boundary layer is thinner and the rate of heat transfer is reduced. Figure 4 shows the temperature profile for different values of Radiation Parameter Na. From this figure it is noticed that an increase in the values of Na results a decrease in the temperature profiles. The effect of Grashof number Gr on velocity is presented in Figure 5. It is observed that an increase in Gr leads to a rise in the velocity boundary layer. Figure 6 shows the velocity profile for different values of modified Grashof number. From this figure it is observed that an increase in the values of modified Grashof number Gm results in increase in the velocity profiles. Figure 7 shows the velocity profiles for different values of radiation parameter Na. From this figure it is notice that velocity decreases with increase in Na. Figure 8 revels the effect of Prandtl number Pr on the velocity

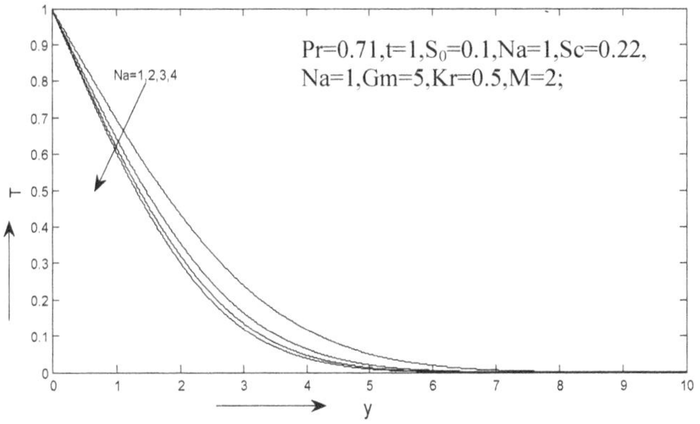


**Figure 4:** Effect of Radiation parameter on Temperature.

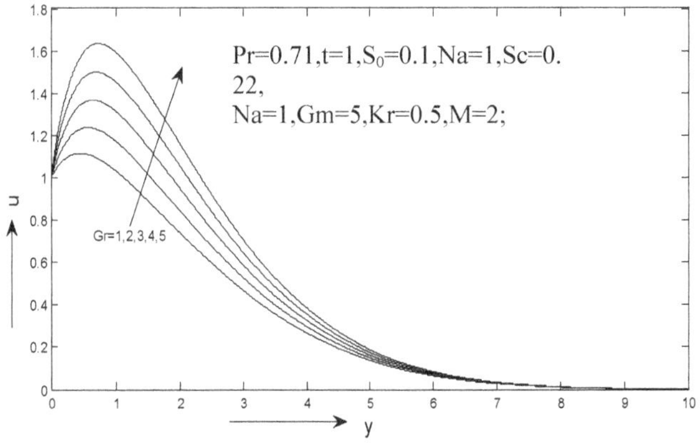


**Figure 5:** Effect of Grashof number on Velocity.

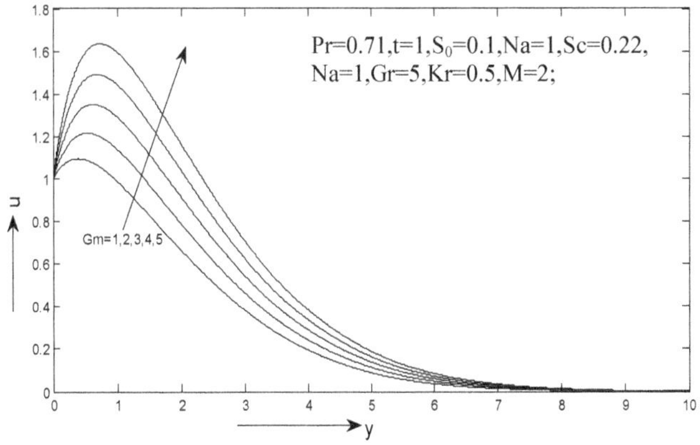


**Figure 6:** Effect of modified Grashof number on Velocity.

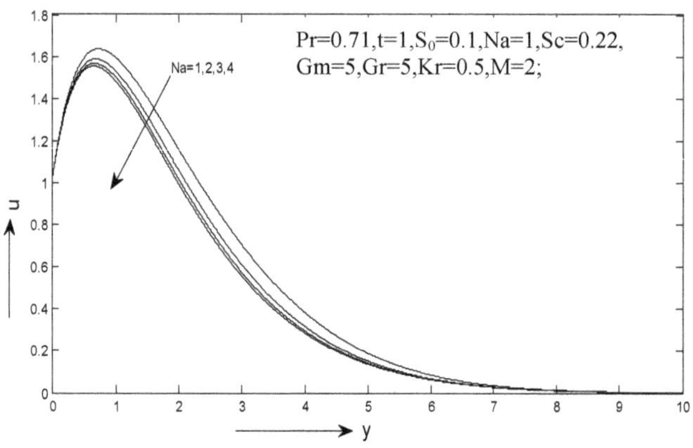


**Figure 7:** Effect of Radiation parameter on Velocity.

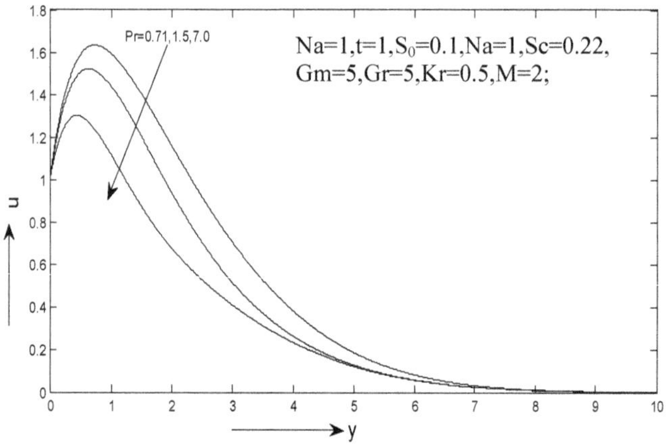


**Figure 8:** Effect of Prandtl number on Velocity.

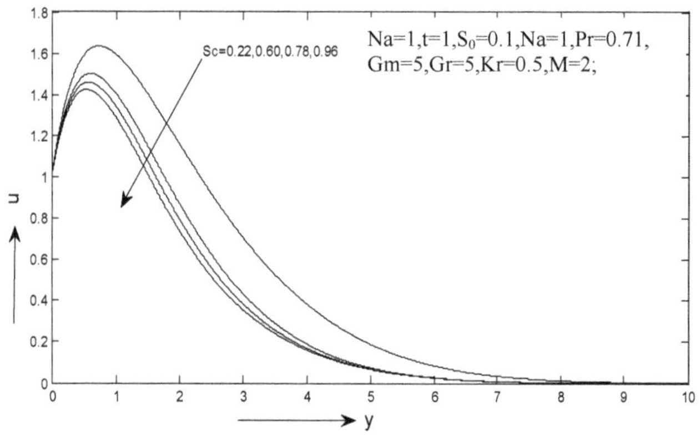


**Figure 9:** Effect of Schmidt number on Velocity.

profile. It is evident from the figure that the velocity decreases with an increase in Pr. Figure 9 illustrates the velocity profiles for different values of Schmidt number Sc. It has been observed that the velocity decreases with increase in Sc. Figure 10 shows the velocity profiles for different values of Soret number $S_0$. It was found that an increase in the value of $S_0$ leads to an increase in the velocity distribution across the boundary layer. This is true, as the $S_0$ increases, small light molecules and large heavy molecules get separated under a temperature gradient, which intern increases the velocity of a fluid. Figure 11 illustrates the velocity profiles for different values of Hartmann number M. From this figure it is notice that velocity decrease with an increase in Hartmann number M. Figure 12 reveals the effect of time t on the transient velocity profiles. It is evident from the figure that the velocity decreases with increase in t. The velocity profiles are plotted in Figure 13 for various values of permeability parameter $K_r$ From this figure, it is noticed that the velocity increases with the increase in the values of the permeability

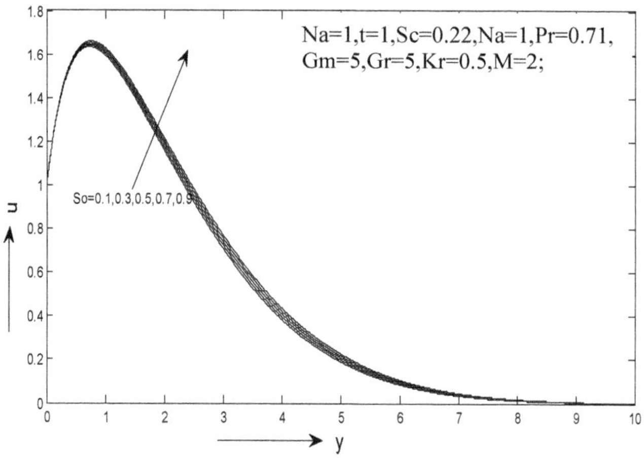


**Figure 10:** Effect of Soret number on Velocity.

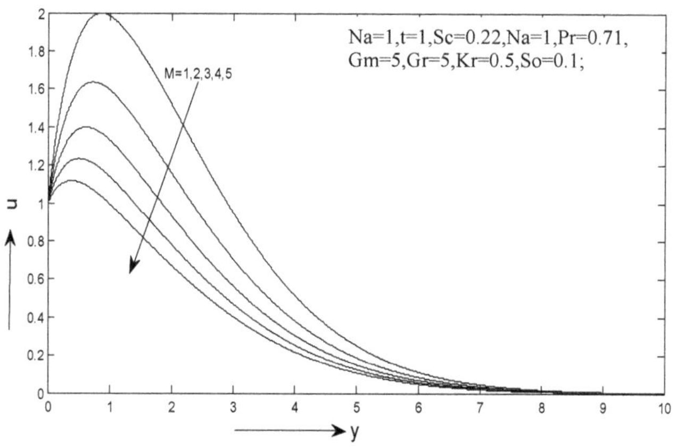


**Figure 11:** Effect of Hartmann number on Velocity.

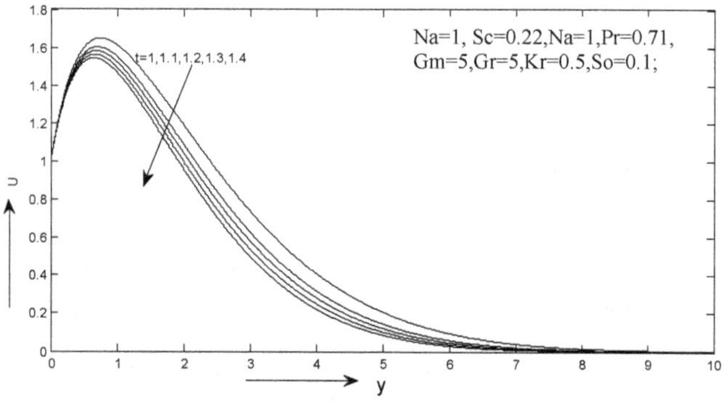


**Figure 12:** Effect of time on Velocity.

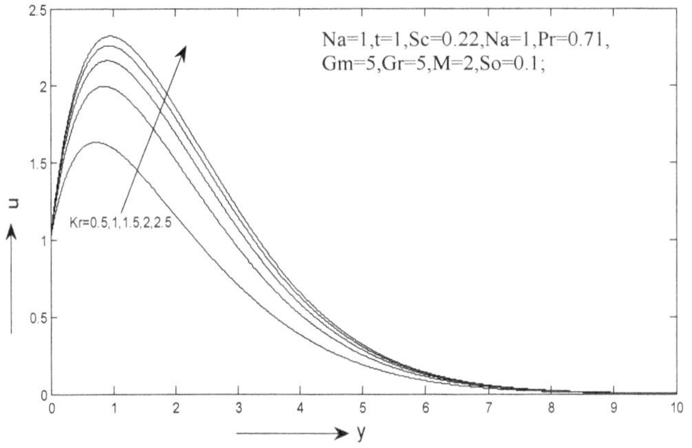


**Figure 13:** Effect of Permeability parameter on Velocity.

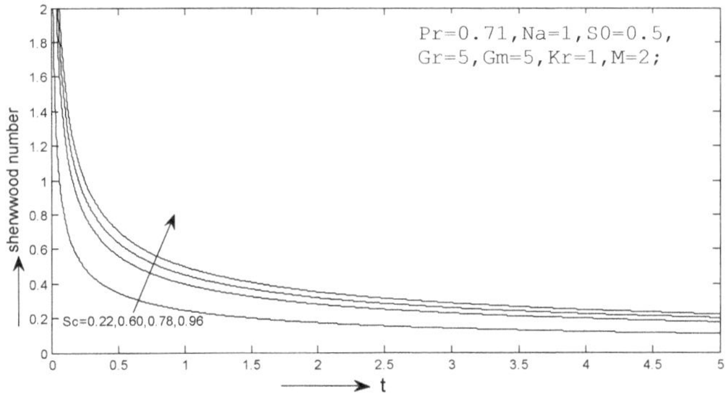


**Figure 14:** Effect of Schmidt number on Sherwood number.

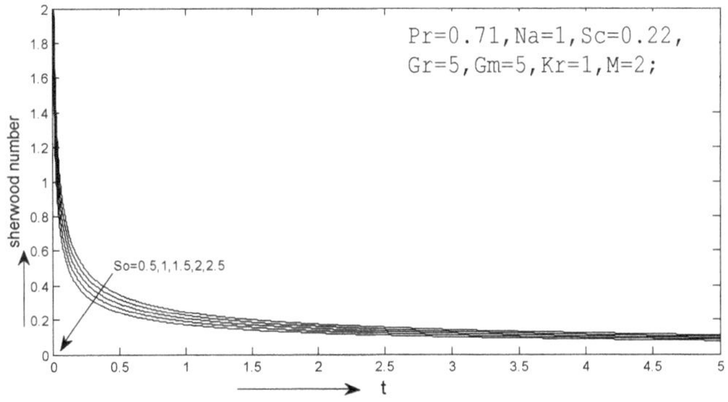


**Figure 15:** Effect of Soret number on Sherwood number.

parameter $K_r$. Physically, an increase in the permeability of porous medium leads to rise, in the flow of fluid during it. When the holes of the porous medium become large, the resistance of the medium may be neglected. A similar approach is noticed with Raju et al. [25]. Figure 14 shows Sherwood number is presented against time t for different values of Schmidt numbers. We observed that Sherwood number increases with increasing Schmidt numbers Sc. Figure 15 shows the Sherwood number (Sh) on the porous plate for different values of Soret number $S_0$. The result display that an increase in the value of $S_0$ results an decrease in the Sherwood number. Figure 16 presents the variation of the Nusselt number Nu against time t for various values of Prandtl number Pr. From this figure we notice that Nusselt number increases when the values of Prandtl number Pr increase. Figure 17 illustrates the Nusselt number for different values of radiation parameter Na. From this figure it is noticed that Nusselt number increases with increase in Na. Figures 18 and 19 depicts skin-friction against time t for different values of Grashof number Gr and modified Grashof number Gm. From these figures it is notice that Skin-friction increases with an increase Gr and Gm. Figure 20 depict skin-friction against time t for different values of Hartmann number M. It is observed that the skin friction decreases with increase in Hartmann number M. Figures 21-23 depicts skin-friction against time t for different values of Permeability parameter

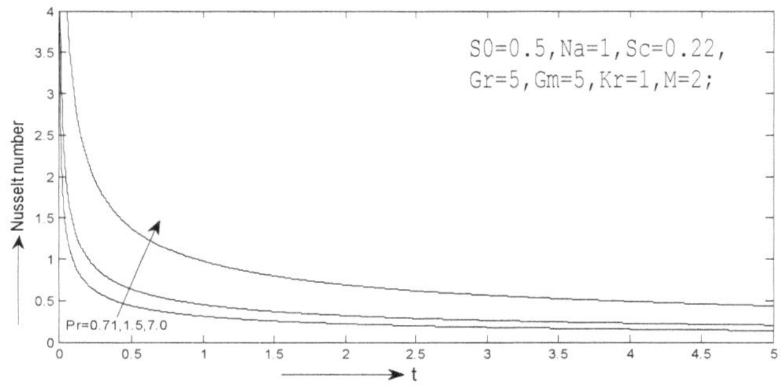


**Figure 16:** Effect of Prandtl number on Nusselt number.

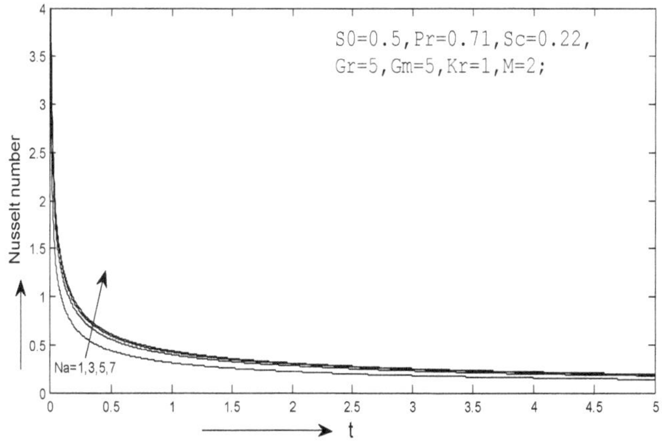


**Figure 17:** Effect of Radiation parameter on Nusselt number.

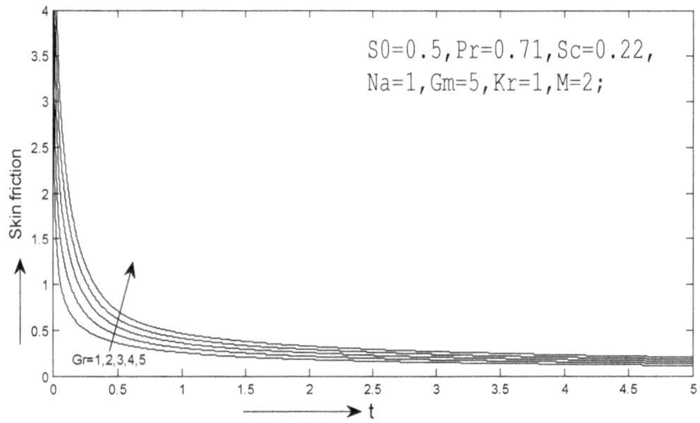


**Figure 18:** Effect of Grashof number on Skin friction.

$K_r$, Schmidt number Sc and Prandtl number Pr. From these figures it notice that skin-friction increases with an increases Kr , Sc and Pr.

## Conclusion

In this paper a theoretical examination has been carried out to study thermal diffusion on MHD free convective radiating flow more an impulsively started vertical plate embedded in a porous medium. Solutions for the model has been derived by using Laplace transform technique. Some conclusions of the study are as follow:

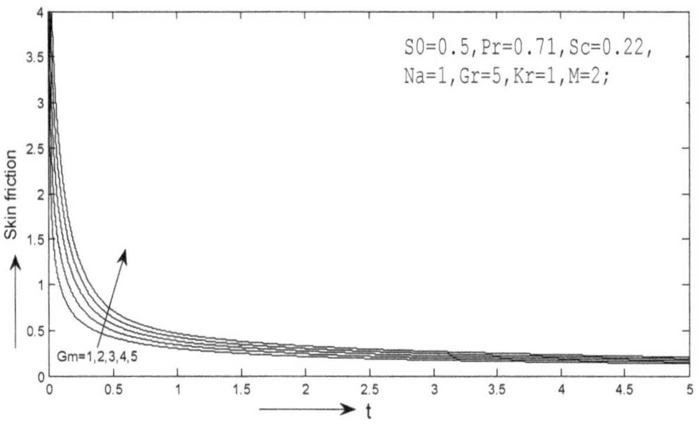


**Figure 19:** Effect of modified Grashof number on Skin friction.

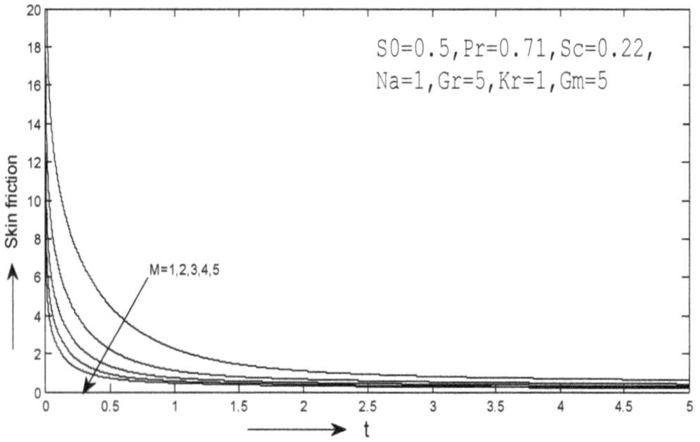


**Figure 20:** Effect of Hartmann number on Skin friction.

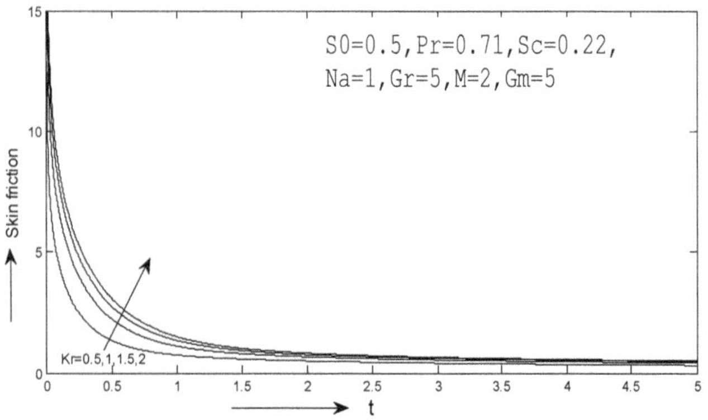


**Figure 21:** Effect of permeability parameter on Skin friction.

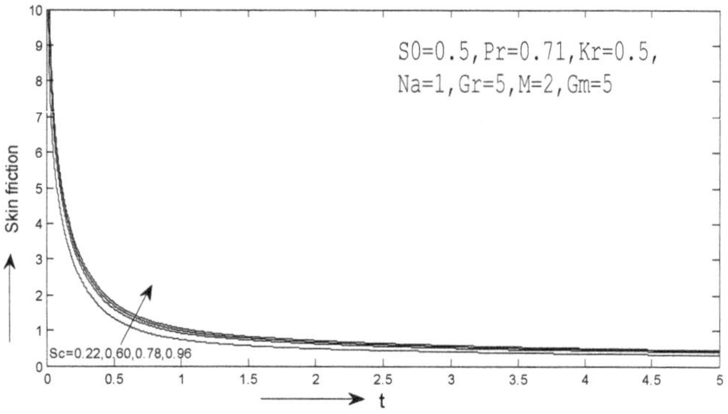


**Figure 22:** Effect of Schmidt number on Skin friction.

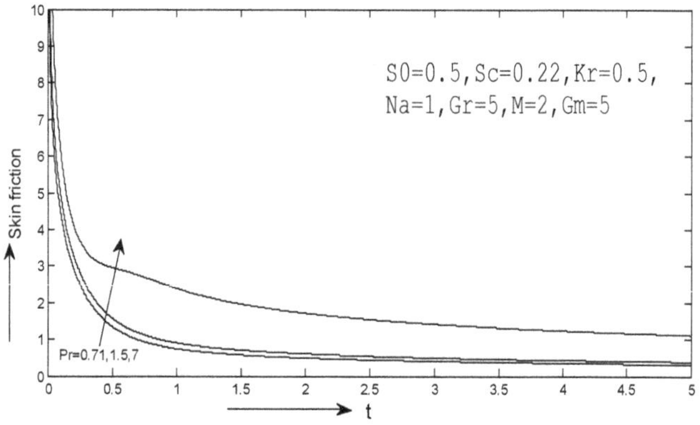


**Figure 23:** Effect of Prandtl number on Skin friction.

- Concentration distributed is observed to decrease with increase in Schmidt number and it increases with increase in Soret number.
- Temperature decreases with increase in Pr and Na.
- Velocity increases with increase in Gr, Gm, $S_0$ and Kr while it decreases with increase in Na, Pr, Sc, M and t.
- Sherwood number increase with increase in Sc and decrease with increase $S_0$.
- Nusselt number increases with increase in Pr and Na.
- Skin-friction increases with an increase in Gr, Gm, Kr, Sc and Pr and decreases with increase in M.

### References

1. Bharat KS, Nityananda S (2015) The Effects of mass transfer on MHD free convective radiation flow over an impulsively started vertical plate embedded in a porous medium.Journal of Applied analysis and Computation 5: 18-27.
2. Ahmed S, Batin A (2013) Convective laminar radiating flow over an accelerated vertical plate embedded in a porous medium with an external magnetic field. IJET 3: 66-72.
3. Chamka AJ (2013) Thermal radiation and buoyancy effects on hydro magnetic flow over an accelerating permeable surface with heat source or sink. IJHMT 38: 1699-1712.
4. Ahmed A, Kalith K, Zueco J (2014) Non-linear magneto hydrodynamic flow over an impulsively started vertical plate in a saturated porous regime Laplace and Numerical approach. J of Engg Physics and Thermo physics 87: 1169-1182.
5. Ravikumar V, Raju MC, Raju GSS Chamkha AJ (2013) MHD double diffusive

and chemically reactive flow through porous medium bounded by two vertical plates. International Journal of Energy & Technology 5: 01-08.

6. Palani G, Abbas IA (2009) FreeConvection MHD flow with thermal radiation from an impulsively-started vertical plate. Nonlinear Analysis: Modelling and Control 14: 73-84.

7. kumar RV, Raju MC ,Raju GSS (2012) Heat and mass transfere on MHD flow of viscous fluid through non-homogeneous porous medium in presence of temperature dependent heat source. International Journal of Contemporary Mathematical sciences 7: 1597-1604.

8. Chen CH (2004) Heat and mass transfer in MHD flow by natural convection from a permeable, inclined surface with variable wall temperature and concentration. Act Mechanica 172: 219-235.

9. Ravikumar V, Raju MC, Raju GSS (2014) Combined effects of heat absorption and MHD on convective Rivlin-Ericksen flow past a semi-infinite vertical porous plate with variable temperature and suction. Ain Shams Engineering Journal 5: 867-875

10. Ahamed S, Abdul B, Chamkha AJ (2015) Numerical/Laplace transform analysis for MHD radiating heat/mass transport in a Darcian porous regime bounded by an oscillating vertical surface. Alexandria Engineering Journal 54: 45-54.

11. Kumar AGV, Varma SVK (2011) Thermal radiation and mass transfer effects on MHD flow past a vertical oscillating plate with variable temperature effects variable mass diffusion. Int J Eng 3: 493-499.

12. Hossain MA, Takhar HS (1996) Radiation effect on mixed convection along a vertical plate with uniform surface temperature. Heat Mass Transfer 31: 243-248.

13. Ibrahim SM, Reddy NB (2013) Similarity solution of heat and mass transfer for natural convection over a moving vertical plate with internal heat generation and a convective boundary condition in the presence of thermal radiation, viscous dissipation, and chemical reaction. ISRN Thermodyn 5: 01-10.

14. Kumar PP, Trilochan B (2015) Analytical solution of MHD free convective flow through porous media with time dependent temperature and concentration. Walailak J Sci & Tech 12: 749-762.

15. Das SS, Satapathy A, Das JK, Panda JP (2009) Mass transfer effects on MHD flow and heat transfer past a vertical porous plate through a porous medium under oscillatory suction and heat source. Int J Heat Mass Tran 52: 5962-5969.

16. Siegel R, Howell JR (2002) Thermal radiation heat transfer. Taylor and Francis Group.

17. Seth G, Nandkeolyar S, Ansari MS (2013)Effects of thermal radiation and rotation on unsteady hydro magnetic free convection flow past an impulsively moving vertical plate with ramped temperature in a porous medium. Journal of Applied Fluid Mechanics 6: 27-38.

18. Das K, Jana S (2010) Heat and mass transfer effects on unsteady MHD free convection flow near a moving vertical plate in porous medium. Bull Soc Math Banja Luka 17: 15-32.S

19. Ravikumar V, Raju MC, Raju GSS, Varma SVK (2013) Magnetic field effect on transient free convection flow through porous medium past an impulsively started vertical plate with fluctuating temperature and mass diffusion. International Journal of Mathematical Archive 4: 198-206.

20. Mamtha B, Raju MC, Varma SVK (2015)Thermal diffusion effect on MHD mixed convection unsteady flow of a micro polar fluid past a semi-infinite vertical porous plate with radiation and mass transfer. International Journal of Engineering research in Africa 13: 21-37.

21. Harinath Reddy S, Raju MC, Reddy K (2015) Unsteady MHD free convection flow of a kuvshinski fluid past a vertical porous plate in the presence of chemical reaction and heat source/sink. International Journal of Engineering Research in Africa 14: 13-27.

22. Ravikumar V, Raju MC, Raju GSS (2015) Theoretical investigation of an unsteady MHD free convection heat and mass transfer flow of a non-Newtonian fluid flow past a permeable moving vertical plate in the presence of thermal diffusion and heat sink. International Journal of Engineering Research in Africa 16: 90-109.

23. Reddy GM, Reddy BN (2011) Mass transfer and heat generation effects on MHD free convection flow past an incline vertical surface in a porous medium. J of A FM 4: 07-11.

24. Senapati N, Dhal RK (2011) Magnetic effects on mass and heat transfer of hydrodynamics flow past an oscillating vertical plate in presence of chemical reaction AMSE B-2: 60-66.

25. Raju KVS, Reddy TS, Raju MC, Satya Narayana PV ,Venkataramana S (2013) MHD convective flow through porous medium in a horizontal channel with insulated and impermeable bottom wall in the presence of viscous dissipation and joule heating Ain Shams Eng J 5: 543-551.

# 13

# Numerical Method for One-Dimensional Convection-diffusion Equation Using Radical Basis Functions

**Su LD[1,2]*, Jiang ZW[2] and Jiang TS[2]***

*[1]North-Eastern Federal University, Belinskogo, Yakutsk, Russia*
*[2]Department of Mathematics, Linyi University, Linyi, P.R. China*

**Abstract**

In this paper, the meshless method is employed for the numerical solution of the one-dimensional (1D) convection-diffusion equation based on radical basis functions (RBFs). Coupled with the time discretization and the collocation method, the proposed method is a truly meshless method which requires neither domain nor boundary discretization. The algorithm is very simple so it is very easy to implement. The results of numerical experiments are presented, and are compared with analytical solutions to confirm the good accuracy of the presented scheme.

**Keywords:** Meshless method; Radical basis function (RBF); Numerical solution; Convection-diffusion equation

## Introduction

Whenever we consider mass transport of a dissolved species (solute species) or a component in a gas mixture, concentration gradients will cause diffusion. If there is bulk fluid motion, convection will also contribute to the flux of chemical species. Therefore, we are often interested in solving for the combined effect of both convection and diffusion.

The convection-diffusion equation is a combination of the diffusion and convection (advection) equations, and describes physical phenomena where particles, energy, or other physical quantities are transferred inside a physical system due to two processes: diffusion and convection [1].

The general convection-diffusion equation has the following form [2,3]:

$$\frac{\partial u(\mathrm{x},t)}{\partial t} = \nabla\cdot(D\nabla u(\mathrm{x},t)) - \nabla(vu(\mathrm{x},t)) + R(\mathrm{x},t).$$

In the above equation, four terms represents transient, convection, diffusion and source term respectively. Where *u(x, t)* is the variable of interest (species concentration for mass transfer, temperature for heat transfer), D is the diffusivity (also called diffusion coefficient), $V$ is the average velocity that the quantity is moving, *R(x, t)* is source term represents capacity of internal sources, $\nabla$ represents gradient and $\nabla\cdot$ represents divergence.

This paper is devoted to the numerical computation of the one-dimension

(1D) convection-diffusion equation:

$$u_t(x,t) + \alpha u_x(x,t) + \beta u(x,t) = \varepsilon u_{xx} + f(x,t),\ a \le x \le b,\ 0 \le t, \tag{1.1}$$

With the initial conditions:

$$u(x,0) = h(x), a \le x \le b, , \tag{1.2}$$

And Dirichlet boundary conditions:

$$u(a,t) = g_0(t), u(b,t) = g_1(t), 0 \le t \tag{1.3}$$

Where $\alpha$, $\beta$ and $\varepsilon$ are known constant coefficients, $h(x)$ and $g_i(t)$ $(i=0,1)$ are known continuous functions.

Recently, much attention has been given to the development, analysis, and implementation of stable methods for the numerical solution of the convection-diffusion equations (see [4] and the reference therein). Jim Douglas, et al. [5] combine definite element and finite difference methods based on the method of characteristic for solving the convection-diffusion problems. Chen and Hon [6] consider the 2D and 3D Helmholtz and convection-diffusion equation using boundary knot method. The meshless local Petro-Galerkin method for convection-diffusion equation was considered in [7]. A new finite difference method described by Ram P. Manohar and John W. Stephenson [8].

In this article, we present a numerical scheme to solve the convection-diffusion equation using the collocation method with Radial Basis Function (RBF). The results of numerical experiments are presented, and are compared with analytical solutions to confirm the good accuracy of the presented scheme.

In last 25 years, the radial basis functions (RBFs) method is known as a powerful tool for scattered data interpolation problem. The use of RBFs as a meshless procedure for numerical solution of partial differential equations is based on the collocation scheme. Because of the collection technique, this method does not need to evaluate any integral. The main advantage of numerical procedures which use RBFs over traditional techniques is mesh-less property of these methods. RBFs are used actively for solving partial differential equations. The examples see [9-11]. In the last decade, the development of the RBFs as a truly meshless method for approximating the solutions of PDEs has drawn the attention of many researchers in science and engineering [12-14]. Meshless method has become an important numerical computation method, and there are many academic monographs are published [15-17].

The layout of the article is as follows: In section 2, we introduce the collocation method and apply this method on the convection-diffusion

***Corresponding author:** Ling-De Su, Department of Mathematics, North-Eastern Federal University, Belinskogo, Yakutsk, Russia
E-mail: sulingde@gmail.com

equation. The results of numerical experiments are presented in section 3. Section 4 is dedicated to a brief conclusion. Finally, some references are introduced at the end.

## The Collocation Method with Radical Basis Function

### Radial basis function approximation

The approximation of a distribution *u(x)*, using RBF may be written as a linear combination of N radial functions, usually it takes the following form:

$$u(x) \approx \sum_{j=1}^{N} \lambda_j \phi(x, x_j) + \psi(x), \text{ for } x \in \Omega \subseteq R^d \tag{2.1}$$

Where $N$ is the number of data points, $x=(x_1, x_2 \ldots x_d)$, $d$ is the dimension of the problem, the $\lambda$'s are coefficients to be determined and $\varphi$ is the radial basis function. Eq. (2.1.1) can be written without the polynomial $\psi$. In that case, $\varphi$ must be unconditional positive definite to guarantee the solvability of the resulting system (e. g. Gaussian or Inverse Multi quadrics). However, $\psi$ is usually required when $\varphi$ is conditionally positive definite, i. e, when $\varphi$ has a polynomial growth towards infinity. We will use the Multi quadrics (MQ), which defined as:

$$\text{MQ: } \phi(x, x_j) = \phi(r_j) = \sqrt{r_j^2 + c^2},\ c > 0 \tag{2.2}$$

Where $r_j = \|x - x_j\|$ is the Euclidean norm. Since $\varphi$ given by (2.2) is $C^\infty$ continuous, we can use it directly.

If $P_q^d$ denotes the space of $d$-variate polynomial of order not exceeding than $q$, and letting the polynomials $(P_1, P_2, \ldots, P_m)$ be the basis of $P_q^d$ in $R^d$, then the polynomial $\psi(x)$ in Eq. (2.1) is usually written in the following form:

$$\psi(x) = \sum_{i=1}^{m} \xi_i P_i(x_j), \tag{2.3}$$

Where $m = (q-1+d)!/(d!(q-1)!)$. To get the coefficients $x=(\lambda_1, \lambda_2 \ldots \lambda_N)$, and $(\xi_1, \xi_2, \cdots, \xi_m)$, the collocation method is used. However, in addition to the $N$ equations resulting from collecting Eq. (2.1.1) at $N$ points and extra $m$ equations are required. This is ensured by the $m$ conditions for Eq. (2.1),

$$\sum_{j=1}^{N} \lambda_j P_i(x_j) = 0,\ i = 1, 2, \cdots m. \tag{2.4}$$

In a similar representation as Eq. (2.1), for any linear partial differential operator $\ell$, $\ell u$ can be approximated by:

$$\ell u(x) \approx \sum_{j=1}^{N} \lambda_j \ell \phi(x, x_j) + \ell \psi(x). \tag{2.5}$$

### The convection-diffusion equation

Let us consider the 1D convection-diffusion equation Eq. (1.1), with the initial conditions Eq. (1.2) and the Dirichlet boundary conditions Eq. (1.3).

First, let us discretize Eq. (1.1) according to the following $\theta$-weighted scheme:

$$\frac{u(x,t+\tau) - u(x,t)}{\tau} + \alpha \cdot \nabla u(x,t) = \theta\left(\varepsilon \cdot \nabla^2 u(x,t+\tau) - \beta \cdot u(x,t+\tau)\right) + (1-\theta)\left(\varepsilon \cdot \nabla^2 u(x,t) - \beta \cdot u(x,t)\right) + f(x,t), \tag{2.6}$$

where $0 \le \theta \le 1$, and $\tau$ is the time step size, and $\nabla u = \frac{\partial u}{\partial x}$, using the notation $u^n = u(x, t^n)$ where $t^n = t^{n-1} + \tau$, we get:

$$\begin{aligned}&(1 + \theta \cdot \beta \cdot \tau) \cdot u^{n+1} - \theta \cdot \tau \cdot \varepsilon \cdot \nabla^2 u^{n+1} = \\ &(1 - (1-\theta) \cdot \beta \cdot \tau) \cdot u^n + (1-\theta) \cdot \tau \cdot \varepsilon \cdot \nabla^2 u^n - \alpha \cdot \tau \cdot \nabla u^n + \tau \cdot f^n\end{aligned}. \tag{2.7}$$

Assuming that there are $N-2$ interpolation points, $u^n(x)$ can be approximated by:

$$u^n(x) \approx \sum_{j=1}^{N-2} \lambda_j^n \phi(r_j),\ j = 1, 2, \cdots, N-2. \tag{2.8}$$

To guarantee the positive definition, here we use the following approximation:

$$u^n(x_i) \approx \sum_{j=1}^{N-2} \lambda_j^n \phi(r_{ij}) + \lambda_{N-1}^n x_i + \lambda_N^n, \tag{2.9}$$

Where $r_j = \|x - x_j\|$ is the Euclidean norm. The additional conditions due to Eq. (2.4) are written as:

$$\sum_{j=1}^{N-2} \lambda_j^n = \sum_{j=1}^{N-2} \lambda_j^n x_j = 0. \tag{2.10}$$

Writing Eq. (2.9) together with Eq. (2.10) in a matrix form we have:

$$[u]^n = A[\lambda]^n, \tag{2.11}$$

Where $[u]^n = [u_1^n \ u_2^n \ \cdots \ u_{N-2}^n \ 0 \ 0]^T$, $[\lambda]^n = [\lambda_1^n \ \lambda_2^n \ \cdots \ \lambda_N^n]$ and $A = [a_{ij},\ 1 \le i, j \le N]$ is given by:

$$A = \begin{pmatrix} \phi_{11} & \cdots & \phi_{1(N-2)} & x_1 & 1 \\ \vdots & \ddots & \vdots & \vdots & \vdots \\ \phi_{(N-2)1} & \cdots & \phi_{(N-2)(N-2)} & x_{N-2} & 1 \\ x_1 & \cdots & x_{N-2} & 0 & 0 \\ 1 & \cdots & 1 & 0 & 0 \end{pmatrix}. \tag{2.12}$$

Assuming that there are $p < N-2$ internal points and $N-2-p$ boundary points, then the $N \times N$ matrix $A$ can be split into: $A = A_d + A_b + A_e$, where

$$\begin{aligned} A_d &= [a_{ij} \ \ for \ \ (1 \le i \le p, 1 \le j \le N) \ \ and \ \ 0 \ \ elsewhere] \\ A_b &= [a_{ij} \ \ for \ \ (p+1 \le i \le N-2, 1 \le j \le N) \ \ and \ \ 0 \ \ elsewhere]. \\ A_e &= [a_{ij} \ \ for \ \ (N-1 \le i \le N, 1 \le j \le N) \ \ and \ \ 0 \ \ elsewhere] \end{aligned} \tag{2.13}$$

Using the notation $\ell A$ to designate the matrix of the same dimension as $A$ and containing the elements $\hat{a}_{ij}$ where $\hat{a}_{ij} = \ell a_{ij}$, $1 \le i, j \le N$, then Eq. (2.2.1) together with the boundary conditions Eq. (1.3) can be written in matrix form as:

$$B\lambda^{n+1} = C\lambda^n + \tau \cdot f^n + [G]^{n+1}, \tag{2.14}$$

where

$$\begin{aligned} C &= (1 - \beta(1-\theta)\tau) A_d + (1-\theta)\tau\varepsilon\nabla^2 A_d - \alpha\tau\nabla A_d, \\ B &= (1 + \beta\theta\tau) A_d - \theta\tau\varepsilon\nabla^2 A_d + A_b + A_e, \\ [G]^{n+1} &= [0 \ \cdots \ 0 \ g_{p+1}^{n+1} \ \cdots \ g_{N-2}^{n+1} \ 0 \ 0]^T, \\ [f]^n &= [f_1^n \ \cdots \ f_p^n \ 0 \ \cdots \ 0]^T. \end{aligned} \tag{2.15}$$

Eq. (2.14) is obtained by combining Eq. (2.6), which applies to the domain points, while Eq. (1.3) applies to the boundary points. Together with the initial condition Eq. (1.2) and Eq. (2.14), we can get all $\lambda$'s, thus we can get the numerical solutions.

Since the coefficient matrix is unchanged in time steps, we use the $LU$ factorization to the coefficient matrix only once and use this factorization in our algorithm.

**Remark:** Although Eq. (2.14) is valid for any value of $\theta \in [0,1]$, we will use $\theta = \frac{1}{2}$ (The famous Crank-Nicolson scheme).

## Numerical Examples

In this section, we present several numerical results to confirm the efficiency of our algorithm for solving the 1D convection-diffusion equation.

### Example 1

In this example, we consider the convection-diffusion Eq. (1.1) in $[0,1]$ with $\alpha = 0.1$, $\beta = 0$, $\varepsilon = 0.01$, with the boundary conditions:

$$u(0,t) = \exp(0.11t), \quad u(1,t) = \exp(-1+0.11t), \quad t > 0,$$

And the initial condition

$$u(x,0) = \exp(-x) \quad 0 \le x,y \le 1,$$

Then the analytical solution of the equation is $u(x,t) = \exp(-x + 0.11t)$. The right side functions $f(x,t) = 0$.

We use $MQ$ radical basis function for the computation, the $L_\infty$, $L_2$ and RMS errors and Root-Mean-Square (RMS) of errors are obtained in Table 1 for t = 0.1, 0.3, 0.5, 0.7 and 1.0 with time steps $\tau = 0.001$ and $dx = 0.001$.

The space-time graph of analytical and numerical solutions for t=1 are given in Figure 1. Note that we cannot distinguish the exact solution from the estimated solution in Figure 1.

| T | 0.1 | 0.3 | 0.5 | 0.7 | 1.0 |
|---|---|---|---|---|---|
| $L_\infty$ -errors | $1.547\times10^{-5}$ | $2.371\times10^{-4}$ | $3.599\times10^{-4}$ | $4.639\times10^{-4}$ | $5.909\times10^{-4}$ |
| $L_2$ -errors | $6.788\times10^{-5}$ | $1.787\times10^{-4}$ | $3.453\times10^{-4}$ | $3.453\times10^{-4}$ | $4.372\times10^{-4}$ |
| RMS- errors | $6.778\times10^{-5}$ | $1.784\times10^{-4}$ | $3.448\times10^{-4}$ | $3.448\times10^{-4}$ | $4.365\times10^{-4}$ |

**Table 1:** For T=0.1, 0.3, 0.5, 0.7 and 1.0 with time steps T=0.001 and *dx*=0.001.

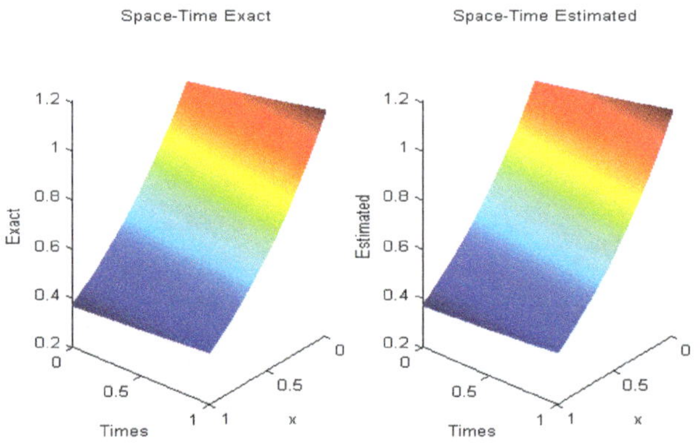


**Figure 1:** Space-time graph of the exact and estimated solutions with T=1 for Example 1.

| T | 0.1 | 0.25 | 0.5 | 0.75 | 1.0 |
|---|---|---|---|---|---|
| $L_\infty$ -errors | $8.855\times10^{-5}$ | $1.031\times10^{-4}$ | $1.245\times10^{-4}$ | $1.501\times10^{-4}$ | $1.811\times10^{-4}$ |
| $L_2$ -errors | $5.929\times10^{-5}$ | $6.882\times10^{-5}$ | $8.305\times10^{-5}$ | $1.002\times10^{-4}$ | $1.208\times10^{-4}$ |
| RMS- errors | $5.842\times10^{-5}$ | $6.781\times10^{-5}$ | $8.183\times10^{-5}$ | $9.870\times10^{-5}$ | $1.191\times10^{-4}$ |

**Table 2:** Numerical Errors at different times with T=0.001 and *dx*=0.001 for Example 2.

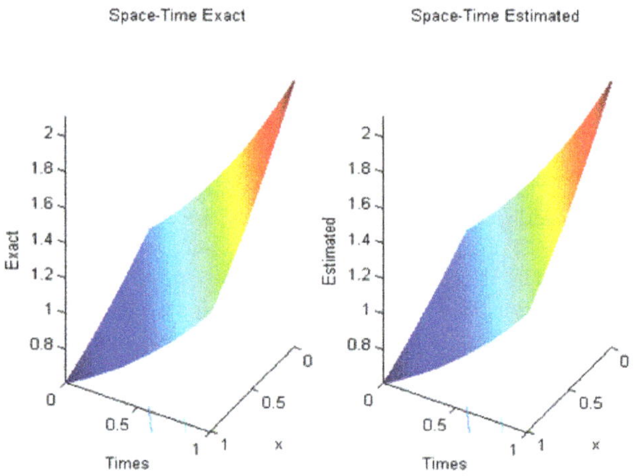


**Figure 2**: The space-time graph of exact and numerical solutions with T=1 for Example 2.

### Example 2

In this example, we consider Eq. (1.1) with $\alpha = 1$, $\beta = 0$, $\varepsilon = 1$ and the boundary conditions:

$$u(0,t) = \exp(\frac{3t}{4}), \; u(1,t) = \exp\left(\frac{-2+3t}{4}\right),$$

And the analytical solution of the equation is given as:

$$u(x,t) = \exp\left(\frac{-2x+3t}{4}\right),$$

We get the initial conditions from the exact solution. The right side functions $f(x,t) = 0$.

The $L_\infty$, $L_2$ and RMS errors and Root-Mean-Square (RMS) of errors are obtained in Table 2 for T=0.1, 0.25, 0.5, 0.75 and 1.0 with time steps $\tau = 0.001$ and $dx$ 0.01.

Similar to the previous example, the space-time graph of analytical and estimated solutions for t=1 are presented in Figure 2.

### Example 3

We consider the convection-diffusion equation Eq. (1.1) with $\alpha = -1$, $\beta = 10$ and $\varepsilon = 1$ in the interval $[0,1]$, the exact solution is given as $u(x,t) = t\sin(\pi x)\exp(-\pi^2 t)$. The boundary conditions are:

$$u(0,t) = 0, \; u(1,t) = 0, \; t \ge 0,$$

The right side functions of $f(x,t) = [(1+10t)\sin(\pi x) - \pi t\cos(\pi x)]\exp(-\pi^2 t)$, and we extract the initial conditions from the exact solution.

These results are obtained with $dx = 0.001$, $\tau = 0.001$. Similar to the previous examples, the $L_\infty$ and $L_2$ error and RMS errors for t =0.5, 0.75, 1.0, 1.25 and 1.5 are presented in Table 3.

| T | 0.5 | 0.75 | 1.0 | 1.25 | 1.5 |
|---|---|---|---|---|---|
| $L_\infty$ -errors | $1.547\times10^{-5}$ | $2.112\times10^{-6}$ | $2.472\times10^{-7}$ | $2.673\times10^{-8}$ | $2.757\times10^{-9}$ |
| $L_2$ -errors | $1.179\times10^{-5}$ | $1.595\times10^{-6}$ | $1.859\times10^{-7}$ | $2.007\times10^{-8}$ | $2.066\times10^{-9}$ |
| RMS- errors | $1.177\times10^{-5}$ | $1.593\times10^{-6}$ | $1.856\times10^{-7}$ | $2.104\times10^{-8}$ | $2.063\times10^{-9}$ |

**Table 3:** Numerical Errors at different times with T=0.001 and *dx*=0.001 for Example 3.

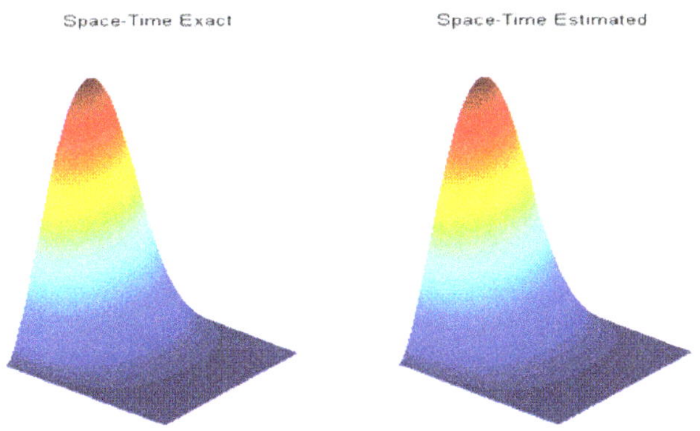


**Figure 3:** Space-time graph of the exact and estimated solutions with T=1 for Example 3.

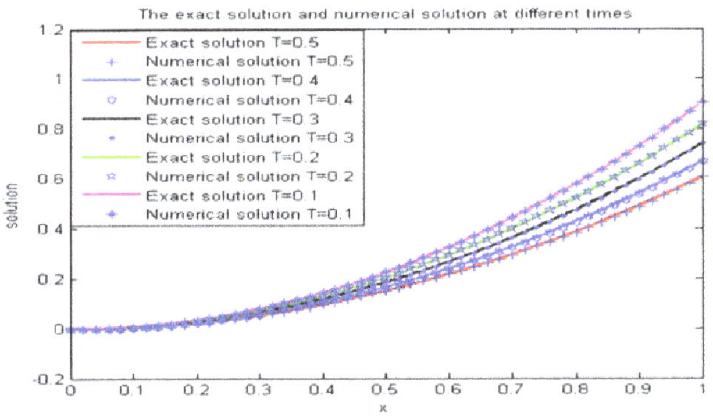


**Figure 4:** The exact and estimated solutions at different times for Example 4.

The space-time graph of analytical and numerical solution for t=1 is presented in Figure 3.

## Example 4

In this example, we consider the convection-diffusion equation in *[0,1]* has the following form:

$$\frac{\partial u}{\partial t}+\frac{\partial u}{\partial x}+2u=\frac{\partial^2 u}{\partial^2 x}+f(x,t)$$

The right side functions $f(x,t)=(x^2+2x-2)\exp(-t)$, with the boundary condition:

$$u(0,t)=0,\ \ u(1,t)=\exp(-t),\ \ t>0$$

Then the analytical solution of the equation is $u(x,t)=x^2\exp(-t)$, we get the initial conditions from the exact solution.

In this case, we use the radial basis functions MQ for the discussed scheme. These results are obtained for $dx=0.001$, $\tau=0.001$. The graph of analytical and numerical solution for t=0.1, 0.2, 0.3, 0.4 and 0.5 is given in Figure 4. The results obtained show the very good accuracy and efficiency of the new approximate scheme. Note that we cannot distinguish the exact solution from the estimated solution in Figure 4.

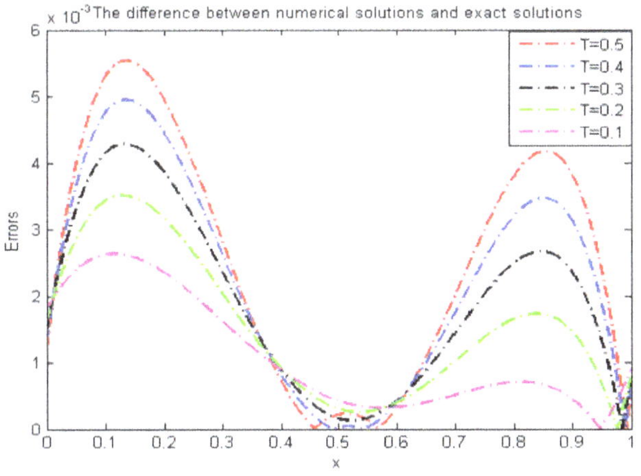


**Figure 5:** The exact and estimated solutions at different times for Example 4.

We also give the difference between exact solutions and numerical solutions in Figure 5.

## Conclusion

In this paper, the collocation method is employed for the numerical solution of convection-diffusion equation based on radical basis functions (RBFs). Coupled with the time discretization, the proposed method is a truly meshless method which requires neither domain nor boundary discretization. The results of numerical experiments are presented, and are compared with analytical solutions confirmed the good accuracy of the presented scheme.

### Acknowledgement

This work was supported by the national Natural Science Foundation of China (Grant Nos.11301252, 11201212) and Applied Mathematics Enhancement Program of Linyi University.

### References

1. Kurganov A, Tadmory E (2000) A New High-Resolution Central Schemes for Nonlinear Conservation Laws and Convection–Diffusion Equations. Journal of Computational Physics 160: 369-383.
2. Stocker T (2014) Introduction to Climate Modelling [M]. Springer.
3. Scott A, Socolofsky, Jirka GH Advective Diffusion Equation.
4. Cockburn B, Shu CW (1998) The Local Discontinuous Galerkin Method for Time-Dependent Convection-Diffusion Systems. Siam J Numer Anal 35: 2440-2463.
5. Douglas J, Russell TF (1982) Numerical Methods for Convection-Dominated Diffusion Problems Based on Combining the Method of Characteristics with Finite Element or Finite Difference Procedures. Siam J Numer Anal 19: 871-885.
6. Chen W, Hon YC (2003) Numerical investigation on convergence of boundary knot method in the analysis of homogeneous Helmholtz, modified Helmholtz and convection–diffusion problems. Int J Numer Meth Engng 56: 1931-1948.
7. Lin H, Atluri SN (2000) Meshless Local Petrov-Galerkin (MLPG) Method for Convection-Diffusion Problems. Computer Modeling in Engineering & Sciences 1: 45-60.
8. Manohar RP, Stephenson JW (1984) A single cell high order scheme for the convection-diffusion equations with variable coefficients. Int J for Num Meth in Fluids 4: 641-651.

9. Jiang ZW, Su LD, Jiang TS (2014) A Meshfree Method for Numerical Solution of Nonhomogeneous Time-Dependent Problems. Abstract and Applied Analysis.

10. Kansa EJ (1990) Multiquadrics-a scattered data approximation scheme with applications to computational fluid dynamics-I. Comout Math Appl 19: 127-145.

11. Zerroikat M, Power H, Chen CS (1992) A numerical method for heat transfer problem using collocation and radial basis functions. Int I Numer Meth Eng 42: 1263-1278.

12. Li M, Jiang TS, Hon YC (2010) A meshless method based on RBFs method for nonhomogeneous backward heat conduction problem. Engineering Analysis with Boundary Elements 34: 785-792.

13. Su LD, Jiang ZW, Jiang TS (2013) Numerical solution for a kind of nonlinear telegraph equations using radial basis functions. Communications in Computer and Information Science 391: 140-149.

14. Jiang TS, Li M, Chen CS (2012) The Method of Particular Solutions for Solving Inverse Problems of a Nonhomogeneous Convection-Diffusion Equation with Variable Coefficients. Numerical Heat Transfer, Part A: Applications 61: 338-352.

15. Fasshauer GE (2008) Meshfree Approximation Methods with MATLAB [M]. Illinois Institute of Technology.

16. Liu GR, Gu YT (2005) An introduction to meshfree methods and there programming [M]. Springer.

17. Chen W, Fu ZJ, Chen CS (2014) Recent Advances in Radial Basis Function Collocation Methods [M]. Springer Briefs in Applied Sciences and Technology.

# 14

# The Entropy Production of a Nonequilibrium Open System

**Ming Bao Yu***

*407 Oak Tree Square, Athens, Georgia 30606, USA*

**Abstract**

A nonequilibrium open system is studied in the projection operator formalism. The environment may linearly deviate from its initial state under the reaction from the open system. If the relevant statistical operator of the system is a generalized canonical one, the transport equation, the second kind of fluctuation-dissipation theorem and the entropy production rate of the open system can be derived and expressed in terms of correlation functions of fluctuations of random forces and interaction random forces.

**Keywords:** Non equilibrium open system; Projection operator; Entropy production rate

## Introduction

In the study of nonequilibrium systems different projection operators are introduced to present a macroscopic description of the system in order to simplify the problem [1-6]. In this approach the macroscopic state of the system is determined by expectation values of a set of basis macro variables, and equations of motions for these expectation values, the transport equations, are derived in the projection operator formalism.

When studying a nonequilibrium open system, the influence of the environment upon the open system is one of the important topics in such studies. It has been shown [7] that the influence from the environment comes from two parts: one is the time-rate of the averaged macro variables resulting from the interaction Hamiltonian $H_{SR}$ and the other from an additional influence term, therefore, the influence of the environment can be completely separated from the corresponding closed system.

When the relevant statistical operator of the system is of a generalized canonical statistical operator (GCSO) by which the entropy of the open system is defined, if the environment is a reservoir, then the memory and influence terms in the transport equation can be given in terms of correlation functions of fluctuations of random forces and interacting random forces, and they can be cast into the Volterra equation formalism.

The purpose of the present paper is to generalize the results to the case that the environment is not a reservoir which may linearly deviate from its initial state under the reaction from the open system. We will show that the memory and influence terms can still be expressed in terms of correlation functions of fluctuations of random forces and interaction random forces, but no longer be able to cast into the Volterra equation formalism, so is the entropy production rate of the open system.

The results obtained in this paper are compared with approaches in linear thermodynamics and statistical mechanics, focusing on the entropy production of a nonequilibrium open system, which is local in both space and time. In contract, the entropy generation [8] is also important in the study of nonequilibrium systems, which is global in space and time, being especially useful in cases involving effects of irreversibility. In addition, another important development in physics today is the so-called quantum thermodynamics [9-14] which has extended the thermodynamics study from the macroscopic scale to the nanometer scale, and even down to the single atom and single photon scale. In Section 2, transport equations of the system are briefly reviewed. In Section 3, a GCSO is introduced. The entropy production rate is derived in Section 4. The influence term and its contribution to the entropy production is studied in Section 5. Comparison of the results with well-known approaches is presented in Section 6 and conclusions are drawn in Section 7.

## Transport equations

Consider an open system S under the influence of its environment R. The total system S $\oplus$ R is characterized by Hamiltonian $H=H_S+H_R+\lambda H_{SR}$ and statistical operator (so) $W(t)$. The open system s is described by a reduced statistical operator $\rho(t)=tr_R W(t)$ satisfying $\partial \rho(t)/\partial t=-iL_S \rho(t)+\eta(t)$ with

$$\eta(t)=-i\lambda tr_R [L_{SR} W(t)] \tag{2.1}$$

describing the influence of R upon S, where $L_S X= i\,[HS, X]$, $h=1$.

Suppose we are satisfied with the description of system S at the macroscopic level by expectation values (EVs) of a set of basis macrovariables $\{ A_j, j=1,\ldots,m\}$ of S, such macroscopic description can be realized by a relevant so $\rho_r(t)$ which is picked up by a time-dependent projection operator $\rho(t)$ from $\rho(t)$: $\rho_r(t)= \rho(t)\, \rho(t)$. We may choose the following projection operator as $\rho(t)$ [3]:

$$p(t)X=\left[\rho_r(t)-\sum_k \frac{\partial\rho_r(t)}{\partial<A_k(t)>}<A_k(t)>\right]tr_s X+\sum_k \frac{\partial\rho_r(t)}{\partial<A_k(t)>}tr_s(A_k X) \tag{2.2}$$

Introduce $q(t)=1-p(t)$, $p(t)q(t)=0$, we have [6]

$$\rho(t)= \rho_r(t)+g(t,0)q(0)\,\rho(0)-$$

$$\int_0^t du g(t,u)q(u)iL_s\rho_r(u)+\int_0^t du g(t,u)q(u)\eta(u) \tag{2.3}$$

*with* $\dot{P}(t)\,\rho(t)=0$; $g(t,u)=T_+ \exp\{-i\int_u^t du_1 q(u_1)L_S\}$ $(t>u)$ is a time-ordered evolution operator satisfying

$$\partial g(t,u)/\partial u=ig(t,u)\,q(u)L_S \text{ and } g(t,t)=1.$$

***Corresponding author:** Ming Bao Yu, 407 Oak Tree Square, Athens, Georgia 30606, USA, E-mail: mingbyu@gmail.com

The transport equation for EV $< A_j(t)>=tr_S[\rho(t)A_j]=tr_s[\rho(t)A_j$ takes the form [6]

$$\frac{\partial}{\partial t}< A_j(t)>=\frac{\partial}{\partial t}< A_j(t)>^{(0)}+Y_j(t) \qquad \text{(j=1,2,...,m).}$$

In Heisenberg picture,

$$\frac{\partial}{\partial t}< A_j(t)>^{(0)}=tr_S[\rho_r(t)\dot{A}_j]+tr_S\left[\rho(0)Q(0)G(0,t)\dot{A}_j\right]+\int_0^t dutr_S\left[\rho_r(u)iL_SQ(u)G(u,t)\dot{A}_j\right], \quad (2.5)$$

here the first term gives the organized motion, the second term the initial condition and the third term the disorganized motion or the memory term [4] and

$$Y_j(t)=\int_0^t dutr_S\left[\eta(u)Q(u)G(u,t)\dot{A}_j\right] \quad (2.6)$$

is an additional term describing the external influence from the environment upon the open system; $G(u,t)$ $T\_\exp\left[i\int du_1 L\, Q(u_1)\right]$ (t>u) is an anti-time-ordered evolution operator defined by $\partial G(u,t)/\partial u=-iL_SQ(u)G(u,t)$ and G(t, t)=1; $Q(t)=1-P(t)$, $P(t)$ is the transposed projection operator of P(t) [4],

$$P(t)X=tr_S\left[\rho_r(t)X\right]+\sum_k\left[A_k-< A_k(t)>\right]tr_S\left[\frac{\partial\rho_r(t)}{\partial< A_k(t)>}X\right]; \quad (2.7)$$

Satisfying

$$tr_S\left[X(t)g(t,u)q(u)Y(u)\right]=tr_S\left[Y(u)Q(u)G(u,t)X(t)\right] \quad (2.8)$$

Since $\dot{A}_j=i(L_S+\lambda L_{SR})A_j$ , (2.4) may be written as [7]

$$\frac{\partial}{\partial t}< A_j(t)>_{(0)}=tr_S\left[\rho_r(t)(iL_SA_j)\right]+tr_S\left[\rho(0)Q(0)G(0,t)(iL_SA_j)\right]+\int_0^t dutr_S\left[\rho_r(u)iL_SQ(u)G(u,t)(iL_SA_j)\right]$$

(j = 1,2,...m) , (2.9)

Where

$$\frac{\partial}{\partial t}< A_j(t)>_{(0)}=tr_S\left[\rho_r(t)(iL_SA_j)\right]+tr_S\left[\rho(0)Q(0)G(0,t)(iL_SA_j)\right]+\int_0^t dutr_S\left[\rho_r(u)iL_SQ(u)G(u,t)(iL_SA_j)\right]$$

(2.10a) is the transport equation of the corresponding closed system, i.e. the time rate of EV resulting from $H_S$, the Hamiltonian of the system $S$ itself; and

$$\frac{\partial}{\partial t}< A_j(t)>_{(i)}=tr_S\left[\rho_r(t)(iL_{SR}A_j)\right]+tr_S\left[\rho(0)Q(0)G(0,t)(iL_{SR}A_j)\right]+\int_0^t dutr_S\left[\rho_r(u)iL_SQ(u)G(u,t)(iL_{SR}A_j)\right] \quad (2.10b)$$

is the time rate resulting from the interaction $H_{SR}$ .

The meaning of (2.9) is clear and simple: The transport equation of an open system is the sum of transport equation of the corresponding closed system, the time-rate of the EV due to the interaction Hamiltonian $H_{SR}$ and the additional influence term $Y_j(t)$.

The influence term (2.6) can be written as (j=1,2...,m) (j=1,2...,m), (2.11)

Where $f_j(u,t)=Q(u)G(u,t)\dot{A}_j$ denotes the random force, which may be split into two:

$$f_j(u,t)=f_j^s(u,t)+\lambda f_j^{SR}(u,t), \quad (2.12a)$$

$$f_j^S(u,t)=Q(u)G(u,t)(iL_SA_j),$$

$$f_j^{SR}(u,t)=Q(u)G(u,t)(iL_{SR}A_j) \quad (2.12b)$$

being respectively the random force and interaction random force associated with the time rate of the basis variable $A_j$ due to $H_S$ and $H_{SR}$, respectively. Since the average of the random force over given ensembles vanishes, so $f_j(u,t)=\delta f_j(u,t)$ In the rest of the paper we will no longer distinguish $f_j(u,t)$ from its fluctuation $\delta f_j(u,t)$.

## Generalized canonical statistical operator In order to go steps further

let us assume $\rho_r(t)$ to be a GCSO:

$$\rho_r(t)=e^{-\sum_l\lambda_l(t)A_l}/z_r(t)=tr_S\left[e^{-\sum_l\lambda_l(t)A_l}\right] \quad (3.1)$$

where $\lambda_t$ $(t)$ $(l=1,...,m)$ are conjugate parameters of the basis macrovariables $\{A_l\}$

Making use of the Kubo identity $\left[X,e^Y\right]_-=\int_0^1 d\alpha e^{\alpha Y}[X,Y]\_e^{(1-\alpha)Y}$ we have

$$iL_S\rho_r(t)=-\sum_l\int_0^1 d\alpha e^{-\alpha\sum_l\lambda_l(t)A_l}(iL_SA_l)e^{\alpha\sum_l\lambda_l(t)A_l}\rho_r(t)\lambda_l(t) \quad (3.2a)$$

$$iL_{SR}\rho_r(t)=-\sum_l\int_0^1 d\alpha e^{-\alpha\sum_l\lambda_l(t)A_l}(iL_{SR}A_l)e^{\alpha\sum_l\lambda_l(t)A_l}\rho_r(t)\lambda_l(t) \quad (3.2b)$$

Introducing the generalized quantum correlation function

$$(X(t),Y(u))_r=\int_0^1 d\alpha tr_S\left[X(t)e^{-\alpha\sum_l\lambda_l(u)A_l}Y(u)e^{\alpha\sum_l\lambda_l(u)A_l}\rho_r(u)\right] \quad (3.3)$$

and making use of (3.2), the integrand in (2.5) may be written as

$$tr_S\left[\rho_r(u)iL_Sf_j(u,t)\right]=\sum_l\left(f_j(u,t),iL_SA_l\right)_r\lambda_l(u) \quad (3.4)$$

Since $f_j(u\ t)$ is in the irregular space because of $Q(u)$ while is in the regular space, their correlation is zero, thus

$$tr_S\left[\rho_r(u)iL_Sf_j(u,t)\right]=\sum_l\left(f_j(u,t),f_l^S(u,u)\right)_r\lambda_l(u) \quad (3.5)$$

Where $f_l^S(u,u)$ is given by (2.12b). Same argument will apply to similar cases later.

Therefore Eq.(2.5) takes the form

$$\frac{\partial}{\partial t}< A_j(t)>^{(0)}=tr_S\left[\rho_r(t)\dot{A}_j\right]+tr_S\left[\rho(0)Q(0)G(0,t)\dot{A}_j\right]+\sum_l\int_0^t du\left(f_j(u,t),f_l^S(u,u)\right)_r\lambda_l(u) \quad (3.6)$$

here the memory term is expressed in terms of quantum correlation function of fluctuations of random forces. The influence term $Y_j(t)$ will be further analysed in Section 5.

## Entropy production rate

Now define the entropy of the noequilibrium open system through its relevant statistical operator [1,15,16]

$$S(t)=-k_Btr_S\left[\rho_r(t)\ln\rho_r(t)\right] \quad (4.1)$$

where $k_B$ is the Boltzmann constant. The entropy production rate reads [7]

$$\frac{\partial S(t)}{\partial t}\quad k_B\sum\frac{\partial< A\ (t)>}{\partial t}{}_j(t) \quad (4.2)$$

which is the sum of products of transport equations and the conjugate parameters. If assume that the initial state of the system is a GCSO: $\rho(0)=\rho_r(0)$ then the initial term in (3.6) vanishes. Combining (4.2) with (2.4) given by (3.6) and (2.11), we obtain

$$\frac{\partial S_1(t)}{\partial t}=\frac{\partial S_1(t)}{\partial t}+\frac{\partial S_2(t)}{\partial t}+\frac{\partial S_3(t)}{\partial t}, \quad (4.3)$$

the first term resulting from the organized motion in (3.6) reads [7]

$$\frac{\partial S_1(t)}{\partial t} = \lambda k_B \sum_{j=1}^{m} \lambda_j(t) tr_S\left[\rho_r(t) tr_S\left(iL_{SR}A_j\right)\right] \quad (4.4)$$

the second term resulting from the disorganized motion in (3.6) takes the form

$$\frac{\partial S_2(t)}{\partial t} = k_B \sum_{j,l} \int_0^t du\left(f_j(u,t), f_l^S(u,u)\right)_r \lambda_l(u)\lambda_j(t) \quad (4.5)$$

because of (3.5); and the third term resulting from the influence term (2.11) is

$$\frac{\partial S_3(t)}{\partial t} = k_B \sum_j \int_0^t dutr_S\left[\eta(u) f_j(u,t)\right]\lambda_j(t) \quad (4.6)$$

These expressions represent the contributions of each term in the transport equation to the entropy production, respectively. Besides, Eq.(4.4) does not involve $iL_S A_j$, indicating that in the organized motion term $H_S$ contributes nothing to the rate.

## Non-reservoir environment

Now we further analyze the contribution of the influence term $Y_j(t)$. Suppose that the environment R is not a reservoir and may linearly deviate from its initial state under the reaction from S. For simplicity, we assume $H_{SR} = \sum_k \gamma_k A_k B_k$, $A_k$ and $B_k$ respectively pertain to S and R, and they are initially independent:

$W(0) = \rho(0)R(0)$ where $R(t) = tr_S W(t)$.

By $W(0) = \rho(0)R(0)$, we have $W(u) = \rho(u)R(u)$, here $\rho(u) = e^{-i(L_S+\lambda L_{SR})u}\rho(0)$ and $R(u) = e^{-i(L_S+\lambda L_{SR})u}R(0)$. By (2.1), $\eta(u)$ may be written as

$$\eta(u) = -i\lambda < L_{SR}(u) >_{R0} \rho(u) = -i\lambda\left[< H_{SR}(u) >_{R0}, \rho_r(t)\right]_- \quad (5.1)$$

$$< H_{SR}(u) >_{R0} = \sum_k \gamma_k A_k < B_k(u) >_{R0,}$$

$$< B_k(u) >_{R0} = tr_R\left[B_k(u)R(0)\right] \quad (5.2)$$

$< B_k(u) >_{R0}$ determines the evolution of the EV of macrovariable $B_k$.

Since $\partial B_k(t)/\partial t = iL_R B_k(t) + i\lambda L_{SR} B_k(t)$ thus $B_k(t) = e^{iL_R t}B_k(0) + i\lambda\int_0^t due^{iL_R(t,u)}L_{SR}B_k(u)$

For weak interaction, keeping only the linear term in λ, we obtain

$$B_k(t) = e^{iL_R t}\left[1 + i\lambda\int_0^t duL_{SR}(u)\right]B_k(0) + o\left(\lambda^2\right), L_{SR}(u) = e^{iL_R u}L_{SR}e^{iL_R u}$$

$$< B_k(u) >_{R0} = < B_k(u) >_{R0}^{(0)} + B_k(u) >_{R0}^{(1)} \quad (5.3a)$$

$$< B_k(u) >_R^{(\ )} = tr_R\left[R(0)e^{iL\ u}B_k(0)\right], \quad (5.3b)$$

$$< B_k(u) >_{R0}^{(1)} = i\lambda tr_R\left[R(0)e^{iL_R u}\int_0^u du_1 L_{SR}(u_1)B_k(0)\right] \quad (5.3c)$$

being the zeroth and first order terms of the EV of $B_K$ when R linearly deviates from its initial state under the weak reaction from S. By (5.1) and (5.2), the integrand in (2.11) takes the form

$$tr_S\{\eta(u) f_j(u,t)\} = -i\lambda\sum_k \gamma_k tr_S\left\{f_j(u,t)\left[A_k, \rho(u)\right]_-\right\} < B_k(u) >_{R0} \quad (5.4a)$$

$$= -i\lambda\sum_k \gamma_k tr_S\left\{F_j(u,t)\left[A_k(u), \rho(0)\right]_-\right\} < B_k(u) >_{R0} \quad (5.4b)$$

in Schrodinger and Heisenberg pictures, respectively; where $F_j(u,t) = e^{i(L_S+L_{SR})u} f_j(u,t)$ *and* $A_k(u) = e^{i(L_S+L_{SR})u}A_k$. Hence we have the influence term

$$Y_j(t) = -i\lambda\sum_k \gamma_k \int_0^t dutr_S\left\{f_j(u,t)\left[A_k, \rho(u)\right]_-\right\} < B_k(u) >_{R0} \quad (5.5a)$$

$$= -i\lambda\sum_k \gamma_k \int_0^t dutr_S\left\{F_j(u,t)\left[A_k(u), \rho(0)\right]_-\right\} < B_k(u) >_{R0} \quad (5.5b)$$

and its contribution to the entropy production in Shrodinger and Heisenberg pictures:

$$\frac{\partial S_3(t)}{\partial t} = -i\lambda k_B \sum_{j,k} \gamma_k \int_0^t dutr_S\left\{f_j(u,t)\left[A_k, \rho(u)\right]_-\right\} < B_k(u) >_{R0} \lambda_j(t) \quad (5.6a)$$

$$-i\lambda k_B \sum_{j\,k} \gamma_k \int dutr_S\left\{F_j(u,t)\left[A_k(u), \rho(0)\right]_-\right\} < B_k(u) >_R \lambda_j(t) \quad (5.6b)$$

Now consider the case that the initial state of S is given by a GCSO:

$$\rho_r(0) = e^{-\sum_l \lambda_l(0)A_l} / tr_S e^{-\sum_l \lambda_l(0)A_l}. \quad (5.7)$$

By the Kubo identity and the initial condition $\rho(0) = \rho_r(0)$ we have

$$\left[A_k(u), \rho(0)\right]_- = -\sum_l \lambda_l(0)\int_0^1 d\alpha e^{-\alpha\sum_l \lambda_l(0)A_l}\rho_r(0),$$

Thus (5.4) may be written as

$$tr_S\{\eta(u) f_j(u,t)\} = \lambda\sum_l\left(F_j(u,t), i < L_{SR}(u) >_{R0} A_l\right)_{r0}\lambda_l(0) = \lambda\sum_l\left(F_j(u,t), \overset{-SR}{f_l}(u,u)\right)_{r0}\lambda_l(0)$$

here we have conducted argument similar to that leads (3.4) to (3.5), and

$$\overset{-SR}{f_l}(u,u) = Q(u)G(u,u)\left[i < L_{SR}(u) >_{R0} A_l\right] \quad (5.8)$$

is the averaged interaction random force. Thus we obtain

$$Y_j(t) = \lambda\sum_l \int_0^t du\left(F_j(u,t), \overset{-SR}{f_l}(u,u)\right)_{r0}\lambda_l(0) \quad (5.9)$$

$$\frac{\partial S_3(t)}{\partial t} = \lambda k_B \sum_{j,l}\int_0^t du\left(F_j(u,t), \overset{-SR}{f_l}(u,u)\right)_{r0}\lambda_l(0)\lambda_j(t) \quad (5.10)$$

With $\left(X(t), Y(u)\right)_{r0} = \int_0^1 d\alpha tr_S\left\{X(t)e^{-\alpha\sum_l \lambda_l(0)A_l}\rho_r(0)\right\}$

If the open system is initially in an equilibrium state

$$\rho(0) = \rho_r(0) = \rho_{eq} = e^{-\beta_0 H_S} / tr_S\left(e^{-\beta_0 H_S}\right), \quad (5.11)$$

$\beta_0 = 1/k_B T_0$ is the initial inverse temperature of the system, then $\rho_{eq}$ is a special case of (5.7) in which $A_1 = H_S, \lambda_1(0) = \beta_0$ and $A_1 = 0\ (l \geq 2)$. Since $A_1$ is the only basic variable, so $\rho_r(t) = e^{-\beta(t)H_S} / tr_S\left[e^{-\beta(t)H_S}\right]$ and $\beta(t) = 1/k_B T(t)$ is the inverse temperature of S. Because $iL_S H_S = 0$, thus $\dot{H}_S = i\lambda L_{SR}H_S$ and

$$f_1^S(u,t) = 0, f_1(u,t) = \lambda f_1^{SR}(u,t); \quad (5.12)$$

the memory term in (3.6) becomes $\lambda\int_0^t du\left(f_1^{SR}(u,t), f_1^S(u,u)\right)_r \beta(u)$ and vanishes, so

$$\frac{\partial < H_S(t) >^{(0)}}{\partial t} = \lambda tr_S\left[\rho_r(t)(iL_{SR}H_S)\right] \tag{5.13}$$

Besides, as a special case of (5.9) and (5.10), we have

$$Y_1(t) = \lambda^2\beta_0\int_0^t du\left(F_1^{SR}(u,t), \bar{f}_1^{SR}(u,u)\right)_{eq} \tag{5.14}$$

$$\frac{\partial S_3(t)}{\partial t} = \frac{\lambda^2}{T_0}\int_0^t du\left(F_1^{SR}(u,t), \bar{f}_1^{SR}(u,u)\right)_{eq}\beta(t) \tag{5.15}$$

Here $\left(X(t),Y(u)\right)_{eq} = \int_0^1 d\alpha tr_S\left\{X(t)e^{-\alpha\beta_0 H_S}Y(u)e^{-\alpha\beta_0 H_S}\rho_{eq}\right\}$ (5.16)

Finally we obtain the transport equation for the only basis variable $H_S$:

$$\frac{\partial S(t)}{\partial t} = \frac{\lambda}{T(t)}tr_S\left[\rho_r(t)(iL_{SR}H_S)\right] + \frac{\lambda^2}{T_0}\int_0^t du\left(F_1^{SR}(u,t), \bar{f}_1^{SR}(u,u)\right)_{eq}\beta(t) \tag{5.17}$$

Eqs.(5.9), (5.10) and (5.14), (5.15) involve the averaged interaction random force (5.8) which has incorporated the linear deviation of the environment from its initial state.

Now consider the case without a given initial condition. By (3.1) and the Kubo identity, we have

$$i < L_{SR}(u) >_{R0}\rho_r(u) = -\sum_l \lambda_l(u)\int_0^1 d\alpha e^{-\alpha\sum_l \lambda_l(u)A_l}\left[i < L_{SR}(u) >_{R0} A_l\right]e^{-\alpha\sum_l \lambda_l(u)A_l}\rho_r(u) \tag{5.18}$$

With (5.1) and (2.3),

$$tr_S\left\{\eta(u)f_j(u,t)\right\} = -\lambda tr_S\left[f_j(u,t)i < L_{SR}(u) >_{R0}\rho(u)\right] = J_1 + J_2 + J_3 \tag{5.19}$$

Eq.(5.19) is similar to Eq.(22) in [17] where the environment is a reservoir and $< L_{SR} >_{R0}$ is time-independent. In the following, we will follow the argument in [17], however, take into consideration that $< L_{SR}(u) >_{R0}$ is time-dependent. Making use of (5.18) and (3.3), leads to

$$J_1 = -\lambda tr_S\left\{f_j(u,t)i < L_{SR}(u) >_{R0}\rho_r(u)\right\} = \lambda\sum_l \lambda_l(u)\left(f_j(u,t), i < L_{SR}(u) >_{R0} A_l\right)_r \tag{5.20}$$

$$J_2 = \lambda\int_0^u du_1 tr_S\left\{f_j(u,t)i < L_{SR}(u) >_{R0}\tilde{K}_1(u,u_1)\rho_r(u_1)\right\}$$

$$= \lambda\sum_l\int_0^u du_1\lambda_l(u_1)\left(K_1(u_1,u)f_j(u,t), i < L_{SR}(u) >_{R0} A_l\right)_r \tag{5.21}$$

$$\tilde{K}_1(u,u_1) = g(u,u_1)q(u_1)iL_S,$$

$$K_1(u_1,u) = iL_S Q(u_1)G(u_1,u) \tag{5.22}$$

$$J_3 = -\lambda\int_0^u du_1 tr_S\left[f_j(u,t)i < L_{SR}(u) >_{R0} g(u,u_1)q(u_1)\eta(u_1)\right]$$

$$= \lambda^2\int_0^u du_1 tr_S\left\{\rho(u_1)\tilde{K}(u_1,u)\tilde{K}(u,t)\dot{A}_j\right\} \tag{5.23}$$

$$\tilde{K}(u,t) = i < L_{SR}(u) >_{R0} Q(u)G(u,t). \tag{5.24}$$

Substituting (2.3) into (5.23) and repeating the above arguments, we have

$$tr_S\left[\eta(u)f_j(u,t)\right] = \lambda\sum_l\{\lambda_l(u)\left(f_j(u,t), i < L_{SR}(u) >_{R0} A_l\right)_r +$$

$$\int_0^u du_1\lambda_l(u_1)\left(K_1(u_1,u)f_j(u,t), i < L_{SR}(u) >_{R0} A_l\right)_{rl} + \ldots\}$$

$$+\lambda^2\sum_l\int_0^u du_1\lambda_l(u_1)\left(K(u_1,u)f_j(u,t), i < L_{SR}(u) >_{R0} A_l\right)_r$$

$$+\int_0^{u_1} du_2\lambda_1(u_2)\left(K_1(u_2,u_1)f_j(u,t), i < L_{SR}(u_1) >_{R0} A_l\right)_{rl} + \ldots\}$$

$$K(u_1,u) = Q(u_1)G(u_1,u)i < L_{SR}(u) >_{R0}, \tag{5.25}$$

which many be written as

$$tr_S\left[\eta(u)f_j(u,t)\right] = \lambda\sum_l\left\{\lambda_1(u)\left(f_j(u,t), \bar{f}_1^{SR}(u,u)\right)_r + M_1(u,t)\right\} \tag{5.26}$$

$$M_1(u,t) = \sum_{n=1}^{\infty}\lambda^{n-1}\int_0^u du_1\ldots\int_0^{u_{n-1}} du_n\lambda_1(u_n)[(K_1(u_{n-1},u_{n-2})\ldots K(u_1,u)f_j(u,t), \bar{f}_1^{SR}(u_{n-1},u_{n-1}))_r$$

$$+\lambda K(u_n,u_{n-1})K(u_{n-1},u_{n-2})\ldots K(u_1,u)f_j(u,t), \bar{f}_l^{SR}(u_n,u_n)_r] \tag{5.27}$$

here we have had argument similar to that leads (3.4) to (3.5). Thus we obtain the influence term and its contribution to the entropy production:

$$Y_j(t) = \lambda\sum_l\int_0^t du\left\{\lambda_l(u)\left(f_j(u,t), \bar{f}_l^{SR}(u,u)\right)_r + M_1(u,t)\right\} \tag{5.28}$$

$$\frac{\partial S_3(t)}{\partial t} = \lambda k_B\sum_{j,l}\int_0^t du\left\{\lambda_l(u)\left(f_j(u,t), \bar{f}_l^{SR}(u,u)\right)_r + M_l(u,t)\right\}\lambda_j(t) \tag{5.29}$$

If we are satisfied with keeping the linear term of λ in $M_l(u,t)$, then

$$M_l(u,t) = \int_0^u du_1\lambda_l(u_l)\left(K_1(u_1,u)f_j(u,t), \bar{f}_l^{SR}(u,u)\right)_r + o(\lambda^2) \tag{5.30}$$

$$tr_S\left[\eta(u)f_j(u,t)\right] = \lambda\sum_l\left(\phi(u)f_j(u,t), \bar{f}_l^{SR}(u,u)\right)_r + o(\lambda^3) \tag{5.31}$$

$$\phi_l(u) = \lambda_l(u) + \int_0^u du_1\lambda_l(u_1)K_1(u_1,u) \tag{5.32}$$

Therefore we have the approximate expressions

$$Y_j(t) = \lambda\sum_l\int_0^t du\left(\phi_l(u)f_j(u,t), \bar{f}_l^{SR}(u,u)\right)_r \tag{5.33}$$

$$\frac{\partial S_3(t)}{\partial t} = \lambda k_B\sum_{j,l}\int_0^t du\left(\phi_l(u)f_j(u,t), \bar{f}_l^{SR}(u,u)\right)_r\lambda_j(t) \tag{5.34}$$

they are up to $\lambda^2$ by (2.12a). Comparing (5.34) with (4.5), we see clearly that $\phi_l(u)$ plays the role of $\lambda_l(u)$ in the case of corresponding isolate system.

Now we rewrite the results obtained above in the form of special dependent. For simplicity, we focus on the simpler expression (5.34). The entropy production of the open system reads

$$\frac{\partial s(x,t)}{\partial t} = \frac{\partial s_1(x,t)}{\partial t} + \frac{\partial s_2(x,t)}{\partial t} + \frac{\partial s_3(x,t)}{\partial t} \tag{5.35}$$

Where

$$\frac{\partial s_1(x,t)}{\partial t} = \lambda k_B\sum_{j=1}^{m}\lambda_j(x,t)tr_S\left[\rho_r(x,t)(iL_{SR}A_j)\right] \tag{5.36a}$$

results from the organized motion in the transport equation due to $H_{SR}$;

$$\frac{\partial s_2(x,t)}{\partial t} = k_B \sum_{j,l} \int_0^t du \left( f_j(u,t), f_l^S(u,u) \right)_r \lambda_l(x,u) \lambda_j(x,t) \quad (5.36b)$$

from the disorganized motion and

$$\frac{\partial s_3(x,t)}{\partial t} = \lambda k_B \sum_{j,l} \int_0^t du \left( \phi_l(x,u), f_j(u,t) \right)_r \bar{f}_l^{SR}(u,u)_r \lambda_j(x,t) \quad (5.36c)$$

from the influence term, respectively.

## Comparison

In this section, we compare the results obtained in the proceeding sections with the well known approaches in the linear nonequilibrium thermodynamics and statistical mechanics. The time rate of the entropy density *s(x, t)* of a nonequilibrium system takes the form [8,18,19]:

$$\frac{\partial s(x,t)}{\partial t} = \sigma(x,t) - \nabla . j_s(x,t) \quad (6.1)$$

Where $\sigma(x,t) = \sum_i X_i(x,t) J_i(x,t)$ (6.2)

is the entropy production density occurring inside the system which is given in terms of the sum of products of thermodynamic fluxes $J_i(x,t)$ and the conjugate thermodynamic forces $X_i(x,t)$; and $j_s(x,t)$ is the density of entropy flux through the border into the system.

Onsager proposed a linear relationship between the fluxes and forces

$$J_i(x,t) = \sum_k L_{ik} X_k(x,t) \quad (6.3)$$

with reciprocity relations

$$L_{ik} = L_{ki} \quad (6.4)$$

Thus we have

$$\frac{\partial s(x,t)}{\partial t} = \sum_{i,k} L_{ik} X_i(x,t) X_k(x,t) - \nabla . j_s(x,t) \quad (6.5)$$

For the special case considered in Sect.5, the interaction between open system S and its environment R takes the form $H_{SR} = \sum_k \gamma_k A_k B_k$, for example, S and R are composed of different kinds of harmonic oscillator [7]. Such interaction implies no obvious border separating S and R, leading to absence of the divergence term on the right hand side of (6.5). Thus the variation of entropy density results from inner entropy production $\sigma(x,t)$ only:

$$\frac{\partial s(x,t)}{\partial t} = \sum_{i,k} L_{ik} X(x,t)_i X_k(x,t) \quad (6.6)$$

Besides, in the Green-Kubo formalism, the transport coefficients $L_{ik}$ can be expressed in terms of time correlation functions of the time rate of corresponding variables [19,20].

$$L_{ik} \sim \int_0^\infty dt < \dot{A}_i(t) \dot{A}_k(0) >_{eq} \quad (6.7)$$

where the average is taken over an equilibrium ensemble and the Markovian effect is taken into account.

In this paper, we study a nonequilibrium open system whose transport equations (2.9)-(2.11) are nonlinear differential-integral ones. Now let us compare (5.36b) with (6.6). We notice that (5.36b) share the same structure as (6.6) because of the facts : (1) parameters $\{\lambda_j(x,t)\}$( j=1,...m) play the role Thermodynamic forces since they may involve spacial gradients of, e.g., temperature, velocity, chemical potential or electric, magnetic fields, etc,; (2) the random forces (2.12b) involve the time rates of variables because of using projection operator technique and (3) the average is taken over GCSO (3.1) instead of an equilibrium ensemble. As for (5.36c), the contribution of $Y_j(t)$ to the entropy production, in which the free term $\phi_l(x,u)$ in the Volterra equation is indeed a generalization of $\lambda_l(x,u)$ in (5.26b), hence (5.36c) possesses the same structure as (6.6) also. Acoordingly, we see clearly that the entropy production rate (5.35) is a natural generalization of (6.6) where the non-linearity and the non-Markovian effect have been taken into consideration. In addition to the entropy production, the *entropy generation* is another useful tool in the study of nonequilibrium systems [8] and especially useful in the analysis of a process occurring in the system during a period of time $\tau$. It is worth noticing the major differences between the two: the entropy production needs the hypothesis of local equilibrium but the entropy generation does not; the former does not consider the time but the latter introduces the lifetime $\tau$ of the process [8]. The two different approaches are closely related and complementary one to another.

## Conclusion

In the present paper we have studied a nonequilibrium open system in interaction with its environment which may linearly deviate from its initial state under the reaction of the open system. We have shown that if the relevant statistical operator of the system is of the form of GCSO, then the transport equation is given by (3.6) and (5.28) or (5.33). The memory term in (3.6) and the influence term (5.28) or (5.33) can be expressed in terms of quantum correlation functions of fluctuations of random forces and interaction random forces, giving the second kind of fluctuation-dissipation theorem for this nonequilibrium open system. We have also shown that the entropy production rate is given by the sum of products of transport equations and the corresponding parameters. In the organized motion term, $H_S$ contributes nothing to the rate, but $H_{SR}$ does; the contributions of the memory and influence terms are expressed in terms of quantum correlation functions of fluctuations of random forces and interaction random forces. The total entropy production rate is given by the sum of contributions resulting from each term in the transport equation, given respectively by (4.4), (4.5) and (5.29) or (5.34). They are natural generalizations of those for a linear nonequilibrium closed system to a nonlinear open system.

### References


1. Robertson B (1996) Equations of Motion in Nonequilibrium Statistical Mechanics. Phys Rev 144: 151.
2. Li KH (1986) Physics of open systems. Phys Rep 134: 1-85.
3. Kawasaki K, Gunton JD (1973) Phys Rev A 8: 2048.
4. Grabert H (1978) Projection Operator Techniques in Nonequilibrium Statistical Mechanics.J Stat Phys 19: 479.
5. Oppenheim I, Levin RD (1979) Nonlinear Transport Processes: Hydrodynamics. Physica A 99: 383-402.
6. Yu MB (1986) Physica A 137.
7. Yu MB (2008) Influence of environment and entropy production of a nonequilibrium open system. Phys Lett A 372: 2572-2577.
8. Lucia U (2013) Stationary open systems: A brief review on contemporary theories on irreversibility. Physica A 392: 1051-1062.
9. Hatsopoulos GN, Gyftopoulos EP (1976) A Unified Quantum Theory of Mechanics and Thermodynamics. Part I. Postulates. Foundations of Phys 6: 15.
10. Hatsopoulos GN, Gyftopoulos EP (1976) A Unified Quantum Theory of

Mechanics and Thermodynamics. Part IIa. Available Energy. Foundations of Phys 6: 127.

11. Hatsopoulos GN, Gyftopoulos EP (1976) A Unified Quantum Theory of Mechanics and Thermodynamics. Part IIb. Stable Equilibrium States. Foundations of Phys 6: 439.

12. Hatsopoulos GN, Gyftopoulos EP (1976) A Unified Quantum Theory of Mechanics and Thermodynamics. Part III Irreducible Quantal Dispersions. Foundations of Phys 6: 561.

13. Beretta GP (2010) Maximum entropy production rate in quantum thermodynamics. J Phys: Conference Series 237: 1-32.

14. Sciacovelli A, Smith CE, Vonspakovsky MR, Verda V (2010) Quantum Thermodynamics: Non-equilibrium 3D Description of and Unbounded System at an Atomistic Level. Int J of Thermodynamics 13: 23-33.

15. Nettleton RE (2001) Positive definiteness of entropy production in the nonlinear Robertson formalism. J Chem Phys 14: 6007-6013.

16. Jarvis JB (2005) Time-dependent entropy evolution in microscopic and macroscopic electromagnetic relaxation. Phys Rev E 72.

17. MB Yu (1990) Phys Lett A 150: 1-10.

18. Groot SD, Mazur P (1962) Non-Equilibrium Thermodynamics, North-Holland.

19. Martyushev LM, Seleznev VD (2006) The restrictions of the Maximum Entropy Production Principle. Phys Rept 426: 1-45.

20. Kubo R (1957) The Glass Transition: Relaxation Dynamics in Liquids and Disordered Materials. J Phys Soc Jpn 12: 570-586.

# Dynamics of Two Charged Particles in a Creeping Flow

**Hassan HK and Stepanyants YA***

*University of Southern Queensland, Toowoomba, Australia*

**Abstract**

We study the interaction of two charged solid particles in a viscous fluid. It is assumed that the particles move either side-by-side or one after another along the same vertical line under the influence of the buoyancy/gravity force, Coulomb electrostatic force or its modification, and viscous drag force. The drag force consists of two components: the quasi-stationary Stokes drag force and Boussinesq–Basset drag force resulting from the unsteady motion. Solutions of the governing equations are analysed analytically and numerically for the cases of perfect fluid and viscous fluid; the comparison of these two cases is presented.

**Keywords:** Micro-particle; Viscous fluid; Charged particle; Stokes drag force; Boussinesq–Basset drag force creeping flow

**AMS subject classification:** 76D07, 70F05, 45A05, 45-04, 70-04

## Introduction

Despite the fact that the study of particle dynamics in a viscous fluid has a long history, many important problems still remain unsolved [1]. This pertains not only to the collective behaviour of big particle ensembles, but even to interactions of two particles in a complex environment of shear flows or in the presence of solid obstacles, walls or free surfaces. Much research efforts have been undertaken over the years to elucidate the particle dynamics in various situations. Importance of such investigations is backed by many practical applications of mixtures of fluids and suspensions; examples being particle motions in the cooling systems of nuclear reactors and particle-liquid mixtures used for pharmaceutical purposes. Other important examples are transport of particles (dusts and aerozoles) in the atmosphere and oceans (dynamics of suspensions, sand, biological products, etc.).

One of the intriguing problems that has not been solved so far is the dynamics of two inter- acting charged particles in viscous fluid. In recent years a great interest is observed to micro- and nano-particles due to their potential applications in modern biotechnologies and other micro-fluid technologies. Quantitative descriptions of such systems represent a certain challenge not only from the practical, but also from the academic point of view.

In this paper we consider elementary acts of interaction of two charged particles moving in a viscous fluid. We consider the interactions of non-conducting charged particles, when the electric charges are uniformly distributed within the spherical particles, and conducting particles, when the charges can freely move within the particles enhancing and reducing the Coulomb forces [2]. The equations of motion are studied analytically where possible and numerically. The particle dynamics is considered in the creeping flow approximation, that is under the assumption that the Reynolds number is very small, $Re \equiv uR/\nu \ll 1$, where u is particle velocity relative to the fluid, $R$ is particle radius, and ν is fluid kinematic viscosity. For simplicity, we assume that particles are solid and have a spherical shape. We study two configurations of particles: i) when they move side-by-side perpendicular to the line connecting their centers and ii) when they move vertically one after another along the same line.

## Equations of Motion and Problem Formulation

A motion of an individual uncharged particle in a viscous fluid at small Reynolds numbers in the creeping regime has been studied [3]. It has been demonstrated that in the case of a transient flow the influence of Boussinesq–Basset drag (BBD) force [4-6] is very important. It provides different character of particle motions in comparison with the well-known Stokes drag (SD) force [7,8]. In the present paper we consider a motion of two electrically charged particles in different setups taking into account gravity/buoyancy force, electrostatic force and viscous drag forces. The effect of viscosity is taken into consideration through the SD force and BBD force which depends on the motion prehistory [4-6,8]. We also take into account a reciprocal influence of particles on the drag force which depends on the particle configuration [9]. To our best knowledge the combined effect of all these factors were not studied thus far.

As a first step we consider two identical metallic particles with electric charges of the same absolute value (they can be either of like or unlike charges). As has been shown [2], the electrostatic force acting on metallic particles deviates from the classical Coulomb law: at small distances the force is not inversely proportional to the square of the distance between the particle centers. This deviation is important at relatively small distances between the particles while at large distances the electrostatic force asymptotically approaches the classical Coulomb law. An exact expression for the electrostatic force $\mathbf{F}_s$ as derived [2] and its asymptotic Coulomb approximation $\mathbf{F}_c$ are as follows,

$$\mathbf{F}_s(\mathbf{r}_1 - \mathbf{r}_2) = -\frac{q^2}{8\pi\varepsilon}\frac{\mathbf{r}_1 - \mathbf{r}_2}{|\mathbf{r}_1 - \mathbf{r}_2|}\frac{S(\beta)}{R^2}, \tag{2.1}$$

$$\mathbf{F}_c(\mathbf{r}_1 - \mathbf{r}_2) = -\frac{q^2}{4\pi\varepsilon}\frac{\mathbf{r}_1 - \mathbf{r}_2}{|\mathbf{r}_1 - \mathbf{r}_2|^3}, \tag{2.2}$$

where R is the particle radius, q is the value of the electric charge, ε is the permittivity of the medium,

***Corresponding author:** Stepanyants YA, University of Southern Queensland, Toowoomba, Australia; E-mail: Yury.Stepanyants@usq.edu.au

$$S(\beta)=\sum_{n=1}^{\infty}\left\{(-1)^{\kappa_n}\frac{n\coth n\beta-\coth\beta}{\sinh n\beta}\left[\sinh\beta\sum_{n=1}^{\infty}\frac{(-1)^{\kappa_n}}{\sinh n\beta}\right]^{-2}\right\},$$

$$\beta=\ln\left(\frac{|\mathbf{r}_1-\mathbf{r}_2|}{2R}+\sqrt{\frac{|\mathbf{r}_1-\mathbf{r}_2|^2}{4R^2}-1}\right).$$

$\kappa_n=n+1$ for like charged particles and $\kappa_n=0$ for unlike charged particles. The physical reason of deviation of electrostatic force from the classical Coulomb formula is explained by redistribution of electric charges within the conducting spheres as illustrated by Figure 1. When electric charges can freely move within conducting spheres they either attract to each other if spheres unlikely charged or repeal from each other when the spheres likely charged.

Figure 2 shows the dependences of the attractive and repulsive electrostatic forces normalized by $q^2/(16\pi\varepsilon R^2)$ as described by Equation (2.1) and the corresponding Coulomb forces under the same normalization as described by Equation (2.2). As one can see from this figure, corrections to the Coulomb forces become notable only when the distance between the particle centres is less than 4R. The modified attractive force infinitely increases when particles approach each other (see line1 in the Figure 2).

When a solid particle moves in a viscous fluid, it experiences an influence of a drag force [7,8]. In the presence of another particle the drag force modifies and depends on many factors, including particle shapes, distance between them, their reciprocal orientation and velocities [9]. The correction to the quasi-stationary SD force acting on

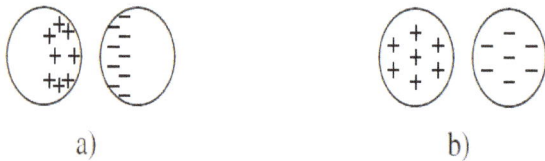

**Figure 1:** Electric charge redistribution within two conducting unlikely charged spheres (a) and two unlikely charged spheres with uniformly distributed carges (b).

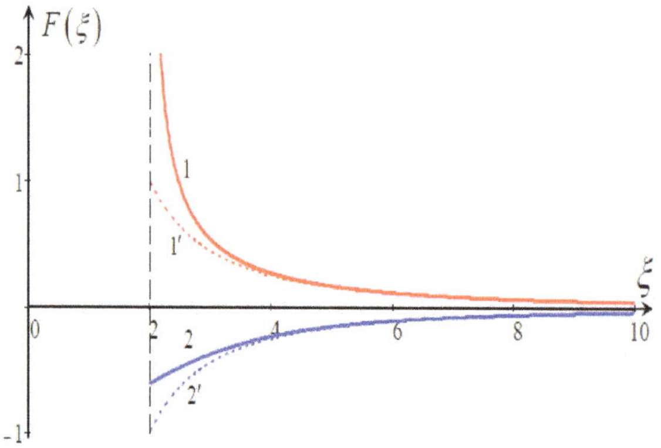


**Figure 2:** (color online) Normalized electrostatic force versus normalized distance, $\xi=x/R$, between two like (line 1) and unlike (line 2) charged particles as per Eq. (2.1). Dashed lines show Coulomb approximations as per Eq. (2.2).

particle in the presence of another particle can be taken into account through the effective viscosity $\nu_{eff}=\nu f(R_1,R_2,\mathbf{r}_1,\mathbf{r}_2,d\mathbf{r}_2/dt)$, where ν is the usual coefficient of dynamic viscosity, and f is a rather complicated function of its arguments. This function gradually reduces to unity when the distance between the particles becomes much greater than their radii.

To the best of our knowledge, modifications of the transient BBD force exerting on a particle in the presence of another particle have not been studied yet. With this in mind, we can consider two notional possibilities: (a) there is no correction to the BBD force due to the presence of another particle, so that the BBD force remains the same as for a single particle; and (b) the correction to the BBD force is described by the same effective viscosity as the SD force. In what follows, both these possibilities will be explored and results will be compared.

Consider further two spherical particles moving in a viscous fluid in the creeping flow regime. The equation of motion of one spherical particle with the added mass effect taken into account is [7,8]:

$$\left(r+\frac{1}{2}\right)\frac{d^2\mathbf{r}_1}{dt^2}=-(r-1)g+\frac{3}{4}\frac{F_s}{\pi R^3\rho}-\frac{9\nu}{2R^2}\left[f\left(R,r_1,r_2,\frac{dr_2}{dt}\right)\frac{dr_1}{dt}+\frac{R}{\sqrt{\pi\nu}}F\left(R,r_1,r_2,\frac{dr_2}{dt}\right)\int_{-\infty}^{t}\frac{d^2r_1}{d\tau^2}\frac{d\tau}{\sqrt{t-\tau}}\right], \qquad (2.3)$$

where $g$ is the acceleration due to gravity, $\rho$ is the particle density, $r$ is the particle-to-fluid density ratio. The added mass effect is taken into account through the coefficient 1/2 in the brackets in the left-hand side of the equation, the first term in the right-hand-side describes the gravity/buoyancy force, second term describes the electrostatic force and the third term describes the total drag force including the SD force (the first term in the square brackets) and the BBD force (the second, integral, term in the square brackets). Function $F\equiv 1$ if the correction to the BBD force is ignored, or $F\equiv f$ if the correction to the BBD force is the same as for the SD force.

The same equation with the indices interchanged holds for the second particle. Subtracting and summing the equations for the individual particles, we obtain

$$\left(r+\frac{1}{2}\right)\frac{d^2(\mathbf{r}_2-\mathbf{r}_1)}{dt^2}=\frac{3}{2}\frac{\mathbf{F}_S}{\pi R^3\rho}-\mathbf{D}_1; \qquad (2.4)$$

$$\left(r+\frac{1}{2}\right)\frac{d^2(\mathbf{r}_2+\mathbf{r}_1)}{dt^2}=-2(r-1)\mathbf{g}-\mathbf{D}_2, \qquad (2.5)$$

where

$$\mathbf{D}_1=\frac{9\nu}{2R^2}\left\{f\left(R,\mathbf{r}_1,\mathbf{r}_2,\frac{d\mathbf{r}_1}{dt}\right)\frac{d\mathbf{r}_2}{dt}-f\left(R,\mathbf{r}_1,\mathbf{r}_2,\frac{d\mathbf{r}_2}{dt}\right)\frac{d\mathbf{r}_1}{dt}+\frac{R}{\sqrt{\pi\nu}}\left[F\left(R,\mathbf{r}_1,\mathbf{r}_2,\frac{d\mathbf{r}_1}{dt}\right)\int_{-\infty}^{t}\frac{d^2\mathbf{r}_2}{d\tau^2}\frac{d\tau}{\sqrt{t-\tau}}-F\left(R,\mathbf{r}_1,\mathbf{r}_2,\frac{d\mathbf{r}_2}{dt}\right)\int_{-\infty}^{t}\frac{d^2\mathbf{r}_1}{d\tau^2}\frac{d\tau}{\sqrt{t-\tau}}\right]\right\}, \qquad (2.6)$$

$$\mathbf{D}_2=\frac{9\nu}{2R^2}\left\{f\left(R,\mathbf{r}_1,\mathbf{r}_2,\frac{d\mathbf{r}_1}{dt}\right)\frac{d\mathbf{r}_2}{dt}+f\left(R,\mathbf{r}_1,\mathbf{r}_2,\frac{d\mathbf{r}_2}{dt}\right)\frac{d\mathbf{r}_1}{dt}+\frac{R}{\sqrt{\pi\nu}}\left[F\left(R,\mathbf{r}_1,\mathbf{r}_2,\frac{d\mathbf{r}_1}{dt}\right)\int_{-\infty}^{t}\frac{d^2\mathbf{r}_2}{d\tau^2}\frac{d\tau}{\sqrt{t-\tau}}+F\left(R,\mathbf{r}_1,\mathbf{r}_2,\frac{d\mathbf{r}_2}{dt}\right)\int_{-\infty}^{t}\frac{d^2\mathbf{r}_1}{d\tau^2}\frac{d\tau}{\sqrt{t-\tau}}\right]\right\}. \qquad (2.7)$$

Below we consider two particular cases of particle configuration when they move (i) side-by- side as sketched in Figure 3a and (ii) one after another as shown in Figure 3b

## Two particles moving side-by-side

Considering the case of two particle moving side-by-side as shown in Figure 3a and assuming that the center of masses of the system does not move in the horizontal direction, we write down the scalar projections of Equations (2.6) and (2.7) onto the horizontal, x, and vertical, z, axes

$$(2r+1)\frac{d^2\xi}{d\theta^2} = -\frac{E_{es}}{2}S_n - f_2(\xi)\frac{d\xi}{d\theta} - \frac{3}{\sqrt{\pi}}F_2(\xi)\int_{-\infty}^{\theta}\frac{d^2\xi}{d\eta^2}\frac{d\eta}{\sqrt{\theta-\eta}}, \quad (3.1)$$

$$(2r+1)\frac{d^2\zeta}{d\theta^2} = -G(r-1) - f_3(\xi)\frac{d\zeta}{d\theta} - \frac{3}{\sqrt{\pi}}F_3(\xi)\int_{-\infty}^{\theta}\frac{d^2\zeta}{d\eta^2}\frac{d\eta}{\sqrt{\theta-\eta}}, \quad (3.2)$$

Where $\xi = \frac{x_2 - x_1}{R}, \zeta = \frac{z}{R}, \theta = \frac{9\nu}{R^2}t, E_{es} = \frac{1}{108\rho}\left(\frac{q}{\pi\nu R}\right)^2, G = \frac{2gR^3}{81\nu^2}$, and

$$S_n = \frac{\sum_{n=1}^{\infty}\left[(-1)^{\kappa_n}\frac{n\coth n\beta - \coth\beta}{\sinh n\beta}\right]}{\sinh^2\beta\left[\sum_{n=1}^{\infty}\frac{(-1)^{\kappa_n}}{\sinh n\beta}\right]^2}.$$

Functions $f_1(\xi)$, $f_2(\xi)$, $f_3(\xi)$, as well as $F_1(\xi)$, $F_2(\xi)$, *and* $F_3(\xi)$ account for the reciprocal influence of particles on the drag forces exerted on them [9]. Functions $f_1$, $f_2$ *and* $f_3$ can be presented in terms of $\xi_1 = 1/(2\xi)$ as follows.

• In the case when two particles move with equal speeds in the same direction along the line connecting their centers:

$$f_1(\xi_1) \approx 1 - 3\xi_1 + 9\xi_1^2 - 19\xi_1^3 + 93\xi_1^4 - 387\xi_1^5 + 1197\xi_1^6 - 5331\xi_1^7 + 19821\xi_1^8 - 76115\xi_1^9 + \frac{2^{20}}{3}\frac{\xi_1^{10}}{1+4\xi_1}. \quad (3.3)$$

• In the case when two particles move with equal speeds on absolute value in the opposite directions along the line connecting their centers:

$$f_2(\xi_1) \approx 1 + 3\xi_1 + 9\xi_1^2 + 19\xi_1^3 + 93\xi_1^4 + 387\xi_1^5 + 1197\xi_1^6 + 5331\xi_1^7 + 19821\xi_1^8 + 76115\xi_1^9 + \frac{2^{20}}{3}\frac{\xi_1^{10}}{1-4\xi_1}. \quad (3.4)$$

• In the case when two particles move with equal speed in the same direction side-by-side in the direction perpendicular to the line connecting their centers and can freely rotate:

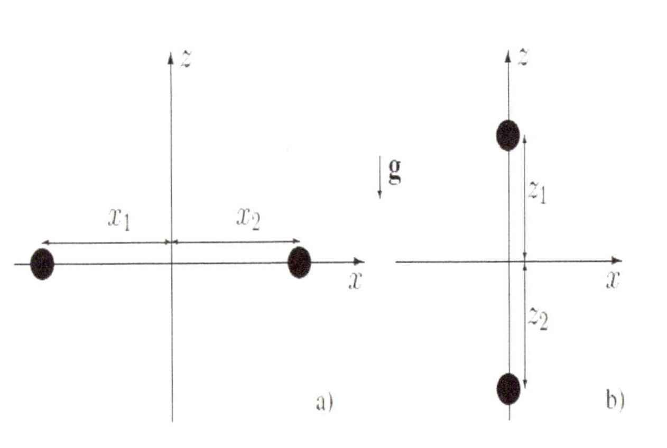


**Figure 3:** Two particles moving side-by-side in the direction normal to the line connecting their centers (a) and one after another along the line connecting their centers (b).

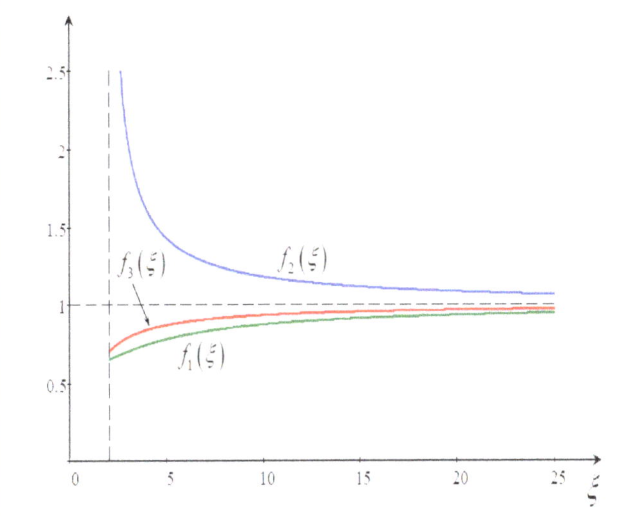

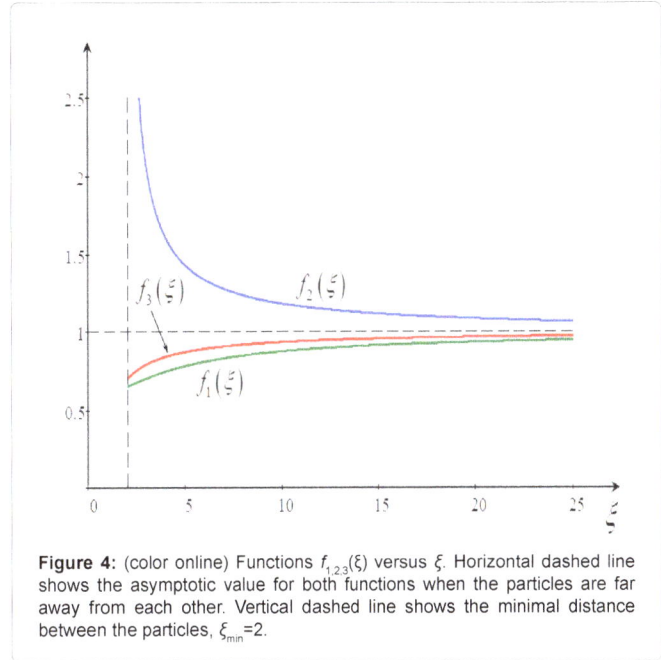

**Figure 4:** (color online) Functions $f_{1,2,3}(\xi)$ versus $\xi$. Horizontal dashed line shows the asymptotic value for both functions when the particles are far away from each other. Vertical dashed line shows the minimal distance between the particles, $\xi_{min}$=2.

$$f_3(\xi_1) \approx 1 - \frac{3}{2}\xi_1 + \frac{9}{4}\xi_1^2 - \frac{59}{8}\xi_1^3 + \frac{273}{16}\xi_1^4 - \frac{1107}{32}\xi_1^5 + \frac{64\xi_1^6}{1+2\xi_1}. \quad (3.5)$$

As have been mentioned above, expressions for functions $F_{1,2,3}(\xi)$ are not known thus far, therefore we will consider two cases, when $F_{1,2,3}(\xi) \equiv 1$ and when $F_{1,2,3}(\xi) \equiv f_{1,2,3}(\xi)$. Functions $f_{1,2,3}(\xi)$ asymptotically approach unity when $\xi \to \infty$, and the corresponding drag force reduces to the drag forces exerted on an isolated particles. However, when $\xi \to 2$, which corresponds to the minimum distance between the particle centers (when the particles touch each other), functions $f_1(\xi)$ and $f_3(\xi)$ go to the finite limits: $f_1(2)=0.647$ and $f_3(2)=0.694$, whereas function $f_2(\xi)$ grows infinitely, $f_2(\xi) \to \infty$ as $\xi \to 2_+$. Physically unacceptable behaviour of the drag forces in this case is the consequence of the approximate character of the formula. Nevertheless, as noted [9], "Hocking states that good agreement on collision efficiencies is obtained with his results and experimental data, so it is apparent that under some conditions the approximate treatment is satisfactory". The dependences $f_{1,2,3}(\xi)$ are shown in Figure 4.

Note that in the Coulomb approximation, when the distance between the particles is much longer than their radii or when the charges of spherical particles are localized in their centers, the first term in Eq. (2.6) takes the simple form $E_{es}/\xi^2$. We will study the particle interaction in both cases, with exact formula for the electrostatic force, $\mathbf{F}_s$ (2.1), and with the Coulomb approximation for the electrostatic force $\mathbf{F}_c$ (2.2).

Equation (3.1) is independent of (3.2) and can be investigated separately. Once its solution is found and $\xi(\theta)$ is determined, Equation (3.2) can be solved then (notice that (3.2) contains $\xi(\theta)$ via functions $f_3(\xi)$ and $F_3(\xi)$).

For computations we used the following values of parameters: water density $\rho=10^3$ kg/m$^3$, water kinematic viscosity $\nu=6.05 \cdot 10^{-7}$ m$^2$/s, particle radius $R=5 \cdot 10^{-5}$ m=50 μ, charge values are equal to $q=1.6 \cdot 10^{-13}$C, water permittivity $\varepsilon=6.954 \cdot 10^{-10}$ F/m, particle-to-water density ratio $r=2.7$ (this corresponds to the aluminium mote). Based on these parameters, the dimensionless parameters are $E_{es}$=3.784 · $10^{-2}$ and $G$=8.272 · $10^{-2}$.

## Particle dynamics in an in viscid fluid

Assuming that two particles with equal charges on absolute value are initially in the rest, let us study first the reference case, when the fluid is perfect and viscosity is absent (formally we put $f_{1,2,3}(\xi)=F_{1,2,3}(\xi) \equiv 0$). Then, Eq. (3.1) in the Coulomb approximation can be solved analytically; solutions for the like and unlike charged particles can be presented in the implicit forms,

$$\theta = \sqrt{\frac{\xi_0(2r+1)}{-2E_{es}}}\left[\frac{\xi_0}{2}\ln\left|\frac{\sqrt{\xi-\xi_0}+\sqrt{\xi}}{\sqrt{\xi-\xi_0}-\sqrt{\xi}}\right| + \sqrt{\xi(\xi-\xi_0)}\right], \quad E_{es} < 0; \quad (3.6)$$

$$\theta = \sqrt{\frac{\xi_0(2r+1)}{2E_{es}}}\left[\xi_0\left(\arctan\sqrt{\frac{\xi}{\xi_0-\xi}} - \frac{\pi}{2}\right) - \sqrt{\xi(\xi_0-\xi)}\right], \quad E_{es} > 0. \quad (3.7)$$

The former solution corresponds to the repulsive case when $\xi > \xi_0$, whereas the latter corresponds to the attractive case when $\xi < \xi_0$.

Solution of Eq. (3.2) without viscosity is trivial – it is simply the motion from the rest with the constant acceleration,

$$\zeta(\theta) = -\frac{G(r-1)\theta^2}{2(2r+1)}.$$

Eliminating θ from the expressions ξ(θ) and ζ(θ), we obtain particle trajectories in both cases of particle repulsion or attraction:

$$\zeta(\xi) = \frac{G(r-1)\xi_0}{4E_{es}}\left[\frac{\xi_0}{2}\ln\left|\frac{\sqrt{\xi-\xi_0}+\sqrt{\xi}}{\sqrt{\xi-\xi_0}-\sqrt{\xi}}\right| + \sqrt{\xi(\xi-\xi_0)}\right]^2, \quad E_{es} < 0; \quad (3.8)$$

$$\zeta(\xi) = -\frac{G(r-1)\xi_0}{4E_{es}}\left[\xi_0\left(\arctan\sqrt{\frac{\xi}{\xi_0-\xi}} - \frac{\pi}{2}\right) - \sqrt{\xi(\xi_0-\xi)2^2}\right]^2, \quad E_{es} > 0. \quad (3.9)$$

Solution of Equation (3.1) beyond the Coulomb approximation can be readily obtained numerically. The equation was integrated by the fourth-order Runge–Kutta method with the fixed integration step using Mathcad-14 software. The infinite sums in Eq. (3.1) were replaced by finite series containing N=200 terms. The results for attractive and repulsive particles are shown in Figure 5 by solid lines 1 and 1′ (trajectories for repulsive particles are labelled by dashed numbers and go to the right, whereas trajectories for attractive particles to the left). In that figure we also show the analytical results obtained in the Coulomb approximation as per Equations. (3.8) and (3.9); they are shown by dashed lines 2 and 2′. Lines 3, 3′ and 4, 4′ pertain to the case of viscous fluid when both viscosity coefficients, for SDF and BBDF, are equally modified by functions $f_2(\xi)$ (3.4) for the horizontal motion and $f_3(\xi)$ (3.5) for the vertical motion (the detailed discussion of the viscosity effect will be presented in the next subsection). Lines 3 and 3′ pertain to the case of exact electrostatic force, and lines 4 and 4′ pertain to the Coulomb approximation. In all cases shown in the figure the particles started to move from the rest when the distance between them was 4 in the dimensionless units. Attractive particle collision occurs when the distance between them ξ=2.

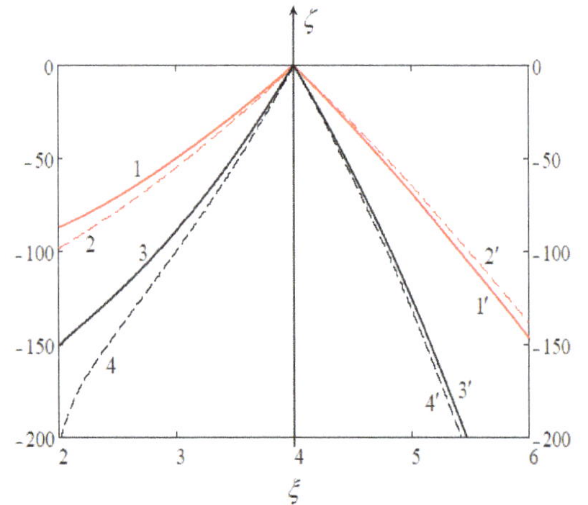

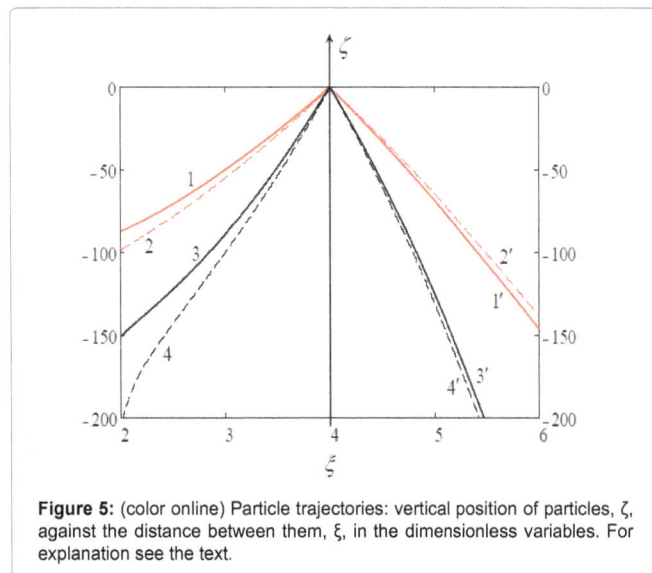

**Figure 5:** (color online) Particle trajectories: vertical position of particles, ζ, against the distance between them, ξ, in the dimensionless variables. For explanation see the text.

As one can see from this figure, the particles collide in a finite time, when they are attracted by each other due to electrostatic force. The collision occurs sooner when the exact electrostatic force is taken into consideration compared to the case of Coulomb approximation. Accordingly, the vertical distance travelled by the particles before they collide is less for the former case compared to the latter case (cf. the trajectories 1 and 2 and 3 and 4 in Figure 5).

The situation is opposite when the particles repulse each other in the perfect fluid: in the case of exact electrostatic force the horizontal motion is slower than in the case of Coulomb approximation. Therefore, trajectory 1′ in the former case lies below the trajectory 2′ in the latter case. The repulsive particles move away from each other to infinity. The observed motion is the direct result of the difference between the exact electrostatic forces in comparison with the Coulomb force. The exact force is larger than the Coulomb force for the attractive particles, but smaller than the Coulomb force for the repulsive particles (Figure 2). When the distance between particles becomes large, the exact electrostatic force quickly reduces to the Coulomb force (Figure 2). However, if particles start moving at a relatively small distance between them, the time lag of trajectory 2′ relative to trajectory 1′ still occurs.

In the viscous fluid repulsive particles move faster in horizontal direction when the Coulomb approximation is used. Therefore, trajectory 4′ lies below the trajectory 3′ which corresponds to exact electrostatic force. This is, apparently, a consequence of a complicated character of modified BB drag with the variable viscosity coefficient $\nu_{eff}$. We will revert to this issue in the next subsection.

Figure 6 illustrates the variation of relative particle velocity in horizontal direction with the distance between them. When particles attract each other their relative speed at the moment of collision (at ξ=2) is higher when the exact electrostatic force is considered then in the case of Coulomb approximation (cf. lines 1 and 2 in Figure 6). When particles repeal from each other, their relative speed varies with distance almost equally both in the case of exact electrostatic force and in the Coulomb approximation; therefore lines 1′ and 2′ are practically indistinguishable.

In viscous fluid horizontal speed of particles always greater when the exact electrostatic force is used in comparison with the Coulomb force (cf. lines 3 and 4, as well as lines 3′ and 4′. Notice that the vertical scale for the viscous case represented by lines 3, 4, 3′, and 4′ (shown on the right) is 20 times greater than in the inviscid case (shown on the left) represented by lines 1, 2, 1′, and 2′.

## Particle dynamics in viscous fluid

The description of particle dynamics becomes much more

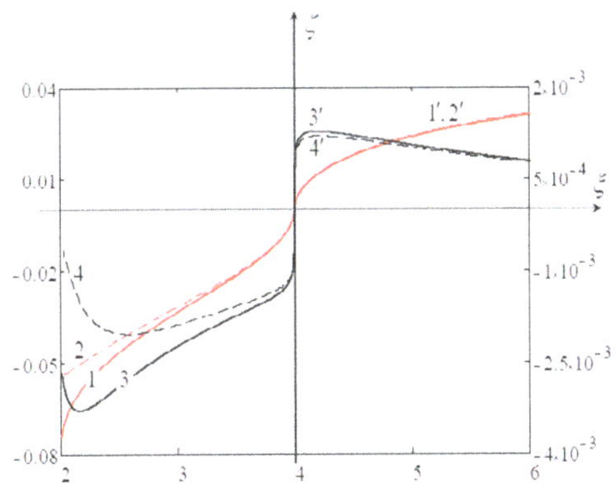


**Figure 6:** (color online) Relative horizontal velocity of particles against the distance between them in the dimensionless variables. Curve numbering corresponds to Figure 5. The vertical scale for the viscous case represented by lines 3, 4, 3′, and 4′ is shown on the right.

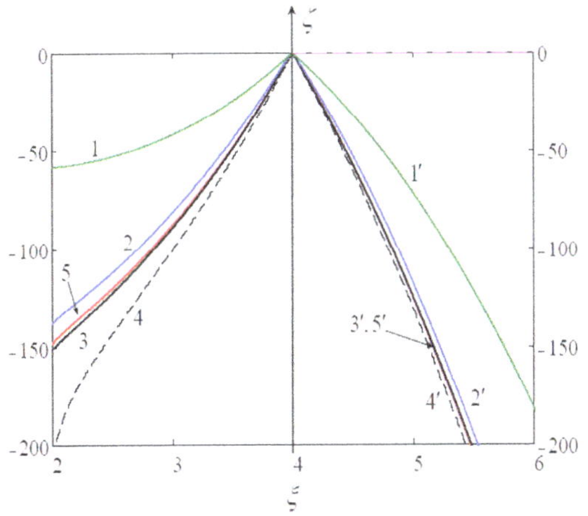


**Figure 7:** (color online) Particle trajectories in viscous fluid with different models of viscosity. For explanations see the text.

complicated when the viscosity is taken into account. The simplest case is the motion of uncharged particles with $E_{es}=0$. Consider first the case when the initial distance between the particles is so large that functions $f_{1,2,3}(\xi)$ and $F_{1,2,3}(\xi)$ can be replaced by unities. In such case the set of equations (3.1)–(3.2) can be solved analytically [3], however, the solution is quite cumbersome. Here we only present the universal asymptotic form of solution for large θ assuming that particles commence motion with zero vertical velocities $v(0) \equiv d\zeta / d\theta|_{\theta=0} = 0$, but with non-zero relative horizontal velocity $u(0) \equiv d\xi / d\theta|_{\theta=0} = u_0$,

$$\xi_{as}(\theta) = (1+2r)u_0\left(1 - \frac{3}{\sqrt{\pi\theta}}\right), \quad u_{as}(\theta) = \frac{3}{2}\frac{2r+1}{\sqrt{\pi}}\frac{u_0}{\theta^{3/2}}. \tag{3.10}$$

$$\zeta_{as}(\theta) = -G(r-1)\left[2(4-r) - 6\sqrt{\frac{\theta}{\pi}} + \theta\right], \quad v_{as}(\theta) = -G(r-1)\left(1 - \frac{3}{\sqrt{\pi\theta}}\right) \tag{3.11}$$

If the distance between the particles is not large enough, then functions $f_{1,2,3}(\xi)$ and $F_{1,2,3}(\xi)$ cannot be replaced by unities. None of Equations (3.1) and (3.2) is integrable in this case. Equations (3.1) and (3.2) were integrated numerically by means of Fortran code using the fourth-order Runge–Kutta scheme and the standard RKGS subroutine in Gill's modification. The infinite sums in Equation (3.1) were replaced by finite series with N=400 terms. The numerical code has been tested against the exact analytical solutions [3] and demonstrated quite reliable results. Examples of numerical solutions of Equations (3.1)–(3.2) with different models of viscosity are presented in Figure 7.

Lines 3, 3′ and 4, 4′ are the same as in Figure 5, i.e. line 3 represents the trajectory of attractive particles when exact electrostatic force is considered and viscosity coefficients are modified in accordance with Equations (3.4) for the horizontal motion and (3.5) for the vertical motion. Line 4 represents the trajectory when the Coulomb approximation is used for electrostatic force. Lines 3′ and 4′ represent the trajectories for the repulsive particles with exact electrostatic force and in the Coulomb approximation correspondingly. As one can see, in the repulsive case trajectories 3′ and 4′ are fairly close to each other, whereas in the attractive case the difference between them is quite noticeable at small distances between the particles. In what follows we consider only exact electrostatic force.

Line 1 (1′) represents particle trajectory in the attractive (repulsive) case when only SD force is taken into consideration with the constant viscosity, i.e. when the influence of another particle is ignored, as well as influence of the BBD force.

Line 5 (5′) represents particle trajectory in the attractive (repulsive) case when only SD force is taken into consideration with the modified viscosity as per Equations (3.4) and (3.5) when the influence of the BBD force is ignored.

Line 2 (2′) represents particle trajectory in the attractive (repulsive) case when the SD force is taken into consideration with the modified viscosity as per Equations (3.4) and (3.5), whereas the viscosity coefficient for the BBD force is assumed constant $F_{2,3}=1$.

As one can see from these graphics, the model with only SD force with constant viscosity pro-videos the results which significantly differ from the results of other models with variable viscosity and BBD force. In the meantime, lines 5 and 5′ are very close to lines 3 and 3′ correspondingly.

This indicates that the BBD force does not play a significant role in comparison with SDF in such motions and hence can be neglected. Figure 8 illustrates the variation of relative particle velocity in the horizontal direction with a distance between them. Line labels in this figure correspond to labels of trajectories in Figure 7. Only the vertical

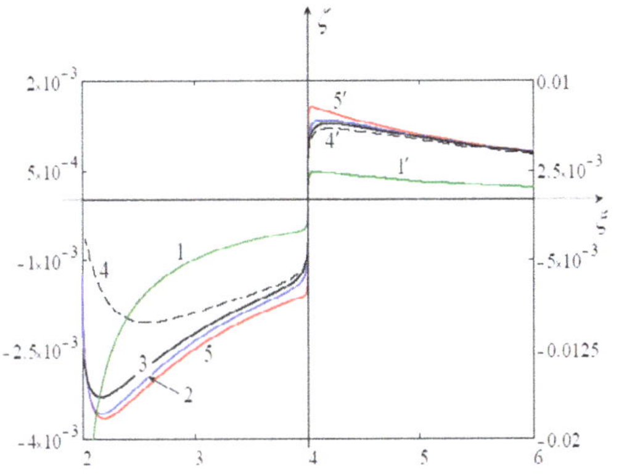


**Figure 8:** (color online) Relative horizontal velocity of particles against the distance between them in the dimension less variables. Curve numbering corresponds to Figure 7. The vertical scale for lines 1 and 1′ are shown on the right. Labels for lines 2′ and 3′ are not shown; these two lines are very close to each other and are disposed between the lines 4′ and 5′.

scale of line 5 (but not line 5'!) is 10 times compressed in comparison to all other lines.

Observe that qualitatively the difference between the exact and Coulomb cases is similar to that shown in Figure 5. Namely, for the attractive particles the horizontal motion is faster when the exact electrostatic force is considered, while for the repulsive particles the horizontal motion is a bit faster when the Coulomb electrostatic force is used.

It is interesting that at the initial stage of motion the relative horizontal speed increases very rapidly, then the speed continue increasing, but with the moderate rate, and then after reaching a maximum value it decreases due to strong influence of drag force correction $f_2(\xi)$ caused by the close presence of the second particle. If such correction is ignored, then the relative horizontal speed monotonically increases until particle collision (cf. line 5 with other lines 1-4 in Figure 7).

As have been noted, in the attraction case the relative horizonatl speed is high for the case of exact electrostatic force in comparison to Coulomb approximation. It is also interesting to note that both the drag-correction factor $f_2(\xi)$ and electrostatic force infinitely increase when the attractive particles approach each other (when $\xi \to 2$). However, the influence of variable viscous term prevails over electrostatic force resulting in the speed deceleration at moment of collision.

## Two Particles Moving One After Another

Consider now the case when two particles move one after another as shown in Figure 3b). Equations of motion in the scalar dimensionless form follow again from Equations (2.3):

$$(2r+1)\frac{d^2\xi}{d\theta^2} = -\frac{E_{es}}{2}S_n - f_2(\xi)\frac{d\xi}{d\theta} - \frac{3}{\sqrt{\pi}}\int_{-\infty}^{\theta}\frac{d^2\xi}{d\eta^2}\frac{d\eta}{\sqrt{\theta-\eta}}, \quad (4.1)$$

$$(2r+1)\frac{d^2\zeta}{d\theta^2} = -G(r-1) - f_1(\xi)\frac{d\zeta}{d\theta} - \frac{3}{\sqrt{\pi}}\int_{-\infty}^{\theta}\frac{d^2\zeta}{d\eta^2}\frac{d\eta}{\sqrt{\theta-\eta}}, \quad (4.2)$$

now $\zeta=(z_1+z_2)/2R$ is the dimensionless coordinate of the mass center, and other dimensionless quantities are defined after Equation (3.2). Function $f_1(\xi)$ is defined in Equations (3.3). Equation (4.1) describes time variation of the relative distance between the particles; it is exactly the same as Eq. (3.1, whereas Eq. (4.2) slightly differs from Eq. (3.2) due to replacement of function $f_3(\xi)$ by function $f_1(\xi)$. The difference between these two functions is not too big, as one can see from Figure 4, therefore solution of the set (4.1, 4.2) does not differ too much from the solution of the set (3.1, 3.2).

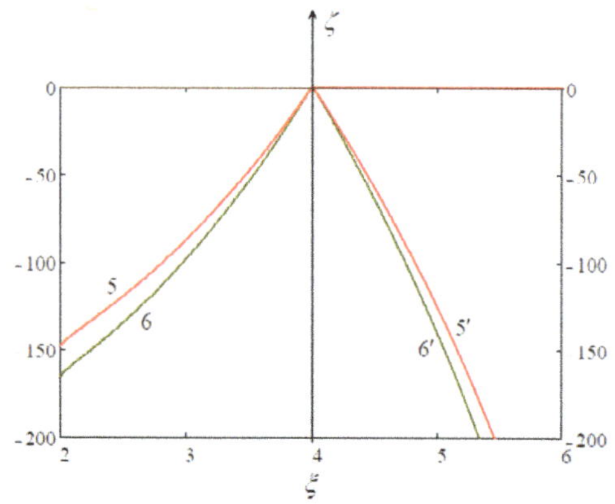

Figure 9: (color online) Particle trajectories moving side-by-side (lines 5 and 5') and vertically one after another (lines 6 and 6') in a viscous fluid. For the detailed explanations.

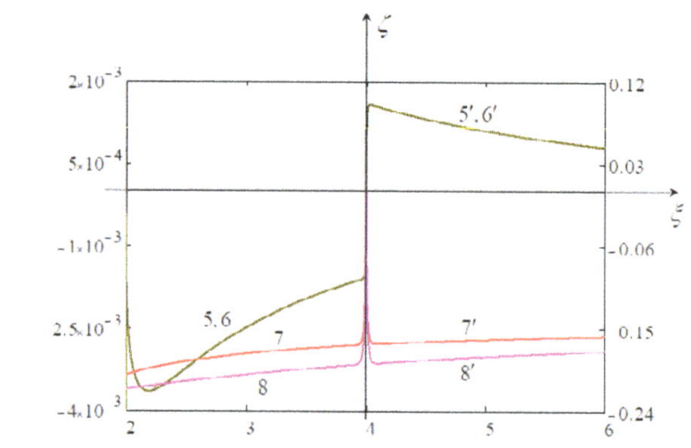

Figure 10: (color online) Relative velocity and the velocity of mass center of two particles against the distance between them in the dimensionless variables. Lines 5, 6 and 5', 6' pertain to the relative velocity of attractive and repulsive particles. Lines 7, and 7' pertain to the side-by-side vertical motion of two particles, and lines 8, and 8' pertain to the vertical motion of two particles one after another. For the detailed explanations see the text. The vertical scale for lines 7, 8, 7', and 8' is shown on the right.

In Figure 9 we present a comparison of two trajectories when the exact electrostatic force was used with the modified viscosity coefficient of SDF only, whereas BBDF was taken with the constant coefficient. Lines 5 and 5' pertain to the case when particles move side-by-side, and lines 6 and 6' pertain to the case when particles move vertically one after another. In the later case the drag force for the motion of mass center is less than in the former case ($f_1(\xi) < f_3(\xi)$), therefore the traversed path by mass center before the particle collision in the later case ($z \approx -166.5$) is a bit greater than in the former case ($z \approx -148.5$).

Due to Equations (3.1) and (4.1) are the same, the relative particle velocities are equal in the corresponding cases of side-by-side and vertical motions (see lines 5, 5' and 6, 6' in Figure 10). But the vertical motion of mass centers in these two cases is slightly different due to the difference in Equations (3.2) and (4.2) (see lines 7, 7' and 8, 8' in Figure 10).

## Discussion and Conclusion

We have considered the dynamics of two unlike and like charged particles in viscous fluid in the creeping flow approximation. Relative particle dynamics have been studied under different models of electrostatic force acting between the particles: the force between two conducting spheres as derived by Saranin [2] and Coulomb's force between point-like particles. Two types of viscous drag forces were taken into consideration: the quasi-stationary Stokes drag force and the transient Boussinesq–Basset drag force. Different models of viscous drag forces were analysed, when the viscosity coefficient is constant, like in the case of a single particle, and when it is modified due to the presence of a second particle. Comparison of particle trajectories under the influence of all these forces were studied and compared with the case of inviscid fluid.

Using the typical value of parameters (see the paragraph before the subsection 3.1.), we obtain that two aluminium micro-particles of a radii 50 μ approaching each other from the distance 200 μ traverse 7.5 mm in the vertical direction before the collision. The maximal relative velocity between the particles is ~ 0.44 mm/s, and their vertical velocity attains ~ 2.28 cm/s.

Results obtained can be useful for the development of control methods of micro- and nano-particle dynamics in viscous fluids in application to technological processes and medicine [10].

## References

1. Shoji M (2004) Studies of boiling chaos: A review. Int J Heat Mass Transfer 47: 1105-1128.
2. Saranin VA (1999) On the interaction of two electrically charged conducting balls. Phys Usp 42: 385-390.
3. Stepanyants YA, Yeoh GH (2009) Particle and bubble dynamics in a creeping flow. Eur J Mech B/Fluids 28: 619-629.
4. Gorodtsov VA (1975) Creeping motion of a drop in a viscous fluid. J Appl Mech Tech Phys 16: 865-868.
5. Lovalenti PM, Brady JF (1993) The force on a bubble, drop, or particle in arbitrary time-dependent motion at small Reynolds number. Phys Fluids A 5: 2104-2116.
6. Kim S, Karrila SJ (2005) Micro hydrodynamics: Principles and Selected Applications. Dover Publ Inc Mineola, New York.
7. Batchelor JK (1970) An Introduction to Fluid Dynamics. Cambridge University Press, Cambridge.
8. Landau LD, Lifshitz EM (1988) Hydrodynamics. (4thedn), Nauka, Moscow.
9. Happel J, Brenner H (1983) Low Reynolds Number Hydrodynamics. Kluwer Academic, London.
10. Sarvazyan A, Ostrovsky L (2009) Stirring and mixing of liquids using acoustic radiation force. JASA 125: 3548-3554.

# Numerical and Experimental Investigations on the Meshing Model Choice of a NACA2415 Airfoil Wind Turbine Placed in an Open Wind Tunnel

**Zied Driss[1]*, Tarek Chelbi[1], Walid Barhoumi[1], Ahmed Kaffel[2] and Mohamed Salah Abid[1]**

[1]*Laboratory of Electro-Mechanic Systems (LASEM), National School of Engineers of Sfax (ENIS), University of Sfax, B.P. 1173, Road Soukra 3038, Sfax, Tunisia, Africa*
[2]*University of Maryland College Park, MD 20742, USA*

**Abstract**

In this work, we are interested on the numerical and experimental investigations of a NACA2415 airfoil wind turbine placed in an open wind tunnel. The study of the meshing effect on the numerical results was developed using a commercial CFD code based on the resolution of the Navier-Stokes equations in conjunction with the standard k-ε turbulence model. These equations were solved by a finite volume discretization method. The developed numerical results are compared with experimental results to choose the adequate meshing.

**Keywords:** Wind turbine; NACA2415; Wind tunnel; Meshing; Modeling; CFD

## Nomenclature

$C$: Width, m

$C_{1\varepsilon}$: Constant of the k-ε turbulence model, dimensionless

$C_{2\varepsilon}$: Constant of the k-ε turbulence model, dimensionless

$C_{\mu}$: Constant of the k-ε turbulence model, dimensionless

$F_i$: Force components, N

$G_k$: Production term of turbulence, $kg.m^{-1}.s^{-3}$

$k$: Turbulent kinetic energy, $J.kg^{-1}$

$L$: Length, m

$p$: Pressure, Pa

$R$: Rotor radius, m

Re: Reynolds number

$t$: Time, s

$u_i$: Velocity components, $m.s^{-1}$

$u_i'$: Fluctuating velocity components, $m.s^{-1}$

$V$: Magnitude velocity, $m.s^{-1}$

$x_i$: Cartesian coordinate, m

$x$: Cartesian coordinate, m

$y$: Cartesian coordinate, m

$z$: Cartesian coordinate, m

$\beta$: Wedging angle, °

$\varepsilon$: Dissipation rate of the turbulent kinetic energy, $W.kg^{-1}$

$\mu$: Dynamic viscosity, Pa.s

$\mu_t$: Turbulent viscosity, Pa.s

$\rho$: Density, $kg.m^{-3}$

$\sigma_k$: Constant of the k-ε turbulence model

$\sigma_\varepsilon$: Constant of the k-ε turbulence model

$\delta_{ij}$: Kronecker delta function, dimensionless

## Introduction

To prepare a truly sustainable development, the community recommends increasing the share of renewable resources for electricity generation. The production of electricity by wind turbines is playing a major role. In this context, many scientists have examined the effects of wind turbines parameters design. For example, Hirahara et al. [1] developed a unique and very small wind turbine with a diameter of 500 mm and a small aspect ratio for wide use in urban space. The basic performance was tested for various free stream and load resistance. The airflow around the turbine was investigated using a particle image velocimetry (PIV). Wright and Wood [2] showed that the acceleration and deceleration of the rotor at speeds below its controlled maximum speed for a range of wind speeds were calculated and compared with data. Schreck and Robinson [3] showed that wind turbine blade aerodynamic phenomena can be broadly categorized according to the operating state of the machine, and two particular aerodynamic phenomena assume crucial importance. At zero and low rotor yaw angles, increasing rotation determines blade aerodynamic response. At moderate to high yaw angles, dynamic stall dominates blade aerodynamic. The main goal of the Mirzaei et al. [4] investigation was to understand the flow field structure of the separation bubble formed on NLF-0414 airfoil with glaze-ice accretions using CFD and hot-wire anemometry and comparing these results with previous researches performed on NACA 0012 airfoil. Hu et al. [5] showed that Coriolis and centrifugal forces play important roles in 3D stall-delay. At the root area of the blade, where the high angles of attack occur, the effect of the Coriolis and centrifugal forces is dominant. Thus, it shows apparent stall-delay phenomenon at the inner part of the blade. However, by increasing the Reynolds number, the separation position has a stronger effect than by increasing the Coriolis and centrifugal forces. Sicot et al. [6] investigated the aerodynamic properties of a wind turbine airfoil.

***Corresponding author:** Zied Driss, Laboratory of Electro-Mechanic Systems (LASEM), National School of Engineers of Sfax (ENIS), University of Sfax, B.P. 1173, Road Soukra 3038, Sfax, Tunisia, Africa, E-mail: zied.driss@enis.tn

Particularly, they studied the influence of the inflow turbulence level (from 4.5% to 12%) and of the rotation on the stall mechanisms in the blade. A local approach was used to study the influence of these parameters on the separation point position on the suction surface of the airfoil, through simultaneous surface pressure measurements around the airfoil. Tahar Bouzaher [7] studied the flow around a NACA2415 airfoil, with an 18° angle of attack, and flow separation control using a rod. It involves putting a cylindrical rod-upstream of the leading edge- in vertical translation movement in order to accelerate the transition of the boundary layer by interaction between the rod wake and the boundary layer. The rod movement is reproduced using the dynamic mesh technique and an in-house developed UDF (User Define Function). Results showed a substantial modification in the flow behavior and a maximum drag reduction of 61%.

Despite the numerous papers already devoted to this subject, studying the aerodynamic characteristics of the horizontal axis NACA2415 airfoil type wind turbine is still needed. The present work is concerned to study the meshing effect on the numerical results developed using the software "Solid Works Flow Simulation". The developed numerical results are compared with experimental results conducted on an open wind tunnel to choose the adequate numerical model.

## Experimental Device

The considered wind turbine is a horizontal axis with a NACA2415 airfoil type placed in the test section of the wind tunnel. The wind tunnel is an open type and provides a stable and uniform air flow in the test section through a downstream vacuum. The compounds of the wind tunnel are presented in Figure 1. The wind turbine consists of three adjustable blades of a length L=110 mm and a width C=45 mm. The rotor radius of the turbine is equal to R=157 mm. The wedging angle is measured between the rotation plane of the wind turbine and the chord; it's equal to β=30° (Figure 2). To develop various experimental tests required at the laboratory scale, we used a specific instrumentation [8-10] (Table 1).

### Anemometer

To determine the velocity profiles in different directions pre-selected in the test section of the wind tunnel, the anemometer type AM 4204 has been used (Figure 3). This anemometer measures wind speed in different positions with a range variation between 0.2 and 20 $m.s^{-1}$ and a resolution reaching 0.1 $m.s^{-1}$. The different characteristics of this anemometer are summarized in Table 2.

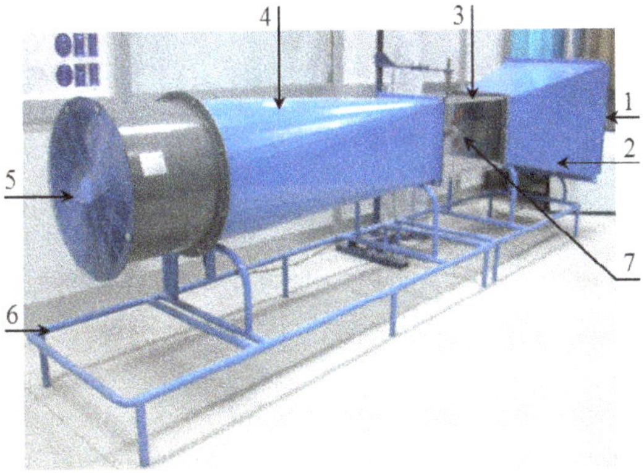


**Figure 1:** Compounds of the open wind tunnel.

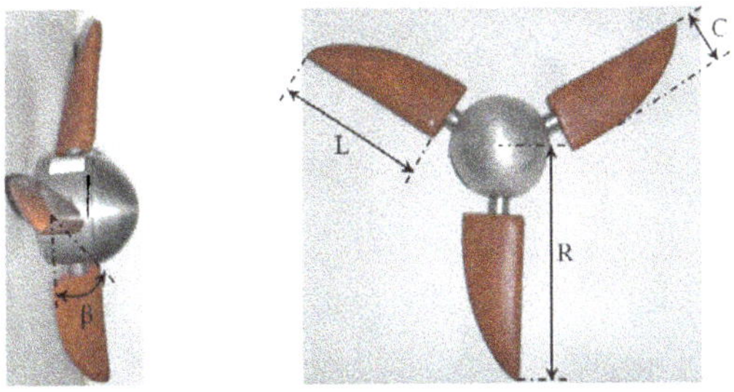


**Figure 2:** Geometric parameters of the wind turbine.

| | |
|---|---|
| 1 | Tranquilization chamber |
| 2 | Collector |
| 3 | Test section |
| 4 | Diffuser |
| 5 | Ventilation chamber |
| 6 | Support |
| 7 | Wind turbine |

**Table 1:** Compounds of the open wind tunnel.

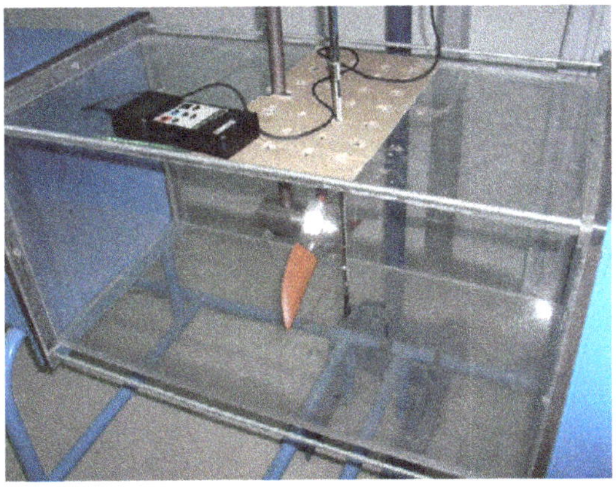

**Figure 3:** Wind speed measuringusing the anemometer type AM 4204.

| Description | Anemometer type AM 4204 |
|---|---|
| **Maker** | Lutron |
| **Probe type** | telescopic |
| **Measurement parameters** | Air velocity, temperature, gas flow |
| **Resolution** | Air velocity 0.1 $m.s^{-1}$<br>Temperature 0.1°C |
| **Precision** | Air velocity 5%<br>Temperature ± 0.8°C |
| **Measuring range** | Velocity 0.1 $m.s^{-1}$<br>Temperature from -20°C to +70°C |

**Table 2:** Characteristics of the anemometer.

### Tachometer

Figure 4 presents the tachometer type CA1725. This device is used for measuring the rotation speed of the rotor. The optical sensor can provide results without disrupting the movement of the rotor. Tachometer features are summarized in Table 3.

## Numerical Model

The software "Flow Simulation" is based on solving Navier-Stokes equations with a finite volume discretization method. The technique

**Figure 4:** Tachometer type CA1725.

| Description | Tachometer type CA1725 |
|---|---|
| **Speed range** | 6 to 100000 tr/min |
| **Resolution** | 0.0006 to 6 according size |
| **Precision** | $10^{-4}$ reading ± 6 points |
| **Supply** | 9 V |
| **Autonomy** | 250 steps of 5 min with optical sensor |
| **Dimension** | 21 × 72 × 47 mm |
| **Weight** | 250 g |

**Table 3:** Characteristics of the tachometer.

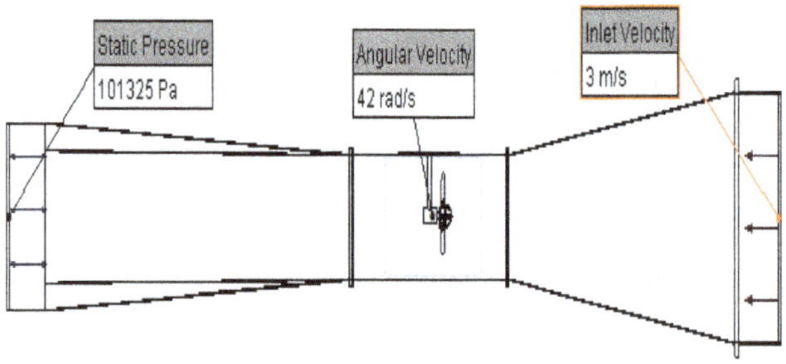


**Figure 5:** Boundary Conditions.

consists in dividing the computational domain into elementary volumes around each node in the grid; it ensures continuity of flow between nodes. The spatial discretization is obtained by following a procedure for tetrahedral interpolation scheme. As for the temporal discretization, the implicit formulation is adopted. The transport equation is integrated over the control volume [10-13].

## Computational domain and boundary conditions

Using the software "Solid Works", the computational domain is defined by the interior volume of the wind tunnel blocked by two planes: the first one is in the tranquillization chamber entry and the second one is in the exit of the diffuser (Figure 5). The boundary conditions are very important and require a lot of precision for an exact description of the problem in order to get satisfactory results. The velocity inlet, measured in the tranquilization chamber, is equal to V=3 $m.s^{-1}$. The static pressure of the air flow through the drive section is made at the atmospheric conditions. For this reason, the pressure outlet is set equal to p=101325 Pa. Around the wind turbine, a rotating area was considered with an angular velocity equal to $\Omega$=42 $rad.s^{-1}$. In these conditions, the Reynolds number is equal to Re=265403. For the initial conditions, we choose null values for all parameters of the computational domain.

## Navier-stokes equations

The equations that govern the movement of fluids are often called the Navier-Stokes equations. Most of these equations involve a process of time-averaging. When turbulence is included, the transported quantity is assumed to be the sum of an average and a fluctuating component [10-11].

The continuity equation is written as follows:

$$\frac{\partial \rho}{\partial t}+\frac{\partial(\rho\, u_i)}{\partial x_i}=0 \qquad (1)$$

The momentum equations are written as follows:

$$\frac{\partial(\rho\, u_i)}{\partial t}+\frac{\partial(\rho\, u_i u_j)}{\partial x_j}=-\frac{\partial p}{\partial x_i}+\frac{\partial}{\partial x_j}\left[\mu\left(\frac{\partial u_j}{\partial x_j}+\frac{\partial u_j}{\partial x_i}-\frac{2}{3}\delta_{ij}\frac{\partial u_i}{\partial x_i}\right)\right]+\frac{\partial(-\rho\, \overline{u_i' u_j'})}{\partial x_j}+F_i \qquad (2)$$

To close the system of equations, we have used the k-ε turbulence model. This model has been used in different anterior works and satisfactory results were obtained [12-15]. It consists on the transport equation of the turbulent kinetic energy k and the transport equation of the dissipation rate of the turbulent kinetic energy ε:

$$\frac{\partial(\rho\, k)}{\partial t}+\frac{\partial(\rho\, u_i\, k)}{\partial x_i}=\frac{\partial}{\partial x_j}\left[\left(\mu+\frac{\mu_t}{\sigma_k}\right)\frac{\partial k}{\partial x_j}\right]+G_k-\rho\varepsilon \qquad (3)$$

$$\frac{\partial(\rho\, \varepsilon)}{\partial t}+\frac{\partial(\rho\, u_i\, \varepsilon)}{\partial x_i}=\frac{\partial}{\partial x_j}\left[\left(\mu+\frac{\mu_t}{\sigma_\varepsilon}\right)\frac{\partial \varepsilon}{\partial x_j}\right]+C_{1\varepsilon}\frac{\varepsilon}{k}G_k - C_{2\varepsilon}\rho\frac{\varepsilon^2}{k} \qquad (4)$$

The hypothesis introduces also the turbulent viscosity $\mu_t$. This term is derived from both k and ε, and involves a constant taken from experimental data:

$$\mu_t=\rho C_\mu \frac{k^2}{\varepsilon} \qquad (5)$$

To summarize the solution process for the k-ε model, transport equations are solved for the turbulent kinetic energy k and the dissipation rate of the turbulent kinetic energy ε. The solutions for k and ε are used to compute the turbulent viscosity $\mu_t$. Using these results, the Reynolds stresses can be computed for substitution into the momentum equations. Once the momentum equations have been solved, the new velocity components are used to update the turbulence generation term. This process is repeated until the convergence of the solutions [12-15].

## Meshing

The goal of this section is to demonstrate various meshing capabilities of "Flow Simulation" allowing us to better adjust the computational mesh to the problem at hand. Although the automatically generated mesh is usually appropriate, intricate problems with thin or small, but important, geometrical and physical features can result in extremely high number of cells, for which the computer memory is too small. In such cases flow simulation options allow us to manually adjust the computational mesh to the solved problem's features to resolve them better. The geometry can be resolved reasonably well. However, if we generate the mesh and zoom in a thin region, we will see that it may still unresolved. In order to resolve these regions properly, we will use the local initial mesh option. The local initial mesh option allows us to specify an initial mesh in a local region of the computational domain to better resolve the model geometry and flow peculiarities in this region. The local region can be defined by a component of the assembly, disabled in the component control dialogue box, or specified by selecting a face, edge or vertex of the model. Local mesh settings are applied to all cells intersected by a component, face, edge, or a cell enclosing the selected vertex. The local mesh settings do not influence the basic mesh but are basic mesh sensitive: all refinement levels are set with respect to the basic mesh cell. To refine the mesh only in a specific region and avoid excessive splitting of the mesh cells in other parts of the model, we apply a local initial mesh at the component surrounding this region. The component is created specially to specify the local initial mesh. The settings on the narrow channels tab control the mesh refinement in the model's flow passages. Characteristic number of cells across a narrow channel box specifies the number of initial mesh cells (including partial cells) that flow simulation will try to set across the model's flow passages in the direction normal to solid/fluid interface. If possible, the number of cells across narrow channels will be equal to the specified characteristic number. Otherwise, it will be close to the characteristic number. If this condition is not satisfied, the cells lying in this direction will be split to satisfy the condition.

In this application, we are interested on the study of the mesh resolution's effect. In fact, we are going to change the size of the mesh and compare the results with the values of the velocity collected from the test section obtained experimentally. Figures 6 and 7 show the 2D

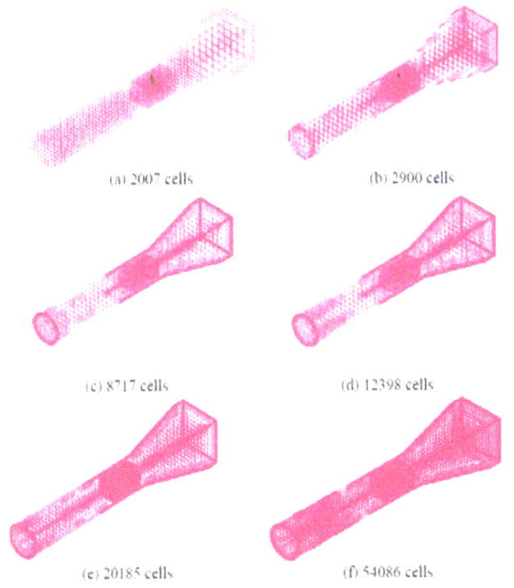


**Figure 6:** 3D view of the meshing.

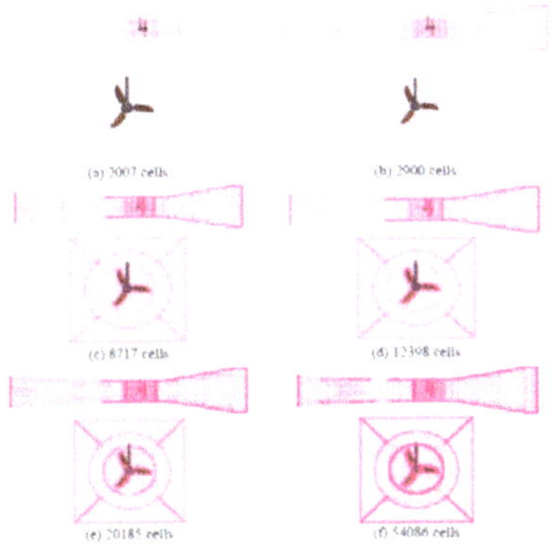


**Figure 7:** 2D view of the meshing.

| Cases study | Cells number | Resolution time (h:m:s) | Velocity (m.s$^{-1}$) | |
|---|---|---|---|---|
| | | | Numerical | Experimental |
| 1 | 2007 cells | 0:1:50 | 17.47 | 13.94 |
| 2 | 2900 cells | 0:2:00 | 16.18 | |
| 3 | 8771 cells | 0:9:32 | 15.15 | |
| 4 | 12398 cells | 0:13:41 | 14.88 | |
| 5 | 20185 cells | 0:21:37 | 15.97 | |
| 6 | 54086 cells | 1:07:27 | 13.34 | |

**Table 4:** Mesh selection criterium.

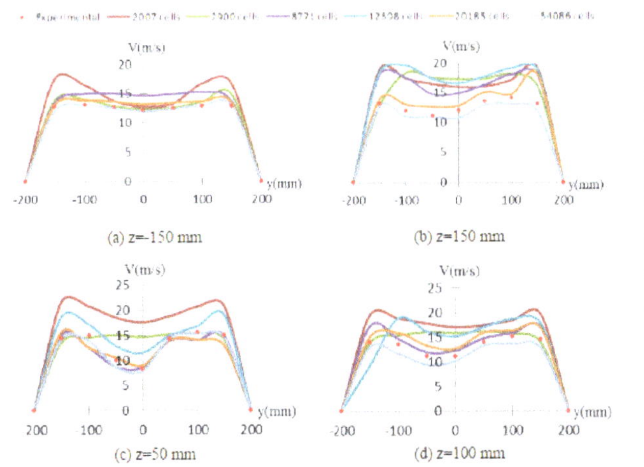


**Figure 8:** Velocity profiles.

and the 3D views of the meshing. In particular, we have chosen to study six meshes. The first case to be treated corresponds to a cell size of 200 mm. The second case corresponds to a cell size of 100 mm. The third case corresponds to a cell size of 20 mm. The fourth case corresponds to a cell size of 15 mm. The fifth case corresponds to a cell size of 10 mm. The latter one corresponds to a cell of 5 mm. In these cases, the number of cells is respectively equal to 2007, 2900, 8717, 12398, 20185 and 54086 cells; which corresponds to a coarse mesh in the first case and a refined mesh in the sixth case.

## Comparison Between Numerical and Experimental Results

In this work, computer investigations are carried out to study the flow field developing around a NACA2415 airfoil wind turbine. The models tested are implemented in the software "Solid Works Flow Simulation" which uses a finite volume scheme. The numerical results are compared with experiments conducted on an open wind tunnel to validate the numerical results. This will help improving the aerodynamic efficiency in the design of packaged installations of the NACA2415 airfoil type wind turbine. Table 4 presents the resolution time and the velocity value measured in the test section for the treated cases. According to these results, it has been noted that the velocity value obtained for the fifth case is the closest to the experimentally measured value for the point repered by the intersection of the planes defined by x=50 mm, y=50 mm and z=100 mm. Also, it has been observed that the time resolution increases with the decrease of the size of mesh cells. Later in the work, it is proposed to use the fifth mesh. This choice leads us to a better result with regards to the precision and the spent time.

Figure 8 shows the different profiles of the average velocity for different cells size. It presents the superposition of the numerical results gathered from the software "Solid works Flow simulation" and the experimental results taken by the anemometer. The velocity profiles are chosen for different directions situated in the test section. The considered directions are defined by the intersection of the plane x=0 mm with the planes z=50 mm, z=100 mm, z=150 mm and z=-150 mm. The different velocity profiles seem to have the same appearance. However, the velocity values depend on the cell size. Indeed the greater the cell size gets the more the gap between numerical and experimental results is large. The best result regarding precision and time is found to be a cell of 5 mm size. In this case, the gap is about 4%. Regarding to these results, the numerical model can predict the aerodynamic results with a good agreements.

## Conclusion

In this work, we are interested on the study of the meshing effect on the numerical results of a horizontal axis wind turbine with a NACA2415 airfoil type. Particularly, we have changed the cells size and we have compared the numerical results with the values of the velocity collected experimentally from an open wind tunnel to choose the adequate numerical model. The different velocity profiles seem to have the same appearance. However, the velocity values depend on the cell size. Indeed the greater the cell size gets the more the gap between numerical and experimental results is large. The best result regarding precision and time is found to be a cell of 5 mm size. In the future, we intend using the particle image velocimetry laser (PIV) system to determine the local characteristics.

### References


1. Hirahara H, Hossain MZ, Kawahashia M, Nonomura Y (2005) Testing basic performance of a very small wind turbine designed for multi-purposes. Renewable Energy 30: 1279-1297.
2. Wrigh AK, Wood DH (2004) The starting and low wind speed behaviour of a small horizontal axis wind turbine. Journal of Wind Engineering and Industrial Aerodynamics 92: 1265-1279.
3. Schreck SJ, Robinson MC (2007) Horizontal Axis wind turbine blade aerodynamics in experiments and modeling. IEEE Transactions on Energy Conversion 22: 61-70.
4. Mirzaei M, Ardekani MA, Doosttalab M (2009) Numerical and experimental study of flow field characteristics of an iced airfoil. Aerospace Science and Technology 13: 267-27.
5. Hu D, Hu O, Du Z (2006) A study on stall-delay for horizontal axis wind turbine. Renewable Energy 31: 821-836.
6. Sicot C, Devinant P, Loyer S, Hureau J (2008) Rotational and turbulence effects on a wind turbine blade: investigation of the stall mechanisms. Journal of Wind Engineering and Industrial Aerodynamics 96: 1320-1331.
7. Tahar Bouzaher M (2014) Numerical study of flow separation control over a NACA2415 airfoil, world academy of science, engineering and technology. International Journal of Mechanical, Aerospace, Industrial and Mechatronics Engineering 8: 782-785.
8. Driss Z, Damak A, Karray S, Abid MS (2012) Experimental study of the internal recovery effect on the performance of a Savonius wind rotor. Research and reviews: Journal of Engineering and Technology 1: 15-21.
9. Damak A, Driss Z, Abid MS (2013) Experimental investigation of helical Savonius rotor with a twist of 180°. Renewable Energy 52: 136-142.
10. Driss Z, Mlayeh O, Driss D, Maaloul M, Abid MS (2014) Numerical simulation and experimental validation of the turbulent flow around a small incurved Savonius wind rotor. Energy 74: 506-517.
11. Frikha S, Driss Z, Hagui MA (2015) Computational study of the diffuser angle effect in the design of a waste heat recovery system for oil field cabins. Energy 84: 219-238.
12. Mabrouki I, Driss Z, Abid MS (2014) Numerical Study of the hydrodynamic structure of a water savonius rotor in a test section. Jordan Journal of Mechanical and Industrial Engineering 8: 127-136.
13. Chtourou W, Ammar M, Driss Z, Abid MS (2014) CFD prediction of the turbulent flow generated in stirred square tank by a rushton turbine. Energy and Power Engineering 6: 95-110.
14. Ammar M, Chtourou W, Driss Z, Abid MS (2011) Numerical investigation of turbulent flow generated in baffled stirred vessels equipped with three different turbines in one and two-stage system. Energy 36: 5081-5093.
15. Driss Z, Bouzgarrou G, Chtourou W, Kchaou H, Abid MS (2010) Computational studies of the pitched blade turbines design effect on the stirred tank flow characteristics. European Journal of Mechanics B/Fluids 29: 236-245.

# On the Exact Values of Daubechies Wavelets

**Hajji MA***

*Department of Mathematical Sciences, United Arab Emirates University, United Arab Emirates*

**Abstract**

In this paper we propose an algorithm for the calculation of the exact values of compactly-supported Daubechies wavelet functions. The algorithm is iterative, performing a single convolution operation at each step. It requires solving, at the first step only, a linear system of a relatively small size. The novelty of the algorithm is that once the values at dyadic points at a certain level *j* are calculated they do not need to be updated at the next step. We find that this algorithm is superior to the well-known cascade algorithm proposed by Ingrid Daubechies. This algorithm can serve well in wavelet based methods for the numerical solutions of differential equations. The algorithm is tested on Daubechies scaling functions as well as Daubechies coiflets. Comparison with the values obtained using the cascade algorithm is made. We found that the cascade algorithm results converge to ours.

**Keywords**: Daubechies wavelets; Daubechies coiflets; Multiresolution analysis; The cascade algorithm; Convolution

## Introduction

The widespread interest in wavelets and their applications started in the 1980s after the breakthrough made by Daubechies [1,2] in constructing the first orthogonal compactly-supported wavelets with arbitrary regularity. Since then many researchers from different fields of science and engineering jumped into the world of wavelets with different intentions. Some were interested in ways to apply wavelets in their fields and others were interested in developing new theories and generalizations.In the engineering side, wavelets have found great success in signal processing such as in the analysis of sound patterns and image processing [3-6]. Wavelets have also found great success in the design and efficient implementation of numerical algorithms for the solution of differential equations [7-16].

Wavelet collocation based methods for solving different equations require knowledge of the values of the wavelet basis elements at collocation points. However, many of the available wavelets, such as the well-known Daubechies compactly-supported wavelets, do not have explicit formulas. Instead, they are defined recursively by refinement equations. In many applications having accurate values of the wavelet bases functions is very important in obtaining accurate solution to the problem. In [2] Daubechies described an algorithm known as the cascade algorithm for computing approximate values of the compactly-supported scaling and wavelet functions with arbitrary high precision. This algorithm works as a refinement scheme. At each step approximately twice as many values are computed, values at odd dyadic points $2^{-j}(2k+1)$ are computed for the first time and values at even dyadic points $2^{-j}(2k)$ are refined from the previous step. In the long run, i.e., as $j\to\infty$, the cascade algorithm produces the exact values. In this work, we propose an algorithm to calculate the *exact* values of Daubechies scaling and wavelet functions. The proposed algorithm avoids the refinement step in the cascade algorithm. Moreover, our algorithm, at each step computes only values at odd dyadic points $2^{-j}(2k+1)$. The values at even dyadic points $2^{-j}(2k)$ have already been computed at the previous step and no need for refinement because the values are exact.

The paper is organized as follows. In section 2, we briefly review compactly-supported scaling and wavelets functions and some related properties. In section 3, we outline the cascade algorithm in [2]. In section 4, we describe the proposed algorithm. In section 5, we apply the proposed algorithm to Daubechies scaling functions and compare our results to those obtained by the cascade algorithm. Finally, we conclude by some remarks in section 6.

## Preliminaries

Wavelet basis is a doubly-index family of $L^2(R)$ functions, $\psi_{j,k}$, $j,k\in Z$, defined by

$$\psi_{j,k}(x) = 2^{j/2}\psi(2^j x - k), \tag{1}$$

where $\psi(x)$ is the mother wavelet defined in terms of a mother scaling function $\varphi(x)$ via the refinement equation:

$$\psi(x) = \sqrt{2}\sum_k g_k\phi(2x-k). \tag{2}$$

The mother scaling function, $\varphi(x)$, is itself defined recursively by the refinement equation:

$$\phi(x) = \sqrt{2}\sum_k h_k\phi(2x-k). \tag{3}$$

The coefficients $h_k$ and $g_k$ are known, in the language of signal processing, as the low- and high-pass filter coefficients, respectively. For orthogonal wavelets, they are related by $g_k = (-1)^k h_{1-k}$.

The scaling function $\varphi$ generates an orthogonal multiresolution analysis (MRA) [17] which is an increasing sequence of subspaces $V_j$ of $L^2(R)$ (approximation spaces) with the following properties

- $\bigcap_j V_j = \{0\}$ and $\overline{\bigcup_j V_j} = L^2(R)$.
- $f(x)\in V_j \Leftarrow f(2x)\in V_{j-1}$.
- $f(x)\in V_j \Leftarrow f(x-2^{-j}k)\in V_j, \forall k\in Z$.
- $\{\phi(x-n), n\in Z\}$ is an orthonormal basis of $V_0$, where $\int\phi(x)dx\neq 0$.

A consequence of the above MRA sructure are the following:

***Corresponding author:** Hajji MA , Department of Mathematical Sciences, United Arab Emirates University, United Arab Emirates
E-mail: mahajji@uaeu.ac.ae

• The set of functions $\{\varphi_{j,k}(x), k\in Z\}$ is an orthonormal basis for $V_j$, where $\varphi_{j,k}(x)=2^{j/2}\varphi(2^j x-k)$.

• Associated with each $V_j$ there is a *wavelet* space $W_j$, its orthogonal complement in $V_{j+1}$, i.e., $V_j \oplus W_j = W_{j+1}$.

• An orthonormal basis for $W_j$ is the set $\{\psi_{j,k}(x), k\in Z\}$.

• $\bigoplus_{j\in Z} W_j = L^2(R)$.

Let $P_j$ and $Q_j$ be the orthogonal projections onto $V_j$ and $W_j$, respectively. Any function $f \in L^2(R)$ can be approximated by a function $f_j \in V_j$ through

$$f \approx f_j = P_j(f) = \sum_k s_k^j \phi_{j,k}, \qquad s_k^j = \langle f, \phi_{j,k} \rangle = \int_{-\infty}^{\infty} f(x)\phi_{j,k}(x)dx$$

Similarly, the orthogonal projection of $f$ onto $W_j$ gives

$$Q_j(f) = \sum d_k^j \psi_{j,k}, \qquad d_k^j = \langle f, \psi_{j,k} \rangle = \int_{-\infty}^{\infty} f(x)\psi_{j,k}(x)dx$$

By the structure of the MRA ($V_j = V_{j-1} \oplus W_{j-1}$), we have

$$P_j(f) = P_{j-1}(f) + Q_{j-1}(f).$$

The coefficients $s_k^{j-1}$ and $d_k^{j-1}$ at a lower scale are computed from $s_k^j$ via Mallat's algorithm [17]:

$$s_k^{j-1} = \sum_n h_{n-2k} s_n^j, \tag{4}$$

$$d_k^{j-1} = \sum_n g_{n-2k} s_n^j, \tag{5}$$

and conversely $s_k^j$ are reconstructed via

$$s_k^j = \sum_n h_{k-2n} s_n^{j-1} + \sum_n g_{k-2n} d_n^{j-1}. \tag{6}$$

For compactly-supported wavelets, the sums in (2) and (3) are finite:

$$\phi(x) = \sqrt{2}\sum_{k=0}^{L-1} h_k \phi(2x-k), \tag{7}$$

$$\psi(x) = \sqrt{2}\sum_{k=0} g_k \phi(2x-k), \tag{8}$$

with $g_k = (-1)^k h_{L-k-1}$ so that both and $\psi$ are supported in [0, $L$-1] .The integer $L=2M$ where $M$ is the number of vanishing moments of $\psi$:[2]

$$\int_{-\infty}^{\infty} x^m \psi(x)dx = 0, \qquad 0 \le m \le M-1.$$

## The Cascade Algorithm

The cascade algorithm is an iterative scheme proposed by Daubechies explanied in [1-3] to calculate approximate values of the scaling and wavelet functions, $\varphi$ and $\psi$, at rational dyadic points $x=2^{-m}k$ . In order to compare this algorithm to our proposed algorithm, it is worthwhile to describe it.

The cascade algorithm is based on the key fact that the scaling function $\varphi$ is the unique function satisfying

$$\langle f(x), \phi(x-n) \rangle = \delta_{0,n}, \tag{9}$$

$$\langle f(x), \psi_{j,k}(x) \rangle = 0, \qquad \forall\, j \ge 0, k \in Z. \tag{10}$$

It is also based on the fact that $2^j\varphi(2^j x)$ is an approximate $\delta$-function as $j\to\infty$ in the sense of the following proposition [2].

**Proposition 1** *If f is a continuous function. Then for any* $x\in R$,

$$\lim_{j\to\infty}\int_{-\infty}^{\infty} f(x+y)2^j\phi(2^j y)dy = f(x). \tag{11}$$

Moreover if $f$ is uniformly continuous, then the convergence in (11) is uniform as well, and If $f$ is Hölder continuous with exponent $\alpha$, i.e., $|f(x)-f(y)| \le C|x-y|^\alpha, \forall\, x,y \in R$, then the convergence in (11) is exponentially fast in $j$, i.e.,

$$|f(x) - \int_{-\infty}^{\infty} f(x+y)2^j\phi(2^j y)dy| \le C2^{-j\alpha}. \tag{12}$$

A consequence of the above proposition is the following.

**Lemma 2** *For any dyadic rational* $x=2^{-m}k$,

$$\phi(2^{-m}k) = \lim_{j\to\infty} 2^{j/2} \langle \phi, \phi_{j,2^{j-m}k} \rangle. \tag{13}$$

Moreover, if $j$ is sufficiently large, say $j > j_0$, we have the estimate

$$|\phi(2^{-m}k) - 2^{j/2}\langle \phi, \phi_{j,2^{j-m}k}\rangle| \le C2^{-j\alpha}, \tag{14}$$

Where $C$ and $j_0$ depend on $m$ and $k$.

Proof By Proposition 1, we have

$$\phi(k2^{-m}) = \lim_{j\to\infty}\int_{-\infty}^{\infty}\phi(2^{-m}k+y)2^j\phi(2^j y)dy$$

$$= \lim_{j\to\infty} 2^{j/2}\int_{-\infty}^{\infty}\phi(u)2^{j/2}\phi(2^j u - 2^{j-m}k)du$$

$$= \lim_{j\to\infty} 2^{j/2}\langle \phi, \phi_{j,2^{j-m}k}\rangle$$

which proves (13). Estimate (14) follows from (12).

Lemma 2 suggests that at any $j$ -level, $\varphi(2^j x)$ can be approximated by

$$\phi(2^{-j}k) \approx 2^{j/2}\langle \phi, \phi_{j,k}\rangle, \tag{15}$$

with the error $|\phi(2^{-j}k) - 2^{j/2}\langle \phi, \phi_{j,k}\rangle| \le C2^{-j\alpha}$ (see estimate (14)).

For $j,k\in Z$ , define the coefficients $c_k^j = \langle \phi, \phi_{j,k}\rangle$. Then by (15), we have

$$\phi(2^{-j}k) \approx 2^{j/2}c_k^j. \tag{16}$$

The coefficients $c_k^j$ can be reconstructed recursively using Mallat's algorithm (6) starting at the scale $j$=0. At scale $j$=0, by (9), we have

$$c_k^0 = \delta_{0,k} = \begin{cases} 1, & \text{if } k=0 \\ 0, & \text{otherwise.} \end{cases} \tag{17}$$

Since $\phi \perp \psi_{j,k}$ for all $j,k \in Z$, $d_k^j = \langle \phi, \psi_{j,k}\rangle = 0$. It follows from (6) that $c_n^j$ are given by

$$c_k^j = \sum_n h_{k-2n} c_n^{j-1}. \tag{18}$$

The cascade algorithm summarizes as follows:

• Start with the sequence $c_n^0 = \delta_{n,0}$ which can be viewed as a first approximation of at the integers.

• For $j \ge 1$ compute $c_k^j$ recursively via (18). The sequence $c_k^j$ gives the approximation of $\varphi$ at the $j$ -level dyadics $x = k/2^j$: $\phi(k/2^j) \approx 2^{j/2}c_k^j$.

We note that if the length of the filter $h$ is $L$ then the length of the sequence $c_k^j$ is $2^j(L-1)-(L-2)$. At every step $j$, the algorithm computes for the first time approximations of $\varphi$ at the odd dyadics $x=(2k+1)/2^j$ and refines the approximations at the even dyadics $x=(2k)/2^j = k/2^{j-1}$ which were obtained at the previous $j$-1 step. To be more precise at step $j$, the algorithm computes a total of $2^j(L-1)-(L-2)$ values (the length $c_k^j$) of which $2^{j-1}(L-1)$ values comprise the new approximations (at the odd dyadics) and the rest ($2^{j-1}(L-1)-(L-2)$) values constitute refinements of the old approximations obtained at the previous $j$-1 step.

Before closing this section, we would like to mention that at any scale $j$ the cascade algorithm does not cover all $j$ -level dyadic points in the interval $[0, L-1]$; it only covers dyadics $_{k=0,1,\ldots,2^j(L-1)-(L-2)}$ for $k = 0,1,\ldots,2^j(L-1)-(L-2)$. Since there are $(2^j(L-1)+1)$ $j$-level dyadics in $[0, L-1]$, the $(L-3)$ dyadic points $x = L-1-k/2^j$ for $k = 0,1,\ldots,L-4$, are not covered; some of them, but not all, will be covered at next step $j+1$.

## The Proposed Algorithm

The algorithm we propose in this paper covers all dyadics at any given scale *j*. It also yields the exact values of the scaling and wavelet functions $\varphi$ and $\psi$ at dyadic points $x = 2^{-j}m,\ j,m \in Z$. Since $\psi$ is given in terms of $\varphi$, it is suffices to describe the algorithm for . The algorithm is based on finding the exact values of $\varphi$ at the integers $n = 1,2,\ldots,L-2$ (the 0 -level dyadics). These are found by solving an eigenvalue problem. Once the values of $\varphi$ at the integers are obtained, at each subsequent step, the algorithm performs a single convolution operation to calculate the values of $\varphi$ at only odd-dyadics. The algorithm is explained in the following.

Since $\varphi$ is supported in $[0, L-1]$, we have, by continuity, $\varphi(x) = 0$ for $x \le 0$ and $x \ge L-1$. Let

$$\Phi^{(0)} = [\phi(1)\ \phi(1)\ \phi(2)\ \ldots\ \phi(L-2)]^T \quad (19)$$

be the column vector containing the values of $\varphi$ at the integers. Then, according to the refinement equation,

$$\phi(x) = \sqrt{2}\sum_{k=0}^{L-1} h_k \phi(2x-k), \quad (20)$$

$\Phi^{(0)}$ satisfies the linear system

$$\Phi^{(0)} = A\Phi^{(0)} \quad (21)$$

where *A* is an $(L\text{-}2)\times(L\text{-}2)$ matrix with entries $a_{ij},\ 1 \le i,j \le L-2$, given by

$$a_{ij} = \begin{cases} \sqrt{2}h_{2i-j}, & \text{if } 0 \le (2i-j) \le L-1, \\ 0. & \text{otherwise.} \end{cases} \quad (22)$$

Equation (21) suggests that $\Phi^{(0)}$ is an eigenvector of *A* corresponding to the eigenvalue $=1$. The existence of the eigenvalue $\lambda=1$ is justified by the following proposition.

**Proposition 3** *Suppose that there exists a continuous solution $\varphi$ to (20). Then $\lambda=1$ is an eigenvalue of the matrix A in (22) and $\Phi^{(0)}$ is an associated eigenvector.*

Proof Let $\Phi^{(0)}$ be as in (19). Then by (20) $\Phi^{(0)}$ satisfies (21). If $\Phi^{(0)}=0$ (the zero vector), then $\varphi(n)=0$ for $k \in \mathbb{N}, l \in \mathbb{Z}$. This implies, using (20), that $\varphi\ (2^{-k}l)=0$ for all $k \in \mathbb{N}, l \in \mathbb{Z}$. Then by continuity of $\varphi$, this would imply that $\varphi=0$ which is a contradiction. It follows that $\Phi^{(0)} \neq 0$ and consequently $\Phi^{(0)}$ is an eigenvector of *A* associated to the eigenvalue $\lambda=1$.

Since $\varphi$ satisfies $\sum_l \phi(x-l) = 1$ for any $x$, $\sum_{n=1}^{L-2}\phi(n) = 1$. Then the vector $\Phi^{(0)}$ is equal to the normalized eigenvector of *A* corresponding to $\lambda=1$. Precisely, $\Phi^{(0)}$ is the unique solution to the $(L\text{-}1)\times(L\text{-}2)$ nonhomogeneous system

$Bx=b$ (23)

where

$$B = \begin{bmatrix} A - I_{(L-2)} \\ \text{—} \\ 11\ldots1 \end{bmatrix} \quad b = \begin{bmatrix} 0 \\ \vdots \\ 0 \\ 1 \end{bmatrix},$$

and $I_{(L\text{-}2)}$ is the identity matrix of order $(L\text{-}2)$.

Next, once the values of $\varphi$ at the integers are known, we apply the refinement equation to find the values of $\phi$ at the *odd* half integers $x = \frac{2n-1}{2}, n = 1,2\ldots,L-1$,

$$\phi((2n-1)/2) = \sqrt{2}\sum_{k=0}^{L-1} h_k \phi(2n-k-1), \qquad n = 1,2,\ldots,L-1. \quad (24)$$

Note that we do need to calculate $\varphi$ at the *even* half integers $x = \frac{2n}{2}$ as they were computed in $\Phi^{(0)}$. Let $\Phi^{(1)}$ be the column vector containing the values of $\varphi$ at the odd 1 -level dyadics $x = \frac{2n-1}{2}, n = 1,2\ldots,L-1$. Then (24) can be viewed as a convolution operation followed by downsampling by 2,

$$\Phi^{(1)} = conv(\sqrt{2}h,\ \Phi^{(0)}) \downarrow 2, \quad (25)$$

where "conv" denotes convolution and ↓ denotes downsampling by 2.

Let $\Phi^{(j)}\ j \ge 1$, be the vector of length $2^{j-1}$ $(L\text{-}1)$ containing the values of $\varphi$ at the odd *j*-level dyadics, i.e.,

$$\Phi^{(j)} = [\phi(1/2^j)\ \phi(3/2^j)\ \ldots\ \phi(L-1-1/2^j)]. \quad (26)$$

A careful examination of the refinement equation (7) gives

$$\Phi^{(j)} = conv(\tilde{h}^j,\ \Phi^{(j-1)}), \qquad for\ \ j \ge 2, \quad (27)$$

where $\tilde{h}^j$ is an "upsampled" version of the vector $\sqrt{2}h$, obtained by inserting $(2^{j-2}\text{-}1)$ zeros between every two successive entries in $\sqrt{2}h$. Explicitly,

$$\tilde{h}_k^j = \begin{cases} \sqrt{2}h_m, & if\ \ k = m2^{j-2}, 0 \le m \le L-1, \\ 0, & otherwise. \end{cases} \quad (28)$$

The length of the vector $\tilde{h}^j$ is equal to $2^{j-2}(L\text{-}1)+1$. Note that the length of $^{(j)}$ given by the convolution formula (27) is the sum of the lengths of $\tilde{h}^j$ and $\Phi^{(j-1)}$ less one, i.e., $(2^{j-2}(L-1)+1)+(2^{j-2}(L-1))-1 = 2^{j-1}(L-1)$, and it agrees with the length of $\Phi^{(j)}$ given by formula (26).

Our proposed algorithm summarizes in the following steps:

1. Compute the values of $\varphi$ at the integers($x$=1, 2,…,$L$-2)given by the normalized eigenvector of *A* corresponding to the eigenvalue 1 and collect them in a vector $\Phi^{(0)}$.

2. Convolve $\Phi^{(0)}$ with the vector $\sqrt{2}h$ and downsample by 2 to get $\Phi^{(1)}$. This gives the values of $\varphi$ at the odd 1-level dyadics $(x = \frac{2n-1}{2}, n = 1,2,\ldots,L-1)$.

3. For $j \ge 2$, compute the values of $\varphi$ at the odd *j*-level dyadics by convolving of the vector $\tilde{h}^j$ with the vector $\Phi^{(j-1)}$ where, again, $\Phi^{(j-1)}$ is the vector containing only the values of $\varphi$ at the odd (*j*-1)-evel dyadics.

The values of the wavelet function $\psi(x)$ at the any *j*-level dyadics can be computed using the relation

$$\psi(x) = \sqrt{2}\sum_{=0} g\ \phi(2x\ \ k) \quad (29)$$

All of the steps in the proposed algorithm are computationally trivial except perhaps for the first one, where an eigenvector of *A* needs to be found. The matrix *A* being sparse, however, this is not very difficult. Using the normalization of $\varphi$, $\sum_{n=1}^{L-2}\phi(n) = 1$, the sought eigenvector $\Phi^{(0)}$ can be obtained as the solution of (23).

As a comparison between the two algorithms we note the following:

- Our proposed algorithm is similar to the well-known cascade algorithm in that the computations of both algorithms involve convolutions, except for the first step in our proposed algorithm, where an relatively small size linear system has to be solved.
- The first step in our algorithm is the most expensive but it is the

crucial one because it yields the exact values at the integers from which everything else is derived.

• The cascade algorithm begins by making an initial guess for $\varphi(x)$ (a Dirac delta function), whereas our algorithm begins by computing $\varphi(x)$ at the integers (0-level dyadics).

• A clear advantage of our algorithm is that (i) it provides the exact values and (ii) once the initial step has been performed, at every subsequent step, only the values of $\varphi(x)$ at the odd dyadics need to be computed, the even dyadics at step $j$ being the odd dyadics at step $j$-1. In contrast, the cascade algorithm requires refinement of $\varphi(x)$ at the even dyadics.

## Numerical Results

We have implemented the proposed algorithm in Matlab and tested it to produce the values of Daubechies' scaling functions as well as Daubechies' coiflets. Plots of Daubechies scaling functions db4 and db6 ($M$=4 and $M$=6 ) as well as Daubechies' coiflets coif2 and coif4 ($M$=2 and $M$=3), obtained by our algorithm, are displayed in Figures 1 and 2.

We compared our numerical values with the ones obtained using the cascade algorithm which is implemented in Matlab under the function "wavefun". Samples of the numerical values obtained are displayed in Tables 1 and 2. Table 1 displays the values of db4 at the integers obtained by the cascade algorithm for $j$=5,10,15, 20 and the ones obtained by our algorithm using only $j$=0, i.e., the solution of the system (23). Table 2 displays selected values of db6 at the $j$=5 level dyadics obtained by the cascade algorithm for $j$=5,10,15, 20 and the ones obtained by our algorithm using $j$=5. The results clearly show that the cascade algorithm results converge to ours as $j$ tends to infinity.We remark that one has to iterate the cascade algorithm for larger value of $j$ to obtain accurate results at a lower $k$ level dyadics. For instance, from Table 2, we needed to iterate the cascade algorithm until $j$=20 to obtain as closer results to the ones obtained by our algorithm with only $j$=5.

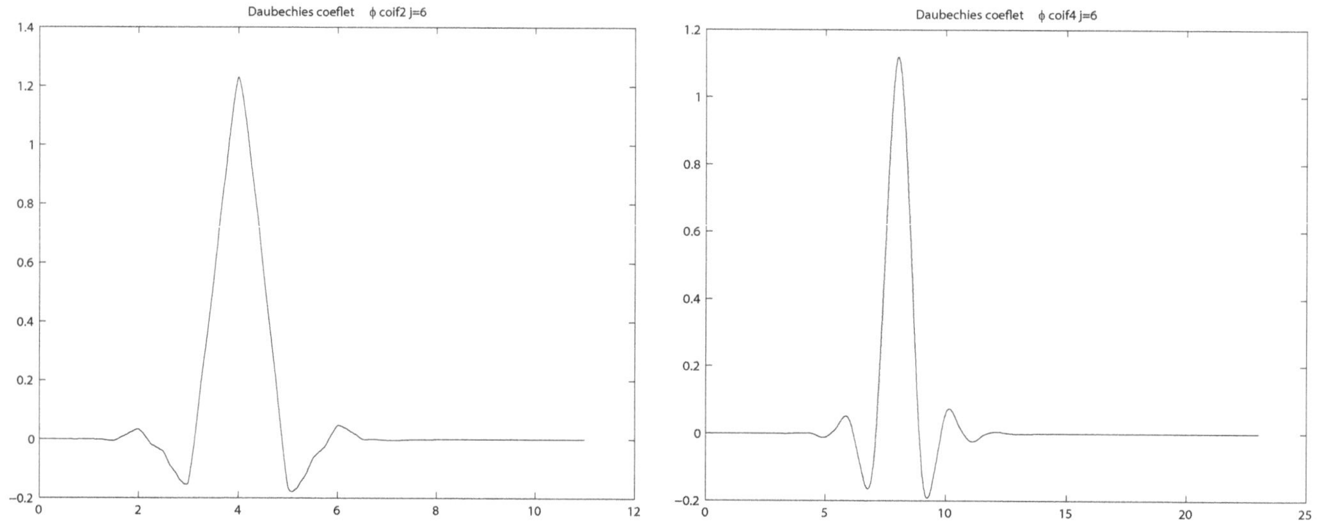


**Figure 1:** A plot of Daubechies db4 and db6.

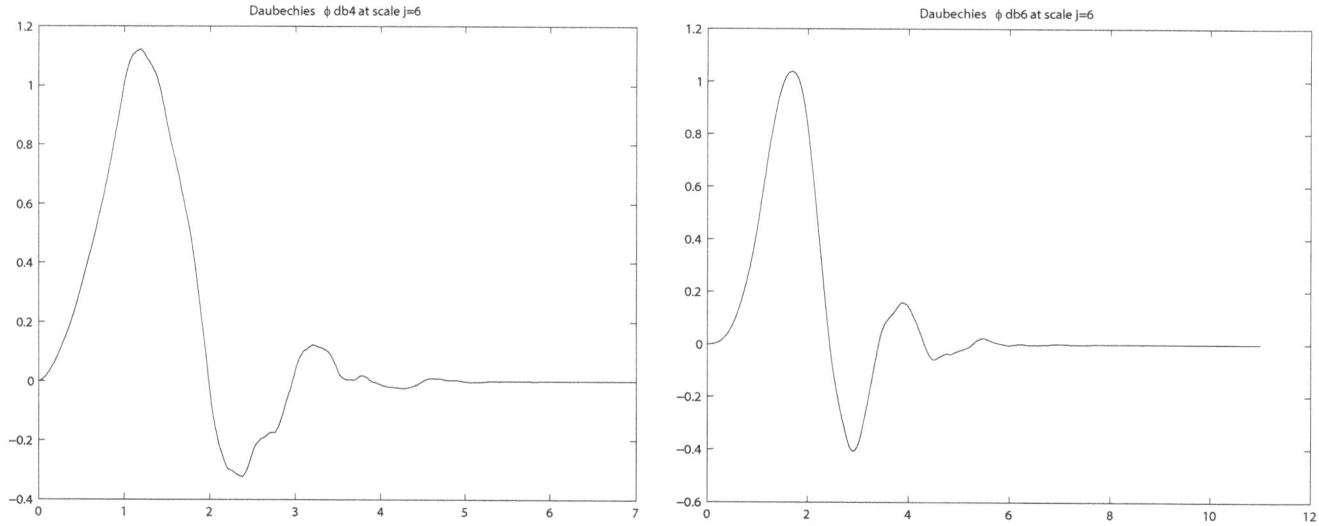


**Figure 2:** A plot of Daubechies coiflets coif2 and coif4.

| Values of Daubechies' scaling function db4 | | | | | |
|---|---|---|---|---|---|
| * $x$ | The cascade algorithm | | | | Our algorithm |
| | $j=5$ | $j=10$ | $j=15$ | $j=20$ | $j=0$ |
| | 1.00747364 | 1.00717932 | 1.00717027 | 1.00716998 | 1.00716997 |
| | -0.03432410 | -0.03385159 | -0.03383741 | -0.03383696 | -0.03383695 |
| | 0.03983843 | 0.03961669 | 0.03961065 | 0.03961046 | 0.03961046 |
| | -0.01180456 | -0.01176501 | -0.01176437 | -0.01176435 | -0.01176435 |
| | -0.00120268 | -0.00119824 | -0.00119796 | -0.00119795 | -0.00119795 |
| | 0.00001926 | 0.00001883 | 0.00001882 | 0.00001882 | 0.00001882 |

**Table 1:** Values of db4 at the integers.

| Values of Daubechies' scaling function db6 | | | | | |
|---|---|---|---|---|---|
| * $x$ | The cascade algorithm | | | | Our algorithm |
| | $j=5$ | $j=10$ | $j=15$ | $j=20$ | $j=5$ |
| $\frac{2}{2^5}$ | 0.0004331 | 0.0002749 | 0.0002707 | 0.0002705 | 0.0002705 |
| $\frac{22}{2^5}$ | 0.1696324 | 0.1623220 | 0.1620967 | 0.1620896 | 0.1620894 |
| $\frac{42}{2^5}$ | 0.8333298 | 0.8208984 | 0.8205067 | 0.8204944 | 0.8204941 |
| $\frac{62}{2^5}$ | 0.9002129 | 0.9135170 | 0.9139223 | 0.9139350 | 0.9139354 |
| $\frac{82}{2^5}$ | -0.1721720 | -0.1591060 | -0.1586933 | -0.1586804 | -0.1586800 |
| $\frac{119}{2^5}$ | 0.1241885 | 0.1214656 | 0.1213838 | 0.1213812 | 0.1213811 |
| $\frac{159}{2^5}$ | -0.0268479 | -0.0277469 | -0.0277748 | -0.0277757 | -0.0277757 |
| $\frac{199}{2^5}$ | 0.0014295 | 0.0014226 | 0.0014213 | 0.0014212 | 0.0014212 |
| $\frac{209}{2^5}$ | -0.0022161 | -0.0023020 | -0.0023046 | -0.0023047 | -0.0023047 |
| $\frac{259}{2^5}$ | 0.0000263 | 0.0000270 | 0.0000270 | 0.0000270 | 0.0000270 |

**Table 2:** Values of db6 at selected level dyadic.

## Conclusion

In this paper, we have presented an efficient algorithm for the computation of the exact values of refinable functions, in particular Daubechies' scaling and wavelet functions. Our motivation for this work stems from the need for more accurate point values of the widely used Daubechies wavelets. This certainly will be useful in the numerical solutions of differential equations where wavelets are being used. Our proposed algorithm produces the exact values whereas the well-known cascade algorithm produces approximate values. What is good about the proposed algorithm is that, at each step, it computes values only at odd dyadics. The cascade algorithm, however, at each step calculates new values and refines old ones. The only expensive step in our algorithm is the first one where we need to solve a relatively small size linear system. This first step is the crucial one in that it provides us with the exact values (to machine precision) at the integers from which the rest is derived.We believe that having exactly values of wavelet functions will give better results in wavelet based numerical schemes for the solution of differential equations.

### References

1. Daubechies I (1988) Orthonormal bases of compactly supported wavelets. Comm Pure and Appli Math 41: 909-996.
2. Daubechies I (1992) Ten Lectures on Wavelets, Society for Industrial and Applied Mathematics. Philadelphia PA
3. Daubechies I, Lagarias JC (1991) Two-scale difference equations. I. Existence and global regularity of solutions. SIAM J Math Anal 22: 1388-1410.
4. Martinet RK, Morlet J, Grossmann A (1987) Analysis of sound patterns through wavelet transforms. Internat J Patterm Recognition and Artificial Intelligence 1: 273-301
5. Antoni, M Barlaud M Mathieu P, Daubechies I (1992) Image coding using wavelet transform. IEEE Trans Image Proc 1: 205-220.

6. Devore RA, Jawerth B, Lucier BJ (1992) Image compression through wavelet transform coding. IEEE Trans Inf Th 38: 719-746.

7. Xu JC, Shann WC (1992) Galerkin-wavelet methods for two-poiny boundary value problems. Numer Math 63: 123-144.

8. Karami A, Karimi HR, Moshiri B, Maralani PJ (2008) Investigation of Interpolating Wavelets Theory for the Resolution of PDEs. Int J Contemp Math Sci 3: 1017-1029.

9. Cai W , Wang J (1996) Adaptive multiresolution collocation methods for initial boundary value problems of Nonlinear PDEs. SIAM J Numer Anal 33: 937-970.

10. Vasilyev OV, Paolucci A (1996) Dynamically adaptive multilevel wavelet collocation method for solving partial differential equations in a finite domain. J Computational Physics 125: 498-512.

11. Vasilyev OV, Paolucci S (1997) Aast adaptive wavelet collocation algorithm for multidimensional PDEs. J Computational Physics 138: 16-56.

12. Bertoluzza S, Naldi G(1996) A wavelet collocation method for the numerical solution of partial differential equations. Applied and Computational Harmonic Analysis 3: 1-9.

13. Alam JM, Kevlahan NKR, Vasilyev OV (2006) Simultaneous space–time adaptive wavelet solution of nonlinear parabolic differential equations. J Computational Physics 214: 829-857.

14. Vasilyev OV, Kevlahan NKR (2005) An adaptive multilevel wavelet collocation method for elliptic problems. J Computational Physics 206: 412-431.

15. Vasilyev OV, Kevlahan NKR (2005) An adaptive wavelet collocation method for fluid-structure interaction at high Reynolds numbers. SIAM J Sci Comput 26: 1894-1915.

16. MA Hajji, Melkonian S, Vaillancourt R (2004) Two-dimensional wavelet bases for partial differential operators and applications. In Advances in Pseudo-Differential Operators, Birkhäuser Basel, USA.

17. Mallat S (1989) Multiresolution approximations and wavelet orthonormal bases of $L^2(R)$ .Trans Amer Math Soc 315: 69–88.

# A Viscosity Hypothesis – That the Presence or Absence of Viscosity Separates Relativistic and Quantum Systems Based on the Simplest Possible Theory of Everything

**Lawrence M***

*Maldwyn Centre for Theoretical Physics, Cranfield Park, Burstall, Suffolk, UK*

**Abstract**

A simple framework for our universe in which the basic constituents act as a background upon which actions by composite particles, composed of those same constituents in motion, act and whose presence or absence from volumes give rise respectively to relativistic and quantum systems. Where the background exists, all composite particles experience energy loss in motion due to viscosity and a maximum velocity and where the background does not exist, there is no energy loss in motion and no maximum velocity. The framework is based on the simple premises of the one size of fundamental building block, the meon, two types of energy, one composite loop form of particle and only three dimensions. Composite loops formed from the unit meon building blocks during different inflation events produce different sizes of fermions, nucleons and atoms, but produce a type of universe with symmetries similar to ours as the inevitable outcome of a successful inflation event. The rate of expansion after a big bang is a function of the size of the electron formed during inflation and that size defines whether the expansion will eventually succeed or fail.

Key paradoxes are shown not to be paradoxes. This framework explains what energy and inertia are, how positive-only mass arises, spin units of ½ h, electrons with 720o of rotation, charge unit sizes, why particles have internal magnetic moments, the second law of thermodynamics and the arrow of time, where there is a maximum speed for particles, why stable states exist, why tired light may reduce the need for dark energy or the size of the universe, why there is no matter/anti-matter imbalance, what dark matter is likely to be, the physical reality underlying zero point energy, why physics fails nowhere and why there is only one universe with threefold symmetry within our nucleons.

**Keywords**: Universe, quantum mechanics; Inflation; Expansion; Black hole; Symmetry; Loops; Stacks; Chain star; Meon; Lepton; Anomalous magnetic moment; Quark; Electron; Viscosity; Dark Matter; Theory of Everything; Dark energy; Arrow of time; Second law of thermodynamics; Viscosity redshift; Prequark; Steady state

## Introduction

Due to size constraints the abstract above was kept short and the following is a more detailed account of what this paper will cover:

It is shown that starting from a very simple set of premises it is possible to construct a framework with properties significantly similar to those of our currently observed universe in which the basic constituents act as a background upon which actions by composite particles, composed of those same constituents in a different dynamic state, act and whose presence or absence from specific volumes give rise respectively to relativistic and quantum systems. The conclusion is that where the background, as described, exists then all composite particles experience energy loss in motion and a maximum velocity and where the background does not exist, there is no energy loss in motion and no maximum velocity.

The background consists of adjusted-Planck volume zero mass black holes which are each comprised of a pair of particle and anti-particle (named meon and anti-meon respectively) whose motions when wholly or partially merged are rotational, vibratory and translational. When unmerged, requiring the same amount of energy every time, the result is a pair of particle and anti-particle chasing each other with one-sixth opposite sign electron charge each. The pair can chase other pairs to form composite loops. Three-pair loops are our normal matter with other size loops, plus specific non-symmetric three-pair loops, being dark matter.

The foundation is based on the simple premises of the one size of fundamental building block, the meon, two types of energy, one composite form of particle and only three dimensions. The loops of three-pairs are the leptons and quarks. Shear viscosity is the reason for a maximum speed through the background medium when particles travel through it, for the arrow of time where all motion loses unrecoverable energy and for an additional redshift of photons. That photons travel at a terminal velocity against the local background defined as the local speed of light shows that there is internal chasing between meon and anti-meon. Where no background is present, the result is non-locality with motion above $c$ and a quantum mechanical environment.

The motion of the meons and anti-meons around the loops can be split into the balancing of mass energies and the resultant motion of charge energies. For every mass energy that a meon or composite particle has, it will always have a similar amount of charge energy, and the same, but opposite, for each meon, anti-meon or anti-particle energy. An analysis of dimensionality, deeper than mass, length and time, shows that the product of volume and shear viscosity is constant,

***Corresponding author:** Lawrence M, Maldwyn Centre for Theoretical Physics, Cranfield Park, Burstall, Suffolk, UK, E-mail: lawrence@maldwynphysics.org

so where all meons have the same volume they are all affected equally by the background shear viscosity and composites will lose rotational rate proportional to distance travelled almost regardless of loop frequency.

The nature of what is currently described as energy is shown to be both a counting phenomenon and a vector property. How what is described as the energy for the big bang was generated and where it is being used currently is shown to affect our understanding of time, mass, light speed, normal and dark matter and the background within which matter exists.

The overall movement of what is currently described as energy starts from the background state of the universe, goes through the formation of loops and gradually returns to the background through the action of the viscosity of the background. It is shown that there can be only one underlying set of the laws of physics and these laws will be the same everywhere and fail nowhere. Composite loops formed from the same meon building blocks during different inflation events can produce different sizes of fermions, nucleons and atoms, but a type of universe with symmetries similar to ours is the inevitable outcome of a successful inflation event. The presence of observable failed inflation events randomly within the background allows big bang and a form of steady state system to coexist, with earlier failed inflation events forming seeds for later gravitational accumulation during our successful big bang. The rate of expansion from a big bang is a function of the size of the electron formed during inflation and that size defines whether the expansion will eventually succeed or fail. It is also explained why only positive masses are observed, and why some particle configurations and orbits are stable.

In the framework it is shown that key paradoxes are not paradoxes, with both interpretations correct given a deeper understanding. The ideas proposed here may be viewed as speculative because they start from a prequark framework. This physics beyond the standard model derives its strength from the number of aspects of the universe that it explains, including what energy really is, how mass arises, what inertia is, why particles have spin in units of ½ *h*, why electrons have 720° of rotation, why all electrons have the same charge size, as do nucleon stacks, why particles have internal magnetic moments, why the second law of thermodynamics exists, why there is an arrow of time, why there is a maximum speed for particles, why stable states exist, why tired light may reduce the need for dark energy or the size of the universe, why there is no matter/anti-matter imbalance, what dark matter is likely to be, what is likely to be the physical reality underlying zero point energy, why physics is the same everywhere and breaks down nowhere and why we have only one universe with threefold symmetry within nucleons in matter.

This paper is not written from an arrogant perspective. The author does not suggest that everything in the paper is unquestionable, only that the current interpretations are open to reinterpretation and that what is included here is one possible solution which happens to solve many paradoxes and provides answers to many otherwise unexplainable observations. It does not doubt one single mathematically accepted observation, although it provides reasons to interpret some of them differently.

The relativistic and quantum realms are understood from fundamental differences [1] to be inconsistent. Even discounting the different treatments of time, size and position, the wave versus particle treatments are difficult to explain [2]. Non-locality has been shown [3] and yet nothing is supposed to exceed local light speed [4].

The solution to the many paradoxes between relativity and quantum mechanics starts by understanding that there are two entirely different environments within the universe – one where a 'background' of basic merged fundamental particle and anti-particles exist against which motion of composites takes place, and one where the background is excluded. This paper will show how and why this happens and how the motion of the photon is effectively a proof of this difference and of the existence at the most basic level of negative fundamental mass, although not of negative mass of composite particles or with the properties usually ascribed to it.

The quest to explain everything has progressed in fits and starts. Each new addition to understanding has been placed within the existing jigsaw of knowledge, either replacing a piece seen no longer to fit or adding to the whole which needs to be solved.

Unfortunately it appears that we have been using the jigsaw pieces to try to solve the wrong picture. They may fit together in places, but do not show the correct picture overall. History has permanently nailed so many pieces in the wrong place that it is difficult for any other picture to be contemplated as a replacement.

What this paper sets out to do is to show what the replacement picture should look like. It shows that many of the pieces are perfectly valid, but just stuck in the wrong place. And it shows that as we delve deeper into nature the mathematics needed get simpler.

As will be explained, what we describe as 'mass' is the frequency of rotation of the loop composites, which is always measured as positive, and all loop composites always have a balance of internal energies and seek to balance their external energies to total zero overall. This is why some configurations of composites in atoms or orbits are stable and shows why we need to reinterpret what we mean by energy as well as inertia.

The fundamental particle pairs that, as will be shown, comprise photons, are affected by the shear viscosity of the background universe equally, almost regardless of the photon frequency. This viscosity also provides a maximum speed of travel for all particles, which we call light speed c. The same viscosity affects the fundamental particle pairs that comprise non-photon composites, providing an arrow of time as they lose rotational rate. Where there is no background, photons are not affected by viscosity and non-locality exists, along with all quantum mechanical effects.

The paper connects the very large to the very small and provides a logical and the simplest possible foundation for a theory of everything.

## Significance and Objectives

The paper sets out to show where relativistic and quantum mechanical systems can exist. At the smallest end of the scale is a simple foundation out of which can be made the particles that we can observe and the basis for those that we cannot. Based on those most simple foundations automatically appear the symmetries we observe and the explanation of why we can only observe a fraction of what exists in the universe. The significance of the framework proposed can be seen in just two different scale examples explained later –contributions to magnetic moments in the charged leptons and how and why failed inflation events occur and what they look like afterwards to an observer.

Based on the simple foundations proposed, as published previously [5], and as corrected with the missing tables [6], the relationships between properties, such as mass, velocity and energy, and the dynamics of the foundation particles, the underlying laws of physics in any part of the universe are simple and cannot be any different. However, the actual sizes of properties are the result of the local inflation event in which the loop composites are formed. In this framework there can be no singularities since the building blocks are the densest particles possible.

The significance of this paper is that it solves many conflicting issues across physics from the very small to the very large, using only very simple foundations. It also shows how viscosity separates the relativistic and quantum environments.

## Basics

The framework used in the foundation equations here is based on the Triple-adjusted Planck Units (TAPU) set of units and property values established in the foundation paper [5] and as corrected [6].

The main relationships and principles will be outlined briefly. All the equations here use only Planck values, unless specifically mentioned otherwise. The Planck, or adjusted-Planck, values are call 'maximal' in that they represent either the largest (eg velocity, $c$) or smallest (eg distance, $L_p$) that is possible for that property.

The analysis is in three stages, where $X_x$ represents any property, such as mass, length, time etc:

a) Eliminating $2\pi$ for simplicity from the generally accepted Planck equations denoted $X_p$ to give Adjusted Planck Units (APU), denoted $X_o$

b) Eliminating $G$ to give Double-Adjusted Planck Units (DAPU), denoted $X_*$

c) Eliminating $h$ to give the final Triple-Adjusted Planck Units (TAPU), denoted $X_T$

The most basic two formulae for defining a Planck unit sized system are the gravitational force equation $F = G\,M^2/L^2 = Q^2c^2/L^2$ and the quantum angular momentum equation $h=M\,C\,L$. The normal usage of the latter is to define a Planck mass $M_p$ and Planck Length $L_p$ such that $\hbar = M_p c\ L_p$ and $M_p = \sqrt{\hbar c / G}$. Unfortunately this introduces the $2\pi$ factor in many equations, where it serves only to confuse.

The preferred definition, to be used here as a starting point, is to define the system without the $2\pi$ factor, which is the first adjustment. This adjustment is split equally between the mass and length units.

Initially the APU mass $M_o$ and APU length $L_o$ are related by $h=M_o\,c\,L_o$ and is defined to be

$$M_o = M_p\sqrt{2\pi} = \sqrt{h\ c\ /G},$$

$$L_o = L_p\sqrt{2\pi}$$

$$Q_o = Q_p\sqrt{2\pi}.$$

In the second stage, however, to achieve the right relationship between $M$ and $L$ in property space, as described below, requires looking at the force equation at the same time.

Rearranging to give $F\ L^2 = G\ M^2 = (Q\ c)^2$ provides the simple relationship that the APU mass $M_o$ and APU charge $Q_o$ are related such that $M_o\sqrt{G} = Q_o c$. Since the latter equation does not include $L_o$ it is not immediately apparent that compared with the Planck properties $M_p$ and $L_p$ there is a need to adjust both by the factor $\sqrt{G}$ in addition to the $\sqrt{2\pi}$ factor, so that now $h = M_* c\ L_* = (M_p\sqrt{2\pi G})c\,(L_p\sqrt{2\pi/G})$ if the latter factors are distributed in the same way as $\sqrt{G}$.

It is possible now to define the second adjustment such that

$$M_* = M_o\sqrt{G} = M_p\sqrt{2\pi G},\ Q_* = Q_o = Q_p\sqrt{2\pi}$$

$$L_* = L_o/\sqrt{G} = L_p\sqrt{2\pi/G}$$

with $h=M_*cL_*$, the basic DAPU units, where $Q_*$ is the DAPU charge. This is the maximum charge based on symmetry with the maximum mass and is not the electron charge, which is considered later.

The result is the foundation of a DAPU property set and units based on

$h=M_*cL_*$

and

$$F_*L_*^2 = M_*^2 = Q_*^2c^2 = h\,c$$

which was the objective of the second adjustment, in that the formulae exclude $G$. The dimensionality of $G$ will be shown to be zero later.

This is the most basic set of Planck properties that can be devised using only two universal constants $h$ and $c$. However, as shown in the third adjustment stage, this is not the minimum number of constants required to establish relationships between the properties.

The relationship between $M_*$ and $Q_*$ is simply $M_*=Q_*c$ with the deeper relationships $M_* = \sqrt{h\,c}$ and $Q_* = \sqrt{h/c}$.

The subsuming of $G$ within the mass and distance units eliminates the difference between gravitational and inertial masses, since there is no longer any purely gravitational mass.

The subsuming of $G$ within the APU mass $M_o$ to produce the DAPU mass $M_*$, and the APU length $L_o$ to produce the DAPU length $L_*$ would seem to ignore the units of $G$, effectively treating $G$ as being without units. This is not the case since $G$ has units of $m^3kg^{-1}\,S^{-2}$, but it is necessary to show that, based on Planck sizes, these units cancel completely to leave only a ratio.

A consideration of the standard laws of nature and the fundamental constants through a form of dimensional analysis shows that if each property at its maximal Planck size is assigned an appropriate dimensionality, every fundamental constant, other than $c$, will have a total dimensionality of zero, or to state the reverse – every property that has dimensionality of zero is a fundamental constant.

The dimensionalities of the main SI, NSI, APU, DAPU or TAPU properties in terms of a hypothetical dimension $Y$ that emerge from the consideration are: $M_*=Y^{+1}$

Mass $M_*=Y^{+1}$ Velocity $c=Y^{+2}$

Length $L_*=Y^{-3}$ Energy $E_*=Y^{+5}$

Charge $Q_* = Y^{-1}$, Time $T_* = Y^{-5}$, $h=Y^0$ $G=Y^0$

The units of $G$ are $m^3kg^{-1}\,S^{-2}=Y^{-9}Y^{-1}\,Y^{+10}=Y^0$ dimensionality and $h$ has units of $m^2kg\,S^{-1}=Y^{-6}Y^{+1}\,Y^{+5}=Y^0$ dimensionality. So the units of both $h$ and $G$ are actually irrelevant because they represent fundamental constants with zero dimensionality. Similarly Boltzmann's constant has units of $J\,K^{-1}=Y^5Y^{-5}=Y^0$dimensionality as well.

Thus adjusting the APU mass to the DAPU mass, and APU length to DAPU length, involves only multiplying or dividing by the ratio $\sqrt{|G|}$ as a dimensionless number, and does not affect the dimensionality of the units of mass, charge or length, other than changing the sizes of the base Planck mass, charge and distance units. This stretches the current property space into the more symmetric DAPU property space which does not affect the current property space topology at all and which is different to treating $G$ to be equal to one.

It is now necessary to make the final third adjustment to produce the most simple definitions possible of mass and charge, that is the TAPU definitions

$$M_T = M_* / \sqrt{h} = \sqrt{c}$$

$$Q_T = Q_* / \sqrt{h} = 1 / \sqrt{c}$$

$$L_T = L_* / \sqrt{h} = 1 / \sqrt{c^{-3}}$$

and to show their simple relationships to all other properties through a new ratio $\vartheta = \sqrt{c / d} = \sqrt{2 \pi c / \alpha}$.

The base formulae are now:

$$1 = M_T c \, L_T$$

and

$$F_T L_T^2 = M_T^2 = Q_T^2 c^2 = c$$

The new TAPU sets are based around the $X_T$ set $M_T = \sqrt{c}$ and $Q_T = 1 / \sqrt{c}$.

One new relationship to emerge from the dimensional analysis is that the product of Shear Viscosity $\eta^*$ of dimension $Y^{+9}$ and Volume V, of dimension $Y^{-9}$ is equal to $h$ in DAPU, so is a constant. Thus any system where all component fundamental particles, like the meons, have the same volume will experience the same effect due to shear viscosity when travelling in the background environment. As will be shown later, this means that the red shift of a photon is dependent almost entirely on the distance travelled by the individual meons rather than the frequency of the two loops that comprise the photon. Tired light is a distance phenomenon, not frequency dependent except at very high frequencies.

It is also worth noting how the current equation relating energy and time, instead of position and momentum in the original Heisenberg relationship [7], in APU was $E_o T_o = \hbar$ and now becomes $E_T T_T = 1$ in TAPU, and that our original target base has now been reached in terms of manipulating and simplifying formulae for use in explaining the framework in our type of universe.

Our 'type' of universe is one with 3-fold symmetry in nucleons, atoms and components, dark matter outweighing normal 3-fold symmetrical matter and three spatial dimensions. These are all inextricably linked in the 'chain then loop' proposals described below using only the one type of foundation particle/anti-particle.

## Foundations

The foundation is a volume which is composed of a merged particle and its anti-partner. When completely merged, nothing is observable from outside that volume.

The particle, which will be termed a 'meon' has positive properties which may be described as fundamental mass and fundamental charge, although these are not necessarily what we understand as normal mass and charge, as will be explained. The anti-partner has negative properties. The meon and anti-meon always appear together as a pair, so the term 'pair' will mean exactly that - but only when the two are completely unmerged. When the two are partly merged, they will be termed a zero-mass black hole 'ZMBH', of adjusted Planck volume. Figure 1 shows the three states of the meons - ZMBHs, pairs and individual.

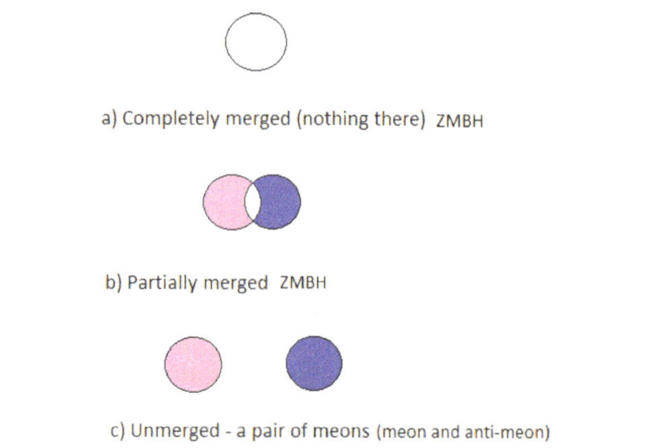


**Figure 1:** Meons and ZMBHs.

The sizes and strength of the actions of mass and charge are equal and are the same for the meons, and anti-meons, as explained in the previous foundation paper [5] in terms of the dimensionality of physical properties. What the dimensionality section shows is that Planck's constant $h$, the gravitational constant $G$ and permeability $u_-$ are constants because they have zero dimensionality - they do not depend on any other values. They are actually ratios set by the use of SI units. This paper uses triple-adjusted SI units, adjusted as explained, so that there is no need to include $h$ or $G$ in formulae. However, for clarity $h$ is used occasionally to emphasis what the normal formulae would look like, but its value in these formulae is equal to 1 in new SI units (TAPU). So the strength of a mass field and charge field of equal fractional Planck value is the same.

Myriad ZMBHs are what our universe is composed of, initially and as the 'background' within which all relativistic events occur. Where there is no background, as will be explained, there are no relativistic effects. The ZMBHs spin, move, rotate and vibrate to transfer forces - these are the force carriers, not the bosons. Composites, loops made from unmerged pairs, affect the ZMBHs and the ZMBHs affect the composite loops.

There is only one size of ZMBH, which is the triple-adjusted Planck unit (TAPU) size and can be defined simply as size equal to the unit 1 for fundamental mass, fundamental charge and volume. When completely merged, there is an empty volume of 1, with no properties observable. The ZMBH background is both a continuum, overlapping everywhere when completely merged, and yet indivisible in that the unmerged constituents cannot be divided. Only in certain circumstances can there be volumes without a ZMBH background, as will be explained later.

To unmerge a ZMBH into meon and anti-meon takes a specific amount of energy every time. The result is a meon and anti-meon, each spinning about its own internal axis (from now on this motion will be defined as 'twisting' to differentiate it from the motion of meons around loops, as explained below, which will be called rotation and the charge-angular momentum of the meons in any loop which is described as the spin of the loop and which is what is usually described as the spin of a fermion). It is proposed as a foundation assumption that the subsequent twisting motion of the meons against the background ZMBHs generates one-sixth the electron charge $q/6 = \sqrt{\alpha / 2\pi} Q/6$, where $Q$ is the TAPU Planck fundamental charge, previously shown as $Q_T$, with sign dependent on i) the spiral orientation of twist versus direction of motion and ii) the original meon fundamental charge. The twist mass energy equal and opposite to the charge energy has $s/6 = \sqrt{\alpha/2\pi}M/6$ where M is the TAPU Planck fundamental mass, previously shown as $M_T$. The motion of the pair is of one chasing the other, as if they were trying to remerge, which will be explained in the section on the hierarchy of zeroes of total energy.

The fundamental charge Q and the one-sixth electron charge $q/6$ are not necessarily the same property, but are assumed to act in the same way and this is shown to be reasonable in the later analysis of mass and charge currents for loops where half of the circumferential charge current arises from the differential dynamics of meons and anti-meons around an electron loop.

In an event where ZMBHs are being unmerged in large quantities to form pairs, which is the start of a big bang, the result of many pairs mixing and chasing each other is the formation of chains. The chains can have any length, but will be broken and reformed many times. The likelihood is that a chain will catch onto its own tail to form a loop. Like the chains, the loops can be any length but the greatest probability will be for the shortest lengths. Assuming the smallest loop to have two pairs, the loops will be 2, 3, 4, 5 and greater number of pairs in length, with the shorter ones more prevalent. In each loop the meon pairs chase, as will be described later in the consideration of the structure of the electron, so that there is always alternating meon and anti-meon along the chain that forms a loop. Figure 2 shows a loop composed of three pairs in motion, chasing each other, at angular frequency *w*.

When originally unmerged, the pair generates twist charges +q/6 and –q/6, with the sign of charge on each being dependent on the initial direction of travel of the pair, although the only charge combination initially for any pair is +-. Subsequent mixing of pairs will result in only four possible twist charge combinations for any pair ++, +-, -+ and --.

Due to the assumed existence of only one fundamental size of meon with one size of energy necessary to separate out a pair from being a ZMBH, then all composite loops, formed from pairs, will have charges in multiples of units of zero or $\pm 1/3\ q$. This will be the case for all loops formed in whatever manner, including failed inflation events described later, with the maximum loop charge being the product of the number of pairs in the loop and $\pm 1/3\ q$ or zero.

## Loop identities

Taking the loops comprised of three pairs, the 3-loops, the identity of the loops can be defined by the total size of their electronic charge and their internal symmetries – meaning where in the loop the one-sixth electron charges sit in relation to each other. The possible charge sizes for 3-loops are 0, 1/3, 2/3 and 3/3*q* electron charge, each positive or negative apart from the zero charge case.

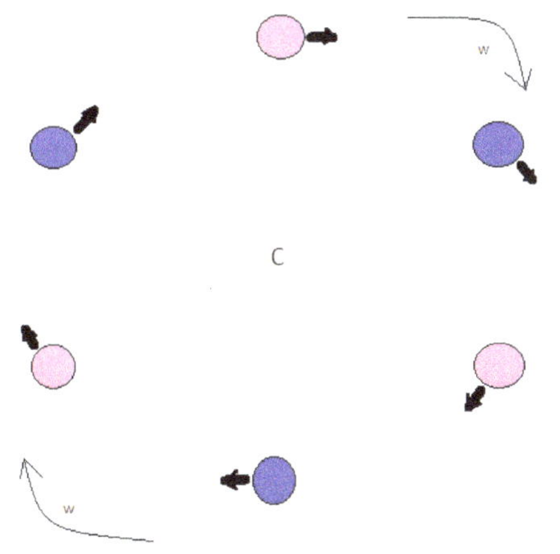


**Figure 2:** Three meon pairs chasing each other in a loop around C at angular frequency w.

The motion of the meons and anti-meons in chasing around the loop can be considered as the loop itself rotating about a central axis perpendicular to the plane of rotation of the meons/anti-meons. Loops always rotate.

The loop with zero total electronic charge (neutrino or anti-neutrino dependent on definition) can be symmetric internally with positive and negative one-sixth charges alternating around the loop, which is actually a symmetric form of hidden 3-fold asymmetry. Asymmetry here refers to pairs in a loop and will be used later more generally to differentiate between symmetric lepton loops and asymmetric quark loops when in stacks. The zero charge loop can also be 2-fold symmetric with three positive one-sixth charges facing three negative one-sixth charges across the loop. The zero charge loops can also be completely asymmetric with no symmetry. As will be explained more later, where loops do not have 3-fold symmetry they are dark matter, unable to interact with 3-fold symmetric loops, or stack, except via gravity and charge. So non-symmetric zero charge loops will not, in usual terminology, 'feel the weak force' and it may appear that there are more flavours of neutrino than the expected three – those which can stack when they are 3-fold symmetric.

The 3/3*q* charge loop can only be perfectly symmetric, with each meon twisting to generate one-sixth charge of the same sign. Again this is a symmetric form of hidden 3-fold asymmetry.

The 1/3*q* charge loop can only be asymmetric with 2-fold asymmetry or non-symmetric, as is the case for the 2/3*q* charge loop. It should be obvious that the fractional charge asymmetric 3-loops are what we call quarks and the symmetric loops, including the asymmetric zero charged loops, are what we call leptons, with all eight 3-loop charge permutations being our fermions.

Further stable structures can be made from combining numbers of these 3-loops if the result can be made rotationally symmetric along their axis of rotation. The most obvious form is a stack of loops with all axes of loop rotation coincident and planes of rotation parallel, like a stack of dishes, requiring that their rotational asymmetries be balanced overall to be stable.

This is fine for the symmetric zero and 3/3*q* charged loops because they can exist separately. However, the asymmetric loops need to be stacked so that their asymmetries cancel each other, otherwise the stack will not be stable. This requires a stack of three 3-loop quarks, a 3-stack, whose asymmetries are at 120 degrees offset to each other when viewed along the mutual axis of rotation of the loops. This is the basis for our three-colour quark colour force framework. The need for symmetry in a stack does not preclude the existence of symmetric loops within a stack. Since the symmetric leptons have hidden 3-fold asymmetry there are good reasons to conclude that a balanced asymmetric 3-stack requires 3-loop symmetric end 'caps' to effectively hide the asymmetric loops from the local environment.

## Loop masses

What is described as the mass of the loops is derived not from the underlying meons, but from their motion around the loop. Each meon should have energy $E_{M+} = (\gamma - 1)Mc^2$ in the loop, where *M* is the

TAPU Planck fundamental mass ($M+$ for a positive fundamental mass meon and $M-$ for a negative fundamental mass anti-meon) and $\gamma$ the relativistic velocity factor for the meon in its motion around the loop. Each anti-meon should have opposite energy $E_{M-}=-(\gamma-1)Mc^2$. So the total energy due to motion of fundamental mass is zero. This is the case for all energies within the loop where even the one-sixth $q$ twisting charge energy is a balance for the twist mass energy.

These two simple equations are actually too simple, although they show the concept nicely. It is necessary to consider all the energies at work, which includes the twist energy on the mass energy side, so the actual equations are:

$$E_{M+} = (\gamma_+ - 1)(M \pm a)c^2 \text{ and}$$

$$E_{M-} = -(\gamma_- - 1)(M \pm a)c^2 \text{, where}$$

$$\alpha = s/6 = \sqrt{\alpha / 2\pi M} / 6$$

and can be either added to or deducted from the fundamental mass energy. The same equations apply to the charges of the meons, giving

$$E_{Q+} = (\gamma_+ - 1)(Q \mp b)c^3 \text{ and}$$

$$E_{Q-} = -(\gamma_- - 1)(Q \mp b)c^3 \text{ where}$$

$$b = q/6 = \sqrt{\alpha / 2\pi Q} / 6 \cdot$$

The $\gamma_+$ here represents the velocity factor for each meon. These four equations are the main ones as shown later that are needed to help resolve the relative magnetic moments of all loops, plus the circumferential and radial electric fields in motion. As previously mentioned the framework used is that the positive meon has positive fundamental mass and charge, and the negative the reverse.

The total mass energy of a chain would differ depending on its charge. The electron chain would have total $q$ charge and $sc^2$ twist energy, as explained later, and the neutrino zero of each, with quarks fractional in each. But in a loop, each meon or anti-meon needs the same mass angular momentum to remain in a stable loop. So the meons and anti-meons adjust their radii of rotation around the loop so that each has the same size angular momentum, positive or negative. This will be explained more later, but the effect means that each $M_+$ has $h^+$ and each $M_-$ has $h^-$, where $h^\pm$ is not necessarily the same as $h$, in the loop, totalling zero overall. Now the sum of the loop mass energies, through the differential rotational velocities, even though all rotate at frequency $w$, is zero.

This is equivalent to a deflection of some flat field which might be termed 'space-time' except for the quantisation of time into the loops so there is no equivalent background time for ZMBHs. In deflection terms, the depths of all deflections affect each other, with a maximum set by the deflection of a single twisting meon or anti-meon. But for a loop, it is effectively the area of the loop which sets the amount of deflection of space – what we term the 'mass' of the loop. The charge of the loop is also a factor as a multiplier but does not change the discussion here.

It is the interaction of one type of energy in one meon or loop with the same type of energy in another meon or loop which underlies their mutual interaction, even though all meons and loops always have a total energy equal to zero. With only fundamental mass and fundamental charge energies, this means there must be deflections of independent flat fields for both charge and mass.

The energy of a loop is also defined by its frequency of rotation $w_{loop}$ such that the simple equation should be

$$E_{loop} = (\gamma - 1)hw_o \cong \frac{1}{2}hw_{loop}$$

Where $w_o$ is the Planck angular frequency. However, in view of the fractional $a$ and $b$ energies, this is not quite right. In order for the loop to remain stable, regardless of the mass angular momentum of any meon, as mentioned earlier, the size of each angular momentum should be the same size, positive or negative, so

$$E_{loop} = \sum \pm(\gamma - 1)hw_o \cong \sum \frac{1}{2}hw_{loop} = 0$$

As will be shown later for the electron analysis, although each meon has the same angular frequency $w$, the actual angular momentum of every meon in all loops except the symmetric neutrinos may be either $h$ or $h^\pm=1.0017h$. To keep open either possibility, $h^\pm$ will continue to be used in the paper, although it could prove to be that $h^\pm=h$. So although the mass energies sum to zero overall for all loops, as do the angular momenta, the relationship between energy and angular frequency is not the simple version. This also applies to the mass formula for a meon.

The motional energy of the meons can also be written simply for fundamental mass as

$$E_{loop} = \sum(\gamma - 1)(\pm M \pm a)c^2 \cong \sum \frac{1}{2}(\pm M \pm a)v_{loop}^2$$

Where $v_{loop}$ is the average meon velocity around the loop and is less than $c$. However, in this instance, the energy depends on how many of the meons have positive or negative $a$ energies and their actual velocities differ depending on whether the $a$ is positive or negative. This will be explained more below.

In the case of a loop with zero overall charge, the $a$ energies sum to zero and the equation becomes the simple version

$$E_{loop} = \sum \pm(\gamma - 1)Mc^2 \cong \sum \pm \frac{1}{2}Mv_{loop}^2 = 0$$

So the two methods of describing the meon motional fundamental mass energy are interchangeable - with appropriate care. What we describe as the mass or energy of a loop is just the size of the two balancing positive and negative energies, even though they always sum to zero.

The same equations apply to the meon motional fundamental charge energies, replacing $M$ by $Qc$, and the energy sizes are identical, adjusted only by $c$ for charge when working in TAPU units. These equations can also be used as the base of the charge current which is one of the components of the magnetic moments of the loops as will be shown below.

The summing of the fundamental $M$ and $Qc$ energies, for every loop, to zero overall means that what we describe as 'energy' is really just a counting mechanism relating the frequency of loop rotation to the depth, and direction, of deflection of each mass or charge field. Even relative motion will not make the energy sum anything different to zero overall. However, the maximum deflections by a loop overall will be limited to be same as that of an isolated twisting meon, thus providing a maximum loop frequency equal to the TAPU Planck frequency, with maximum meon relative velocity of $c$. The actual maximum energy balance depends on how much mass angular momentum any meon has in total.

## Loop properties and dynamics

Whilst the sum of the meon motional $\Sigma(\pm M\pm a)c^2$ fundamental mass and twist energies can be described as the loop mass, the sum of the meon motional $\Sigma(\pm Q\pm b)c^3$ fundamental charge and one-sixth electromagnetic charge energies is the spin energy of the loop – in both cases although they sum to zero, we usually consider only one side of the zero total balloon or balance. If we consider the 3/3 $q$ charged 3-loop, this is the electron and the simple equations imply that the mass energy of the electron is identical to its spin energy where

$$E_{loop} = \sum \pm(\gamma - 1)Mc^2 = \sum \pm(\gamma - 1)hw_o$$

$$\cong \sum \pm\frac{1}{2}Mv_{loop}^2 <=> oneside = m_{loop}c^2 \cong \frac{1}{2}hw_{loop}$$

These energies are not usually connected in this way because the spin of the electron is taken as ½ $h$, missing out the $w_{loop}$ factor. The more complex equations show that each loop needs to be considered meon by meon, as will be shown for the leptons in Tables 1 and 2, but the relationship between mass and frequency remains with $h^{\pm}$ angular momentum.

This identification of the mass of the loop as based on the frequency of rotation of the meons around the loop shows why the property called the 'fundamental mass' $M$ of a meon is not the same as the 'mass' of a loop. The latter will always be a positive frequency, whereas the former may be positive or negative and may not correspond to the normal interactions expected of masses. The assumption here for meon fundamental mass interactions is that same type of masses attract, opposites chase, as explained later.

What this loop framework shows is that the simple spin of a loop is $1h$, and the ±½ belongs to the frequency, so the loops have twice the frequency expected at normal energies. This is discussed in the section on the electron below and is why electrons appear to have 720 degrees of rotation, shown by the electron spin g-factor being slightly more than 2 rather than 1.The factor 2 also appears in the section on the electron showing how the charge current arises both from offset rotational radii of meons with their fundamental charges as well as from the six q/6 charges rotating around the loop.

In this explanation the consideration of the motion of both the $Q$ and $q/6$ charges leads to the non-zero magnetic moment of the charged loops, contributions to the anomalous magnetic moments and an obvious reason in the planar nature of the loops why they can be described as spin up or down and why there cannot be any magnetic monopoles.

Importantly it must be noted that because the mass of any loop is proportional to its rotational frequency rather than the underlying number of meons in the loop, the length of the loop, however many pairs it contains, is independent of the mass of the loop. The same is the case for the spin of the loop, so all single loops (including all fermions) are spin ± ½, in accepted usage.

So a 3-loop of radius x has a mass of y, and a 4-loop of radius x also has a mass of y and both are spin ± ½. However, the magnetic moment of non 3-loops will be different because the net current flow, radial and circumferential electric fields will be different. This gives a way of identifying different pair-number dark matter loops from identical charge and mass normal matter loops.

The simple dynamics of meons in a loop is given by

$h = Mv_{loop}r_{loop}$ and $v_{loop} = r_{loop}w_{loop}$

so the smaller the physical size of the loop, the larger is its energy, but the extra $a$ and $b$ energies make the actual equations more complex. It is always the case though, that $v=rw$ for all meons individually.

Apart from the charge and gravitational formulae, these few

| Type | +M | -M | +Q | -Q | H/S/Any inner/outer | H/S/Any inner/outer |
|---|---|---|---|---|---|---|
| $T_1$ | +a | +a | -b | -b | H + inner | S - outer |
| $T_2$ | -a | -a | +b | +b | S + outer | H - inner |
| $T_3$ | +a | -a | -b | +b | H + inner/Any | H - inner/Any |
| $T_4$ | -a | +a | +b | -b | S + outer/Any | S - outer/Any |

**Table 1:** Pair combinations of *H* and *S* giving types $T_x$.

| Type | e- | e+ | $v_s$ | $v_a$ | d- | d+ | u+ | u- |
|---|---|---|---|---|---|---|---|---|
| $T_{111}$ | ✓ | | | | | | | |
| $T_{222}$ | | ✓ | | | | | | |
| $T_{333}$ | | | ✓ | | | | | |
| $T_{444}$ | | | ✓ | | | | | |
| $T_{344}$ | | | | ✓ | | | | |
| $T_{334}$ | | | | ✓ | | | | |
| $T_{123}$ | | | | ✓ | | | | |
| $T_{124}$ | | | | ✓ | | | | |
| $T_{134}$ | | | | | ✓ | | | |
| $T_{133}$ | | | | | ✓ | | | |
| $T_{144}$ | | | | | ✓ | | | |
| $T_{112}$ | | | | | ✓ | | | |
| $T_{234}$ | | | | | | ✓ | | |
| $T_{122}$ | | | | | | ✓ | | |
| $T_{244}$ | | | | | | ✓ | | |
| $T_{233}$ | | | | | | ✓ | | |
| $T_{223}$ | | | | | | | ✓ | |
| $T_{224}$ | | | | | | | ✓ | |
| $T_{113}$ | | | | | | | | ✓ |
| $T_{114}$ | | | | | | | | ✓ |

**Table 2:** Combinations of types $T_x$ to give the fermions.

equations describe the only laws of physics needed at this fundamental level and will be used in their more complex form to analyse the electron below.

## Matter and anti-matter

Currently the basic assumption on matter/anti-matter asymmetry is that somehow there is an excess of matter over anti-matter and after mutual interaction, only the matter excess survives. This is not the case in the framework proposed here.

Consider a chain of pairs of any number travelling across a theoretically-existing flat surface. The chain encounters an obstacle which deflects the chain either right or left so that it catches its own tail. One version will become a clockwise rotating loop and the other an anticlockwise rotating loop. If the clockwise loop, knocked to the right in this thought experiment, is defined to be 'spin +½' and the anticlockwise as 'spin -½' it is apparent that the spin energies are the same and the mass energies – the loops' rotational rates – are also the same. The charges of the loops will also be the same since the meons have not changed twist orientation. How then to define a matter particle or an anti-matter particle?

The underlying difference can only be investigated by switching the time direction for every property of each meon. This means the initial direction of travel of the chain, the twist orientation of each meon and also the underlying identity of each meon which has to become an anti-meon and vice versa.

So now the meon twisting right hand screw along one spatial direction (forwards), generating negative one-sixth electron charge, will become an anti-meon also twisting right hand screw along the opposite spatial direction (backwards), generating positive one-sixth electron charge. The difference is also that the chain previously deflected right will now form a spin -½ loop instead of the earlier spin +½ loop since the chain travel direction is reversed and the obstacle is in the way of its new path.

However, in both cases, for both deflections, we could choose the opposite framework in which to view the orientation of spin of the loop. So it cannot be that the definition of matter versus anti-matter can include any spatial orientation.

The only property that provides an unambiguous definition that can be used to define matter and anti-matter is the sign of charge of the loop. If the positive charge is matter, then the negative charge of the electron makes it an anti-matter particle, as would be all other negatively charged loops. Thus all neutral atoms are balances of matter and anti-matter, and all atomic photon emission energies will be identical whether the atoms are composed of neutrons and positively charged protons or anti-neutrons and negatively charged anti-protons with balancing electrons or positrons respectively.

The neutrino could then be defined as both matter and anti-matter since it has no overall charge. Even if a specific position for the start of the loop is defined, so that it would be possible to call one matter and the other anti-matter and rotating either loop by 60 degrees would convert one to the other, this would be the same spatial framework change as for spin and so not usable. The anti-loop of a positive charge spin -½ loop is a negative charge spin -½ loop. So a photon, as shown below, being loop and anti-loop rotating in the same sense, is a perfectly balanced composite of matter and anti-matter.

Since unmerging ZMBHs produces a balance of fundamental charges as well as of one-sixth electron charges in the twisting meon and anti-meon pair, there can be no matter/anti-matter imbalance in the universe.

## Quantum and Gravitational Orbits

The existence of both mass and spin energies in all loops, considering just the mass side of energy interactions, implies that the motion of both energies should be included in the gravitational orbital equations of motion, making them identical to the quantum mechanical versions. This is also seen in the energies within the photon, where both mass and spin energies need to be considered in the relativistic deflection of their motion past the Sun, so although we measure 1 *hw* of photon energy, the deflection shows that 2 *hw* are present [8].

With the new TAPU units, explained in the foundation paper [5] showing that both *G* and *h* are dimensionless ratios, (each property *M*, *Q* etc is now in the new units) the new gravitational and quantum orbital formulae at Planck values become:

$$Energy = \frac{MM}{r} = \frac{QQc^2}{r} = Mv^2 = w$$

$$Force = \frac{MM}{r^2} = \frac{QQc^2}{r^2} = \frac{Mv^2}{r} = w / r$$

However, what is measured in a gravitational orbit is still the kinetic mass energy only, since the kinetic energy of spin can only be measured with spin energy, even though the kinetic energy of spin exists.

The spin energy itself only appears in the equation above when the two bodies have mainly aligned spin components – as is the case for small quantum mechanicals system. For randomly spin-aligned bodies, there will be little overall effect.

In the energy equation above, the kinetic side, since $M=Qc$ and there are two different energies in motion, should really be displayed as

$$Mv^2 = \frac{1}{2}Mv^2(mass\,KE) + \frac{1}{2}(Qc)v^2\vartheta(spin\,KE)$$

where $\vartheta$ is a factor for the relative alignment of overall spins in the two bodies, equal to one when both have all loop spins aligned parallel and in the same sense.

This means that the actual total energy of all stable orbits is always zero. That is why the orbits are stable. To move to another stable orbit requires that the energies, potential versus motional, change equally on both sides of the equation. For electron orbitals this is achieved by adding a stack of photon loops of the correct energy onto electron loops or removing them from existing electron-photon stacks and adjusting the orbital size smaller or larger. To be stationary requires the photon double loops of electron and positron to unmerge and stack, probably produced by impact onto the existing stack or single electron loop. Further impact of the appropriate energy may dislodge the stacked electron and positron to reform the photon and chase up to light speed.

In eliminating *G* from the 'old' Planck units, as a dimensionless ratio, it becomes clear that the strength of mass and charge fields are identical. This is the case for fundamental mass, normal mass and charge properties. It is only the relative size of the one-sixth electron twisting charge ($\sqrt{\alpha / 2\pi Q / 6}$) versus the normal range of loop sizes of our fermions ($10^{-20}$~$10^{-23}$ *planck mass*) that makes it appear that gravity is weaker than charge. The use of *G* has hidden the underlying symmetry of the strength of the actions of mass and charge in terms of fractional Planck values, as explained in the foundation paper [5].

The above interpretation implies that curved motion does not result in acceleration - centrifugal forces are the real ones - and that energy is just as much a vector in action as force, it just requires a different interpretation of 'energy', where transformation into a framework of 'outwards and inwards relative to a point' eventually leads to greater clarity, although remembering that total energy is always zero for both charge and mass.

Two cases are worth explaining here. The first case is Newton's bucket which has the liquid contents of a vertically rotated bucket staying inside the container. This is because the force on the liquid is outwards, away from the centre of rotation, and has nothing to do with the mass of the rest of the universe acting on the liquid. When the rotational rate is too low, the liquid will escape because the upward vertical centrifugal force is no longer enough to overcome the downward gravitational force.

The second case is the 'remarkable' stability of a bicycle in motion. It is not remarkable once it is understood that the rotation of the wheels causes an outward force, in the plane of the upright wheel, to act on the wheel tending to keep it upright. When the speed of the bicycle, or the rotational frequency of the wheel touching the ground, is large enough in relation to the bicycle mass, there will be a net outward and upward force tending to return the wheel vertical when tilted against the gravitational force trying to topple the bike.

In the case of a particle moving in an inertial reference frame, it is the case that it moves because it already has energy relative to that frame and that energy is another way of describing the force acting within the particle along its direction of travel. The force required to stop the particle is what we describe as equal to the particle's inertia. It is that if the body already moves, it is because it is acted on by an internal force and has an internal energy, relative to the frame of reference, acting along its direction of travel.

## Beyond Single Loops

Having looked earlier at the single loops that are the fermions, it is necessary to look at combinations of loops, which have been termed 'stacks'. The following analysis looks at rotating and balancing symmetry within loops, meaning just where the charges sit within a consistent framework defined by the identity of a similar starting point for every loop. It is not concerned with the magnetic moments of the loops, other than in terms of the spin ±½ of the loop (magnetic moment up or down), or the radii of rotation of the individual meons since these are secondary to the effect of the charges of the meons. As shown later all meons with $H$ mass factors in loops in a stack where all loops are the same size (mass or radius) will rotate at the same inner radius $r\,(1+g/r)$ and all $S$ mass factors at $r(1+k/r)$ so the larger total charges will have the greatest effects. But it is clear that it is all the mass and charge types that affect the interactions between loops in a stack – the loop deflections (masses), spin energies, s/6 energies, $q/6$ charges, MM and QQ meon/meon or meon/anti-meon interactions across loops. The strong force is a mix of these different variants of charge and mass actions along with the loop structure.

### Stacks and symmetry – where 'chemistry' appears

How do we get 'chemistry' is the question. 'Chemistry' means the formation of nucleons and atoms and their interactions. With 3-loops, a stack of three loops is required for overall balance for three different 3-fold asymmetric loops. Further consideration shows that 2-loops have 2-fold symmetry. 4-loops have 2- and 4-fold symmetry. 5- loops have 2- and 5-fold symmetry. This ignores the non-symmetric loops, which cannot be balanced by any loop except their own anti-loop.

In order to produce overall symmetric stacks requires that the internal symmetries be balanced. Ignoring 2-stacks for the moment, for 4-loops the only stacks that can be made symmetric contain an even number of loops. The same is true for all even-loops, where there is an even number of pairs in the loops.

What we require for chemistry, at the most basic level in forming atoms, is that the largest-charge symmetric loop can be balanced in orbit around an opposite charge central stack so that the total overall charge is zero. In each loop framework the largest charge-loop is always symmetric, as it must be. Thus it can exist on its own, in a stack of one. But in order to be in orbit, it cannot be only the stack charge that must balance with the orbiting loop, but also the total stack spin must balance with the orbiting loop spin. This is evident from the previous formulae where the spin KE is as important as the mass KE.

Each loop has spin ±½ in the current terminology. In a stack, in order to avoid the possibility of adjacent loops merging, they must stack with alternating spin orientations. Where 2-loops, or other even pair number loops, have the same spin orientation they will form bosons.

To maintain balancing symmetries all asymmetric loops in a stack must rotate at the same frequency – the same size, which should mean same mass except for the fractional effect of fractional charges in the asymmetric loops which changes the observable loop frequencies and the loop apparent masses. So a stack of two will have zero total spin. So will all stacks, where nucleons are concerned, which have even numbers of loops.

What this means is that no even-loop charged stacks can have orbiting maximum-charge loops in stable orbit because the even-loop stacks always have zero spin whilst the maximum-charge loops always have simple spin ±½. Only odd-loop charged stacks will have spin ±½ in total and so will balance the spin ±½ of the orbiting maximum-charge loop.

The smallest odd-loop number for which spin ½ stacks can have orbiting spin ½ loops is the 3-loops, with 3-fold symmetry, and these are atoms. This is our matter.

Other odd-loops which can form nucleons with orbital systems are 5- loops, 7-loops etc. But these contain more pairs than the 3-loops and so are less likely to be formed in the first place. The loops that do not form chemistry have no atoms and can have no emission spectra, so are versions of dark matter.

Also likely to be dark matter are the completely non-symmetric 3-loops, which can only balance with their own partner anti-loops, so reducing their likelihood of producing any 3-stacks.

Overall then, chemistry will be most likely with 3-loop systems and the bulk of other systems will not form nucleons or atoms. The larger number of 2-loops formed in preference to our 3-loops will mean that our 3-loop matter will be a smaller fraction of the universe than the sum of the even-loop and other odd-loop systems.

This framework gives rise to multiple loops and composites that react weakly with our 3-loops and their composites. The greatest differentiator will be the charges of many symmetric charged dark matter loops, which will be different to our $3/3q$ electron charge loops and the magnetic moments of these non-3-loops which appear, from their masses, to be normal matter.

Differentiation between loops with different symmetries will be possible because, as shown later, although the meons will rotate at the same H and S radii, and will have similar radial electric field sizes, their circumferential fields will be different in both size and angle to the radial perpendicular. The meons will be increasingly closer together circumferentially with greater symmetry number and there will be more and stronger inter-meon electric fields.

## Stack identities

There are two forms of 2-stack identities, those where each loop rotates in the opposite sense to the other and those where they rotate in the same sense.

**Opposite-rotating stacks**: The opposite rotating stacks can be formed of any loop pair number provided only that the pair number is the same and they can balance. So a 3-loop can stack with any other 3-loop provided the total stack is symmetric and loops have the same frequency *w*. This usually means loop and anti-loop, but could be up quark 3-loop with down quark 3-loop, for example. Also possible is 4-loop with anti 4-loop, etc.

One form of opposite rotating stack that is probably present everywhere is the zero spin, zero charge stack made of a 3-loop electron and 3-loop positron. This 'zeron' is probably what underlies the quantum mechanical effect of 'pair creation'. The stack being hit by a particle with sufficient energy can be separated briefly into the two loops - a pair of opposite charge, opposite spin loops seems to appear from nowhere, but were actually always there.

If space is filled with zerons, alongside the existing underlying ZMBHs and acting as part of the viscosity drag in volumes from which they are not excluded (the background), then at every frequency *w*, centred on every point, there should exist a zeron of apparent energy *hw* as part of a concentric shell of zerons. Each of the loops will have apparent energy *hw*/2, and this may represent the zero point energy of that point - although, as mentioned earlier, the total energy of all loops is always zero. This concentric zeron shell framework may also explain physically the Casimir effect, in that the exclusion of shells beyond two parallel plates will result in excess pressure from the surrounding unaffected concentric shells.

Additionally, the denser gravitational clumping effect of 3-loop chemical composite nuclei may result in the preferential location of 3-loop zerons within the same symmetry volumes. This could effectively squeeze the more abundant 2-loop zerons away from the 3-loop gravitational clumps. However, the same zero point energy would still exist across all zerons.

Because of the extra loop identities in loop systems above 3-loops, for simplicity only the 4-loop systems are considered here to represent these as examples of dark matter composites. The total charge sizes for the possible 4-loops are $0q$, $1/3q$, $2/3q$, $3/3q$, $4/3q$ electron charge, positive or negative. The symmetric 4-loops have zero and $4/3q$ charge. The asymmetric loops are the equivalent of quarks. To form a stable 4-loop 2-stack requires 2-fold or 4-fold symmetry and total charge equal to $4/3q$ electron charge. This is possible with a $1/3q$ 4-loop and a $3/3q$ 4-loop of the same charge for 2-fold symmetry, so this is one form of zero spin 2-stack 4-loop particle that should exist. For 4-stacks, one possible stack combination could be 'quarks' of charge $1/3q$, $-2/3q$, $3/3q$ and $2/3q$ to give $4/3q$ overall.

For 2-loop stacks, there is only 2-fold symmetry and two loops are required in a stack. The 2-loop system consists of $0q$, $1/3q$ and $2/3q$ electron charges. The only way to achieve a stable stack using asymmetric loops is with 2-loop 'quarks' of charge $1/3q$ and $1/3q$ with the same charge sign, which stack then totals the $2/3q$ charge of the symmetric maximum-charge 2-loop. These zero spin stacks should also exist, although in a 3-loop environment where $q=3/3q$ electron charge, such a stable $2/3q$ charged stack will not stably attach to any 3-loop or 3-loop stack to be observed, although it will be deflected appropriately in charge or magnetic fields.

Any length of stack is possible if the constituent loops sum to whole number or zero charges and their overall symmetry is maintained. So considering a proton core of three 3-loop quarks, possible stack additions could be neutrino end caps, quark and anti-quark, electron and neutrino - each of 3-loop and total added spin of zero. The identities of these would be, in order, proton, penta-quark-proton and neutron.

For other sized loops with different symmetries, the same stack lengthening will occur provided the additional loops have the same symmetries.

An extreme example might be the stacking of electrons to form what would be observed to be a heavy electron, with spin ½*h* and integer charge greater than one. This would require extreme pressure to overcome the inter-electron charges. More likely would be the stacking of a single electron with a zeron to look like a heavy electron with total spin ½*h* and a large mass.

**Same-rotating stacks**: Considering the stacks where the loops rotate in the same sense, there are also two forms here. One is where the loops are a loop and its anti-loop. The other where the loops are different - including different symmetry isomers of anti-loops.

Where the loops are loop and anti-loop rotating in the same sense, the result is a photon with spin 1. Because the frequency of any loop is always measured to be positive, which is the base for all loops having positive observable mass and (even though its energy is zero overall, we measure the split balancing part only) the photon will always be measured to have spin +1.

In the photon, the same forces that drive the chasing of a meon pair round inside a loop also drive the chasing of one meon in the loop towards an anti-meon in the anti-loop, or vice versa.

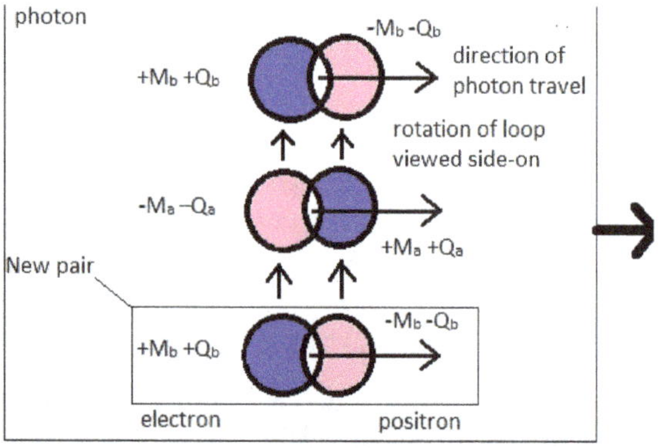


**Figure 3:** Photon as partially merged electron and positron loop showing chase forces around the loops and between meons in each loop (loop rotational axia in plane of paper). Each new pair is like an original ZMBH but in motion with in the loops.

Figure 2 shows the forces driving meons around a loop and Figure 3 shows the same force action between opposite meon types in a partially merged loop and anti-loop that form a photon.

The result is acceleration of the loops along their mutual axis of rotation up to the velocity at which the chasing force is equal to the effect of the viscosity of the background field of ZBMHs and zerons.

This maximum speed is what we call the local speed of light, *c*. This effect can be seen in an experiment [9] where the acceleration of photons is interpreted as being due to them having positive and negative effective masses within crystals. Actually it is a very clever way of stacking photons so that the chasing from meon to anti-meon internally also occurs externally – as if the photons have become six meon/anti-meons short chains rotating about their axis of motion. This is because a photon contains six of the shortest possible chains – the two loops' original meons in one loop and the anti-meons in the other loop, and vice versa, acting as new pairs chasing each other –and adding more pairs to the chain can be done by stacking other photons in front or behind the first photon.

Where the viscosity is great, when the ZMBH and zeron numbers are denser near other loops or large masses (planets, stars) then the actual speed of light will be lower than when the ZMBH and zeron field is less dense. But in each location the velocity is the fastest possible and defined as *c*. That photons travel at a terminal velocity defined as the local speed of light shows that there is internal chasing between meon and anti-meon from loop to loop. This can only be explained if there are negative and positive fundamental masses which chase.

The energy used in matching the background viscosity in order to travel at *c* reduces the rotation rate of the meons in the two loops that comprise the photon. This is tired light and, apart from at very high rotational frequencies, is proportional directly to the distance travelled by the meons, and thus by the photon, almost regardless of photon frequency.

The spiral distance $D_\gamma$ travelled over time *t* by each meon in a photon with frequency *w* is given by $D_\gamma = \sqrt{1+w/w_T}\,ct$ where $w_T$ is the TAPU Planck frequency. Where $w \ll w_T$ then the meons in different energy photons will each have experienced virtually the same viscosity red shift. The path difference travelled by each meon, comparing a gamma ray at around $10^{24}$ Hz and visible light around $10^{14}$ Hz, is the difference between $\sqrt{1+10^{-18}}\,ct$ and $\sqrt{1+10^{-28}}\,ct$ respectively – not significant given the uncertainty in the emission point of each.

So the redshift observed in any photon has to take account of this extra viscosity redshift factor. This implies that the size of dark energy may need to be radically reassessed, to the extent that possibly the rate of expansion of the universe is not accelerating at all, if all the excess redshift were due to tired light. Alternatively, the universe, or our successful big bang event, may actually be failing and we are contracting - if the viscosity red shift is larger than the contraction blue shift, or the expansion could be zero and the main red shift is caused by tired light. Overall, it is likely that the observed redshift is a mix of factors including the tired light effect.

This explanation of photons losing frequency to the background has been simplified, as will be explained later, because it is the emission shell (a volume of no-background) that loses the energy and the photon provides it as it skips around the shell.

Using multiple bootstrapping methodologies to measure distances to light emitting objects and comparing these results should allow the existence of the tired light effect to be confirmed and its size calculated within the overall redshifts observed.

The viscosity effect is rather like a very diffuse aether, whose red shift effects can be easily observed only over light years, except that the direct effects are not length contraction in a specific direction, just redshift in all directions. Regardless of the expansional motion of our successful inflation event against the background ZMBHs, because of the velocity of all photons at maximum local speed *c* the redshift factor (frequency loss) at any point will be the same in all directions.

Photon loops, when measured from the point of emission, can expand and contract in radius due to external frequency transfer, physical interaction or internal frequency balancing between component electron and positron loops, and are stable at any radius. However the fermions, having maximum radii locked in during inflation are only stable at those radii, but can decrease radii, increasing energy, when external motion adds energy to the loops, in a stationary frame of reference. They can also increase in radii as they lose energy due to the background viscosity, explained below.

Where the two loops rotating in the same sense are not loop and anti-loop, the result is a spin 1 boson. These are not force carriers, but are composites likely with high velocities. These also have to obey the total symmetry rules and can be composed of any loop pair number, provided both are the same, so that the total symmetry is stable.

For non-photon loops, the rotation of the loops will be opposed by the viscosity of the background ZMBHs. This should slow the rotation of the loops – effectively reduce their frequencies and what are observed to be their masses. However, it is likely that the continual stacking and unstacking of photon rings provides the extra frequency boost to return a 'weak' loop to its preferred size (frequency or mass). In this way, photons and loops lose energy to the background ZMBH and zerons by spinning, moving or otherwise providing them with energy equal to that lost by the action of the ZMBH and zeron background viscosity. The rotational rates which we describe as energy are being gradually returned to the background ZMBHs from where they originated in a big bang event. The preferred sizes for the non-photon loops are those locked in during inflation. It may be possible to observe the mass loss in non-photon rings by isolating a specific highly accurately measured mass from any light sources and leaving it for long enough – then measure it in the absence of light sources. The rate of loss should be the same as that experienced in the tired light effect for photons, although reversion to normal size when exposed to photons of suitable frequency will be very quick.

Whilst a photon is more likely to be a single electron and positron ring rotating together, it is possible that neutrino and anti-neutrino, or quark and anti-quark could also form a photon double loop. And it is even more likely that longer stacks of photons could form because each pair partially merged across two loops could be similarly merged with another photon of the same size. The actual length of a stack of photons may be the wrong way to describe such a photon. A better description may just be the number of double loops presents.

It is even possible that a stack of separated photons, each chasing the one ahead as if it were a chain, could catch onto its own tail. Then the 'mass' of that super-loop would depend on the rotational frequency of the photons as they travel around the super-loop. Given that the photons would be travelling at local *c*, the 'mass' of the super-loop would be the adjusted-Planck mass regardless of how many photons formed that super-loop.

## Strong and Weak Forces

So far the description of stacks has not directly mentioned strong forces, only forces or energies due to charge and mass. That is because there is no different strong force. What looks like a strong force is only the actions of charge and mass, both fundamental and normal, by loop systems when close together. At the intra-stack distances, the actions are largely from meons in one loop to meons/anti-meons in the adjacent loop in the stack. At inter-stack distances, where two stacks are similarly aligned adjacent, the same is likely also the case. It is the loop nature of the composites that introduces preferential separations. At stable intra-loop and inter-loop distances the forces of attraction and repulsion due to charge and gravity (of equal strength for equal adjusted Planck fractional values, and via their various types) are balanced.

The weak force is not mentioned because it is the effect of the physical replacement of one loop in a stack by another. For example, a neutron stack (5 or 7 3-loops with a core of three quarks and caps of neutrinos and an electron) can have its electron loop knocked out of the stack by the appropriate size impact of an incident neutrino, resulting in a proton. The effect is random and depends on the density and energy of incident neutrinos.

The reason why the electron and neutrino can exist within a neutron stack is because they are symmetric already, and have hidden 3-fold asymmetry, and because their components, the meons and anti-meons, are large. The actual physical size of the loop, which we call its mass, is not of consequence as to whether it can exist inside a nucleus or not.

However, there is a preferred size of symmetric loop to enable better stacking, so the electron swaps frequency for magnetic moment. As shown by the symmetric charged members of loop families, the electron family can change physical size/mass in exchange for magnetic moment between preferred family sizes. Magnetic moment is just another way of describing the mass or charge energy of the loops.

Interaction between separated loops is complex. For every loop there is an extra chasing force towards or away from, another loop due to mass-chasing, which would not be expected. The effect may be masked by charge interaction, or be zero if the loops are at unchanging separation, as in a stable orbit. The effect is dependent on the difference between the two loop frequencies, the initial direction of travel of the two and the mutual relative orientation of their planes of rotation. Mostly the effect is to introduce an effective vibration of the loops as the chase direction between meon/anti-meon changes. But there may be an overall net chasing driving the meon/anti-meon, and their respective loops, along the lines between them. The effect is most efficient and obvious in the case of neutrino/anti-neutrino or electron/positron comprising a photon.

## Where the Electron Got Its Size from – Inflation

We probably owe our existence to the size of the electron. Had our original successful big bang been less inflative, the electron would have been physically smaller, so of larger mass, and our inflation event could have failed.

In the great unmerging event that preceded our big bang, where chains formed, broke and eventually formed loops, at some point those loops interacted physically to hit each other. The result was a sudden drop in speed of some meons with a resulting huge increase the physical size of the loops, of which they were components, to conserve angular momentum $h$ for the meons in the loops (when considered in DAPU units and ignoring $h^{\pm}$, as explained later).

In a three dimensional universe these 'inflated' loops would have been quickly aligned along three axes to provide three different rates of inflation, a different one along each axis. Even a loop being off axis would become on-axis with sufficient inflation, producing flatness and standardisation of fermion loop sizes with preference for remaining within the three family sizes. Three dimensions are needed as a minimum, because it is not possible to overlap or merge objects of only two dimensions, and more than three are not necessary.

Given a suitable mix of the amount of inflation along the three axes, it is possible to produce eleven fermion masses in the right range out of the twelve we observe. So four fermions and three dimensions of inflation probably represent all that is needed to produce the three fermion families with their physical sizes/masses.

It is possible to assume that inflation only affects the loops where meon and anti-meon are not merged – single fermion loops and not photons where each meon/anti-meon pair are effectively a reformed ZMBH. This means that the single fermion loops would inflate, but not the photons.

The effect would be that the single fermion loops would inflate so that their centres were further away from the origin of the initial unmerging event than the chains which formed photons at near Planck energy. This would allow the fermion single loops to be ahead of the photons in expansion until the photons caught up. At that point, the energy (frequency) of the photons interacting with the fermions would result in the formation of fermion then nucleon stacks and the subsequent reemission of the photons to continue their outward expansion, some now ahead in the expansion of the fermion/nucleon mix.

Expansion is used here to mean simply the outward motion of the components of the initial inflation across the background of generally stationary ZMBHs. The vector nature of energy, as represented by half the zero balance, means that a symmetric expansion also has zero energy of outward motion in total.

## Failed Inflation Events and Chain Stars

The paper sets out in part to provide a logical framework for using the simplest possible assumptions to build a universe with the laws, properties and symmetries which we observe. These assumptions do not lead to anything which can exist outside our universe, since everything that emerges is made of components from within our universe.

This does not preclude different inflation events. However, these will be within our universe. So if there is only one size of building block, no expansion of space can be taking place, part of what we see as expansion is red-shift in photons.

Where these events failed, they may be observable and where successful, they also may be observable, but not yet recognised, or beyond our observable horizon but still part of our universe, not any form of external multiverse.

Where multiverse and black hole event horizon models are some of the 'most speculative' [10] physical theories, what is proposed here is considerably less speculative, but is still speculative from a completely different viewpoint to those normally used in multiverse or black hole speculations.

In this section the normal understanding of 'energy' will be used in the explanation, although, as previously explained, it should be understood that for every rotational or motional mass energy considered, there is an opposite balancing charge energy. The reason for any change is the different way in which charge and mass energies interact with themselves, rather than each other, from loop to loop or meon to meon in different loops. So the total energy released in unmerging, inflation and expansion is zero. What is retained is the angular momentum of each meon and anti-meon which, in total across all meons and anti-meons, will also total zero.

The inflation of the loops, in the meon speed drop, releases huge amounts of energy (frequency) from the loops which then drive expansion away from the source of the initial collisions by speeding the resultant larger radius (smaller mass) loops outwards. The question is how much is released. If the loops that emerge from inflation are large enough in physical size (small enough in mass/energy) then the expansion will more easily drive the loops outward and their mutual gravitation will be too low to oppose that expansion.

The release of energy is the difference between the small size, approximately the Planck-energy, at which loops formed from the initial chains, and the eventual post-inflation large sizes (small energies) of those loops – the masses of the fermions. This factor for us is a maximum around the inverse of the mass of the electron in adjusted Planck units, or of the order $10^{23}$. The time taken for the meons in an electron loop to adjust loop size from around Planck energy to the electron observed size at velocity *c* will be of the order $10^{-29}$s, the path being across the diameter of the new larger radius electron loop at velocity *c*.

If the speed drop is not large enough in the physical interactions between loops, so that the resultant inflation does not release enough energy for expansion, the loop sizes will be small and their masses large. Here the energy for expansion may be insufficient eventually to overcome the gravitation of the resultant larger mass loops. This is a 'failed inflation' event, where the balance between inflation and expansion was wrong, and the loops will collapse back towards their starting point under the effect of their large mutual gravitation.

A failed inflation event does not lead to a big crunch though, because the initial unmerging of ZMBHs to form loops has changed the environment. When the loops fall back in to the centre of their expansion, the result is likely to be a 'chain star' where the loops eventually become broken by the strength of gravity into pairs again. Within the chain star are the same processes that occurred after the big bang unmerging event, but without the formation of the first loops at the smallest possible size (greatest energy).

The loop size (mass) and spin energy are both reduced to zero as the loop is stretched, by differential strength of action of gravity across the loop, and its frequency reduces as it descends into the chain star and is finally broken. The charge stays with the pairs in the chain. Inside the chain star, the entering chain will get broken, mixed, form, reform continuously until a photon double loop of sufficiently large energy (smallest size) escapes. As it does so, it will lose a lot of its energy. This is how a chain star evaporates and where it gets its temperature from and how the previously entering loops' information escapes in a different form.

A chain star chaotically makes and breaks the loops that fall into it. It emits only symmetric loops in the form of high energy photons that have enough energy to escape the gravity of the chain star as low energy photons. Chain stars are what we describe as black holes. The chain stars reduce in physical size as they successfully create high energy loop/anti-loop pairs that are emitted as photons [11]. The photons do not need to be 3-pair size and will be any number of pairs, although 2-pairs will dominate as dark matter photons.

Failed inflation events themselves will have gradations of end results. For the 'heaviest' loops formed, it will be difficult for the loops to escape mutual gravitational collapse and a chain star will be the result. But for less heavy loops, maybe near in size but not as light as our fermions, the result could look like a galaxy with the results of expansion almost balanced by the effects of mutual gravitation. So a scale of failed inflation events could exist, dependent also on whether the initial inflation event contained any rotational effect. This latter effect may make a galaxy appear to have stars with greater velocities than normal mechanics would expect and might be viewed as needing surrounding dark matter to exist.

Additionally, it is not clear that a failed inflation event with a slightly larger 'electron' mass would allow its larger scale atomic emission spectra to be differentiated from either a faster moving galactic source or one at smaller distance, given the accuracy of spectral observation. As the fermion mass sizes increase with greater failure effect, at some point the difference should become observable.

The photon emission spectra from such a 3-fold symmetric system with greater 'electron' mass would be proportional to the 'electron' mass, it would appear to be more energetic and so blue shifted when viewed in isolation to the other redshift components.

Assuming that unmerging events happen randomly in the background ZMBH space, then many black holes or galaxies will have been formed before our successful inflation, and also since then. They are embedded in the background which is the foundation of our universe. What we see as an excess of early black holes, or the early formation of stars or galaxies because of those black holes [12], may be just a reflection of the rate of failed inflations. If we could differentiate between failed inflation black holes or galaxies and subsequent black holes and galaxies formed from excess density gas clouds, we could make an estimate of the rate of unmerging events.

Another aspect that may allow the accuracy of this framework of black holes as failed big bangs would be the discovery of unarguably conjoined galaxies or black holes [13], where the redshift of each is markedly different without any alternative explanation. One redshift would represent the relatively stationary background mixed with tired light from it, whilst the other has a different redshift representing its local rate of expansion mixed with the same tired light factor as the first. Finding a number of these would allow the tired light factor to be separated from the overall redshifts. Effectively the lower redshift represents the partly stationary background tired light factor and the difference will be mostly the motional expansion redshift. The number of such real conjoined systems within a given volume would provide an estimate of the rate of failed big bangs.

The universe cannot be expanding due to inflating space between galaxies because that would imply that the base unit of space was expanding. And that cannot be the case since all ZMBHs are the same adjusted TAPU Planck size, so expansion is just motion versus the background of ZMBHs.

This represents a mixture of both the big bang and steady state theories of the universe. The steady state here is confined to random big bang events that either fail or succeed, instead of the continual creation of matter between expanding space. The result is that the universe could be far older than our inflation event suggests and could be much smaller. Furthermore, earlier failed inflation events could have formed seeds for later gravitational accumulation during our own big bang, providing a quicker route to star and galaxy formation.

Black holes also have another two roles in the universe. One is that they convert asymmetric loops into symmetric ones, and in the process reduce the amount of loops showing mass within the universe, so reducing the gravitational effect of all loops with observable mass in the universe. This has two effects. One is that the expansion of the universe may accelerate as the overall gravitational effect reduces, which might look like the acceleration of expansion or dark energy [14]. The other is that there will appear to be more photons than would otherwise be expected with increasing discrepancy over time [15]. This is because more bodies, symmetric and asymmetric loops will have been swallowed by the black holes and turned into exiting photons. The symmetry generating effect arises as matter, anti-matter and dark matter loops are broken when they enter the black holes and they can only re-emerge as symmetric photons, having sufficient energy to escape, as mentioned above.

The current quantum pair creation explanation for emissions from, and evaporation of, black holes is not possible in this framework. Zerons, which underlie pair creation, already exist and become observable given sufficient impact. If zerons were close to a black hole, as if acting as an event horizon, they would be stretched and swallowed by the hole. So pair creation as the primary source of black hole emissions is not likely.

The second other role of black holes is in dark matter transformation. The breaking and reforming of loops shifts the ratio of dark matter to normal matter because the same probabilities of reforming chains of different pair number will make many more 2-pair photons than 3-pairs. Dark matter photons will be preferentially formed and emitted, although they will not have any effect on the charged lepton's $g$-2 value since they have the wrong symmetry [16].

So the current ratio of dark matter to normal may have been significantly changed by black holes over time, especially if the galaxy under observation is a failed inflation event from before our own big bang. The dark photons may either not be observable to our 3-pair detectors or may have been broken into loop and anti-loop amongst other dark matter loops. With enough black holes and galaxies emitting dark photons, there will seem to be no reason for the extra ionisation of gas clouds [17].

The consequent effect of dark matter transformation would be to reduce the amount of normal matter surrounding the black hole and leave the galaxy looking as if it does not have enough matter to retain its matter in orbit around itself.

So it may be that the amount of inflation sufficient to produce a small enough electron, and our other fermions, is what separated our successful inflation from a failed inflation. It is also possible that there is a range of possible small loop sizes that could be successful in overcoming gravitational collapse, but outside the parameters necessary for successful chemistry. Maybe our electron size is the lowest limit for both successful inflation and successful chemistry.

It is possible that the symmetry of a failed inflation event affects the final outcome. With perfect symmetry there would be no rotation of the expansion away from the event and the resulting deflation would be rotationless. The result could be a simple black hole. However, an asymmetric expansion could involve the rotation of the whole number of loops formed, which would delay the deflation and possibly enable the formation of a rotating galaxy.

For loops entering a black hole, the breaking of the loop is where time stops for that loop. Only loops have our time and our time only formed when loops did. For different big bangs, successful or unsuccessful, normal or dark matter, the only time the loops know is their own rotational rate. In such a system, without overall rotation of the universe, all centres of rotation are equal and there is no preferred central frame of reference.

## Laws of Physics are Identical – Always and Everywhere

The loop sizes are the only variables in the chemistry of the universe, because the one-sixth charge will always be the same since it represents the energy required to unmerge all foundation ZMBHs into meon and anti-meon pairs.

These failed inflations are within our universe. Using the ZMBHs as the only source of building blocks for composite loops, it is not possible to have any universes 'outside' ours. There may be other unmerging events, inflations and big bangs, but they are using our foundations, within our three dimensions. Other successful inflation events may be so rare that there have been none, could have occurred beyond our visible horizon or not have been recognised as such yet.

As shown above, any inflation events, failed or otherwise, will produce 3-loop systems with potential for chemistry and dark matter with the same approximate relative starting ratio of matter to dark matter as ours. What will differ is exactly what size the loops end up as. This will affect the sizes of atoms, photon emission lines etc but not how gravity and charge work, the size of one-sixth charges and three dimensions. All underlying physical laws will be the same in all 'universes', if that is what inflation events are called, whilst they evolve to be either failed or successful.

The idea that physics breaks down within black holes, forming singularities, is wrong and is based on the idea that energy/matter is packets of amorphous waves or particles.

If the foundation of loops is TAPU Planck size meon and anti-meon, then they are the densest particles possible in the universe. There is no possible assembly of particles which could break meons apart. So a black hole, or chain star in the preferred description, is far less dense than any meon. And if the universe is populated by and constructed from unbreakable meons, then physics does not break down anywhere. And that is why the foundations of the universe, as merged volumes of meon and anti-meon, are called ZMBHs – zero mass black holes. Everything we are made of and can observe is composed of the densest micro black holes, and massive black holes do not contain singularities.

The question of whether there is an event horizon around a black hole is the wrong question. The loops are stretched until they break back into chains as they approach the massive black hole. However, different loops break at different distances from the black hole, so the break point is smeared, without a specific horizon and depends largely on the strength gradient of the gravitational field – a factor dependent on the size of the black hole. The larger the size, the lower the gradient and the more smeared the horizon.

And the question of how time is affected as the hole is approached can be understood – as the loops elongate on their way towards the black hole, due to the differential gravitational field on each meon in the loop, they slow in frequency and their own time slows until it ceases to exist when the loops break. For both the loops and any observer of those loops, the time/energy of those loops is no longer observable.

When the loops break, the identity of the loops is lost and the angular momentum of the meons in the loops has been transferred to the hole. That internal momentum was what is described as the mass and spin of the loop, so the mass and spin are transferred, as is the overall loop charge.

What happens inside the hole is that the pairs in the chain, that was a loop, retain their twists and if they later exit the hole as part of a photon, they will show the same fractional electron charge that they have always had. So although the loop identity disappears, the later emergence of any pairs from that broken loop will see some of that identity reappear, although within a different loop. So information is conserved, but at a lower level than the loops.

Also worth mentioning is the lack of need for any Higgs mechanism. Mass is the size of the loops and is a measure of their frequency of rotation. The Higgs is just a boson composed of an even number of loops. The loop framework does not require anything further, although when producing a framework for the successful modelling of the magnetic moments of the loops and nucleons there is a further relationship between the size of magnetic moment of loops and their masses and charges, as described later. There may be many more matter bosons yet to be discovered at high energies, but they are simply longer stacks.

## Hierarchy of Zeroes of Total Energy

What drives the states of matter that we observe? It is a preference for having the least energy possible in the simplest state, using 'energy' here in its accepted meaning.

The state of the stationary ZMBHs could be said to be a quadruple of zeroes of total energy ('ZOTEs'). Not only do they have no motion if considered in their own frame of reference, but for fundamental energies, each charge energy is balanced by an opposite charge energy, each mass energy is balanced by an opposite mass energy, and each charge energy is balanced versus a mass energy. Even if they rotate, spin or vibrate, these motions can be reference framed away. This is the preferred state for two merged meons. Once unmerged, they would prefer to get back to that 'perfect' state.

When unmerging a ZMBH, two ZOTEs are lost and each meon or anti-meon is left with only a balance of charge versus mass energies, although it has added a ZOTE for one-sixth electron charge energy versus mass twist energy. The pair is driven to regain its lost ZOTEs, due to charge balancing charge and mass balancing mass. So they chase each other to try to remerge.

The formation of a chain does not change the situation for the pair, but the chain latching onto its own tail produces another ZOTE. This is the balancing of the motional energy of the pairs with their spin energy, their mass and spin, from the point of view of the loop combination. This could be reference framed away by considering the loop to be stationary and then the meons would be missing their drive to remerge. So the loop framework conserves that remerging drive and adds the mass versus spin ZOTE.

The next level is the stacking of asymmetric loops, which is a form of ZOTE in that the result is stable, so the imbalanced energies are hidden. However, the stack will have an odd number of loops in our threefold symmetric matter and so the next ZOTE will be to balance out the spin energy of the stack by forming a stable composite with a suitable symmetric loop in orbit. Additional charged nucleon stacks in a core will require additional loops in orbit and any lack of the latter will drive the formation of compounds with other atoms which have an excess or deficit of these loops in orbit, as another ZOTE is formed.

Even at gravitationally dominated levels, the orbiting of two bodies will prefer the formation of a ZOTE – it is just that so far the spin KE of the loops in the orbiting bodies has been not been considered correctly in the energy equations. This means that the zero total energy involved in being in a gravitationally stable orbit has not been understood.

Whilst photons are considered as having energy, this composite is again simply another form of ZOTE. For an electron and positron, the mutual formation of a photon provides almost the original quadruple of being in a nearly reformed ZMBH. Each meon and anti-meon, merged across both loops, has the quadruple ZOTE, but additionally there are the twist balance and the mass versus spin ZOTEs present. For the loop overall, its motion produces another ZOTE because all the motional energies balance, but this could be reference framed away in any case.

It can be said generally that all interactions are preferentially driven towards ZOTE states in some form. Where particles or systems are not in ZOTE states, they will not be stable until they reach such states, whether by motional, potential or other energy gain or loss.

It is not clear how pairs could eventually remerge into the original pure quadruple ZOTE state because they have the additional twist ZOTE, having gained it on unmerging. However, this does suggest that the hierarchy of ZOTEs may partially underlie the second law of thermodynamics. The direction of progress from ZMBH to gravitational system may be from low number of ZOTEs to higher number, perhaps with increasing entropy in each ZOTE number that the particle or system has, with the reverse direction requiring an energy input.

Interaction between loops close enough together is by physical interaction or by the loops' mass and charge fields that transfer loop frequency from one loop to another. The total frequency of the loops involved, adjusted for potential, magnetic or other positional energy effects, will remain the same. This is the basis of conservation of energy.

Another method of conservation of energy is the change in loop size/frequency of electron into muon, for example, balanced by the change in magnetic moment, gravitational, kinetic or other energies.

Because loop frequency is also a measure of the effective temperature of a loop, the inability of a lower frequency loop to speed up a higher frequency loop provides another physical the basis for the second law of thermodynamics.

And underlying all motion is the loss of frequency by all loops when they move across the background viscosity.

## How to Uncover the Dark Matter Framework

Amongst the dark matter particles should be opposite sense rotating variants of symmetric 4- or other even-loop zero spin particles which when separated would have charges of $4/3q$ electron charge, and the same for spin 1 bosons split apart. The magnetic moments of non-3-loop particles will be different to 3-loop moments, which will enable their identification.

Another proof is in the photons emitted by black holes, whether from failed inflations or otherwise. The photons emitted must be symmetric, but can be of any loop pair number. But it is only the 3-loop photons that will stack successfully with electrons to boost them between orbitals because they have 3-fold symmetry. So photons that arrive at detectors with suitable energies but which do not boost electrons, or that when separated have symmetric and stable 'electron' and 'positron' emerge with fractional electron charges, will show that the loop hierarchy is correct.

It is also possible that other successful inflation events have not yet been recognised as such. The failed ones will probably be embedded within other matter systems by now. But there may be examples of large black holes, chain stars, which have no surrounding matter and no other explanation for why they exist alone, but could be observed in the search for MACHO objects [18].

## Time and Quantum Mechanics

Quantum mechanics has not been mentioned much so far, but the identification of fermions as loop entities underlies their wave nature. So loops are the basis for all gravitational interactions and are the simplest form of quantum gravity, in that each loop is its own quanta of gravity.

All loops carry their own time with them, as the inverse of their frequency. In a non-rotating, or slowly rotating, universe this means that all loops have frequencies (their masses) which are absolute in the framework of the universe, but which are relative to each other. However, in the framework of the ZMBHs, there is no observable time because the loops, which we use to observe, cannot observe ZMBH rotation or vibration. So there are two different motional types of time in the universe. One which the loops exist within and one for ZMBHs. The latter time may also be as relative between ZMBHs as it is for loops relative to each other. The only thing that can be said is that the viscosity of the background ZMBHs and zerons, where they exist within a volume, effectively transferring energy to them from the loops, provides an unambiguous arrow of time. No event can be reversed because all loop components in the event have lost energy to the background, so the loop has lost frequency.

There is one further level of time, but this is the equivalent of either a stationary ZMBH space where there is no motion, rotation or vibration or the complete absence of background ZMBHs within a volume. This is the complete absence of time – in the latter, no viscosity affects the motion of loops and so there is no velocity limited to *c*. This state is what seemingly underlies most quantum mechanical effects. The absence of ZMBHs from a volume is effectively outside our relativistic universe, but part of the whole universe.

The emission of a photon should actually be considered as the creation of an expanding spherical shell within which the photon moves outwards from its source. The shell is an absence of ZMBHs within which the photon can move instantaneously and randomly within the shell to any point in the shell, called 'skipping'. It is the shell that at all points moves against the ZMBH viscosity outside, like the bow wave of a boat leaving a depression in its wake, with the photon as the indicator of the loss of energy against that drag. When the photon is observed, the effect of all the stars and galaxies over which that portion of the shell has moved is reflected in the photon.

The instant before the photon is observed it could have been situated at the farthest point of the shell from the observer and then skipped around the shell to be observed, travelling at above *c*. If it did not skip into the right place, it would not have been observed. So the path of a photon from emission to an observer's eye is a completely random walk around the expanding shell with the only fixed points at emission and observation.

For example, whether one or two slits are open before a photon is observed hitting a detector does not matter to the photon until it is actually observed. One gate can be closed after the photon should have passed through, but the result will be what the shell experiences [19].

The non-local effect provides what is called superposition of particles. Here it is not a shell, but a tunnel without ZMBHs or zerons, through the background ZMBHs and zerons, which links the particles. The particles skip randomly and almost instantaneously between the tunnel extremes because there is no viscosity in the tunnel and the observation of one particle immediately destroys the tunnel and locks the other particle where it was at that instant. The effect appears to be superposition because the particles are exchanging position through the tunnel too frequently to observe. The entanglement of single or multiple loops opens up a ZMBH-free tunnel between the entangled entities and is the basis for non-locality.

Another example is the orbitals of electrons around atoms. Here the shell, within which no background exists, may be in multiple pieces around the atom, but the electron can skip within the shell parts without viscosity and instantaneously from any point to any other point within the shell parts. There is no time spent in motion between the shell parts and the electron can be anywhere in the totality of the shell at any time. It is possible, when considering the same sort of split between expanding orbital shell and loop position for photon travel, that it is the whole orbital that retains the action of the charge within it regardless of the actual instantaneous position of the electron. This could be experimentally investigated for non-spherically symmetric orbitals, although the extremely fast skipping would make separating the two distributions difficult.

It is possible that the charges initially generated by the twisting of the meons against the background would no longer be generated in tunnels where there are no background ZMBHs. This would allow a loop and anti-loop to have no charge effect of each other as they move along the tunnel. Possibly the meon twist energy would be balanced by translational motion along the tunnel until the tunnel is destroyed and the charges re-emerge due to the background flooding in. Or possibly the charge energy may be used instead to keep the tunnel open until the system is disturbed.

Another possibility is that separated parts of a whole orbital are joined by tunnels, although this would mean that the electrons could exist in volumes (the tunnels between parts) which they should not from a probability analysis. This is not the preferred solution, but the alternative is that orbital parts are joined in some higher dimension, which is also not a supported solution.

Another description may be that the superposition of the two loops that comprise a photon create a bond between them that could be seen as a tunnel joining them through the background, keeping the background out of the way, so that they can move along the tunnel, swapping places randomly at above light speed, because there is no background to slow them. The distance we measure between the two loops does not matter because the tunnel effectively means that they are always in contact, because their travel time is almost instantaneous.

## The Action Modes of ZMBHs

The framework here does not require bosons as force carriers. Instead the background ZMBHs transmit forces in five different modes of action across space. The five 'space' actions of the ZMBHs can be split as follows:

A Magnetic Flux Line - The ZMBHs, as a pair, are partially unmerged and spin about their symmetric axis of rotation, without any external motion of the pair. The field between the meons is both electric and gravitational, but since there is no relative motion between the pair, external gravitational chasing does not occur, they are in balance. The sum of the mass-like part is zero, as is the charge-like part, and all that remains is the rotating electric field. This rotating electric field generates a magnetic field along the line of the rotational axis which does not extend far beyond the ZMBH pair. This is one building block of a flux line. A continuous line of these ZMBH pairs will form a complete flux line, started by the magnetic field produced by a loop or any magnet and finishing as a complete circle on the opposite side. The rate of rotation and ZMBH pair separation define the strength of the field at that particular point.

B Electric Field Line – attractive charges - The ZMBH pair is partially unmerged and is not rotating or moving externally. Since it is not rotating and the meons have no relative velocity, there is no magnetic field or any external gravitational chasing. The electric field lines up between two opposite charges, with alternating ZMBH pairs, each separated or partially merged with the next in line, forming an electric field line.

C Electric Field Line – repulsive charges - The ZMBH pair is partially unmerged and is not rotating or moving externally. It does vibrate along the line between charges, centred on the stationary meon with charge opposite to the external charges. The same-charge meon vibrates through the opposite-charge meon and its slight excess energy relative to the opposite-charge meon, due to its relative motion, is enough to maintain the pair in position. It is also possible that the moving meon orbits around the non-moving anti-meon, or vice versa, which would maintain meon to anti-meon separation and avoid external gravitational chasing between them. The vibrational or orbital frequency would depend on the strength of the field at that point and the orientation of rotation would have to alternate along the chain to keep the average total magnetic field zero.

D Gravity – Because the loop has zero energy overall due to mass or spin energies, it is the meons and anti-meons in the loop themselves which interact with the ZMBHs. Since an isolated meon is the densest particle possible, it will represent a maximum action of gravity (or charge). A positive meon in a loop will be moving at the loop rotational velocity and the negative meon in an external ZMBH pair will chase it, with its own partner doing the same to it. This will extend out in an external chain-like structure with alternating meon and anti-meon ZMBH pair, latched onto the meon in the loop. The meon-to-anti-meon separation/merger will represent the strength of gravity at that point – a form of deflection acting on any other loop that approaches. An external chain based on a loop anti-meon will have the reverse orientation, although at great enough distances the difference may not be observable. This action is called dragging – where the meons and anti-meons in the loop drag chains of ZMBHs around as the loop rotates. This whirlpool-like effect extends out to an influence distance, beyond which the swirling no longer enables one loop to identify the orientation of another loop, only its frequency of rotation which represents its mass.

E Spin – The ZMBH interaction is identical to that for gravity, except that it is the charge of the loop meons and anti-meons which attracts the ZMBH pair. The size of effect is the same and the chains formed are the same. However, at separations greater than the interaction distance, because the orientation information is no longer available, the interacting loops cannot react to each other's spin energies. Where the gravitational chains provide a continuous gradient of gravity along the chain towards the loop meons or anti-meons (attraction), the same ZMBH chains due to charge provide no overall charge gradient along a chain. So the interaction of loops due to spin is because physically their whirlpools interact.

The action of two whirlpools with similar rotational orientations is to combine, whereas the action of two oppositely rotating whirlpools is to repel, both when inside the influence distance. The relative angle of the whirlpool planes of rotation will also affect the strength of interaction with a minimum when the planes are at right angles. Although the overall charge along each chain is zero, there will be an alternate electric field extending outward along the chain from each loop meon and anti-meon which will be non-zero and larger closer to the loop meon or anti-meon.

This explains why the spin-spin energy interaction is very small in large bodies with randomly oriented spins, but is large in aligned spin systems. It also shows that the kinetic energy of the spin energy in any motional system needs to be considered in exactly the same way as the mass kinetic energy. And that the size of mass energy and spin energy is the same in any loop.

## The Electron

The next level up from single fermion loops was considered earlier, as were the wider implications at the largest scale. Now the foundations of the composite nature of fermions and the basic structure of loops can be analysed in more detail.

As shown earlier the two sets of equations governing the motion of meons around a ring can be shown as

$$E_{M+} = (\gamma_+ - 1)(M \pm a)c^2$$

$$E_{M-} = -(\gamma_- - 1)(M \pm a)c^2$$

$$a = s/6 = \sqrt{\alpha / 2\pi} M / 6$$

$$E_{Q+} = (\gamma_+ - 1)(Q \mp b)c^3$$

$$E_{Q-} = -(\gamma_- - 1)(Q \mp b)c^3$$

$$b = q/6 = \sqrt{\alpha / 2\pi} Q / 6 \cdot$$

We set $(\gamma_+ - 1) \cong \frac{1}{2} v^2 (1 + p/v)^2$ at $v<c$ and $v(1+p/v)=r(1+g/r)w$ for the positive meon, with $(\gamma_- - 1) \cong \frac{1}{2} v^2 (1 + f/v)^2$ at $v < c$ and $v(1+f/v)=r(1+k/r)w$ for the negative meon. Here $v$ and $r$ are the simple values that would be the case if there were only M energies, not the additional $a$ and $b$ mass and charge twist energies and the factor $p$, $k$, $f$ and $g$ are small changes in the velocity and radius of rotation respectively of the meons.

Because a meon pair, as considered here, must rotate at the same frequency $w$, it can be shown that $k/r=f/v$ and $g/r=p/v$.

So now the four equations become

$$E_{M+} = \frac{1}{2} Mvrw(1 \pm a/M)(1 + g/r)^2$$

$$E_{M-} = \frac{1}{2} Mvrw(-1 \pm a/M)(1 + k/r)^2$$

$$E_{Q+} = \frac{1}{2}Qcvrw(1 \mp b/Q)(1+g/r)^2$$

$$E_{Q-} = \frac{1}{2}Qcvrw(-1 \mp b/Q)(1+k/r)^2$$

## Mass and charge currents

To find the part of the magnetic moment due to the motion of the charges around the loop requires the simple formula relating the actual charges and area

$$\mu_{act} = IA = Q_{act}\pi r_{act}^2 / t$$

$$= Q_{act}(w/2\pi)\pi r_{act}^2 = \frac{1}{2}Q_{act}v_{act}r_{act}$$

Comparing this result with the four formulae shows that the equations for μ are contained within the energy formulae, using the appropriate actual values for each meon.

Considering firstly the symmetric electron. It is composed of three identical pairs of meons. In order to have an overall negative charge of size equal to one electronic charge, it must have all $b$ of negative charge. This means that all $a$ must be positive. We can consider just one pair and multiply the result by three.

The charge current for one pair can be calculated as

$$\mu_{Qe-(1)} = \frac{1}{2}Qvr(1-b/Q)(1+g/r)^2 + \frac{1}{2}Qvr(-1-b/Q)(1+k/r)^2$$

This is not the only contribution to the magnetic moment because there are also electric fields, in the frame of reference of the rotating loop, across the loop to the opposite meon and both towards the next in line, and from the following meon. Figure 4 shows the relative positions.

Similar to the charge current is the mass current, which for the three electron pairs will each be

$$\mu_{Me-(1)}/3 = \frac{1}{2}Mvr(1+a/M)(1+g/r)^2 + \frac{1}{2}Mvr(-1+a/M)(1+k/r)^2$$

This is not to say that there is a magnetic moment due to the meon mass energies in motion, only that all equivalent motions of mass or charge can use the same equations. If there were no $a/M$ energy factor,

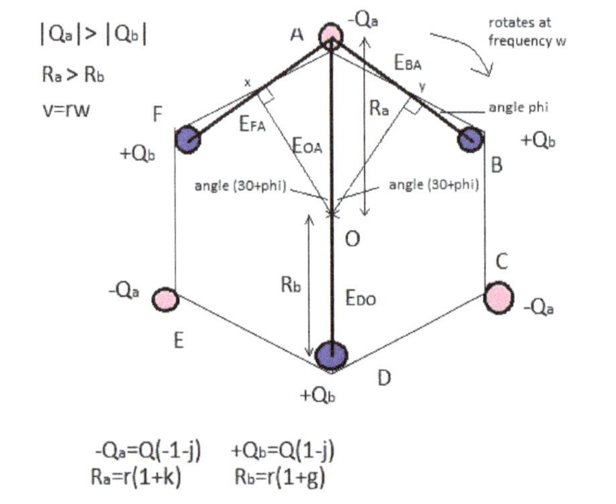


**Figure 4:** Electron loop with charges $-Q_a$ at Ra and + $Q_b$ at $R_b$ each from center of rotation O.

then $k$ and $g$ would be zero and the simple equation for one meon would be

$$\mu_{M+} = \frac{1}{2}Mvr = \frac{1}{2}h$$

Note that the relative radii in the meon equation have swapped relative to the charge equation, as has the sign of $a/M$.

Because $a/M$ and $b/Q$ are both ratios and of the same size, they can be simplified to $|a/M|=|b/Q|=j$.

Since the mass angular momenta for each meon in a loop must be the same, this means that the mass current for the loop must be zero, so the mass current equation for three pairs becomes

$$\mu_{Me-}/\frac{3}{2}Mvr = (1+j)(1+\frac{g}{r})^2 - (1-j)(1+k/r)^2 = 0$$

The solution for this relates $k$ and $g$ but does not pin down either yet. Replacing $(1+j)=H$, meaning 'Huge or larger' and $(1-j)=S$, meaning Smaller, the relationships are

$$k/r = (g/r)\sqrt{H/S} + (\sqrt{H/S} - 1)$$

$$g/r = (k/r)\sqrt{S/H} + (\sqrt{S/H} - 1)\cdot$$

The use of $H$ and $S$ is important in the simplification of the possible combinations of different radii of rotation of the meons within loops.

**A possible simplifying assumption**: Although the value of $h^{\pm}$ could be different to $h$, as has been used so far, it is also possible to consider that each meon, even when at $(1+g/r)r$ or $(1+k/r)r$ has in total still only $h$ angular momentum, as it would if there were no $a/M$ energy factor. So individually $(1+j)\left(1+\frac{g}{r}\right)^2 = 1$ and $(1-j)\left(1+\frac{k}{r}\right)^2 = 1$ would be the case.

The result would be that $g$ must have a negative value, representing a distance inside the 'no $a/M$' radius $r$ and further that the charge current equation for one pair would now become

$$\mu_{Qe-(1)} = \frac{1}{2}Qvr(1-j)/(1+j)$$

$$+\frac{1}{2}Qvr(-1-j)/(1-j)$$

$$= \frac{1}{2}Qvr\left[\frac{(1-j)}{(1+j)} - \frac{(1+j)}{(1-j)}\right]$$

$$\mu_{Qe-(1)} = \frac{1}{2}Qvr4j/(1-j^2)$$

And for three pairs gives the pleasing result that the effect of the offset rotational radii is slightly more than double the effective $q$ charge current. Instead of six $q/6$ charges providing just one $q$ charge current, the offset adds more from the meon fundamental charges, since the effect is

$$_{Qe\ (3)}/\frac{}{2}Qvr = -2\left(\frac{}{}\right)/(1-\frac{}{72})$$

This does not eliminate the need to consider other values for $h^{\pm}$, and the other effects providing magnetic fields, but is interesting in itself, because it pins the angular momentum of every meon, regardless of rotational orbit, at $h$. However, the possible alternate values of $h^{\pm}$ will be considered in the rest of the paper.

If the value of $h^{\pm}$ were equal to $h$, it would follow that the same formulae would apply to all charged lepton families for circumferential charge currents, meaning the same adjustment factors $k$ and $g$ relative to the different radii of the electron, muon and tauon loops, although each would be different absolute adjustment distances.

**Scale of magnetic moments**: The relative scale of this moment needs to be considered because the calculations here are all based on internal loop velocities, radii and charges, whereas the usual meaning of the magnetic moment of the electron relates to the external velocity and radius of an electron orbiting in a magnetic field.

As shown, the charge current for three pairs in the electron can be recast in the form

$$\mu_{Qe-(3)} = \frac{1}{2}h^{\pm}(12j)/(1-j^2)c$$

where $h^{\pm}$ means the value of angular momentum appropriate for the radius of rotation of the meon under consideration, which may not be $h$.

For a 'standard loop' where there would be no twist energies present then $j=k=g=0$ with $h^{\pm}=h$. The standardised magnetic moment of any loop would be $\mu_{Qs} = \frac{1}{2}h/c$ and the electron standardised anomalous factor due to charge current would be

$$A_{e-(MQ)} = \mu_{Qe-(3)} / \mu_{Qs} = -2(\frac{h^{\pm}}{h})(\frac{q}{Q})/(1-j^2)$$

To convert this into the orbital framework requires the multiplication of this factor by the ratio

$$A_{(MQ>mq)} = \left(\frac{Q}{q}\right)\left(\frac{m_e}{M}\right)$$

So that in terms of an electron orbiting at velocity $v_{orb}$ and radius $r_{orb}$ with mass $m_e$, with $h=m_e v_{orb} r_{orb}$ and

$$\mu_{qe} = -\frac{1}{2}qv_{orb}r_{orb}$$

Then the value of standardised internal electron charge current as an anomalous magnetic moment in the usual orbiting framework will be

$$A_{e-(mq)} = -2(\frac{h^{\pm}}{h})(\frac{me}{M})/(1-j^2)$$

This value is too small to affect the 12 decimal places of current accuracy of the electron anomalous magnetic moment. The case for the electric fields is analysed below.

**Mixing rotational radii in loops:** For each meon pair there are only four different outcomes of $H$ and $S$ for the meon mass $M$ as shown in Table 1, reverting to using $a$ and $b$ to define the different energy components. The first outcome $H+$ $S-$ is called $T_1$ and has the H+ positive mass meon at radius $(r+g)$, the inner radius of the two. So it has a larger positive mass factor but rotates at a smaller radius. For the charge side, this same meon has the smaller charge also at the smaller distance $(r+g)$. So this pair may have a balance of mass current but they have an excess of negative charge current since the larger charge orbits at the larger radius. So this pair of $T_1$ type represents one-third of the electron loop. By symmetry, the type $T_2$ pair makes up one-third of the positively charged positron.

Type $T_3$ has a mix of charges on each meon of the pair, with both meons showing $H$ sized mass current. These will rotate at the same $(r+g)$ radius when they are mixed in with a loop that requires equal angular momenta, but otherwise if all three pairs are $T_3$, in a symmetric neutrino, they can take any radius.

Type $T_4$ has a mix of charges on each meon of the pair, with both meons showing $S$ sized mass current. These will rotate at the same $(r+k)$ radius when they are mixed in with a loop that requires equal angular momenta, but otherwise if all three pairs are $T_4$, in a symmetric neutrino, they can take any radius.

It is by mixing types within a loop that the quarks and asymmetric neutrinos can be made. The 20 mixes of types that are necessary to form leptons and quarks are shown in Table 2.

What is apparent is that in every loop, other than the symmetric neutrinos, there are only two radii at which meons rotate. All meons with $H$ as the mass factor will rotate at $(r+g)$ and all meons with $S$ as the mass factor will rotate at $(r+k)$ when $(r+k)>(r+g)$, and it is possible, as shown earlier, that $g$ could have a negative value.

Although this simplifies the possible structure of each loop, it does require each meon and its relationship with all other meons in the loop to be considered. This means mainly the two adjacent meons and the one opposite. In a simple non $a$, $b$ system the same sign meon interactions could be ignored because the actions of the foundation sizes of mass $M$ and charge $Q$ would cancel each other out. But this is not the case, although the interactions are much smaller – there is for the symmetric leptons no electric field between same sign meons, but there is for the other three opposite sign meons in the loop.

Table 3 shows the breakdown of the only 14 type combinations possible that form different loops. This is not the same as comparing the positional charge symmetry from one loop to another loop, which is the basis of balancing loops in a stable stack. In the latter case, the relative positions of each meon matter rather than the $H$ and $S$ combinations, so a down quark $d^-$ composed of $T_1$ $T_4$ $T_4$ pairs in order round the loop is a different 'colour', meaning physical rotational phase, to another down quark composed of $T_4$ $T_1$ $T_4$ pairs in order. In the table, the order is not necessarily important for defining the properties of the loop, although it would be for phase (colour) and balance, as will be analysed when considering stack symmetry. The 14 different type combinations are split into 7 for leptons and 7 for quarks.

## Electric fields between meons

What the only two radii of rotation do simplify is the relative angles that the circumferential electric fields can take to the perpendicular to the radial line between two meons as shown in Figure 4. If the meons rotate at the same radius, the angle will be zero. For every other combination of $H$ and $S$ radii, the angle will be the same size $\phi$ outward or inward from the centre for each meon. This is because the circumferential electric field between any two different-radius adjacent meons have different velocities at each point along the line from $v(1+k/r)$ to $v(1+g/r)$, or vice versa. This affects the magnetic field generated when considered in the external frame of reference in which the loop is observed to rotate at $w$.

| Type | e- | e+ | $v_s$ | $v_a$ | d- | d+ | u+ | u- |
|---|---|---|---|---|---|---|---|---|
| Any | | | $T_3$ $T_4$ | $T_{34}$ | | | | |
| r(1+g) only | | | $T_3$ | | | | | |
| r(1+k) only | | | $T_4$ | | | | | |
| r(1+k) and r(1+g) | $T_1$ | $T_2$ | | $T_{123}$ $T_{124}$ | $T_{12}$ $T_{14}$ $T_{13}$ $T_{134}$ | $T_{12}$ $T_{24}$ $T_{23}$ $T_{234}$ | $T_{23}$ $T_{24}$ | $T_{13}$ $T_{23}$ |

**Table 3:** Radius combinations of types $T_x$ in the fermions.

The angle $\varphi$ is given by $tan\varphi = cos30(k-g)/(1+\frac{1}{2}(k+g))$

There are two different electric fields to consider. The first is circumferential and the second radial.

**Circumferential fields**: The two circumferential fields $E_{FA}$ and $E_{BA}$ in Figure 3 have differential velocities at each meon and along the line between them because the positive charges $+Q(+1-j)$are at $R_b=(r+g)$ which is less than the $R_a=(r+k)$ of the $-Q(+1+j)$ negative charges. But there is a zero of velocity along the line of each field, where the off-direction of velocity of the fields at each point becomes zero. These points, where the fields lines are perpendicular to the radial lines are at x and y, and these points are at angles of $(30+\phi)$ to the charge at A.

Note that here in the formulae is considered positive, outward beyond *r*, but, as seen earlier (and as drawn in the figure), could have a negative value, dependent on the value of $h^{\pm}$.

So each E field is split into fractional fields $E_{FX}$ and $E_{XA}$ and into $E_{BY}$ and $E_{YA}$. And either side of points x and y along the two fields, the velocity relative to the field lines act in different directions, either outwards or inwards relative to the centre. This means that the B fields generated act in the same direction, in Figure 4 it would be up out of the plane of the paper.

**Radial fields**: There is one radial field acting between meons at D and A, but in the same way that the circumferential fields were split into two due to the presence of a zero of velocity this field needs to be split. So $E_{DA}$ is split into two fractional radial fields $E_{DO}$ and $E_{OA}$ which are affected by the perpendicular angular velocity acting at each point, which increases outwards from zero at the centre O. Once again, due to the different velocity components, both E fields result in B fields up out of the plane of the paper in Figure 4.

However, dependent on the mix of *Q* and *q/6* charge on each meon being considered, the zero of motion may not coincide with the zero of electric field. The net magnetic effect will be in line with the expected sign, but will be larger or smaller due to the non-zero or reverse electric field respectively near the centre of rotation.

The result is that all six E fields have the same formulae, differentiated only by the charge and distance sizes and the limits of integration along the fields. The formulae can be used for all loops, using masses, charges, radii and limits, but need to consider each set of relationships between any meon and all the others in the loop. For asymmetric loops there are no shortcuts, although the small number of possible adjacent meon to meon relationships will keep the total number of different values low.

Radial and circumferential fields will exist in neutrinos, even though there is no charge current in total, so they will have a very small anomalous magnetic moment.

So far this consideration of the electron has looked mainly at relative radial effects, although *k* and *g* also represent absolute values circumferentially when considering relative angles. When looking at the families of fermions, the difference between the relative and absolute effects will probably give rise to the differences in anomalous magnetic moments.

It is not yet clear whether the internal effects described here are sufficient to provide the total anomalous magnetic moments of the charged leptons, or whether they are adjustments currently beyond the capabilities of experimental justification.

**Total magnetic moment formulae:** The end result should be a formula containing all the internal components generating magnetic moment in any loop. The basis here is specific to the charged leptons, but simply changing the input parameters of charge and radius will provide any other loop magnetic moments – even for dark matter loops. The difference that makes the formula so powerful is the reliance on actual ring frequency – which means the mass of the loop. So each loop in the lepton family has a different magnetic moment.

Unfortunately the analysis, basically the equation for the electric field between meons needing integration dependent on velocity at each point is beyond the author's capabilities at this time. Hopefully this will be rectified in due course.

**Solving for the actual values:** Once the integrations have been managed, there would be two simultaneous equations to solve with only two unknowns, the fractional radial distances *k* and *g*. The solution would be found by reduction method because of their complexity. One solution produces $h^{\pm}=1.0017h$, using unsatisfactory integration results, so is discounted, but does suggest an upper limit since it was based on coincident zeros of field and velocity, whereas offset zeros would reduce the value. The power of the total formula should be in arriving at adjustments to, or the observed magnetic moment of, the electron to an accuracy of 12 decimal places.

## Discussions

This long and convoluted paper has unfortunately been necessary to understand how so many issues can be explained based on such simple foundations.

The hypothetical existence of meons, and their twisting motions, may seem far-fetched. But the number of paradoxes solved are a testament to its power. And that this comes about using only one type of particle and its anti-partner, only two types of energy, one size other than 1 and three dimensions of space makes this explanation the most simple that could ever be imagined for a theory of everything.

These are a few of paradoxes already explained.

a) How can nothing give rise to something? The answer is zero mass black holes that unmerge into meons.

b) How can particles appear to be simultaneously in two places at once - in entanglement or superposition? The answer is that they can`t, and only appear that way because the skipping frequency is too high for us to measure in an environment without background and velocity above *c*.

c) How do quantum mechanics and relativity fit together, when the latter requires a time component which the former does not? The answer is that classical and quantum mechanics have not recognized the zero energy of stable orbital states, although in different ways, that underlies the independence of classical and quantum energies from time and that all particles have zero total energy. Relativity is a result of viscosity limiting motional speeds of meons and losing energy (frequency) from loops to the background ZMBHs whilst quantum mechanics arises when loops move where there are no ZMBHs and the zero energies are more obvious. So viscosity separates the relativistic and quantum environments.

d) Why is the action of charge so strong compared with gravity? The answer is that the strength of action of both is identical. It is the relative size of the charges generated by the meons in a loop when compared with the loop 'masses' that gives the appearance of different strengths. The gravitational constant also serves to confuse the situation.

e) In calculating energies, orthodoxy interprets particles as points which results in infinite interaction energies that have to be what is quaintly described as `renormalized`, effectively dividing everything by infinity, to obtain reasonable numbers. With loops, every loop has a definite non-zero size as do all meons, so renormalisation is not required.

f) The most interesting paradox is that which exercised Bohr and Einstein so much. Put simply, is the universe spooky acting at a distance with no reality until measurement or is it clockwork composed of real particles? The answer surprisingly is that both are correct! The difference is that each refers to different levels of meon structure. The spooky action at a distance and lack of reality until observation are aspects of the absence of viscosity and the existence of zero energy orbitals and shells on which loops are seemingly randomly appearing during skipping which results in probabilistic observations. The clockwork interpretation applies to the individual meons, which are real particles. Unfortunately we have no access to their non-probabilistic motions because we can only observe using probabilistic loops. So Bohr was right at the level we can observe, whereas Einstein was correct in describing the underlying particulate form of nature. The framework here enables paradoxes such as this to be true in both interpretations – it requires only having an appreciation of the deeper prequark nature of the universe for understanding.

g) Time and its apparent non-reversible nature also is no longer paradoxical in the loop framework. Before loops formed, there was no time that we can observe - since all our measurement of time is based on using loops to observe. So time is a construct of the structures formed by meons and is not an extra dimension. That the physical size, the mass, of a ring is linked to a measurement of time is fundamental in the loop framework - they are just two different ways of describing the loop size. This is what underlies Heisenberg`s uncertainty principle. For every loop the simple product of the momentum of each meon and its frequency of rotation around the loop is its energy ±½ *hw* at low energy in DAPU form. However, this result is more accurately written as $(\pm\frac{1}{2}w)h$, since the ½ belongs to the angular frequency *w*, as the relativistic expansion of $(\gamma\text{-}1)w_o$, and the *h* to the meons and anti-meons. This means that the energy of a loop due to its mass energy alone is half the expected value, but twice the angular momentum expected. But when spin energy of the same size, but opposite type, is added, the total content is twice expected the value.

So the really the product of energy and time (inverse angular frequency) in DAPU is always *h*. Measuring any two associated properties like energy and time for any loop will always result in the value *h*. For particles composed of multiple loops, the value will be a multiple of *h*, so the minimum value is *h*. Because every meon and loop has zero total energy, it should be possible to ignore the time content for reverse situations, except that the background viscosity always deducts energy and ensures no process can be reversed without needing the addition of energy. In TAPU form, these values are all just ±1, and Heisenberg`s equation becomes $E_T T_T=1$

h) The `twins` paradox is one that is not truly a paradox, but is worth understanding from a loop perspective. The two twins are separated, one travels elsewhere at high speed and the other doesn`t. When they meet again, after the traveller returns, one is older than the other. The round trip by the fast twin involves no vector energy difference if the start and end points are coincident, but what has changed is the phase difference between the loops that comprise the bodies of the twins. The motion of the fast twin`s loops has shifted the phase from being identical to being different. This is the equivalent of introducing a time difference. So initially identical loops that then have phase differences introduced are no longer the same and a time difference has been added.

i) The paradox that suggests that normal physics does not apply inside stellar black holes can be seen in the loop framework to be false. Since normal physics is actually based on loops composed of the most dense black holes possible, then normal physics applies inside stellar black holes. What gets destroyed on the way into large black holes is not the meons, but the loops – being broken into chains.

j) One of the predictions of the standard model is that there is an energy at which all forces become equal. As has been shown, this can never be the case since the mass of a loop and the electronic charge are generated in different ways. However, there is an equivalent for loops, in that the spin interactions depend on the loop frequency (mass) and relative spin orientation. At very high loop frequencies, where the relativistic effect becomes much greater than ½ *hw*, the parallel spin orientations become indistinguishable and the loop identities only distinguishable by charge. If these are the energies at which stacks initially formed, then only the overall charge of the stacks would have mattered in the local environment, provided the individual loops were the correct isomers to fit symmetrically within a particular stack to provide stability. So regardless of whether the loops in a stack were electron/positron, quark or neutrino, they would have been of equal energy, differentiated only by charge.

k) Wave-particle duality is another paradox of sorts. How can a particle be both wave and particle? As has been shown, a loop is exactly that. It has both a frequency of rotation and what we observe as a mass. These two aspects are just different sides of the same coin. But note that there is no need for any Higgs boson to provide mass to loops.

l) Matter and antimatter continue to exercise the standard model. How can they be created in equal amounts and yet then some mechanism manages to destroy all the antimatter and leave just a small amount of matter only. The answer in the loop framework is that all categories of normal loop, whether electrons, positrons, neutrinos, antineutrinos or quarks and antiquarks, and dark matter equivalents of each pair number, were created in equal numbers when the unmerging of zero mass black holes turned into our Big Bang. Which loop is the matter and which is its antimatter partner depends on the definition, but only charge can be used as the differentiator.

m) In the model produced here, the universe is both continuous and particulate. The particulate parts are the meons themselves, and the loop formed by them, with the continuous part the ZMBHs in their overlapping, rotating and vibrating state. The later provide values for charge and mass action at all values between one and zero on a random basis.

On a more specific discussion of magnetic moments – because the overall paper is so long, it is left for others to calculate the moments of the quarks and stacks. However, they will find that the size of mass observed for quarks alone and in stacks is a function of not only on the frequency of loop rotation, but also on the charge of the loop, which relates to the actual values of *s* for each loop.

It is also evident that quantum effects such as superposition ought to occur at what has been previously considered classical levels. Possibly it is only the complexity of the objects being considered that has stopped such effects being observed so far.

On the subject of energies and the wider implications of ZMBHs unmerging and failed inflation events, whilst multiverses may be an exciting area of theory, the speculation involved is more far-fetched than that proposed here with only one single universe. There is no need for 10, 11, 13 or 26 dimensions hidden from view, although the ideas of loop quantum gravity, M-theory and loop string theory are not too far away from this pre-fermion loop framework. They only require, at the basic level, the appreciation that a pre-fermion framework using actual particles rather than strings, membranes or solid doughnuts solves both upward combinations, like nucleons formed from stacks through asymmetric balancing, and downward combination into ZMBHs (strictly, the reverse in unmerging) which provide a form of modern very diffuse aether, with the background and loops each influencing each other.

One of the most interesting targets would be to be able to estimate the proportion of 3-loops versus non-3-loops to see what percentage each should make of the total loop population, compared to the observed matter/dark matter ratio. This would help estimate whether there would still be room for other ideas on the observed acceleration of expansion or rotational rates in galaxies. Unfortunately the proportion calculation is more complex than it appears at first sight, given that the starting point is an unknowable number of pairs and of pre-existing chain stars that have been re-sorting the population throughout the life of the universe. However, it may be possible using that current split, given a suitable equation for the conversion rate, to estimate how long the steady state existed before our big bang, or at least a minimum time.

Another observable target should be the untangling of the tired light effect from observed redshifts, which would strongly support the composite nature of the photons, where all component meons in any frequency of photons emitted by one object and observed by another travel almost exactly the same distance from emission to observation regardless of the loop (photon) energies, except at very high frequencies. The discovery of unarguably conjoined galaxies with significantly different redshifts would represent a background failed big bang event before our successful one physically adjacent to part of our expansion, leading to a direct calculation that the difference in redshift is due to the difference in expansion velocity at that location versus background motion.

Intriguingly it may also be possible to observe the background viscosity effect by the simple act of putting an object in a box - although not a cat. If left long enough without photon interaction, it may be possible to observe some temporary mass loss in the object [20]. Unfortunately the act of taking the object out of the box may well quickly allow the reduced-frequency loops to be recharged back up to their normal frequency by photons. And there are many other experimental issues that will complicate such an observation - local gravity changes, radioactive decay, atmospheric pressure differences, humidity etc

The ideas proposed here may be viewed as speculative because they start from a prequark framework. This physics beyond the standard model derives its strength from the number of aspects of the universe that it explains, including how mass arises, what inertia is, why particles have spin in units of ½ $h$, why all electrons have the same charge size, as do nucleon stacks, why particles have magnetic moments, why the second law of thermodynamics exists, why there is an arrow of time, why there is a maximum speed for particles, why stable states exist, why tired light may reduce the need for dark energy, why there is no matter/anti-matter imbalance, what dark matter is likely to be, what is likely to be the physical reality underlying zero point energy, why physics is the same everywhere and breaks down nowhere and why we have only one universe with threefold symmetry within nucleons in matter.

The weakness of some aspects of the proposals is drawn from the accepted interpretations that can only be reinterpreted correctly in the loop framework. These include the existence of adjusted Planck mass and density meons, the proposal on different screw motions of twist aligned along meon direction of travel generating different sign of one-sixth electron charge, the re-emergence of centrifugal forces as the direct expression of two outward energies in circular motion instead of centripetal acceleration due to curved motion, vector energy, the resultant adjustment of orbital energy equations and the re-emergence of a very diffuse and novel form of aether as the background of ZMBHs. These aspects may prove hard to persuade doubters about, but the simplicity and limited extent of the starting foundations, the logic of what can be built and how much it resembles the universe that is observed provide strong arguments in its favour.

Once it is accepted that the speed of light is a terminal velocity against the viscosity of the local environment, then the chase action between different mass types and the existence of negative fundamental mass are both supported.

Because all meons and composites composed of meons have zero total energy at all times, the mathematics currently employed to describe the energies or interactions of systems is insufficient at the foundation level and a simpler mathematics is required. Physics gets simpler as analysis gets deeper.

## Conclusions

Using the simple foundation of a universe composed only of ZMBHs of one size volume, which split into mirror meon and anti-meon to form chains then loops - which are our fermions and dark matter particles and the only composite particle form - it is possible to construct many of the major aspects of the universe as we observe it. And at every stage, every particle has zero total energy by our current definition. This shows that at a basic level current physics does not understand what energy is. This paper provides an explanation and direction on many other issues.

Inflation can be seen as the process by which the high energy initial loops lose energy to fund expansion and the balance between the amount of inflation and resultant loop masses decides whether the inflation and expansion event succeeds or fails. We see the failed inflations as chain stars, some of our black holes. Multiple big bangs within the background ZMBHs link the big bang theory to a form of steady state theory.

Given the loop framework and internal loop symmetries, there can be only one loop pair length, the 3-loop, which, as the shortest odd number pair loop, is the main source of any chemistry. The actual chemistry values will depend on the specific inflation rates along the only three spatial dimensions that exist. So the underlying physical laws of all loop systems will be the same, but the actual values of the interactions will depend on the specific rates of inflation of each inflation event.

The only conclusion that can be drawn is that with loop systems a type of universe with symmetries similar to ours is the inevitable outcome of a successful inflation event, even though the details may differ due to the different size of the fermions created in each.

It is also the case that there will always be two different types of volume which will exist within the universe. One will have a background of ZMBHs and zerons which provide viscosity and a terminal speed of travel. The other will not have a background and will have no terminal velocity of travel.

Using the adjusted Planck size and density meons in this pre-fermion framework ensures that physics will be the same everywhere and break down nowhere. And the existence of the viscosity of the background ZMBHs provides an arrow of time and a method of reassessing whether the universe is expanding at the rate currently accepted.

The final piece of the jigsaw in rearranging the picture of physics is the production of part of the anomalous magnetic moments of the loops. By showing how these contributions are produced in the leptons, the whole ZMBH-to-loop framework is underpinned.

Such a framework as described offers so many new ways of reinterpreting our current understandings that it deserves to be considered on a wider basis, providing possible solutions to open problems and direction for future research.

This paper also presents new ways of understanding the relationships between properties whilst undermining the current interpretation of where the quantum and classical worlds diverge. The novel insights and predictions include:

i. Physics is the same everywhere and breaks down nowhere. There are no singularities.

ii. There is only one universe, whose base fundamental components are ZMBHs of one size and two energy types that sum to zero.

iii. There are only two sizes in the universe, other than the loop sizes ('masses') which were locked-in by inflation, which are the TAPU Planck size of the meons and the fine structure constant, a function of the energy needed to unmerge ZMBHs.

iv. There is no beginning or end to the universe. ZMBHs have always existed and all loop and unmerged pair energies will eventually return to the background ZMBHs. The ZMBH background is both a continuum and the source of indivisibles.

v. The laws of physics can be no different anywhere because the maximal values of all properties are powers of $c^{1/2}$, or $c^{1/2}$ and the fine structure constant α. Loop sizes define the size of interactions but not the relationships between properties.

vi. If a loop is not passing through the background of ZMBHs, it is not limited to c and will not lose energy due to viscosity so exists in a quantum mechanical environment..

vii. Viscosity of the background ZMBHs underlies relativity, the arrow of time, electric charge generation and the second law of thermodynamics.

viii. ZMBHs unmerging enable loops, boson stacks, nucleon stacks and atoms. Nothing thus produces something, although the total energy is always zero.

ix. The absence in some volumes of the background ZMBHs underlies quantum mechanics and non-locality.

x. Matter and anti-matter are created equally. All stable systems have equal quantities because the only differentiator is the sign of charge.

xi. All meons and loops and everything composed of loops have total energy equal to zero. It is how the two types of energy in any particle interact with the same type in another that determines the result.

xii. Charge and gravitation have equal strengths of interaction when considered in fractional Planck terms in TAPU form.

xiii. Only two forces exist, due to mass and charge. Actions of the strong force are due to the loop nature of interactions between meons in adjacent loops, and the other energies in those loops. The displacement of loops in stacks by collision is the weak 'force'. The colour force is the balancing of asymmetric loops in a stack to produce rotational symmetry along the stack.

xiv. Energy is a counting mechanism. What we call the 'mass energy' of a loop is its component meons' rotational rate and is equal in size and opposite in type to the spin energy of the loop.

xv. Quantum mechanics and relativity co-exist within loops and which is observed depends on whether the background ZMBHs interact with the loops or not.

xvi. Superposition is the skipping of loops around tunnels in the background ZMBH space at frequencies too high to observe because the loops travel above *c*. Once one entangled loop is observed the tunnel or shell closes and the other loop is stranded.

xvii. Tunnels through the background ZMBH space which enable travel at speeds above c are the basis of quantum non-locality.

xviii. Elimination of *h* and *G* shows that size is not what differentiates gravitational from quantum systems. The energy equations in both systems are the same when the kinetic energy of spin is accounted for.

xix. General relativity requires time because it depends on the frequencies of loop rotations. Quantum mechanics does not require time because its non-local effects are outside the background ZMBH space.

xx. A loop is both a wave and a composite particle underlying wave-particle duality.

xxi. Time for particles composed of loops did not exist before loops formed. Time exists mainly in loops and when a loop breaks as it falls into a black hole it loses all time and reverts to a chain.

xxii. Inertia is the vector mass energy that a particle has in an external frame of reference.

xxiii. Twist charges occur in units of 1/6 electron charge because it takes the same amount of energy to unmerge a ZMBH into a meon and anti-meon pair.

xxiv. Normal matter is loops of three pairs. Dark matter is mainly loops with other than three pairs.

xxv. The speed of light is the maximum local velocity at which a meon can travel against the background ZMBHs, balancing viscosity forces against the mass chasing force between meons in the two photon component loops.

xxvi. Stable states exist as multiple levels of zero energy balance. All systems prefer states of zero total energy.

xxvii. Viscosity red shift requires the rethinking of how much, or whether, dark energy exists and the size and age of our big bang.

xxviii. ZMBHs are the force carriers, not the bosons. The background is rather like a very diffuse form of aether with loops acting on the background and the background acting on the loops. Overall the relativistic interaction is driven by the viscosity of the background on the meons.

xxix. Threefold symmetry in normal matter arises because there are three meon pairs in normal matter loops.

xxx. Chemistry arises because of the need to balance loop stack spin by the orbiting of the largest charge symmetric loop of the same pair number.

xxxi. The volume of dark matter exceeds that of normal matter because loops with less than three pairs are easier to make and black holes convert symmetric and asymmetric loops into mainly symmetric dark matter photons.

xxxii. The mass of normal and dark matter loops will be the same as ½ hw, simply. Their spin energies will all be ½ hw as well, but the magnetic moments will depend on the number of pairs in a loop.

xxxiii. Many black holes and galaxies are failed inflation events. Isolated black holes with no surrounding matter would prove that they were such events.

xxxiv. The unit size of meons means that the universe cannot be expanding in the accepted sense of all distances increasing. The observance of expansion in this sense is due at least partially to the viscosity of the background producing a red shift in photons which has not yet been taken into account.

xxxv. The big bang and (a form of) steady state theories can coexist, with failed inflation events appearing randomly as isolated black holes or galaxies and earlier such events acting as gravitational seeds for our big bang expansion.

xxxvi. Where two conjoined galaxies have different red shifts, one will be the result of a failed inflation in the 'stationary' ZMBH background and the difference in red shifts will represent the net expansion at that point in space.

xxxvii. Black holes are symmetry filters, sucking in asymmetry and emitting only symmetric photons.

xxxviii. Inflation along the only three dimensions locked in the three family sizes of the fermion loops.

xxxix. Negative fundamental mass exists with the chase interaction between opposite sign fundamental masses, and attraction between same sign fundamental masses.

xl. Fundamental charge sizes exist, as shown by their contribution to the anomalous magnetic moments of the loops.

xli. The fundamental constants $h$ and $G$ have zero values for dimensionality and can be eliminated from all equations by appropriate adjustment of SI units because they are only dimensionless ratios.

xlii. There are only three spatial dimensions because there are only three families of fermions and no evidence exists of any more.

xliii. There are three levels of time – outside the ZMBH stationary background, which has no time, ZMBH motion/rotation/vibration and loop time.

xliv. A loop has a magnetic moment when it has mixed twist charges on the meons which give different meon radii of rotation. The loop will have balanced mass currents, but net charge current and internal electric fields producing magnetic fields due to loop rotation, the latter even in neutrinos.

xlv. All meons have only two possible radii of rotation in asymmetric loops. In symmetric neutrinos, the radii are the same and can be any size, which enables neutrinos to adjust size and frequency easily.

xlvi. Slower loops cannot speed up faster ones.

xlvii. Pair creation is the temporary un-stacking of a zeron.

xlviii. Zero point energy is multiple concentric shells of zerons at every point in space.

xlix. Only loops with odd pair number can produce chemistry because the net spin of the nucleon stack can be balanced by the spin of the orbiting maximum charge loop.

l. Bohr and Einstein were both correct. They referred to different levels of meon structure without being aware that there were different levels.

li. The twins' paradox is not really a paradox. The relative motion between similar loops results in a locked-in loop phase difference which is a time difference.

lii. All observable particles, leptons and quarks and dark matter all have the same loop structure and the same unit observable sizes based on zero or $q/3$ charge, and $s$ twisting (1/2 $h$ spin), energies.

liii. To correctly understand the relationships between properties the fundamental constant $G$ needs to be split equally between both mass and distance properties and $h$ equally between both mass and charge, and distance properties.

liv. The size of loop mass energy and spin energy are equal, but the energies are opposite types.

lv. Fermions are composed of three meons and three anti-meons to give all fermion charges and ½ h spin.

lvi. Nucleons are stacks of loops, each loop rotating opposite to its adjacent loop.

lvii. Asymmetric neutrinos do not 'feel the weak force' in usual terminology because they are not 3-fold symmetric and so could appear to be more flavours of neutrino than the three already established.

lviii. A photon is effectively six ZMBHs reformed with chasing between meons in the two loops whose force is balanced against the background viscosity and the lost energy reduces the frequency as a red shift.

lix. Non-photon loops will also lose energy via viscosity in rotation, so 'mass' will be lost in the absence of photons which would otherwise refuel those loops.

lx. An isolated black hole is probably a failed inflation event.

lxi. Black holes transform loops preferentially to dark matter photons which may explain the observation of excess gas ionisation.

lxii. The physical electron loop size is possibly the largest possible (smallest mass) to produce a successful inflation event and possibly defines the limit between success and failure for a big bang and the subsequent rate of expansion or contraction.

lxiii. Because all meons and composites composed of meons have zero total energy at all times, the mathematics currently employed to describe the energies or interactions of systems is insufficient at the foundation level and a simpler mathematics is required. Physics gets simpler as analysis gets deeper.

## References


1. Aldrovandi R, Pereira JG, Vu KH (2005) Gravity and the Quantum: Are they Reconcilable? Conference Proceeding of Quantum Theory: Reconsideration of Foundations-3.Vaxjo University, Sweden.
2. Afshar SS, Flores E, McDonald KF, Knoesel E (2007) Paradox in Wave-Particle Duality. Found Phys 37: 295-305.
3. Bernhard C, Bessire B, Montina A, Pfaffhauser M, Stefanov AS (2014) Wolf Non-Locality of Experimental Qutrit Pairs. Journal of Physics A: Mathematical and Theoretical 47: 424013.
4. Ellis GFR (2007) Note on Varying Speed of Light Cosmologies. Gen Rel Grav 39: 511-520.
5. Lawrence M (2016) How SI Units Hide the Equal Strength of Gravitation and Charge Fields. J Phys Math.
6. www.maldwynphysics.org.
7. Heisenberg W (1930) The Physical Principles of Quantum Theory. University of Chicago Press, USA.
8. Albert E (1936) Lens-Like Action of a Star by the Deviation of Light in the Gravitational Field. Science 84: 506-507.
9. Wimmer M, Regensburger A, Bersch C, Miri MA; Batz S, et al. (2013) Optical diametric drive acceleration through action–reaction symmetry breaking. Nature Physics 9: 780-784.
10. Barrau A, Fourier UJ, Grenoble (2008) Physics in the multiverse: an introductory review. Cern Courier 47: 13-17.
11. Kimura M, Isogai K, Kato T, Ueda Y, Nakahira S, et al. (2016) Repetitive patterns in rapid optical variations in the nearby black-hole binary V404 Cygni. Nature 529: 54-58.
12. Wu XB (2015) An ultraluminous quasar with a twelve-billion-solar-mass black hole at redshift 6.30. Nature 518: 512-515.
13. Halton ARP (1966) Atlas of Peculiar Galaxies. Publ Pasadena: California Inst. Technology.
14. Paál G (1992) Inflation and compactification from galaxy redshifts? ApSS 191: 107-124.
15. Kollmeier JA,Weinberg DH, Oppenheimer BD, Haardt F, et al. (2014) The Photon Underproduction Crisis. The Astrophysical Journal Letters.
16. Brookhaven National Laboratory (2015) Searching for signs of a force from the 'dark side' in particle collisions. ScienceDaily.
17. Kollmeier JA, Weinberg DH, Oppenheimer BD, Francesco H (2014) The Photon Underproduction Crisis. The Astrophysical Journal Letters 798: L32.
18. Alcock C (2000) The MACHO Project: Microlensing Results from 5.7 Years of LMC Observations. Astrophys J 542: 281-307.
19. Manning AG, Khakimov RI, Dall RG, Truscott AG (2015) Wheeler's delayed-choice gedanken experiment with a single atom. Nature Physics.
20. Georgia Institute of Technology. Official Kilogram Losing Mass: Scientists Propose Redefining It As A Precise Number Of Carbon Atoms. ScienceDaily, 21 September 2007. P. Mohr Recent progress in fundamental constants and the International System of Units Third workshop on Precision Physics and Fundamental Physical Constants (6 December 2010).

19

# Inertia effect on free convection over horizontal surface in a porous medium with a variable wall heat flux and variable wall temperature containing Internal heat generation

**Ferdows M[1]* and Liu D[2]**

[1]Department of Applied Mathematics, University of Dhaka, Bangladesh
[2]Department of Engineering and Science, Louisiana Tech University, USA

### Abstract

The analysis is performed for the steady-state boundary layer flow with free convective heat transfer over a horizontal surface embedded in a fluid-saturated porous medium subject to variable wall heat flux (WHF) and variable wall temperature (VWT). In modeling the flow inertia, no slip boundary conditions and exponentially decaying internal heat generation (IHG) are taken into account. The similarity equations are solved numerically and their solutions are dependent on the problem parameters G, m, λ. The influence of such parameters on flow characteristics such as velocity, temperature profiles, the skin friction coefficient, and the Nusselt number are thoroughly discussed. It is found that the flow profiles along with physical parameters significantly altered the rate of heat transfer and induced more flow within the boundary layer than that of without IHG.

**Keywords:** Inertia; Nusselt number; Rayleigh number

## Introduction

Numerous phenomena involve free convection driven by internal heat generation. Free convection, as a process of vital significance, has been of interest to researchers in both industry and academia, such as a recent numerical work by Liu et al. [1]. Two important modelling of free convection are related to nuclear energy and combustion, specifically in the development metal waste from spent nuclear fuel and the storage of spent nuclear fuel, see Horvat et al. [2], Jahn and Reinke [3]. A couple of recent papers have been devoted to the subject of similarity solutions for free with internal heat generation in porous media for several geometric configurations, see Bagai [4], Postelnicu and Pop [5], Postelnicu et al. [6,7]. Several recent papers, such as Magyari et al. [8,9], Mealey and Merkin [10], Merkin [11] throw new lights on the internal heat generation in porous media, remaining in the frame of similar solutions.

Turning now to the non-Darcy formulations, with focus on vertical and horizontal configurations in porous media, we present below a short review of the literature dedicated on these topics.

**a) Vertical configuration**

- Thermal dispersion, which becomes prevalent in the non-Darcy regime, and viscous dissipation effects are studied by Murthy [12]) for a vertical flat plate configuration, with inertia effect included in the momentum equation. Constant wall temperature conditions were considered in that study.
- Nonsimilarity solutions for the non-Darcy mixed convection from a vertical flat plate embedded in a saturated porous medium for a power-law variation of the wall temperature were reported by Kodah and Duwairi [13].
- A number of non-Darcian effects, including the high-flow-rate inertia were analysed by Hung and Chen [14] for a free convection along a impermeable vertical plate embedded in a thermally stratified fluid-saturated porous medium. Variable heat flux conditions are imposed in that study.

**b) Horizontal configurations**

- The presence of inertia (Forchheimer form-drag) in the steady free convection boundary layer over an upward-facing horizontal embedded in a porous medium was studied by Rees [15].
- This work was extended later by Hossain and Rees [16] by taking the surface temperature as a power-law variation in terms of the distance from the leading edge.
- Duwairi et al. [17] analyzed the combined free and forced convection regimes within a non-Darcy model for a horizontal configuration.

In view of the above literature survey, it seems that the effect of inertia on free convection from a horizontal surface embedded in a porous medium with internal heat generation was not considered till now, at our best knowledge. Two cases are considered in the paper: variable wall heat flux and variable wall temperature. Distinctive from the previous quoted studies where inertia effects have been considered, in the present paper there are obtained similarity solutions, in the spirit of studies dealing with internal heat generation.

## Analysis

Consider the natural convection , laminar, two dimensional boundary layer flow over horizontal surfaces in porous media with internal heat generation of wall temperature $Tw$ and uniform ambient temperature $T\infty$,where $Tw > T\infty$. The x-coordinate is measured along the surface and the y-coordinate normal to it. Under the Boussinesq and boundary layer approximations, the basic equations are continuity equation, momentum equation and energy equation

$$\frac{\partial^2\psi}{\partial y^2}+\frac{K'}{\upsilon}\frac{\partial}{\partial y}\left[\left(\frac{\partial\psi}{\partial y}\right)^2\right]=-\frac{g\beta\beta K}{\upsilon}\frac{\partial T}{\partial x} \qquad (1)$$

***Corresponding author:** Ferdows M, Professor, Department of Mathematics, University of Dhaka, Bangladesh, E-mail: ferdows@du.ac.bd

$$u\frac{\partial T}{\partial x}+v\frac{\partial T}{\partial y}=\alpha\frac{\partial^2 T}{\partial y^2}+q''' \quad (2)$$

where the stream function Ψ was introduced through the relationships

$$u=\frac{\partial\psi}{\partial y}\quad v=-\frac{\partial\psi}{\partial x} \quad (3)$$

such as the continuity equation

$$\frac{\partial u}{\partial x}+\frac{\partial v}{\partial y}=0 \quad (4)$$

Is identically satisfied.

## Case of variable wall heat flux

The boundary conditions for the model are

$\upsilon=0$, $q_w(x)=ax^m$, at $y=0$ (5a)

$u=0$, $T=T_\infty$, as $y\to\infty$ (5b)

We now introduce the following dimensionless variables

$$\eta=\frac{y}{x}\left(Ra_x\right)^{1/4}\dot{e}\psi=\alpha\left(Ra_x\right)^{1/4}f(\eta),\ (\eta)=\left(T-T_\infty\right)\frac{\left(Ra_x\right)^{-1/4}}{q_w x/k} \quad (6)$$

Where the local Rayleigh number is defined as

$$Ra_x=\frac{g\beta q_w(x)Kx^2}{k\upsilon\alpha} \quad (7)$$

We then consider the internal heat generation (IHG) of the form

$$q'''=\frac{\alpha q_w(x)x}{k}\cdot\frac{\left(Ra_x\right)^{3/4}}{x^2}e^{-\eta} \quad (8)$$

Equations (1) and (2) become

$$f''+2Gf'f''+\frac{m-2}{4}\eta\theta'+\frac{3m+2}{4}\theta=0 \quad (9)$$

$$\theta''+\frac{m+2}{4}f\theta'-\frac{3m+2}{4}f'\theta+e^{-\eta}=0 \quad (10)$$

Where

$$G=\frac{\alpha K'}{\upsilon}\cdot\frac{\left(Ra_x\right)^{1/2}}{x} \quad (11)$$

Is the inertia parameter. The corresponding boundary conditions are

$f(0)=0$, $\theta'(0)=-1$ (12a)

$f'(\infty)=0$, $\theta(\infty)=0$ (12b)

Quantities of engineering interest are the skin friction coefficient and Nusselt number, defined as

$$C_{fx}=\frac{\tau_w}{\rho U_c^2/2}\quad \tau_w=-\mu\left(\frac{\partial u}{\partial y}\right)_{y=0} \quad (13)$$

$$Nu_x=\frac{hx}{k},\ h=\frac{q_w}{T_w-T_\infty} \quad (14)$$

where $U_c$ is a characteristic velocity, taken here as

$$U_c=\frac{\sqrt{\upsilon\alpha}}{x} \quad (15)$$

Performing the calculations, we get

$$C_{fx}\left(Ra_x\right)^{-3/4}=-f''(0) \quad (16)$$

$$Nu_x\left(Ra_x\right)^{-1/4}=\frac{1}{\theta(0)} \quad (17)$$

## Case of variable wall temperature

1. The boundary conditions in this case are

$\upsilon=0$, $T=T_w(x)=T_\infty+Ax^\lambda$, at $y=0$ (18a)

$u=0$, $T=T_\infty$, as $y\to\infty$ (18b)

We now introduce the following dimensionless variables

$$\eta=\frac{y}{x}\left(Ra_x^*\right)^{1/4},\ \psi=\alpha\left(Ra_x^*\right)^{1/4}f(\eta),\ \theta(\eta)=\frac{T-T_\infty}{T_w-T_\infty}\left(Ra_x^*\right)^{-1/4} \quad (19)$$

Where the local Rayleigh number is defined as

$$Ra_x^*=\frac{Kg\beta\left(T_w-T_\infty\right)x}{\upsilon\alpha} \quad (20)$$

Now, the internal heat generation term is taken in the following form

$$q'''=\alpha\left(T_w-T_\infty\right)\cdot\frac{\left(Ra_x^*\right)^{3/4}}{x^2}e^{-\eta} \quad (21)$$

in order to obtain similarity solutions.

The problem becomes, in transformed variables,

$$f''+2Gf'f''+\frac{\lambda-3}{4}\eta\theta'+\frac{5\lambda+1}{4}\theta=0 \quad (22)$$

$$\theta''+\frac{\lambda+1}{4}f\theta'-\frac{5\lambda+1}{4}f'\theta+e^{-1}=0 \quad (23)$$

That must be solved along the boundary conditions

$f(0)=0$, $\theta(0)=1$ (24a)

$f'(\infty)=0$, $\theta(\infty)=0$ (24b)

The expression of the skin friction coefficient remains the same, so that

$$C_{fx}\left(Ra_x^*\right)^{-3/4}=-f''(0) \quad (25)$$

While for the Nusselt number we get

$$Nu_x\left(Ra_x^*\right)^{-1/2}=-\theta'(0) \quad (26)$$

## Numerical Analysis and Results

The two sets of boundary value problems (9), (10), (12) and (22-24) were solved by two methods.

a) A well-known finite difference method, namely the Keller-box method, incorporated in our own code

b) Using the *dsolve* routine from MAPLE [18].

The results given by these methods were compared and in each case the maximum value of η, say $\eta_{max}$, was adjusted in order to conciliate the outputs and get grid independence in case a). As a rule, the dynamic and thermal boundary layers become thicker as *G* increases and also the convergence of the numerical procedure is achieved with more difficulty. Basically, $\eta_{max}$ was raised till 20, when *G* was increased. Results are reported separately for the two cases considered in this study: Figures 1-6 are variable surface heat flux (VHF), while Figures 7-12 for variable wall temperature boundary conditions (VWT). The variations of the skin friction and Nusselt number with the inertia

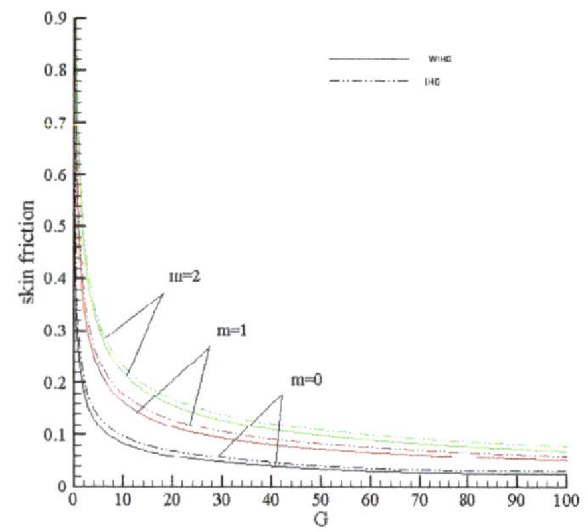

Figure 1: Skin friction variation with the inertia parameter G, VHF case.

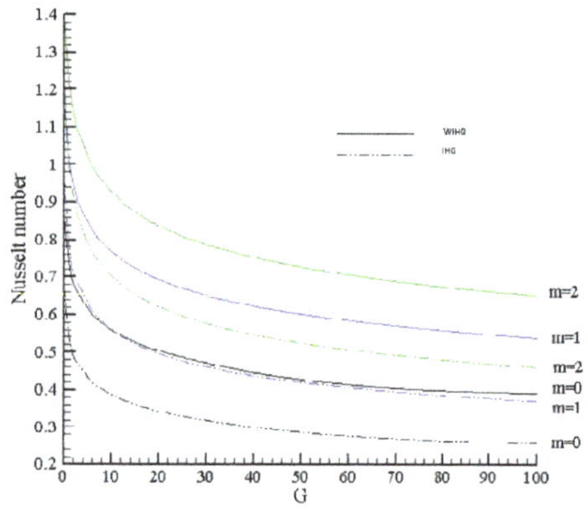

Figure 2: Nusselt number variations with the inertia parameter *G*, VHF case.

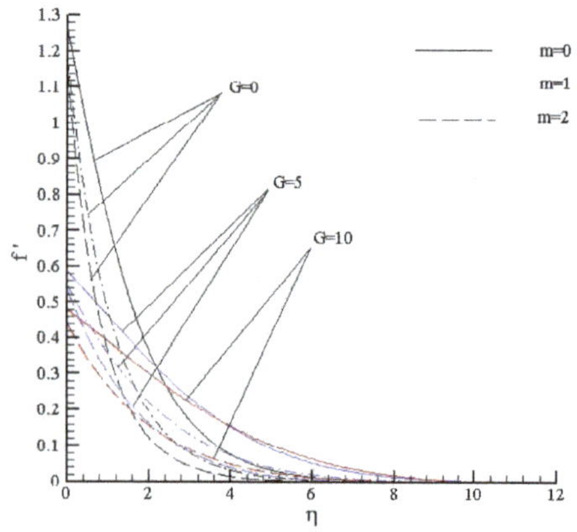

Figure 3: Velocity profiles in VHF case, without internal heat generation, for several values of *G* and *m*.

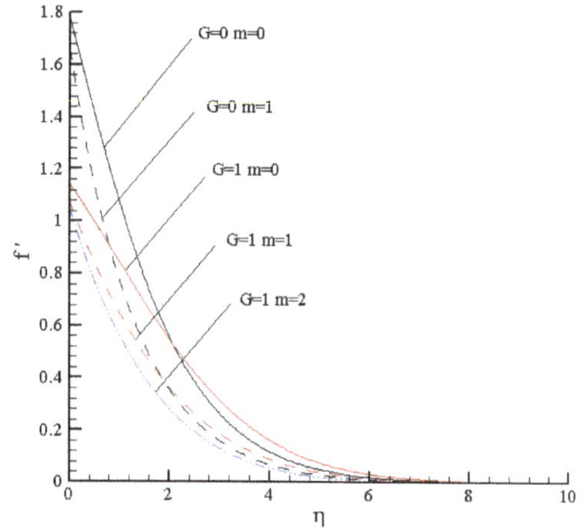

Figure 4: Velocity profiles in VHF case, internal heat generation, for several values of *G* and *m*.

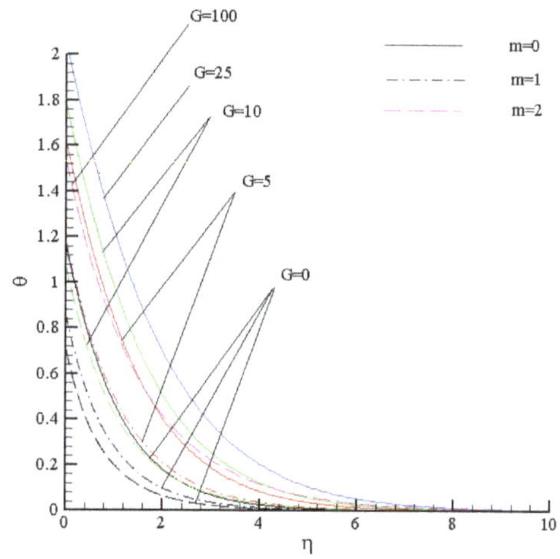

Figure 5: Temperature profiles in VHF case, without internal heat generation, for several values of G and *m*.

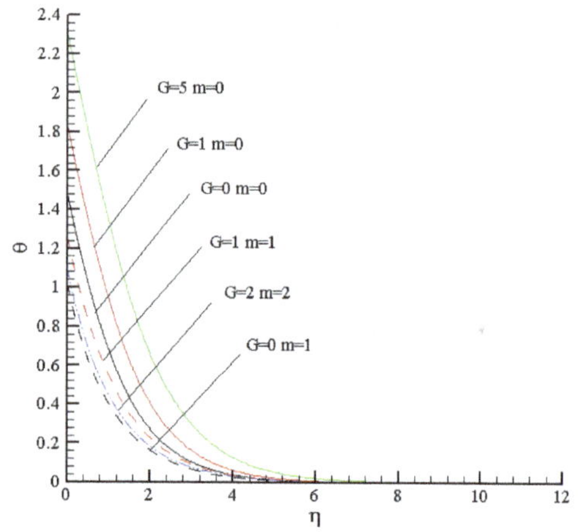

Figure 6: Temperature profiles in VHF case, internal heat generation, for several values of G and *m*.

parameter are shown in Figures 1 and 2 respectively considering with and without IHG. From Figure 1, it is seen that in both the cases (with and without IHG) the skin friction is more significant and induced more flow than that of without IHG, hence give rise to the velocity gradient leads to the increase of the skin friction. On the other hand, one can easily remark from Figure 2 that the Nusselt number with IHG is less intensive than the presence of IHG.

Figures 3-6 presents the solutions of the dimensionless velocity and temperature profiles for several values of G and m considering with and without IHG. We can see that the velocity profiles decreases with increasing G (and m). The reverse happens as one move far away from the surface indicating that the effects of the parameters are more dominant near the surface. Further it is also seen that the momentum boundary layer thickness has influenced greatly by IHG. From Figures 5 and 6, it is seen that the temperature profiles increases and tends to zero at the edge of the thermal boundary layer satisfying the boundary condition θ()=0. Also the thermal boundary layer thickness increases when the effect of IHG is considered. The dimensionless velocity and temperature profiles as well as ski friction coefficient and Nusselt number for some values of G and λ in case of VWT are presented in Figures 7-12. We can draw the same conclusions as in the case of VHF.5.

## Conclusions

At our best knowledge, the available studies on free convective boundary layers in porous media with inertia effects included in non-Darcy models do not report variations of the skin friction and Nusselt number with G, due to the fact that only non-similarity solutions have been obtained. In such circumstances, the authors concentrated themselves on the variation of these two quantities along the non-

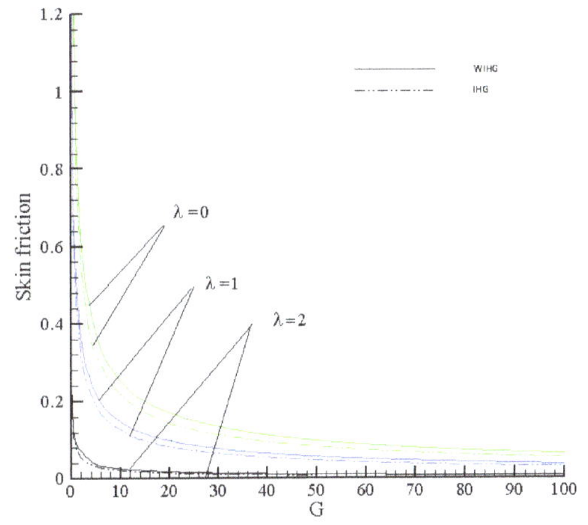


**Figure 7:** Skin friction variation with the inertia parameter *G*, VWT case.

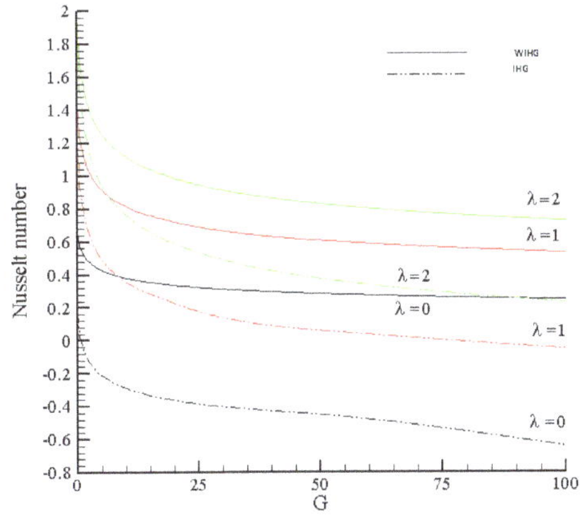


**Figure 8:** Nusselt number variation with the inertia parameter *G*, VWT case.

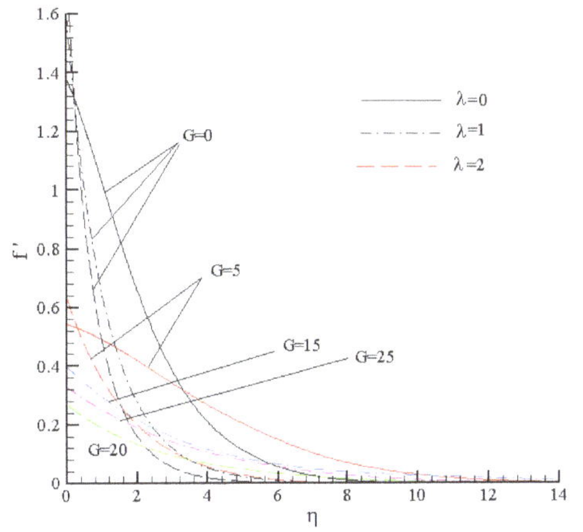


**Figure 9:** Velocity profiles in VWT case, without internal heat generation, for several values of *G* and λ.

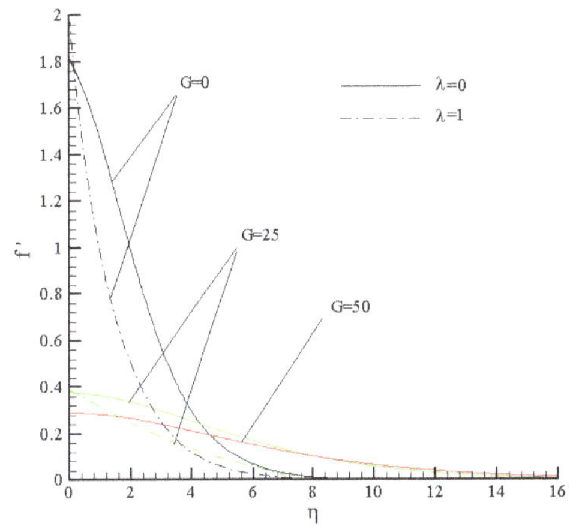


**Figure 10:** Velocity profiles in VWT case, with internal heat generation, for several values of *G* and λ.

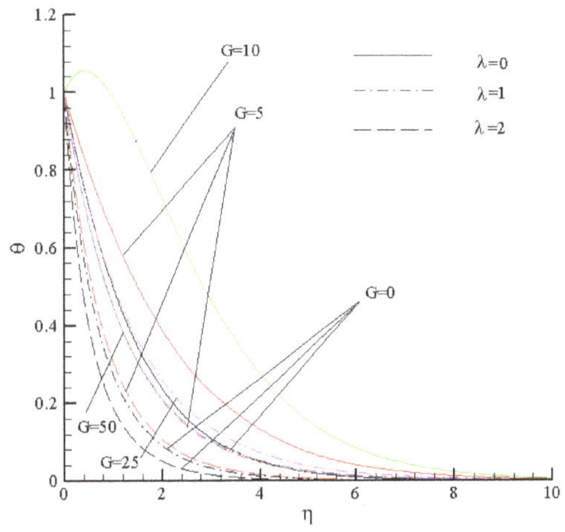


**Figure 11:** Temperature profiles in VWT case, without internal heat generation, for several values of *G* and λ.

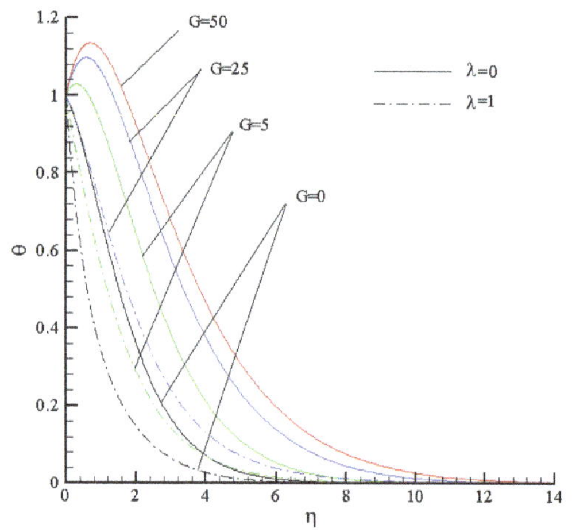


**Figure 12:** Temperature profiles in VWT case, with internal heat generation, for several values of *G* and λ.

similar variable (in stream wise direction). We are here in another situation, due to the existence of similarity solutions, so that we found that for both cases, VHF and VWT, a maximum value $G$=100 are enough to achieve an almost constant value of the skin friction and Nusselt number.

## References


1. Wang Y, Liu D, Zhang H (2014) Spectral Nodal Element Simulation of Conjugate Heat and Mass Transfer: Natural Convection Subject to Chemical Reaction Along A Circular Cylinder. International Journal of Computer Science and Application 3: 20-24.
2. Horvat A, Kljenak I, Marn J (2001) Two-dimensional large eddy simulation of turbulent natural convection due to internal heat generation. Int J Heat Mass Transfer 44: 3985-3995.
3. Jahn M, Reinke HH (1974) Free convection heat transfer with internal energies sources: calculation and measurements. 5th Int Heat Transfer Conference, Japan.
4. Postelnicu A, Pop I(1999) Similarity solutions of boundary layer free convection flows with internal heat generation about vertical and horizontal surfaces in porous media. Int Comm Heat Mass Transfer 26: 1183-1191.
5. Postelnicu A, Pop I, Grosan D (2000) Free convection boundary-layer flows over a vertical permeable flat plate in porous media with internal heat generation. Int Comm Heat Mass Transfer 27: 729-738.
6. Postelnicu A, Grosan T, Pop I (2001) The effect of variable viscosity on forced convection over a horizontal flat plate in a porous medium with internal heat generation. Mech Res Comm 28: 331-337.
7. Bagai S (2003) Similarity solutions of free convection boundary layers over a body of arbitrary shape in porous medium with internal heat generation. Int Comm Heat Mass Transfer 30: 997-1003.
8. Magyari E, Pop I, Postelnicu A (2007) Effect of the source term on steady free convection boundary layer flows over an vertical plate in a porous medium Part I. Transport in Porous Media 67: 49-67.
9. Magyari E, Pop I, Postelnicu A (2007) Effect of the source term on steady free convection boundary layer flows over an vertical plate in a porous medium Part II. Transport in Porous Media 67: 189-201.
10. Mealey LR, Merkin JH (2008) Free convection boundary layers on a vertical surface in a heat-generating porous medium. IMA Journal of Applied Mathematics 73: 231-253.
11. Merkin JH (2008) Free convective boundary-layer flow in a heat-generating porous medium: similarity solutions. Q J Mech Appl Math 61: 205-218.
12. Murthy PVSN (1998) Thermal dispersion and viscous dissipation effects on non-Darcy mixed convection in a fluid saturated porous media. Heat Mass Transfer 33: 101-107.
13. Rees DAS (1996) The effect of inertia on free convection from a horizontal surface embedded in a porous medium. Int J Heat Mass Transfer 39: 3425-3430.
14. Hossain MA, Rees DAS (1997) Non-Darcy free convection along a horizontal heated surface. Transport in Porous Media 29: 309-321.
15. Kodah ZH, Duwairi HM (1996) Inertia effects on mixed convection for vertical plates with variable wall temperature in saturated porous media. Heat Mass Transfer 31: 333-338.
16. Hung CI, Chen CB (1997) Non-Darcy free convection in a thermally stratified porous medium along a vertical plate with variable heat flux. Heat Mass Transfer 33: 101-107.
17. Duwair HM, Aldoss TK, Jarrah MA (1997) Nonsimilarity solutions for non-Darcy mixed convection from horizontal surfaces in a porous medium. Heat Mass Transfer 33: 149-156.
18. Maple Software.

20

# Solitonic Model of the Electron, Proton and Neutron

**Sladkov P***

*Independent Researcher, Russia*

**Abstract**

In paper, which is submitted, electron, proton and neutron are considered as spherical areas, inside which monochromatic electromagnetic wave of corresponding frequency spread along parallels, at that along each parallel exactly half of wave length for electron and proton and exactly one wave length for neutron is kept within, thus this is rotating soliton. This is caused by presence of spatial dispersion and anisotropy of strictly defined type inside the particles. Electric field has only radial component, and magnetic field - only meridional component. By solution of corresponding edge task, functions of distribution of electromagnetic field inside the particles and on their boundary surfaces were obtained. Integration of distribution functions of electromagnetic field through volume of the particles lead to system of algebraic equations, solution of which give all basic parameters of particles: charge, rest energy, mass, radius, magnetic moment and spin.

**Keywords:** Structure of elementary particles; Structure of matter; Theory of elementary particles; Electron; Proton; Neutron; Nuclei; Electromagnetic field; Atom; Microcosm; Elementary particles; Fundamental interactions; New theory; New physical theory

## Introduction

In present article alternative (to Standard Model) hypothesis of structure of electron, proton and neutron is suggested. The others elementary particles (except photon and neutrino) are not stable and they are considered as unsteady soliton-similar formations. In series of experiments indirect confirmations of existence of quarks were obtained, for instance in experiments by scattering of electrons at nuclei, performed at Stanford linear accelerator by R. Hofshtadter, look for instance [1]. At that, experiments by elastic and deeply inelastic scattering gave quite different results: in first case take place pattern of scattering at lengthy object, in second case is pattern of scattering at "point" centers, that is interpreted as confirmations of existence of quarks. However what "point" formations appear only in deeply inelastic scattering don't may be an evidence of quarks existence, because to above-mentioned fact may be given and another explanations: in moment of birth of new particles, which take place in deeply inelastic scattering, structure of nucleon change, it sharply diminish in volume, but after appearance of new particles nucleon return to initial state. Or process of birth of new particles occur in "point" volume inside nucleon and these energy "point" centers disappear after completion of process particles birth. And fact that experiments by elastic scattering gave pattern of scattering at lengthy object prove inexistence of quarks in nucleus. In theory of Standard (quarkual) Model come into at least 20 parameters artificially introduced from outside, such as "colour" of particles, "aroma" etc., that is its fundamental demerit. Theoretical work, which is present here, has no demerits of Standard Model, it completely describe structure of elementary particles therefore it can help in discovery new ways of making energy, elaboration perfectly new devices for its production and to achieve progress in such fields as nuclear power engineering, nanotechnology, high-powerful lasers, clean energy and others.

## Rotating Monochromatic Electromagnetic Wave

Let us write down Maxwell's equations in spherical coordinates supposing that:

1) There are no losses;

2) Only $\dot{E}_r$, $\dot{H}_\theta$, $\dot{j}_\phi$, $\dot{\rho}$ are not equal to zero.

$$\frac{1}{r}\left(\frac{\partial}{\partial r}(r\dot{H}_\theta)\right) = \dot{j}_\phi; \quad (1)$$

$$\frac{1}{r}\frac{\partial \dot{E}_r}{\partial \theta} = i\omega\,\mu\,\dot{H}_\phi = 0; \quad (2)$$

$$\frac{1}{r}\frac{\partial \dot{E}_r}{\partial \theta} = i\omega\,\mu\,\dot{H}_\phi = 0; \quad (3)$$

$$\frac{1}{r\sin\theta}\left(-\frac{\partial \dot{H}_\theta}{\partial \phi}\right) = i\,\omega\,\varepsilon\,\dot{E}_r; \quad (4)$$

$$\frac{1}{r^2}\frac{\partial}{\partial r}(r^2\varepsilon\quad \dot{E}_r) = \dot{\rho}; \quad (5)$$

$$\frac{1}{r\sin\theta}\frac{\partial}{\partial \theta}(\sin\theta(\mu\,\dot{H}_\theta)) = 0. \quad (6)$$

Here $r,\theta,\ \phi$ spherical coordinates of the observation point; $\dot{E}_r$ и $\dot{H}_\theta$ - components of the electromagnetic field, $\dot{j}_\phi$ - density of electric current, $\dot{\rho}$ - volume charge density; $\omega$-circular frequency of field alteration $i$ - imaginary unit ε-dielectric permittivity - magnetic permeability Figure 1.

Substituting the expression for $\dot{H}_\theta$ from (2) in (4), we obtain:

$$\frac{\partial}{\partial} + \varepsilon\,\mu\,\omega^2\,r^2\sin^2\theta\,E \ = 0; \quad (7)$$

$$\frac{1}{r^2\sin^2\theta}\frac{\partial^2\,\dot{E}_r}{\partial\phi^2} + \omega^2\,\varepsilon\,\mu\,\dot{E}_r = 0; \quad (7')$$

This is Helmholtz homogeneous equation. Let us designate.

$$\sqrt{\varepsilon\,\mu}\ \omega r\sin\theta = k_1 - \quad (7'')$$

Wave number - General solution of Helmholtz equation:

$$\dot{E}_r = E_0\,e^{-ik_1\phi} + E_0\,e^{ik_1\phi}. \quad (8)$$

***Corresponding author:** Sladkov P, Independent Researcher, Russia
E-mail: sladkovpaul@gmail.com

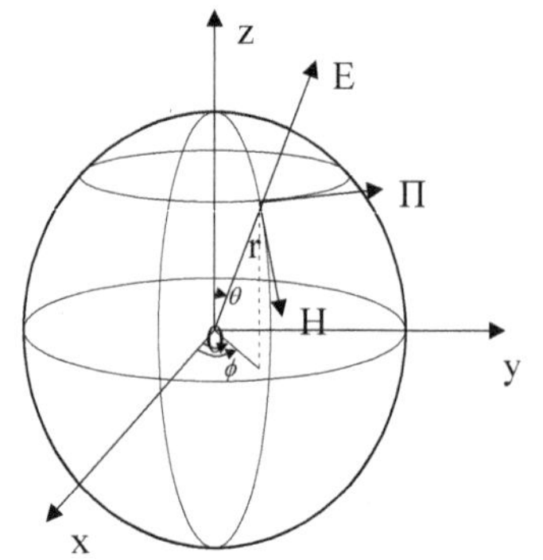

**Figure 1:** Representing the spherical coordinates of the observation point.

This expression describes two waves, moving to meet one another by circular trajectories, along the parallels. Pointing's vector in each point is directed at tangent to the corresponding parallel [2,3].

Let us consider a wave, moving in positive direction $\phi$.

$$\dot{E}_r = E_0\, e^{-ik_1 \phi}\, F(r,\theta); \quad (9)$$

Here

$$k_1\phi = \sqrt{\varepsilon\,\mu}\,(\omega\, r \sin\theta)\,\phi -$$

Wave phase;

$K_1$ - Dimensionless analog of the wave number. If to introduce a wave number of traditional dimension ($\frac{1}{m}$);

$$\beta = \omega\sqrt{\varepsilon\,\mu} = \frac{k_1}{r\sin\theta},$$

The wave phase will be written down as

$$k_1\,\phi = \sqrt{\varepsilon\,\mu}\,\omega\,(r\sin\theta)\,\phi = \beta\, l,$$

Where

$$l = (r\sin\theta)\,\phi -$$

Arc length along the corresponding parallel. In the considered case the wave number is a function of coordinates and frequency. Thus, the wave, which is described, can exist only at availability of spatial and frequency dispersion [4-6]. Dispersion equations will be obtained below, apart from the already found expression (7″).

From expression (2), taking into account (7″) and (9), we have:

$$\dot{H}_\theta = \frac{\sqrt{\varepsilon\,\mu}\,\omega\, r\sin\theta}{\mu\,\omega\, r\sin\theta} E_0\, e^{-ik_1\phi}\, F(r,\theta) = \frac{E_0}{z} e^{-ik_1\phi}\, F(r,\theta)\cdot \quad (9')$$

For actual amplitudes:

$$E_r = E_0\, F(r,\theta) \quad \sin k_1\phi\,; \quad (10)$$

$$H_\theta = \frac{E_0}{z} F(r,\theta)\sin k_1\phi.\cdot \quad (10')$$

Here

$$z = \sqrt{\frac{\mu}{\varepsilon}} -$$

Means characteristic impedance.

The last expressions describe an electromagnetic wave, rotating around axis Z in positive direction $\phi$.Conditions of self-consistency:

1) $z$=constant

2) Along each parallel on the circle length, the integer number of half-waves must be kept within.

$$2\pi\, r\sin\theta = n\frac{\lambda}{2}; \quad (11)$$

Here $\lambda = \frac{v}{f} -$

Wave length, v - phase velocity of wave, f - frequency, n=1,2,3...

Let us consider the case when n=1,

$$2\pi\, r\sin\theta = \frac{\lambda}{2};$$

$$v = 2\omega\, r\sin\theta. \quad (11')$$

Along each parallel, exactly half of wave length is kept within.

Phase velocity of wave is the function of frequency and distance up to the axis of rotation.

$$v = \frac{1}{\sqrt{\varepsilon\,\mu}} = 2\omega\, r\sin\theta\,;$$

$$\varepsilon\,\mu = \frac{1}{4\omega^2\, r^2\sin^2\theta}; \quad (11'')$$

$$z = \sqrt{\frac{\mu}{\varepsilon}};$$

$$\mu = \varepsilon\, z^2; \quad (11''')$$

We are substituting in (11″):

$$z = \frac{1}{2\omega\, r\varepsilon\sin\theta}; \quad (12)$$

$$\varepsilon = \frac{1}{2\omega\, r\, z\sin\theta}\,. \quad (12')$$

From (11″) $\varepsilon = \frac{\mu}{z^2}$;

We are substituting in (12′).

$$\mu = \frac{z}{2\omega\, r\sin\theta}; \quad (12'')$$

$$z = 2\mu\,\omega\, r\sin\theta\cdot \quad (12''')$$

Taking into account (8) and (11″)

$$k_1 = \frac{\omega\, r\sin\theta}{2\omega\quad r\sin\theta} = \frac{1}{2}\cdot$$

Then

$$E_r = E_0 F(r,\theta)\sin\frac{\phi}{2}; \quad (13)$$

$$H_\theta = \frac{E_0}{z} F(r,\theta)\sin\frac{\phi}{2}\cdot \quad (13')$$

Function $\sin\frac{\phi}{2}$ is on valued in angles interval $0 \le \phi \le 2\pi$.

This situation can be interpreted as rotation of spherical coordinate system around axis z in positive direction $\phi$ with angular velocity $\frac{d\phi}{dt}$. Let us find it from the condition:

$$\omega\, t - \frac{\phi}{2} = const.$$

Having differentiated this expression on t, we receive,

$$\frac{d\,\phi}{dt} = 2\omega.$$

At the same time the electromagnetic field, about spherical

coordinate system, is determined by expressions (13) and (13′). Further from (3): as $H_\phi=0$

$$\dot{E}_r(\theta) = const;$$

$$E_r(\theta)=const \tag{14}$$

From equation (6)

$$\frac{\partial}{\partial\theta}\left(\frac{\sin\theta\,(z\,H_\theta)}{2\,\omega r\sin\theta}\right)=0$$

Follows

$$\dot{H}_\theta(\theta) = const;$$

$$H_\theta(\theta)=const \tag{14′}$$

To receive field dependence from $r: E_r(r); H_\theta(r)$, let us find solution of three-dimensional Helmholtz equation in spherical coordinates.

$$\frac{1}{r^2}\frac{\partial}{\partial r}\left(r^2\frac{\partial E_r}{\partial r}\right)+\frac{1}{r^2\sin\theta}\frac{\partial}{\partial\theta}\left(\sin\theta\frac{\partial E_r}{\partial\theta}\right)+\frac{1}{r^2\sin^2\theta}\frac{\partial^2 E_r}{\partial\phi^2}+k^2E_r=0. \tag{15}$$

$E_r$ does not depend from θ, look (14), therefore three-dimensional Helmholtz equation transfers into two-dimensional one.

$$\frac{1}{r^2}\frac{\partial}{\partial r}\left(r^2\frac{\partial E_r}{\partial r}\right)+\frac{1}{r^2\sin^2\theta}\frac{\partial^2 E_r}{\partial\phi^2}+k^2E_r=0. \tag{15′}$$

Let us suppose that

$$k^2=k_2^2+k_3^2,$$

now

$$\frac{1}{r^2}\frac{\partial}{\partial r}\left(r^2\frac{\partial E_r}{\partial r}\right)+\frac{1}{r^2\sin^2\theta}\frac{\partial^2 E_r}{\partial\phi^2}+k_2^2E_r+k_3^2\,E_r=0. \tag{15″}$$

This equation can be satisfied, if

$$\left\{\begin{aligned}&\frac{1}{r^2\sin^2\theta}\frac{\partial^2 E_r}{\partial\phi^2}+k_2^2E_r=0;\\&\frac{1}{r^2}\frac{\partial}{\partial r}\left(\frac{r^2\,\partial E_r}{\partial r}\right)+k_3^2E_r=0.\end{aligned}\right. \tag{16),(17}$$

Thus, initial Helmholtz equation has split into the system of two equations. We substitute in these equations instead of $E_r(r,\phi)=f(r)\,g(\phi)$, (i.e. we are searching the solution as the product of two functions) and divide the first equation by $f(r)$, and the second – by $g(\phi)$. We receive

$$\left\{\begin{aligned}&\frac{d^2g}{d\phi^2}+k_2^2r^2(\sin^2\theta)\,g=0;\\&r^2\frac{d^2f}{dr^2}+2r\frac{df}{dr}+k_3^2r^2f=0.\end{aligned}\right. \tag{18), (19}$$

Equations (16) and (18) are equivalent to equations (7) и (7′), which were received earlier from Maxwell's equations, and

$$k_2=\omega\sqrt{\varepsilon\,\mu}=\frac{\omega}{v}=\frac{1}{2r\sin\theta};$$

$$k_1=k_2\,r\sin\theta=\frac{1}{2}$$

The solution of equation (18) was found earlier, look (13).

$$g(\phi)=\sin\frac{\phi}{2}. \tag{20}$$

Let us copy (19) as:

$$r^2\frac{d^2f}{dr^2}+2r\frac{df}{dr}+k_4^2f=0; \tag{19′}$$

Where

$k_4=k_3r$

19' - centrally symmetric Helmholtz equation. Let us suggest, $k_4=k_3r$

$$k_3=\frac{\omega}{v_r},$$

Where $v_r$- phase velocity of electromagnetic wave in radial direction. As in the central symmetric equation angular dependence is absent, it is logical to assume that

$v_{r=}v=2\omega r\sin\theta$

at $\theta=\frac{\pi}{2}$; i.e.

$v=2\omega r$

$$k_3=\frac{\omega}{v_r}=\frac{1}{2r}; \tag{21}$$

$$k_4=k_3\,r=\frac{1}{2}. \tag{21′}$$

Instead of (19′), we are having

$$r^2\frac{d^2f}{dr^2}+2r\frac{df}{dr}+\frac{1}{4}f=0. \tag{19″}$$

This is Euler equation, it has the solution

$$f=r^{-\frac{1}{2}}(C_1+C_2\ln r). \tag{22}$$

Let us converse expression (22).

$$f=\frac{(C_1+C_2\ln r)}{\sqrt{r}}=\sqrt{\frac{a}{r}}(C_3+C\ln r)=\sqrt{\frac{a}{r}}(1+\ln C_5+\ln r^C)=\sqrt{\frac{a}{r}}(1+\ln C_5\,r^C). \tag{22′}$$

Here $C_1=\sqrt{a}\,C_3$; $C_2=\sqrt{a}\,C$; $C_3=1+\ln C_5$; $a$-value of radius $r$, at which the rotating monochromatic electromagnetic wave ceases to exist, and $E_r=E_0$; $f=1$ hence

$$\begin{aligned}&1+\ln C_5\,a^C=1;\\&\ln C_5\,a^C=0;\\&C_5\,a^C=1;\\&C_5=\frac{1}{a^C}.\end{aligned} \tag{22″}$$

In view of this,

$$f=\sqrt{\frac{a}{r}}\left(1+\ln\left(\frac{r}{a}\right)^C\right)=\sqrt{\frac{a}{r}}\left(1+C\ln\frac{r}{a}\right).$$

Let us designate $C=p$ now

$$f=\sqrt{\frac{a}{r}}\left(1+p\ln\frac{r}{a}\right).$$

Thus, for $E_r$ we are having

$$E_r=E_0\,g(\phi)\,f(r)=E_0\sqrt{\frac{a}{r}}\left(1+p\ln\frac{r}{a}\right)\sin\frac{\phi}{2}. \tag{23}$$

At $r\to\infty$, $E_r=0$; $f=0$.

Really

$$\lim_{r\to\infty}\frac{\ln r}{\sqrt{r}}=\lim_{r\to\infty}\frac{1/r}{1/2\sqrt{r}}=0.$$

So that at alteration of $r$ within the interval from 0 to $a$, $E_r$ would not change its sign, observance of the following requirement is necessary $P\leq 0$

At $r = 0$, $E_r = \infty$; $f = \infty$.

At $E_r = E_0 \sqrt{\frac{a}{r}} (1 + p \ln \frac{r}{a}) \sin \frac{\phi}{2}$;

## System of Equations for Electron

Basing on results of the previous section, let us write down expressions for electromagnetic field inside the electron, assuming that it is concentrated inside the orb of radius $a$

$$E_r = E_0 \sqrt{\frac{a}{r}} (1 + p \ln \frac{r}{a}) \sin \frac{\phi}{2}; \tag{23'}$$

$$H_\theta = \frac{E_0}{z} \sqrt{\frac{a}{r}} (1 + p \ln \frac{r}{a}) \sin \frac{\phi}{2}. \tag{23''}$$

Here $a$ is electron radius, $E_0$ amplitude of electric field intensity at $r=a$; $z=const$ characteristic impedance inside the electron, $P$- unknown coefficient and $P \le 0$

At that the internal electron medium possesses frequent and spatial dispersion, as well as anisotropy. Dispersion equations have the following appearance [7-11]. '

$$v_r = v_\theta = 2\omega r; \tag{24}$$

$$v_\phi = 2\ r \sin\theta \tag{24'}$$

$$z_r = z_\theta = z_\phi = z = const \tag{24''}$$

Here $v_r, v_\theta, v_\phi$ - phase velocity of rotating monochromatic electromagnetic wave in corresponding direction. In viewed case, the electromagnetic wave is being spread only in the direction $\phi$, and we shall need expressions $v_r$ and $v_\theta$ for searching the formulas of dielectric and magnetic permeability, as well as wave numbers of corresponding directions; $z_r, z_{\theta,} z_{\phi,} z$ characteristic impedances inside the electron; $\varepsilon_\varphi$, и, $\mu_\varphi$ were found before, see (12') (12'')

$$\varepsilon_\phi = \frac{1}{2\,\omega\, r\, z \sin\theta} = \frac{1}{v_\phi\, z};$$

$$\mu_\phi = \frac{z}{2\,\omega\, r \sin\theta} = \frac{z}{v_\phi}.$$

In view of (24)(24')(24''), let us write down expressions for $\varepsilon_r$ $\varepsilon_\theta \mu_r \mu_\theta$

$$\varepsilon_r = \frac{1}{2\,\omega\, r\, z};$$

$$\mu_r = \frac{z}{2\,\omega\, r};$$

$$\varepsilon_\theta = \frac{1}{2\,\omega\, r\, z};$$

$$\mu_\theta = \frac{z}{2\,\omega\, r}. \tag{24'''}$$

From considerations and formulas adduced, it follows that dielectric and magnetic permeability are tensor values.

$$\|\varepsilon\| = \begin{vmatrix} \varepsilon_r & 0 & 0 \\ 0 & \varepsilon_\phi & 0 \\ 0 & 0 & \varepsilon_\theta \end{vmatrix} = \begin{vmatrix} \frac{1}{2\,\omega\, r\, z} & 0 & 0 \\ 0 & \frac{1}{2\,\omega\, r\, z \sin\theta} & 0 \\ 0 & 0 & \frac{1}{2\,\omega\, r\, z} \end{vmatrix}.$$

$$\|\mu\| = \begin{vmatrix} \mu_r & 0 & 0 \\ 0 & \mu_\phi & 0 \\ 0 & 0 & \mu_\theta \end{vmatrix} = \begin{vmatrix} \frac{z}{2\,\omega\, r} & 0 & 0 \\ 0 & \frac{z}{2\,\omega\, r \sin\theta} & 0 \\ 0 & 0 & \frac{z}{2\,\omega\, r} \end{vmatrix}.$$

Let us find dimensionless wave numbers.

$$k_\phi = \frac{\omega}{v_\phi} r \sin\theta = \frac{\omega\, r \sin\theta}{2\,\omega\, r \sin\theta} = \frac{1}{2};$$

$$k_\theta = \frac{\omega}{v_\theta} r = \frac{\omega\, r}{2\,\omega\, r} = \frac{1}{2};$$

$$k_r = \frac{\omega}{v_r} r = \frac{\omega\, r}{2\,\omega\, r} = \frac{1}{2}.$$

Thus

$$k_\phi = k_\theta = k_r = \frac{1}{2}.$$

Let us remind that in the viewed case, the electromagnetic wave is spread only in the direction of $\phi$

At $r=0$ we are having a special point:

$$E_r = \infty;\ H_\theta = \infty;\quad \rho = \infty;\quad j_\phi = \infty;\quad \|\varepsilon\| = \infty;\quad \|\mu\| = \infty.$$

Despite of this, all basic electrons' parameters – charge $q$ rest energy $W$, magnetic moment $M$-expressed through integrals by volume from the functions specified above, prove to be finite quantities [12-15]. Look further.

From (5), we find volume charge density inside electron

$$\rho = div\,(\varepsilon_r\, E_r) = \frac{}{r} \frac{\partial}{\partial r} \left[ \frac{r\ \ E\ \ \sin -}{rz} ((\frac{a}{r})^{\frac{1}{2}} + p\, (\frac{a}{r})^{\frac{1}{2}} \ln \frac{r}{a}) \right] =$$

$$= \frac{E_0 \sqrt{a} \sin \frac{\phi}{2}}{2\,\omega\, z r^2} ((\frac{1}{2} + p) \frac{1}{\sqrt{r}} + \frac{p}{2\sqrt{r}} \ln \frac{r}{a}) \cdot \tag{25}$$

Integrating $\rho$ on electron's volume, we shall receive this expression for its charge $q$.

$$q = \int_V \rho\, dV = \frac{E_0 \sqrt{a}}{2\,\omega\, z} \int_0^{2\pi} \int_0^{\pi} \int_0^{a} \frac{\sin \frac{\phi}{2} r^2 \sin\theta}{r^2} ((\frac{1}{2} + p) \frac{1}{\sqrt{r}} + \frac{p}{2\sqrt{r}} \ln \frac{r}{a})\, d\phi\, d\theta\, dr =$$

$$= \frac{4\, E_0 \sqrt{a}}{\omega\, z} (2\sqrt{a}\, (\frac{1}{2} + p) + p\sqrt{a}\ \ln a - 2p\sqrt{a} - p\sqrt{a} \ln a - p\sqrt{0} \ln 0) = \frac{4\, E_0\, a}{\omega\, z}. \tag{26}$$

On the other hand, from the third integral Maxwell's equation, it is possible to find electron's charge as a stream of vector electric induction D through the surface of the orb of radius $a$

$$q = \oint_S \varepsilon_r E_r\ dS = \int_0^{2\pi} \int_0^{\pi} \frac{E_0 \sin \frac{\phi}{2} a^2 \sin\theta}{2\,\omega\, z a} d\phi\, d\theta = \frac{4\, E_0\, a}{\omega\, z}. \tag{26'}$$

As we can see, expressions (26) и (26') are equivalent to each other.

From (1), we obtain expression for current density $j_\phi$

$$j_\phi = \frac{1}{r} \frac{\partial}{\partial r} (r \frac{E_0 \sin \frac{\phi}{2}}{z} (\sqrt{\frac{a}{r}} + p\sqrt{\frac{a}{r}} \ln \frac{r}{a})) = \frac{E_0 \sqrt{a} \sin \frac{\phi}{2}}{r\, z} ((\frac{1}{2} + p) \frac{1}{\sqrt{r}} + \frac{p}{2\sqrt{r}} \ln \frac{r}{a}) \cdot \tag{27}$$

From expressions (25), (27) it is visible that in the interval of change of $r$ from 0 to $a$,$\rho$ and $j_\phi$ once change the sign. It can be explained by the fact that in the viewed structure, the substantial role is played by the rotating monochromatic electromagnetic wave, and the space charge density and electric current density – are auxiliary or even fictitious quantities in the sense that inside the particle there is neither any charged substance nor its motion [16]. Inside the electron, it is not the charge that is the source of electric field, but electric field is the source of the charge. In its turn, it is not the electric current that is the source of magnetic field, but magnetic field is the source of the electric current [17-22]. Thus, a deduction about vector nature of elementary charge can be made.

Now we shall determine electron's rest energy as electromagnetic wave energy inside a particle.

$$W = \int_V w\, dV.$$

Here $w$- is volume density of electromagnetic wave energy,

$w = \frac{\Pi}{v_\phi}$, where

$\Pi$ - Pointing vector,

$$\Pi = \left[E_r\ H_\theta\right],$$

$v_\phi$ -phase velocity of electromagnetic wave in direction of $v\phi$.

$v_\phi = 2\omega r\sin\theta$

$$W = \int_0^{2\pi}\int_0^{\pi}\int_0^{a} \frac{E_0^2 \sin^2\frac{\phi}{2}}{2\,\omega\, rz\sin\theta}\left(\frac{a}{r} + 2p\frac{a}{r}\ln\frac{r}{a} + p^2\frac{a}{r}\ln^2\frac{r}{a}\right) r^2 \sin\theta\ d\phi\, d\theta\, dr =$$

$$= \frac{\pi^2 E_0^2 a}{2\omega z}\left(a + 2\,p\,a\ln a - 2\,p\,0\ln 0 - 2\,p\,a - 2\,p\,a\ln a + p^2 a\left(2 - 0\ln^2 0 + 2*0\ln 0\right)\right) =$$

$$= \frac{\pi^2 E_0^2 a^2}{2\,\omega\, z}\left(1 - 2\,p + 2\,p^2\right). \qquad (28)$$

$$\frac{\pi^2 E_0^2 a^2}{2\,\omega\, z}\left(1 - 2\,p + 2\,p^2\right) = \hbar\,\omega; \qquad (28')$$

Here $\hbar$ is Planck's constant.

We shall be searching electron's magnetic moment in the form of a sum. $M = M_m + M_L$

$M = M_m + M_L$

Where $M_m$- is magnetic moment, created by volumetric current; $M_L$ -magnetic moment, attributed to impulse moment, i.e. to rotation.

$M_L = \gamma L$

Where $\gamma$-gyromagnetic ratio; $L$-impulse moment of electron.

Basing on Barnett effect, we are making a supposition, that the impulse moment, attributed to rotation, creates additional magnetic moment [21,23].

Being aware of the fact that electron's impulse moment is equal $\frac{\hbar}{2}$, from (28′) we find expression for L.

$$L = \frac{\pi^2 E_0^2 a^2}{4\,\omega^2\, z}\left(2p^2 - 2p + 1\right);$$

$$M_L = \frac{\gamma\,\pi^2 E_0^2 a^2\left(2p^2 - 2p + 1\right)}{4\,\omega^2\, z};$$

or $M_L = \gamma\frac{\hbar}{2}$.

Let us calculate $M_m$ as electric current magnetic moment in volume V, relating to axis z by the formula:

$$M_\partial = \frac{1}{2}\int_V \left[r_z\ j_\phi\right] dV.$$

See for instance [3], page 111, where $r_z$ - distance to axis z,

$r_{z=} r\sin\theta$

$$M_\partial = \frac{1}{2}\int_0^{2\pi}\int_0^{\pi}\int_0^{a} \frac{r\sin\theta\, E_0\sqrt{a}\,\sin\frac{\phi}{2}\, r^2\sin\theta}{z\, r^{\frac{3}{2}}} \qquad (29)$$

$$\left(p + \frac{1}{2} + \frac{p^`}{2}\ln\frac{r}{a}\right) d\phi\, d\theta\, dr = \frac{\pi\, E_0\, a^3}{z}\left(\frac{8p+5}{25}\right).$$

$$M = M_\partial + M_L = \frac{\pi\, E_0\, a^3}{z}\left(\frac{8\,p+5}{25}\right) + \gamma\frac{\hbar}{2}. \qquad (29')$$

Or

$$M = \frac{\pi\, E_0\, a^2}{z}\left[a\left(\frac{8p+5}{25}\right) + \frac{\gamma\,\pi\, E_0}{4\,\omega^2}\left(1 - 2p + 2p^2\right)\right]. \qquad (29'')$$

Thus, we have received the system of algebraic equations for electron.

$$\begin{cases} \frac{4\, E_0\, a}{\omega\, z} = -e; & (30) \\ \frac{\pi\, E_0\, a^3}{z}\left(\frac{8p+5}{25}\right) + \frac{\gamma\,\hbar}{2} = -1{,}0011595\,\frac{e\,\hbar}{2\,m}; & (31) \\ \frac{\pi^2 E_0^2\, a^2}{2\,\omega\, z}\left(1 - 2p + 2p^2\right) = \hbar\,\omega; & (32) \end{cases}$$

Here $e$ - charge of electron, $m$- it's mass.

Three equations contain five unknown quantities: $E_0$, $a$, $z$, $p$, $\gamma$ Let us add this system with equations, which we shall receive from boundary conditions.

At $r=a$, $R=a$

$$\varepsilon_r\, E_0 = \varepsilon_0\, E_{внешн.} \qquad (33)$$

In the exterior area, the same as and in the interior area, electric field intensity possesses only radial component. Here R - distance from electron's center to the observation point in the exterior area, $\varepsilon_0$- vacuum dielectric permeability [24].

Further. $H_0 = \frac{E_0}{z} = H_{внешн.}$ (34)

In the exterior area, the same as and in the interior area, magnetic field intensity possesses only meridional component.

It is obvious that

$$\varepsilon_r \geq \varepsilon_0, \qquad (33')$$

then from (33) follows:

$$E_0 \leq E_{внешн.} \qquad (33'')$$

On the other hand it is known that the electric field, having passed through dielectric layer, cannot increase, therefore

$$E_0 \geq E_{внешн.} \qquad (33''')$$

In other words, correlations (33′) (33″) (33‴) will be simultaneously executed only in one case, if

$$\varepsilon_{r=}\varepsilon_0; \qquad (35)$$

$$E_{внешн.} = E_0. \qquad (36)$$

Now under Biot-Savart's law, we are finding magnetic field in the exterior area.

$$B_{внешн.} = \frac{1}{4\pi} \int_V \frac{\left[j_\phi \ R\right] \mu_\phi}{R^3} dV.$$

In last expression we substitute (12″) and (27).

$$B_{внешн.} = \frac{E_0 \sqrt{a}}{4 \pi R^2 z} \int_0^{2\pi} \int_0^{\pi} \int_0^{a} \frac{z \sin \frac{\phi}{2} \left[ (\frac{1}{2} + p) \frac{1}{\sqrt{r}} + \frac{p}{2\sqrt{r}} \ln \frac{r}{a} \right]}{2 \omega r^2 \sin \theta} r^2 \sin \theta \, d\phi \, d\theta \, dr = \quad (37)$$

$$= \frac{E_0 \sqrt{a}}{2 \omega R^2} \left[ (\frac{1}{2} + p) 2\sqrt{a} - p\sqrt{a} \ln a + p\sqrt{a} \ln a - p\sqrt{0} \ln 0 - 2p\sqrt{a} \right] = \frac{E_0 \ a}{2\omega R^2}.$$

$$H_{внешн.} = \frac{B_{внешн.}}{\mu_0} = \frac{E_0 \ a}{2 \mu_0 \ \omega R^2}. \quad (38)$$

At $r=a$ $R=a$

$$H_{внутр.} = \frac{E_0}{z} = H_{внешн.}$$

$$\frac{E_0}{z} = \frac{E_0 \ a}{2 \mu_0 \ \omega a^2} = \frac{E_0}{2 \mu_0 \ \omega a};$$

$$z = 2 \mu_0 \ \omega a. \quad (39)$$

On the other hand, from (24‴)

$$z = 2 \mu_\theta \ \omega r.$$

At $r=a$

$$z = 2 \mu_\theta \ \omega a. \ (39')$$

We substitute in (39).

$$2 \mu_\theta \omega a = 2 \mu_0 \omega a;$$

$$\mu_\theta = \mu_0 \quad (40)$$

Thus, at $r=a$

$$\varepsilon_r = \varepsilon_\theta = \varepsilon_0;$$

$$\mu_\theta = \mu_r = \mu_0; \quad (41)$$

$$v_r = v_\theta = 2\omega a = \frac{1}{\sqrt{\mu_0 \ \varepsilon_0}} = c.$$

Here $c$ - velocity of light, $\omega$=7,7634421*10[20] Hz- Compton circular frequency of electron.

$$a = \frac{c}{2\omega} = 0,1930796 * 10^{-12} (m) \cdot \quad (42)$$

As it is known, atom's radius approximately equals to $10^{-10}$ m, volume of atom - 4,18879*$10^{-30}$ $m^3$. We found, that radius of electron equals to 1,930796*$10^{-13}$ m, volume of electron -3,0150724*$10^{-38}$ $m^3$. That is one electron occupies 0,7197955*$10^{-8}$ from atom's volume and, for example, 100 electrons (as in atoms located at the end of the periodic system) occupy 0,7197955*$10^{-6}$ from atom's volume [8,25].

We substitute (42) в (39).

$$z = \frac{2\mu_0 \omega c}{2\omega} = \sqrt{\frac{\mu_0}{\varepsilon_0}} = 376,73032 (Ohm). \quad (43)$$

Let us solve the system (30), (31), (32), taking into account (42) and (43).

$$\frac{4 \ E_0 \ c}{\omega \sqrt{\frac{\mu_0}{\varepsilon_0}} 2\omega} = -e;$$

$$E_0 = -\frac{\omega^2 \ \mu_0 \ e}{2}. \quad (30')$$

$$\frac{\pi \ E_0}{8\omega^3 \mu_0{}^2 \varepsilon_0} (\frac{8p+5}{25}) + \frac{\gamma \ \pi^2 \ E_0{}^2 (1-2p+2p^2)}{16\omega^4 \mu_0{}^{3/2} \varepsilon_0{}^{1/2}} = -1,0011595 \frac{e\hbar}{2m}. \quad (31')$$

$$\frac{\pi^2 \ E_0{}^2}{8\omega^3 \mu_0{}^{3/2} \varepsilon_0{}^{1/2}} (1-2p+2p^2) = \hbar\omega. \quad (32')$$

We substitute (30′) in (32′).

$$p^2 - p + \frac{1}{2} - \frac{16 \ \hbar \varepsilon_0{}^{1/2}}{\pi^2 e^2 \mu_0{}^{1/2}} = 0;$$

$$p_1 = 4,6747427;$$

$$p_2 = -3,6747427.$$

$P$ must be negative, therefore we select

$$p_2 = p = -3,6747427.$$

We substitute (30′) in (31′)

$$-\frac{\pi \ e}{16\omega\mu_0 \ \varepsilon_0} (\frac{8p+5}{25}) + \frac{\pi^2 \ \gamma \ e^2 \mu_0{}^{1/2}}{64\varepsilon_0{}^{1/2}} (1-2p+2p^2) = -1,0011595 \frac{e \ \hbar}{2 \ m}. \quad (31'')$$

We substitute $p$ meaning in (31″) and find $\gamma$

$$\gamma = -0,2434911 * 10^{12} (\frac{1}{T * s}).$$

From solution of equation (31), it is visible that two components of magnetic moment of electron $M_m$ и $M_L$ are directed to opposite sides and $M_L \succ M_m$

Let us also calculate numerical value of $E_0$ by formula (30′)

$$E_0 = -6,0673455 * 10^{16} (\frac{V}{m}).$$

"Dimensions" of electron for the present are not discovered by experimental way, though precision of measuring is led to $10^{-18}$ m. Within the framework of the model considered it may be explained by the next way: electron is not hard particle with this quantity of vector E, which exist inside it, unlike from proton and neutron, quantity of vector E inside which approximately $10^7$ times as much [26-31].

For positron, the system of equations will take a somewhat different view.

$$\begin{cases} \frac{4 \ E_0 \ a}{\omega \ z} = e; & (44) \\ \frac{\pi \ E_0 \ a^3}{z} (\frac{8p+5}{25}) + \frac{\gamma \ \pi^2 \ E_0{}^2 \ a^2}{4\omega^2 \ z} (1-2p+2p^2) = 1,0011595 \frac{e \ \hbar}{2 \ m}; & (45) \\ \frac{\pi^2 \ E_0{}^2 \ a^2}{2\omega \ z} (1-2p+2p^2) = \hbar \ \omega; & (46) \end{cases}$$

Boundary conditions are the same as for electron. Hence

$$z = \sqrt{\frac{\mu_0}{\varepsilon_0}};$$

$$a = \frac{c}{2\omega} = 0,1930796 * 10^{-12} (m).$$

The system of equations (44), (45), (46) with exactness to a sign, has the same solutions, as the system (30), (31), (32).

$$E_0^{e+} = -E_0^e = 6,0673455 * 10^{16} (\frac{V}{m});$$

$$\gamma_{e+} = -\gamma_e = 0,2434911 * 10^{12} (\frac{1}{T * s});$$

$$p_{e+} = -p_e = -3,6747427.$$

## System of Equations for Proton

By applying reasoning and mathematical calculations of the

previous section in relation to proton, we shall receive the relevant system of equations.

$$\begin{cases} \frac{4E_0\, a}{\omega\, z} = e; & (47) \\ \frac{\pi\, E_0\, a^3}{z}(\frac{8p+5}{25}) + \frac{\gamma\, \pi^2 E_0^{\,2}\, a^2(1-2p+2p^2)}{4\omega^2 z} = -2{,}7928475\frac{e\hbar}{2\,m}; & (48) \\ \frac{\pi^2 E_0^{\,2} a^2}{2\omega\, z}(1-2p+2p^2) = \hbar\,\omega; & (49) \end{cases}$$

Here corresponding letters mean parameters of proton.

Boundary conditions: at $r=a$

$$\varepsilon_r = \varepsilon_\theta = \varepsilon_0;$$

$$\mu_\theta = \mu_r = \mu_0;$$

Hence

$$z = \sqrt{\frac{\mu_0}{\varepsilon_0}};$$

$$a = \frac{c}{2\omega} = 1{,}0515447 * 10^{-16}\ \mathrm{m}.$$

Here: $\omega$=1,425486*$10^{24}$ Hz- Compton circular frequency of proton [32,33].

Solving the system (47), (48), (49), we shall receive

$$E_0 = 2{,}0455794 * 10^{23}\,(\frac{V}{m});$$

$$p = -3{,}6747427;$$

$$\gamma = -2{,}3081218 * 10^8\,(\frac{1}{T*s}).$$

From the solution of equation (48) it is visible that two components of proton's magnetic moment $M_m$ и $M_L$ have identical direction, and $M_L \succ M_m$

Let us write down the system of equations for antiproton explained in [34].

$$\frac{4E_0 a}{\omega\, z} = -e; \quad (50)$$

$$\frac{\pi\, E_0 a^3}{z}(\frac{8p+5}{25}) + \frac{\gamma\, \pi^2 E_0^{\,2} a^2(1-2p+2p^2)}{4\omega^2 z} = 2{,}7928475\frac{e\,\hbar}{2m}; \quad (51)$$

$$\frac{\pi^2 E_0^{\,2} a^2}{2\omega\, z}(1-2p+2p^2) = \hbar\omega. \quad (52)$$

Boundary conditions: at $r=a$

$$\varepsilon_r = \varepsilon_\theta = \varepsilon_0;$$

$$\mu_r = \mu_\theta = \mu_0;$$

Hence

$$z = \sqrt{\frac{\mu_0}{\varepsilon_0}};$$

$$a = \frac{c}{2\omega} = 1{,}0515447 * 10^{-16}\,(m).$$

System of equations (50), (51), (52) with exactness to a sign has the same solutions, as system (47), (48), (49).

$$E_0^{\bar{p}} = -E_0^{p} = -2{,}0455794 * 10^{23}\,(\frac{V}{m});$$

$$\gamma_{\bar{p}} = -\gamma_p = 2{,}3081218 * 10^8\,(\frac{1}{T*s});$$

$$p_{\bar{p}} = p_p = p_e = -3{,}6747427.$$

## System of equations for neutron

$$E_r = E_0\sqrt{\frac{a}{r}}(1+p\ln\frac{r}{a})\sin\phi; \quad (53)$$

$$H_\theta = \frac{E_0}{z}\sqrt{\frac{a}{r}}(1+p\ln\frac{r}{a})\sin\phi. \quad (53')$$

Along each parallel, exactly one wave length is kept within. In this case:

$$v_\phi = \omega\, r\sin\theta;$$

$$\varepsilon_\phi = \frac{1}{\omega\, r\, z\sin\theta}; \quad (54)$$

$$\mu_\phi = \frac{z}{\omega\, r\sin\theta}.$$

$$v_r = 2\omega\, r;$$

$$\varepsilon_r = \frac{1}{2\omega\, r\, z}; \quad (54')$$

$$\mu_r = \frac{z}{2\omega\, r}.$$

$$v_\theta = v_r = 2\omega\, r;$$

$$\varepsilon_\theta = \varepsilon_r = \frac{1}{2\omega\, r\, z}; \quad (54'')$$

$$\mu_\theta = \mu_r = \frac{z}{2\omega\, r}$$

In other words, anisotropy is taking place, $\varepsilon$ and $\mu$ are tensor quantities.

$$\|\varepsilon\| = \begin{vmatrix} \varepsilon_r & 0 & 0 \\ 0 & \varepsilon_\phi & 0 \\ 0 & 0 & \varepsilon_\theta \end{vmatrix} = \begin{vmatrix} \frac{1}{2\omega\, r\, z} & 0 & 0 \\ 0 & \frac{1}{\omega\, r\, z\sin\theta} & 0 \\ 0 & 0 & \frac{1}{2\omega\, r\, z} \end{vmatrix}.$$

$$\|\mu\| = \begin{vmatrix} \mu_r & 0 & 0 \\ 0 & \mu_\phi & 0 \\ 0 & 0 & \mu_\theta \end{vmatrix} = \begin{vmatrix} \frac{z}{2\omega\, r} & 0 & 0 \\ 0 & \frac{z}{\omega\, r\sin\theta} & 0 \\ 0 & 0 & \frac{z}{2\omega\, r} \end{vmatrix}.$$

Here and further, corresponding letters mean parameters of neutron.

Let us find rest energy of neutron.

$$W = \int_V w\, dV = \int_V \frac{[E_r\, H_\theta]}{v_\phi}\, dV =$$

$$= \int_0^{2\pi}\int_0^{\pi}\int_0^{a} \frac{E_0^{\,2}\frac{a}{r}\sin^2\phi}{\omega\, r\, z\sin\theta}(1+2p\ln\frac{r}{a}+p^2\ln^2\frac{r}{a})\, r^2\sin\theta\, d\phi\, d\theta\, dr =$$

$$= \frac{\pi^2 E_0^{\,2} a}{\omega\, z}\left[a + 2p\, a\ln a - 2p0\ln 0 - 2p\, a - 2p\, a\ln a + p^2 a(\ln^2 1 - 2\ln 1 + 2) - p^2 a(0\ln^2 0 - 2*0\ln 0)\right] =$$

$$= \frac{\pi^2 E_0^{\,2} a^2(1-2p+2p^2)}{\omega\, z}. \quad (55)$$

Further. Charge of neutron is equal to zero.

$$q = \oint_S \varepsilon_r E_r\, dS = 0.$$

Really,

$$\int_0^{2\pi}\int_0^{\pi}\frac{E_0 \sin\phi}{2\omega\, az} a^2 \sin\theta\, d\phi\, d\theta = 0.$$

It is obvious that

$$\int_0^{\pi}\int_0^{\pi}\frac{E_0 a \sin\phi \sin\theta}{2\omega\, z} d\phi\, d\theta = -\int_{\pi}^{2\pi}\int_0^{\pi}\frac{E_0 a \sin\phi \sin\theta}{2\omega\, z} d\phi\, d\theta \neq 0.$$

It is logical to assume that

$$\int_0^{\pi}\int_0^{\pi}\frac{E_0 a \sin\phi \sin\theta}{2\omega\, z} d\phi\, d\theta = -\int_{\pi}^{2\pi}\int_0^{\pi}\frac{E_0 a \sin\phi \sin\theta}{2\omega\, z} d\phi\, d\theta = e.$$

Then

$$\frac{2\, E_0 a}{\omega\, z} = e. \quad (56)$$

Magnetic moment for neutron will be searched as the sum:

$M = M_m + M_L$

Where $M_m$ - magnetic moment created by volume current; $M_L$ - magnetic moment, attributed to impulse moment, i.e. to rotation.

$$M_m = \frac{1}{2}\int_0^{2\pi}\int_0^{\pi}\int_0^{a}\frac{E_0\sqrt{a}\, r \sin\theta \sin\phi}{z\, r^{3/2}}\left(p + \frac{1}{2} + \frac{1}{2} p \ln\frac{r}{a}\right) r^2 \sin\theta\, d\phi\, d\theta\, dr = 0;$$

as

$$\int_0^{2\pi}\sin\phi\, d\phi = 0.$$

$$M_L = \gamma\, L,$$

$$M_L = \gamma\frac{\hbar}{2} = M. \quad (57)$$

Now we shall write down the system of equations for neutron.

$$\begin{cases} \dfrac{2E_0 a}{\omega\, z} = e; & (56') \\ \dfrac{\pi^2 E_0^{\,2} a^2}{\omega\, z}(1 - 2p + 2p^2) = \hbar\,\omega; & (55') \\ \gamma\dfrac{\hbar}{2} = -0{,}96623707 * 10^{-26}. & (57') \end{cases}$$

Boundary conditions: at $r=a$

$$\varepsilon_r = \varepsilon_\theta = \varepsilon_0;$$

$$\mu_r = \mu_\theta = \mu_0;$$

Hence

$$z = \sqrt{\frac{\mu_0}{\varepsilon_0}}.$$

From (54) и (54′) follows that

$$\varepsilon_\phi = \varepsilon_r \frac{2}{\sin\theta};$$

and from (54) и (54″)that

$$\mu_\phi = \mu_\theta \frac{2}{\sin\theta}.$$

$$v_\phi = \omega\, a \sin\theta = \frac{1}{\sqrt{\varepsilon_\phi \mu_\phi}} = \frac{1}{\sqrt{\varepsilon_0\mu_0 \dfrac{4}{\sin^2\theta}}} = \frac{c\sin\theta}{2}.$$

So

$$\omega\, a = \frac{c}{2};$$

$$a = \frac{c}{2\omega} = 1{,}0500973 * 10^{-16}\,(m).$$

Here $\omega$=1,4274508*$10^{24}$ Hz- Compton circular frequency of neutron.

Let us solve system (56)(55′)(57′)

$$E_0 = \frac{e\omega^2\sqrt{\dfrac{\mu_0}{\varepsilon_0}}}{c}. \quad (56'')$$

$$E_0 = E_0^n = 4{,}1024444 * 10^{23}\,\left(\frac{V}{m}\right).$$

We substitute (56″) B (55′)

$$p^2 - p + 0{,}5 - \frac{2\,\hbar}{\pi^2 e^2\sqrt{\mu_0/\varepsilon_0}} = 0;$$

$$p_1 = 1{,}8999321;$$

$$p_2 = -0{,}8999321;$$

*P* must be negative, therefore we select

$$p_2 = p_n = -0{,}8999321.$$

From (57′) we find $\gamma$

$$\gamma = -1{,}8324711 * 10^8\left(\frac{1}{T * s}\right).$$

Let us write down the system of equations for antineutron.

$$\begin{cases} \dfrac{2\, E_0 a}{\omega\, z} = e; \\ \dfrac{\pi^2 E_0^{\,2} a^2}{\omega\, z}(1 - 2p + 2p^2) = \hbar\,\omega; \\ \gamma\dfrac{\hbar}{2} = 0{,}96623707 * 10^{-26}. \end{cases}$$

Boundary conditions are the same, as at neutron, hence

$$z = \sqrt{\frac{\mu_0}{\varepsilon_0}};$$

$$a = \frac{c}{2\omega} = 1{,}0500973 * 10^{-16}\,(m).$$

The last system with exactness to a sign has the same solutions, as system (56)(55′)(57)

$$E_0^{\tilde{n}} = E_0^n = 4{,}1024444 * 10^{23}\,\left(\frac{V}{m}\right);$$

$$p_{\tilde{n}} = p_n = -\,0{,}8999321;$$

$$\gamma_{\tilde{n}} = -\gamma_n = 1{,}8324711 * 10^8\left(\frac{1}{T * s}\right).$$

## Conclusion

Within the framework of the model, which is considered, electron, proton and neutron represent a monochromatic electromagnetic wave of corresponding frequency spread along parallels inside the spherical area, i.e. a wave, rotating around some axis. At that along each parallel, exactly half of wave length for electron and proton and exactly one wave length for neutron, is kept within, thus this is rotating soliton. This is caused by presence of spatial dispersion and anisotropy of a strictly defined type inside the particles. In electron vector E is directed to center of particle, that correspond to negative charge, and in proton vector E is directed from center of particle, that correspond to positive charge. Thus, by natural way, all basic parameters of particles are obtained: charge, rest energy, mass, radius, magnetic moment and spin, that is confirmed by mathematical expressions, which are discovered.

## References


1. Sarycheva LI (2000) Structure of matter. Soros educational, Russia.
2. DI Blokhintsev (1983) Principles of quantum mechanics.
3. Born M (1967) Atomic physics.
4. Bredov MM, Rumyantsev VV, Toptygin IN (1985). possibility of measuring the thermal vibration spectrum g ($ omega $) using coherent inelastic neutron scattering from a polycrystalline sample. Inst of Semiconductors, Leningrad.
5. Vinogradova MB, Rudenko OV, Sukhorukov AP (1990) Theory of waves.
6. Vlasov AD, PMurin B (1990) Physical quantities units in science and technology.
7. Vladimirov VS, Jeffrey A, Schroeck FE (1971) Equations of mathematical physics. American Journal of Physics 39: 1548-1548.
8. Kalashnikov SG (1985) Electricity.
9. Kamke E (1976) Reference manual on ordinary differential equations. Campman and Hall/CRC.
10. Landau LD, Lifshits EM (1989) Quantum mechanics. Non-relativistic theory .Physics Today 11: 56-60.
11. Landau LD, Lifshits EM (1973) Field theory.
12. Landau LD, Lifshits EM (1982) Electrodynamics of continuous mediums.
13. Prudnikov AP, Yu A, Brychkov O, Marychev I (1981) Integrals and series. Elementary functions. Gordon and Breach Science Publishers, UK.
14. Taylor B, Parker V, Langerberg D (1991) Fundamental constants and quantum electrodynamics. Doklady Akademii Nauk Ukrain 5: 40-45.
15. Terletskiy YP, Yu P Rybakov (1990) Electrodynamics.
16. Turanyanin D (2004) On interaction of motional masses. Journal of Theoretics.
17. Xu Z, Buehler MJ (2010) Interface structure and mechanics between graphene and metal substrates: a first-principles study. Journal of Physics: Condensed Matter 22: 45-56.
18. Kyriakos AG (2003) The electrodynamics form concurrent to the Dirac electron theory. Physics Essays 16: 365-374.
19. Kyriakos AG (2005) The massive neutrino-like particle of the non-linear electromagnetic field theory.
20. Kyriakos AG (2004) Yang-Mills equation as the equation of the superposition of the non-linear electromagnetic waves.
21. Kyriakos AG (2004) Non-linear Theory of quantized Electromagnetic Field equivalent to the Quantum Field Theory.
22. Ivanov IP (2003) Quark model is not quite correctly. Scientific Russia.
23. Kopelyowicz V Topologic soliton models of baryons and its predictions. Scientific Russia.
24. Diakonov D, Petrov V, Polyakov M Unpolarized and polarized quark distributions in the large-$N_c$ limit Z. Phys Rev D.
25. Skyrme THR (1962) A Non-Linear Field Theory. Nucl Phys 31: 556.
26. Polyakov M (2000) Infinite conformal symmetry in two-dimensional quantum field theory. Eur Phys J 241: 333-380.
27. Nakano T (2003) Evidence for Narrow S=+1. Baryon Resonance in Photo-production from Neutron. Phys Rev Lett.
28. Polyakov MV, Rathke A (2003) on photoexcitation of baryon antidecuplet. Eur Phys J A 18: 691-695.
29. Diakonov D, Petrov V (1985) Baryons as solitons. Elementary particles. Moscow Energoatomisdat.
30. Chemtob M (1985) Skyrme model of baryon octet and decuplet. Nucl Phys B 256: 600-608.
31. Yi SD, Onoda S, Nagaosa N, Han JH (2009) Skyrmions and Anomalous Hall Effect in a Dzyloshinskii-Moriya Spiral Magnet. Physical Review B.
32. Walliser H, Eckart G (1992) baryon resonance as fluctuations of the skyrme soliton. Nuclear physics a 429: 514-526.
33. Diakonov D, Petrov V, Polyakov M (1997) Exotic Anti-Decuplet of Baryons: Prediction from Chiral Solitons. Z Phys A 359: 305-314.
34. Ivanov IP (2006) Last days of Standard Model?

# 21

# Mathematical Issue in Section 2 of 'On the Electrodynamics of Moving Bodies'

**Makanae M***

*Researcher, Representative Free Web College, Japan*

**Abstract**

In Albert Einstein's first published work, 'On the Electrodynamics of Moving Bodies', he introduced a set of equations, namely $t_B - t_A = \frac{\gamma_{AB}}{c-v}$ and $t'_A - t_B = \frac{\gamma_{AB}}{c+v}$ which have the form 'time=distance/velocity', as a part of the conclusion of Section 2. In these equations, Einstein implied that an event in a moving system viewed from within that moving system differs from the same event viewed from a reference stationary system. This perspective became the fundamental basis of the special theory of relativity (STR).

However, considering Sections 1 and 2 of Einstein's paper using practical examples and numerical values, we find that an inconsistency is caused by using 'relative speed' as 'velocity' in the universal equation 'time=distance/velocity'. In the conventional mathematics, only 'mobile speed' is admitted as 'velocity' in 'time=distance/velocity'. This is a pure mathematical issue that should be solved if we continue to use the STR, under the premise that Sections 1 and 2 of 'On the Electrodynamics of Moving Bodies' are correct.

**Keywords:** Special theory of relativity; Mathematics; Lorentz factor

## 1. Introduction

In 1905, Albert Einstein's first published work, 'On the Electrodynamics of Moving Bodies' [1], was released. The theory presented therein was later termed the special theory of relativity (STR). In Section 1 of his paper [1], Einstein defines the concept of 'time'. He then describes a method of confirming the synchronization of two clocks placed at points *A* and *B*.

In Section 2 of his paper [1], based on '*the principle of relativity*' and '*the principle of the constancy of the velocity of light*', Einstein considers the relationships between a moving system and a reference stationary system with the concepts of 'length' and 'time', using the form 'velocity=distance/time'. Both systems have a relationship under the special condition that an observer at rest in the reference stationary system observes an event in the moving system, which moves under parallel translation with uniform velocity with respect to the reference stationary system. Through the consideration in Section 2 of the paper [1], Einstein implies that an event in a moving system viewed within that moving system differs from the same event viewed from a reference stationary system. This perspective became the fundamental basis of STR.

In Section 3 of the paper [1], Einstein develops the expression $\beta = \frac{1}{\sqrt{1-\frac{v^2}{c^2}}}$ which was later called the 'Lorentz factor'. This expression denotes the ratio of the value of 'length' or 'time' between the moving system and the reference stationary system when both systems are under the same special conditions described above.

The expression $\sqrt{1-\frac{v^2}{c^2}}$ on the right side of the Lorentz factor denotes the condition of the length or time in the moving system viewed from the reference stationary system, when the reference value of length or time in the stationary system is defined to be unity. In this expression, *c* denotes the velocity of light, and *v* denotes the velocity of the moving system. The effectiveness of the Lorentz factor for length and time is the starting point for studying STR.

From the above, we can say that, in Section 3 of the paper [1], the Lorentz factor was developed as the core relationship in the STR, while Sections 1 and 2 of the paper [1] were provided as the premises of Lorentz factor.

However; if examining the equations provided in Section 1 and 2 of the paper [1] by using numerical values in practical examples with Einstein's description in the paper [1], problems are caused by using equation 'time=distance/velocity' in a method that had never been considered in the field of mathematics.

## 2. Confirming Equations Provided in Sections 1 and 2 of the Paper [1]

At the beginning of his study in Section 1 of the paper [1], Einstein emphasized that the concept of 'time' is important in the study of physics. Then, he provided a method for confirming the synchronization of two clocks placed at points *A* and *B* using light traveling between the two points:

$$t_B - t_A = t'_A - t_B \tag{1}$$

Here, *t* represents time, $t_A$ is the point in time at which light is emitted from the light source placed at *A*, $t_B$ is the point in time at which the light is reflected by a mirror placed at *B*, and $t'_A$ is the point in time at which the light returns to *A*. The left-hand side of equation (1) represents the time required for the light to 'Go' (from *A* to *B*), and the right-hand side represents the time required for the light to 'Return'

***Corresponding author:** Makanae M, Researcher, Representative Free Web College, Japan, E-mail: edit@free-web-college.com

(from *B* to *A*). Based on the premise that the velocity of light is constant in a vacuum, the right- and left-hand sides of equation (1) are equal. Thus, we can confirm that the synchronization of the two clocks, placed at *A* and *B*, is satisfied. This method became the predominant basis of the thought experiments described in the paper [1]. In the same section of the paper [1], Einstein provided

$$\frac{2\overline{AB}}{t'_A - t_A} = c \qquad (2)$$

The term $\overline{AB}$ was described as the distance between the points *A* and *B*; therefore, $2\overline{AB}$ denotes the distance of the round trip of light between the clock positions at points *A* and *B*. The term $t'_A - t_A$ is described as the time required for the same round trip of light, and c is the velocity of light. In other words, equation (2) corresponds to the form of equation 'distance/time=velocity'. In the first half of Section 2 of the paper [1], the following expression was provided with the explanation that '*where time interval is to be taken in the sense of the definition in Section 1*'.

$$velocity = \frac{light\ path}{time\ interval} \qquad (3)$$

Equation (3) corresponds to the form 'velocity=distance/time'; in other words, the form of equation (2) and equation (3) is established to have the same order. Therefore, the $2\overline{AB}$ term of equation (2) corresponds to the 'light path' of equation (3), while $t'_A$-$t_A$ of equation (2) corresponds to the 'time interval' of equation (3).

Immediately after equation (3), Einstein included the following description:

'*Let there be given a stationary rigid rod; and let its length be l as measured by a measuring-rod which is also stationary. We now imagine the axis of the rod lying along the axis of x of the stationary system of co-ordinates and that a uniform motion of parallel translation with velocity v along the axis of x in the direction of increasing x is then imparted to the rod*'.

Then, the two different '*operations*' were provided to '*inquire as to the length of the moving rod*'.

Operation (a): '*The observer moves together with the given measuring-rod and the rod to be measured, and measures the length of the rod directly by superposing the measuring-rod, in just the same way as if all three were at rest*'.

Operation (b): '*By means of stationary clocks set up in the stationary system and synchronizing in accordance with section 1, the observer ascertains at what points of the stationary system the two ends of the rod to be measured are located at a definite time. The distance between these two points, measured by the measuring-rod already employed, which in this case is at rest, is also a length which may be designated 'the length of the rod*'. Then, Einstein predicted the length of the moving rod as follows:

'*In accordance with the principle of relativity the length to be discovered by the operation (a)—we will call it 'the length of the rod in the moving system'—must be equal to the length l of the stationary rod. The length to be discovered by the operation (b) we will call 'the length of the (moving) rod in the stationary system.' This we shall determine on the basis of our two principles, and we shall find that it differs from l*'. Based on the above conditions, Einstein introduced the equations:

$$t_B - t_A = \frac{\gamma_{AB}}{c - v} \text{ and } t'_A - t_B = \frac{\gamma_{AB}}{c + v} \qquad (4)$$

The form of both equations corresponds to the form 'time=distance /velocity,' and the premises were implied for equation (4) as follows:

1. Two coordinate systems: the moving system is described as a moving rigid rod, and the stationary system is described as the reference frame.

2. The rigid rod travels at a uniform velocity, undergoing parallel translation with respect to the stationary system, along the positive direction of the x-axis.

3. *A* is the point of the end rod, closest to the origin of x-axis, and *B* is the point of the end of the rod, at a distance *l* from *A*.

4. A light source is placed at *A*, and a mirror is placed at *B* to reflect the light in the opposite direction.

5. A clock is placed at each of the points *A* and *B*.

6. The round trip of a ray of light between *A* and *B* is performed according to the formalism contained in equation (1).

7. The velocity of the moving rigid rod is *v*, and *c* is the velocity of light.

8. The point in time at which the light is emitted by the light source is $t_A$.

9. The point in time at which the light is reflected by the mirror is $t_B$.

10. The point in time at which the light returns to *A* is $t'_A$.

11. '$\gamma_{AB}$ *denotes the length of the moving rigid rod - measured in the stationary system*'.

Conditions 4, 5, 8, 9, and 10 are the same conditions as those in equation (1).

When an observer in the reference stationary system observes this round trip of the light, equation (4) holds. The left-hand side of equation (4) corresponds to the Go condition (light leaving *A*), and the right-hand side corresponds to the Return condition (light returning to *A*). Unlike equation (1), the structure of equation representing go and return is different in equation (4).

In total, four equations were provided in Section 1 and 2 of the paper [1]:

$$t_B - t_A = t'_A - t_B \qquad (1)$$

$$\frac{2\overline{AB}}{t'_A - t_A} = c \qquad (2)$$

$$velocity = \frac{light\ path}{time\ interval} \qquad (3)$$

$$t_B - t_A = \frac{\gamma_{AB}}{c - v} \text{ and } t'_A - t_B = \frac{\gamma_{AB}}{c + v} \qquad (4)$$

[Note] We should call (3) an 'expression,' but, for convenience, we refer to it as 'equation.'

## 3. Examining Each Equation Using Practical Numerical Values

We now examine these four equations using numerical values from practical examples. First, we provide the mutual conditions in the four equations. Regarding the velocity of light, we employ 'the principle of the constancy of the velocity of light' as Einstein did, i.e., the velocity of light is always *c*. We assume the velocity of the moving system *v* is half of velocity of light, i.e., 0.5 *c*. The distance between points *A* and *B* is given by $l$. We further assume $1l$ is the distance that light can move in 1s when the system containing points $A$ and $B$ is stationary. Regarding the time required for light to travel from point $A$ (or $B$) to point $B$ (or $A$), this value is given as 1s based on above assumptions when the system containing points $A$ and $B$ is stationary.

### 3.1 Examining equation(1)

The examination of equation (1), namely, $t_B - t_A = t'_A - t_B$, is the simplest.

$$1\text{s} - 0\text{s} = 2\text{s} - 1\text{s} \tag{5}$$

### 3.2 Examining equation (2)

The examination of equation (2), namely $\frac{2\overline{AB}}{t'_A - t_A} = c$, which corresponds to the universal form 'velocity=distance/time,' is also simple. This equation is established by the round trip of light between points $A$ and $B$ viewed within the stationary system that contains points $A$ and $B$. The distance $AB$ is $1l$, so that $2\overline{AB}$ is $2l$. The round trip travel time of $t'_A - t_A = 2$ s. Therefore, equation (2) is given by

$$2l/2\text{s} = c \tag{6}$$

### 3.3 Examining equation (3)

Assuming that the system containing points $A$ and $B$ is stationary and the observer at rest in the same system observes the round trip of light between $A$ and $B$, the observer will obtain the simple results in equation (3), $velocity = \frac{light\ path}{time\ interval}$

- Go interval (the trip of light from $A$ to $B$): $c = 1l/1$ s (7)
- Return interval (the trip of light from $B$ to $A$): $c = 1l/1$ s (8)

On the other hand, assuming that the system containing points $A$ and $B$ moves at half of the velocity of light and that the observer at rest in the reference stationary system observes the round trip of light between $A$ and $B$, we have to assume tangible conditions.

First, we consider the time sequence. In this case, viewed from the reference stationary system, the points $A$ and $B$, corresponding to the locations of the light source and the mirror, respectively, are in motion. In other words, the positions of $A$ or $B$ are variable with the progress of time when viewed from the reference stationary system; thus, we need to specify the time sequence. For this purpose, we describe each position using the following references to time data:

- The position of $A$ at time $t_A$ is denoted as $At_A$.
- The position of $B$ at time $t_B$ is denoted as $Bt_B$.
- The position of $A$ at time $t'_A$ is denoted as $At'_A$.

Using this notation, the description of the Go and Return intervals of the trip of light, viewed from the reference stationary system, can be described as follows:

- Go interval: from $At_A$ to $Bt_B$
- Return interval: from $Bt_B$ to $At'_A$

To show that every element of equation (3) is concerned with the motion of light between $A$ and $B$, we describe equation (3) $velocity = \frac{lightpath}{time\ interval}$ as velocity of light=light path from $At_A$ (or $Bt_B$) to $Bt_B$ (or $At'_A$)/time interval required for light to travel between $At_A$ (or $Bt_B$) and $Bt_B$ (or $At'_A$).

The velocity of light of the above expression is c, because we already employed '*the principle of the constancy of the velocity of light.*' Therefore, using the practical example assumed at the beginning of this section of this paper, we obtain the following results:

Go interval

$$c = \frac{2l}{2s} \tag{9}$$

$$\text{Return interval } c = \frac{l \times 2/3}{2/3s} \tag{10}$$

The process of calculating equations (9) and (10) is given below.

We assume that, when $t$=0 s, the light source placed at point $A$ of the moving system emits light. The time intervals, which are required for the trip of light in the Go interval and in the Return interval, viewed from the reference stationary system, are calculated as follows.

In the Go interval, when $l + vt = ct$, the tip of the light emitted by the light source placed at point $A$ reaches the mirror placed at point $B$. We can transpose $l + vt = ct$ to

$$t = \frac{l}{c - v} \tag{11}$$

In the Return interval, temporarily assuming that the point in time at which the light is reflected by the mirror is 0 s, when $l - vt = ct$, the tip of the light reflected by the mirror reaches point $A$. We can transpose $l - vt = ct$ to

$$t = \frac{l}{c + v} \tag{12}$$

If we use the numerical values for our example in these equations, equation (11) becomes $t = \frac{1l}{0.5c}$, and equation (12) becomes $t = \frac{1l}{1.5c}$. With these calculations, we obtain the time in the Go interval to be 2s, and the time in the Return interval is 2/3 s. From these results, each point in time becomes clear:

- $t_A$ (the point in time at which light is emitted by the light source): 0 s
- $t_B$ (the point in time at which light is reflected by the mirror): 2 s
- $t'_A$ (the point of time at which light returns to $A$): 2/3 s past 2 s

Under '*the principle of the constancy of the velocity of light,*' the light moves at velocity $c$ (i.e., $1c$) in the Go and Return conditions, even if observing from the stationary reference system. Therefore, the light path from $At_A$ to $Bt_B$ (i.e., in the Go interval) becomes $2l$; the light path from $Bt_B$ to $At'_A$ (i.e., in the Return interval) becomes $2/3l$.

From the above, we can obtain the numerical values in equation (3) as shown in the equations (9) $c = \frac{l \times 2}{2\text{s}}$ (in Go interval) and (10) $2/3\text{s} = \frac{l \times 2/3}{c}$ (in Return interval), when the system, which contains the light source and the mirror, is in motion at the velocity $0.5c$ viewed from the reference stationary system.

We can transpose equations (9) and (10) to:

$$\text{Go interval: } 2\text{s} = \frac{l \times 2}{c} \tag{13}$$

$$\text{Return interval: } 2/3\text{s} = \frac{l \times 2/3}{c} \tag{14}$$

Now we should remember the following facts for later discussion in our study.

- Equations (4), (13), and (14) are established under '*the principle of the constancy of the velocity of light*'.
- Einstein's equation (4) $t_B - t_A = \frac{\gamma_{AB}}{c-v}$ and $t'_A - t_B = \frac{\gamma_{AB}}{c+v}$, corresponds to the form 'time=distance/velocity'.
- Equations (13) and (14) correspond to the form 'time=distance/velocity'.
- Equations (13) and (14) are established under conditions similar to that of equation (4).
- The velocity of equation (4) is '$c - v$' (in Go interval) or '$c + v$' (in Return interval).
- The velocity of equation (13) or (14) is just '$c$' in both intervals.

### 3.4 Examining equation (4)

In Einstein's equation (4) $t_B - t_A = \frac{\gamma_{AB}}{c-v}$ and $t'_A - t_B = \frac{\gamma_{AB}}{c+v}$ which correspond to the form 'time=distance/velocity,' the numerical values in the practical example are clear for the time case, because these are the same as the results using equation (11) $t = \frac{l}{c-v}$ and equation (12) $t = \frac{l}{c+v}$.

Therefore, the Go interval $t_B - t_A$ is 2s, and the Return interval $t'_A - t_B$ is 2/3 s as viewed from the stationary reference system.

Regarding the numerical value of $c-v$ or $c+v$, which provided as 'velocity' of 'time=distance/velocity' by Einstein in Go or Return interval, the former becomes $1c-0.5c=0.5c$, and the latter becomes $1c+0.5c=1.5c$, under the assumption that $v$ is $0.5c$.

Regarding the term '$\gamma_{AB}$', which is defined as '*the length of the moving rigid rod - measured in the stationary system*', we should consider the difference between the concepts of '$\gamma_{AB}$' and 'light path'. Usually, if two terms differ, the concept behind the two terms also differs, unless one term is rephrased to have the same meaning as the other term. However, it is difficult to say that '$\gamma_{AB}$' is a rephrasing of the 'light path', even though both terms were provided within the same context in the same section. This is because, in the first place, the rigid rod is a solid body but the light path is not, regardless of whether they belong to the moving system or whether observing them from the reference stationary system.

However, there is a light source placed at point $A$ of the 'light path' as well as at point $A$ of '$\gamma_{AB}$': similarly, a mirror is placed at point $B$ of the 'light path' as well as at point $B$ of '$\gamma_{AB}$'. In other words, the round trip of light between points $A$ and $B$ is incarnated not only in the concept of 'light path' but also in '$\gamma_{AB}$' under the same conditions of Section 2 [1].

Based on the above, there are two possibilities for the relationship between the terminologies '$\gamma_{AB}$' and 'light path':

Possibility 1: '$\gamma_{AB}$' and 'light path' are effectively the same concept with different names.

Possibility 2: '$\gamma_{AB}$' and 'light path' are different concepts, as given by the difference in names.

For the present, we choose Possibility 1 with the reasoning that the round trip of light between points $A$ and $B$ can be incarnated on a moving rod, i.e., the '*light path*' can be incarnated on the moving rod, and considering Einstein's equation (4) under the condition of Possibility 1 in Sections 3, 4, and 5 of our study. Regarding Possibility 2, we will consider in Section 6 of our study.

Under the above choice, we consider the numerical value of '$\gamma_{AB}$' in our practical example. The length of the rod measured by Einstein's operation (a) has a value of $1l$. However, in our examination, we are trying to clarify each value when the system containing points $A$ and $B$ is in motion and is observed from the stationary reference system; thus, we employ Einstein's operation (b). The numerical value of the length of the moving rod measured using operation (b) can be calculated using equations for the Go interval:

$$\gamma_{AB}=(t_B - t_A) \times (c - v) \tag{15}$$

and for the Return interval

$$\gamma_{AB}=(t'_A - t_B) \times (c + v). \tag{16}$$

With the numerical values of $t_B - t_A$=2 s and $t'_A - t_B$=2/3s from equations (11) and (12), in the Go interval, equation (15) becomes $\gamma_{AB}$=2s × 0.5$c$; therefore, $\gamma_{AB}=1l$. Likewise, in the Return interval, equation (16) becomes $\gamma_{AB}$=2/3$s$ × 1.5$c$; therefore, $\gamma_{AB}=1l$.

From the results of these calculations, we can describe Einstein's equation (4), $t_B - t_A = \frac{\gamma_{AB}}{c-v}$ and $t'_A - t_B = \frac{\gamma_{AB}}{c+v}$ as

$$2\text{s} = \frac{1l}{0.5c} \text{ and } 2/3\text{s} = \frac{1l}{1.5c} \tag{17}$$

## 4. Validating the Numerical Values of Each Examination

Let us validate our examination of the four equations in Section 3 of our study. Regarding the equations (5), (6), (7), (8), (9) and (10), these calculations are correct.

Regarding the equation (17), which is the result of the calculation of Einstein's equation (4)

$t_B - t_A = \frac{\gamma_{AB}}{c-v}$ and $t'_A - t_B = \frac{\gamma_{AB}}{c+v}$ the numerical values

$2\text{s} = \frac{1l}{0.5c}$ and $2/3\text{s} = \frac{1l}{1.5c}$ are clearly consistent.

However, considering this result with Einstein's prediction that '*The length to be discovered by the operation (b) we will call 'the length of the (moving) rod in the stationary system'. This we shall determine on the basis of our two principles, and we shall find that it differs from l*' as described in Section 2 of [1], we find an inconsistency. Because the numerical value of '$\gamma_{AB}$' is $1l$ in both the Go and Return intervals in the above results, these values do not '*differ from l*', even though equation (17) was examined using operation (b). Therefore, we should validate Einstein's equation (4) with equation (17) in more detail.

Hereafter, we call Einstein's prediction 'the prediction '*it differs from l*' ' for brevity.

## 5. Validating Equation (4) More Thoroughly with Equation (17)

If we interpret the prediction '*it differs from l*' as that the length of the moving rod viewed from the reference stationary system differ from $1l$, this prediction becomes correct by using the Lorentz factor. In particular, when calculating the length of the moving rod by using the

Lorentz factor, the value is $\sqrt{0.75l}$ when viewed from the reference stationary system in all cases if the velocity of the moving rod is constant at 0.5$c$; i.e., the value of $\gamma_{AB}$ is constant as $\sqrt{0.75l}$ in both interval of the round trip of light. Certainly, $\sqrt{0.75l}$ differs from 1$l$. However, the Lorentz factor was provided as a conclusion of its development process in Section 3 [1]; in other words, the prediction '*it differs from l*', which was provided in the first half of Section 2 [1], and the description of Lorentz factor are distantly discussed in the paper [1].

From the above conditions, it is not clear that '*it*' of '*it differs from l*', which corresponds to the value of the length of the moving rod viewed from the reference system, was controlled by Einstein's equation (4) or by the Lorentz factor.

Even if we neglect this issue, there is another issue which we discuss below.

The numerical values of $\gamma_{AB}$ in the Go interval and Return interval are the same value, namely, 1$l$ (if it was controlled by Einstein's equation (4)) or $\sqrt{0.75l}$ (if it was controlled by Lorentz factor). However, the values of 'light path' in the Go interval and Return interval, which can be incarnated on the moving rod, are different when viewed from the stationary reference system, as shown in equation (9) as 2$l$ and in equation (10) as 2/3$l$. If the measurement is performed by operation (a), the length of rod and the '*light path*' between points $A$ and $B$ are constant as 1$l$ because the values are obtained '*directly by superposing the measuring-rod, in just the same way as if all three were at rest*'. However, the value of $\gamma_{AB}$ cannot be incarnated by operation (a) because the rod defined in $\gamma_{AB}$ is not at rest. Thus, Einstein's equation (4), which contains $\gamma_{AB}$, can only be determined using operation (b).

However, examining the description of operation (b) more thoroughly, we find that '*the observer ascertains at what points of the stationary system the two ends of the rod to be measured are located* '*at a definite time*'. We can interpret this description as follows: the observer at rest in the stationary reference system measures the length of the moving rod '*at a definite time*'. The meaning of measuring the rod '*at a definite time*' is effectively the same as the meaning of measuring the moving rod when this rod temporarily stops. Therefore, if the observer at rest in the stationary reference system measures the length of the moving rod '*at a definite time*', the observer will obtain the value of the length of the rod as 1$l$.

Therefore, we can say that equation (17), which denotes the numerical values of $\gamma_{AB}$ in the Go and Return intervals as 1$l$ using Einstein's equation (4) in operation (b), is correct. In reality, the numerical values of equation (17) were provided at a definite time, namely when the light reaches the mirror or when the reflected light reaches point $A$. However, the numerical values of the 'light path,' which can be incarnated on the moving rod in the Go interval and Return interval, are 2$l$ and 2/3$l$ when viewed from the stationary reference system, if employing '*the principle of the constancy of the velocity of light*'. Thus, Possibility 1, the prediction '*it differs from l*' and operation (b) cannot be assumed in the same context.

## 6. Considering Equation (4) in Possibility 2

Let us consider equation (4) using Possibility 2, in which $\gamma_{AB}$ and the 'light path' are different concepts. The condition of '*the principle of the constancy of the velocity of light*' is still employed.

First, we confirm the form of equation (4) $t_B - t_A = \frac{\gamma_{AB}}{c-v}$ and $t'_A - t_B = \frac{\gamma_{AB}}{c+v}$.

The term '$\gamma_{AB}$'(i.e., the length of the moving rod measured in the reference stationary system) was used as the distance in equation of the form 'time=distance/velocity'. Both $t_B - t_A$ and $t'_A - t_B$ were described as time and denote the time interval of the round trip of light between points $A$ and $B$. $c$-$v$ and $c$+$v$ were provided as the velocity of 'time=distance/velocity'. Therefore, we can say that equation (4) denotes the time in which the light moves between points $A$ and $B$ is equal to the length of the moving rigid rod divided by $c$-$v$ (in the Go interval) or $c$+$v$ (in the Return interval).

Fundamentally, when we want to determine the numerical value for the time in which the light moves between points $A$ and $B$ using 'time=distance/velocity,' the distance should be associated with the motion of light between points $A$ and $B$. However, in equation (4), '$\gamma_{AB}$', which corresponds to 'distance,' was defined by Einstein as the length of the moving rigid rod measured in the stationary system, i.e., this definition does not discuss the motion of light.

We can treat '$\gamma_{AB}$' as the 'light path' using the reason that a round trip of light can be incarnated between points $A$ of $\gamma_{AB}$ and $B$ of $\gamma_{AB}$; however, this assumption belongs to Possibility 1. Therefore, in Possibility 2, we ignore the motion of light when considering $\gamma_{AB}$, regardless of whether the 'light path' can be incarnated between points $A$ of $\gamma_{AB}$ and $B$ of $\gamma_{AB}$.

If we ignore the motion of light in '$\gamma_{AB}$,' equation (17), namely, $2\text{s} = \frac{1l}{0.5c}$ and $2/3\text{s} = \frac{1l}{1.5c}$ correct when employing the interpretation that the meaning of '*the two ends of the rod to be measured are located at a definite time*' using operation (b) is the same as measuring the length of the rod when it temporarily stops. Under this condition, the observer at rest in the reference stationary system measures the length of the rod as 1$l$ in both the Go interval and Return interval. However, the same observer can also measure the length of the rod as 1$l$ using operation (a). Thus, classifying the operation as (a) and (b) become meaningless.

Therefore, we can conclude that Section 2 of the paper [1], which provided '*the principle of the constancy of the velocity of light,*' operations (a) and (b), and equation (4), is inconsistent.

## 7. Considering the Factor that Caused the above Problems

Here, let us consider the factor that caused the problems introduced in Section 5 and 6 of our study. Primarily, 'time=distance/velocity' states that the time in which a certain object moves between two points is equal to the distance that a certain object moves between the two points divided by the velocity with which the object moves between the two points. In other words, if we consider the equation 'time=distance/velocity' as it is used for an object that moves from a certain point to another point, each element contained in this equation should be associated with the motion of the object. From this viewpoint, we call the velocity in this equation the 'mobile speed'. Therefore, if using 'time=distance/velocity' to perceive the motion of light, the above description can be arranged as follows: the time in which light moves between two points is equal to the distance that light moves between two points divided by the velocity with which the light moves between the two points (i.e., the mobile speed of light).

However, the terms $c$-$v$ and $c$+$v$, which correspond to 'velocity' in Einstein's equation (4), are not the mobile speed because the mobile speed of light is always '$c$' if employing '*the principle of the constancy of the velocity of light.*' Einstein did not define to these terms; thus, let us denominate the concepts of '$c$-$v$' and '$c$+$v$' as 'relative speed' using the following reasoning.

Regarding $c$-$v$, under '*the principle of the constancy of the velocity of light*' in the Go interval, the light moves '*relatively*' to the point at which the light source is placed after light is emitted with velocity $c$-$v$ in the positive direction of the x-axis, as viewed from the stationary reference system. In reality, Einstein used the term '*relatively*' and '$c$-$v$' in the following sentence: '*the ray moves relatively to the initial point of k, when measured in the stationary system, with the velocity c-v*' in Section 3 of the paper [1]. (Note that the term '*k*' corresponds to the moving system.) Therefore, we can say that the term $c$-$v$ represents relative speed in the Go interval.

Regarding $c$+$v$; this term represents the reverse situation of the Go interval that the concept of relative speed $c$-$v$ can be established, therefore we can say that the term $c$+$v$ represents the relative speed between the moving system and the reflected light in the Return interval.

## 8. Considering the Factor Causing Einstein to use 'Relative Speed' in 'Time=Distance/Velocity'

From the earlier considerations in our study, we can determine that Einstein's equation (4) states that the time in which the light moves between points $A$ and $B$ is equal to the length of the moving rigid rod divided by the relative speed between the light and the moving system. However, the method, which uses 'time=distance/velocity' in the above form, had never been considered in the conventional law of mathematics. Because the velocity of 'time=distance/velocity' must be the mobile speed, the relative speed $c$-$v$ or $c$+$v$ cannot be used as the mobile speed of light.

Therefore, we trace the factor causing Einstein to use the relative speed $c$-$v$ and $c$+$v$ as 'velocity' in the universal equation 'time=distance/velocity'.

We first confirm the background facts.

1. In the Go interval: when $l + vt=ct$, the tip of the light emitted from the light source reaches the mirror (i.e., is reflected by the mirror). Moreover, $l + vt=ct$ can be transposed to the form in equation (11), $t = \frac{l}{c - v}$.

2. In the Return interval: when $l - vt=ct$, the tip of the light reflected by the mirror reaches point $A$ if we temporarily assume that the point of time at which the mirror reflected the light is 0 s. Then, $l - vt=ct$ can be transposed to the form in equation (12), $t = \frac{l}{c + v}$.

3. Both equations (11) and (12) can be used to calculate the time interval between two points, such as $t_B - t_A$ or $t'_A - t_B$, under the condition that distance is $l$ as constant.

4. The form of equations (11) and (12) corresponds to the form 'time=distance/velocity,' i.e., the term $t$ of the equations (11) and (12) corresponds to 'time', the term $l$ of equations (11) and (12) corresponds to 'distance,' and the terms $c$-$v$ and $c$+$v$ of equations (11) and (12) correspond to 'velocity'.

5. The examination of equations (11) and (12) is correct using practical examples under the condition of a time point at which the tip of light reaches the mirror or point $A$.

In Section 2 of the paper [1], equations (11) and (12) of our study are not provided. However, the conditions given in points 1 and 2 above are tacit facts in Section 2 of the paper [1]; therefore, we can conjecture that Einstein considered and examined equations (11) and (12) when he prepared the paper [1]. If this conjecture is correct, we can infer the factor causing Einstein to use $c$-$v$ and $c$+$v$ as 'velocity' in

'time=distance/velocity' is that he relied on equations (11) and (12) as the basis that $c$-$v$ and $c$+$v$ can be used as the 'velocity' in 'time=distance/velocity'.

## 9. Conclusion

The Lorentz factor $\beta = \frac{1}{\sqrt{1 - \frac{v^2}{c^2}}}$, which is the core premise of the STR, requires '$v$' as one of its elements. The Lorentz factor was established based on the concept of Einstein's equations of time=distance/velocity provided in the paper [1]. However, the 'mobile speed' of light '$c$' is configured for the velocity in

'time=distance/velocity' even if we consider the round trip of light in the moving system viewed from the stationary reference system; in other words, the velocity of the moving system '$v$' does not always affect the velocity of light, regardless of whether the system containing the light source and the mirror moves, if '*the principle of the constancy of the velocity of light*' is employed.

If $v$ does not always affect the velocity of light, the numerical value of $v$ in the Lorentz factor is always zero; thus we can describe $\beta = \frac{1}{\sqrt{1 - \frac{v^2}{c^2}}}$ $\beta = \frac{1}{\sqrt{1 - \frac{0^2}{c^2}}}$ i.e., $\beta$=1/1, the ratio of length or time between the moving system and the reference system is always unity. This means that relativity itself does not affect the physical condition of length or time.

Einstein's STR is the one of the most important and respected theories of modern physics; therefore, all the more reason that the ambiguity that lurks in the premise of the theory should be solved. For this purpose, we need to establish a new law, which admits that the relative speed can be used instead of mobile speed as 'velocity' in the universal equation 'time=distance/velocity'. Had this law been already established, everybody should have learned it in youth because 'time=distance/velocity' is the subject taught in mathematics of high degree of elementary school or junior high school; however, nobody learns it. Therefore, we can declare that the new law has not yet been established. If this issue is neglected and we continue using the Lorentz factor, the STR will become less persuasive someday. This is a purely mathematical issue concerning the human wisdom; thus, we require a coalition of mathematicians to establish the new law.

**Acknowledgments**

I would like to thank Editage and Elsevier for English language editing.

**References**

1. Perrett,W., Jeffery, G.B., On the Electrodynamics of Moving Bodies, English translation of 'Zur Elektrodynamik bewegter Körper', German Das Relativatsprinzip, [4th ed] Tuebner, 1922
2. Einstein A (1905) On the electrodynamics of moving bodies. Annalen der Physik 17: 891-921.

# New Analytical Formulae to Calibrate Well-type Scintillation Detectors Efficiency

**Abbas MI[1*], Ibrahim OA[2], Ibrahim T[1,2] and Sakr M[1,2]**

[1]*Department of Physics, Faculty of Science, Alexandria University, Egypt*
[2]*Department of Physics, Faculty of Science, Beirut Arab University, Lebanon*

**Abstract**

Direct Mathematical calculations for the absolute efficiency of well-type gamma ray scintillation crystals are described. Calculated detection efficiencies are drawn for two commonly used well crystals. Comparisons are made between present works and published Monte Carlo values. The present approach proved quite success in predicting the efficiency of well type detectors providing only the geometry and materials of the system formed of source, detector and shielding as well as the energy of the emitted photons.

**Keywords:** New mathematical method; Gamma-spectroscopy; Absolute efficiency; Well-type detector

## Introduction

Direct mathematical method to calculate the efficiency of radioactive systems was first proposed by Selim and Abbas and then it was used by several authors [1-12]. The comparison between the calculated values of efficiencies and experimental ones provided in these papers shows that the error is very small. This method proves its accuracy and success for many detector geometries; like: coaxial detectors [1-4], parallelepiped detectors [5] and borehole detectors [12]; and types like: semiconductors [4-7] and scintillators [1-3] Besides, the direct mathematical method is built on simple mathematical approach and it does not need long computer programs nor previous experimental calculation of the efficiency of standard source like other Monte Carlo method [13-16]. In this work we used direct mathematical method to predict the efficiency of two well type detectors, in detecting the gamma photons produced by sources placed in their well cavities, over wide range of photon energy for seven different source geometries and positions including point sources with very small dimensions and cylindrical sources. These detectors are scintillators and their active material is Sodium Iodide (NaI). Sodium Iodide is widely used in detecting of gamma photons nowadays because of the high efficiency of these crystals, relatively low cost and availability in different shapes and sizes. In our calculation, the effect of beam attenuation caused by well's walls and bottom is included as well as self-attenuation caused by the material of cylindrical sources.

For each source geometry, a comparison between our calculated value of efficiencies using this approach and reference values obtained from Pomme et al. [15] is given. This comparison shows that our values were in good agreement with Pomme et al. [15], proving the success of our approach in predicting the efficiencies of these detectors over wide range of photon energy.

The arrangement of this paper is as follows: providing a general explanation of the direct mathematical method in section 2, geometric description of well type detector is given in section 3. Detailed explanation of the mathematical perspective used is done. Comparison between calculated and reference values of efficiency is done in seven independent figures. Finally the conclusion is presented.

## Direct Mathematical Method

This method is theoretical based on the following fundamentals: When a photon with certain energy strikes a detector without absorbing walls or windows, the probability for it to interact with this detector and depose energy to be recorded is:

$$f = 1 - e^{-\mu . d} \qquad (1)$$

Where, $\mu$ represents the total attenuation coefficient of the detector active volume, and $d$, is the photon path length traveled through the detector active volume. The factor determining the beam attenuation due the source container and the detector end cap materials is called the attenuation factor, $F_{att}$. So the probability is rewritten as:

$$f_j = F_{att}\left(1 - e^{-\mu . d}\right) \qquad (2)$$

The attenuation factor $F_{att}$, is expressed as:

$$F_{att} = e^{-\sum_i \mu_i \delta_i} \qquad (3)$$

Where, $\mu_i$ is the attenuation coefficient of the $i^{th}$ absorber for a gamma photon with energy $E_\gamma$ and $\delta_i$ is the gamma photon path length through the $i^{th}$ absorber.

## Geometric Description of Well Type Detector

The detector, we calculated its efficiency, is a well type NaI detector of cylindrical shape. This detector is illustrated in Figure 1. The total height of this detector is marked as "T" in the mentioned figure and its total radius "R". This detector contains a coaxial cavity (well) of cylindrical shape also that has a radius r and a height D as shown in Figure 1. A small radioactive source can be placed in the well coaxial cavity. The source might be a small cylinder that can fit in the cavity or an extremely small source (point source). This source contains usually radioactive nuclei that undergo beta or alpha decay, and then emit gamma photons.

***Corresponding author:** Abbas MI, Department of Physics, Faculty of Science, Alexandria University, Egypt, E-mail: mabbas@physicist.net

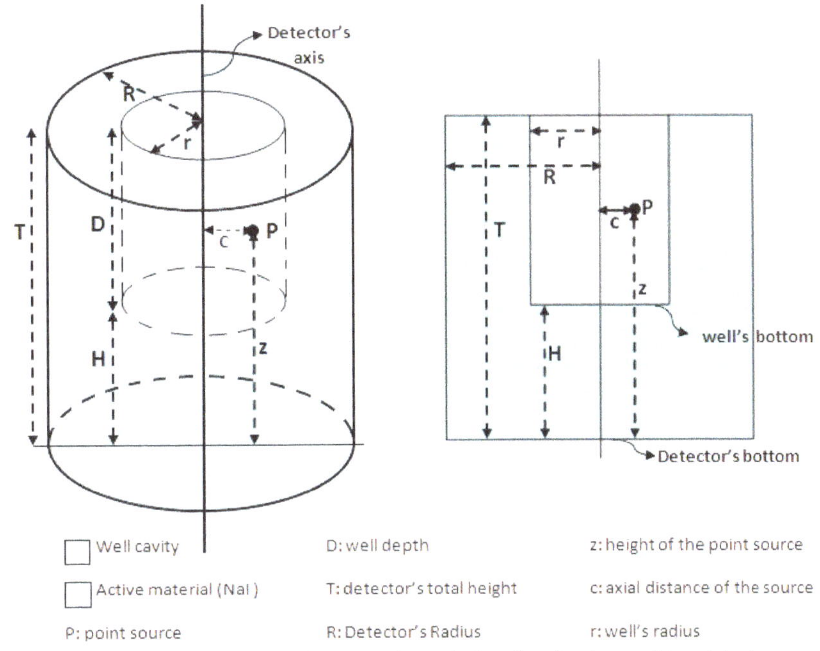

**Figure 1:** Three dimensional sketches and a vertical section view for a well type detector.

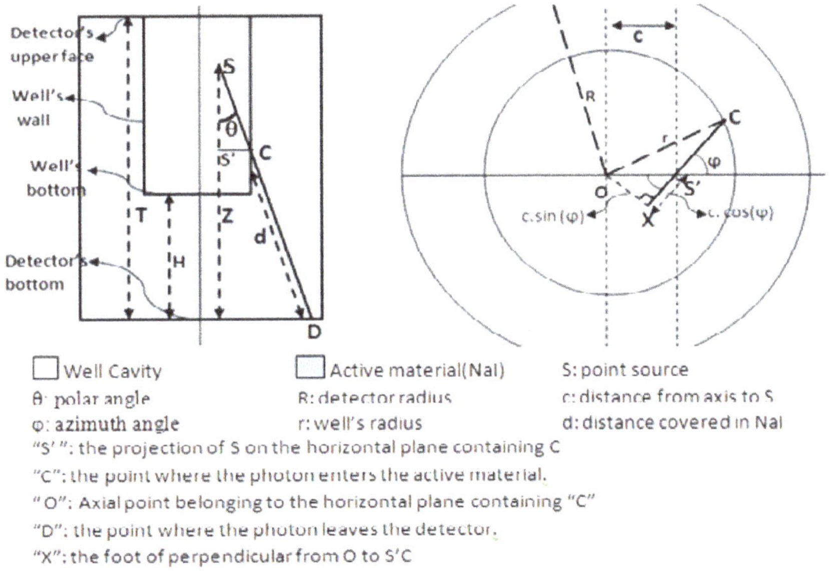

**Figure 2:** Three dimensional schematic diagram of a well type detector (left) and to a horizontal view plane (right) to illustrate the path length function $d_2$.

## Mathematical Perspective

### Path-lengths

There are five different path lengths "faiths" ($d_1$, $d_2$, $d_3$, $d_4$ and $d_5$) can be followed by a photon in our system depending on where the photon enters the active material and leaves it, the photon can enter either from well's bottom or the well's side and leaves either from detector bottom, upper face or outer side.

**If the photon enters the active material from the bottom and leaves from the bottom:** Just by making projection on the axis of the detector, one can derive the function of path length to be:

$$d_1 = \frac{H}{\cos(\theta)} \tag{4}$$

**If the photon enters the active material from well's side and leaves from detector's bottom:** By making projections on a vertical and horizontal planes (Figure 2). We can derive the path length function of the distance covered in the active material (NaI) in terms of the detector and source dimensions shown in Figure 1, as well as the polar and azimuth angle. As shown in Figure 2. If the photon enters from the side of the well cavity and leaves from the detector's bottom, the distance it covered inside the detecting material (NaI) is labeled by CD (d) in the figure, the distance $\overline{CD} = \overline{SD} - \overline{SC}$ where $\overline{SD}$ can be taken from the vertical view (left) to be $z/\cos(\theta)$, and SC is simply $S'C/\sin(\theta)$. The horizontal view to the right of the same figure. Figure 2 shows the triangle OXC right at X, the simple geometry of OXC gives $\overline{CX} = \sqrt{OC^2 - OX^2}$ and $\overline{CS'} = \overline{CX} - \overline{XS'}$; but $\overline{OX} = c.\sin(\varphi)$ and $\overline{XS'} = c.\cos(\varphi)$ and $\overline{OC} = r$. So $CS' = \sqrt{r^2 - (c.\sin(\varphi))^2}$ the path length covered by a photon entering from the well's side and leaving from the detector's bottom will be:

$$d_2 = \frac{z}{\cos(\theta)} - \frac{\pm c.\cos(\varphi) + \sqrt{r^2 - (c.\sin\varphi)^2}}{\sin(\theta)} \quad (5)$$

To discuss the "$\pm$" sign in the equation of path length $d_2$, we consider the projection on a horizontal plane containing the point where the photon enters the active material shown in Figure 3, the photon can enter the active material from the nearer arc shown as dashed arc in the figure or from the farther arc shown as solid arc. If it enters from the nearer arc (left part of Figure 3), the distance will be $\overline{S'A} = \overline{XA} - \overline{XS'}$ and thus $\overline{S'A} = \sqrt{r^2 - (c.\sin(\varphi))^2}$ If the photon enters from the farther arc (right part of the figure), the distance $\overline{S'B}$ in the figure is done by $\overline{S'B} = \overline{XB} + \overline{S'X} = \sqrt{r^2 - (c.\sin(\varphi))^2} + c.\cos(\phi)$

**If the photon enters the active material from well's bottom and leaves from the outer side:** As shown in Figure 4, if the photon enters from the well's bottom and leaves from detector's outer side the distance covered inside the active material is $\overline{AB}$ where $\overline{AB} = \overline{SB} - \overline{SA}$ SA is clearly $f/\cos(\theta)$ and $\overline{SB} = \overline{S'B}/\sin(\theta)$ .To derive S'B with respect to R, and $\varphi$, the horizontal view to the right is done, $S'B = \overline{BX} + \overline{XS'}$ here

$$\overline{BX} = \sqrt{r^2 - (c.\sin(\varphi))^2} \text{ and } \overline{XS'} = c.\cos(\varphi)$$

So $\overline{SB} = \dfrac{-c.\cos(\varphi) + \sqrt{R^2 - (c.\sin\varphi)^2}}{\sin\theta}$ and the path length

function "$d_3$" in this case will be:

$$d_3 = \frac{\pm c.\cos(\varphi) + \sqrt{R^2 - (c.\sin\varphi)^2}}{\sin\theta} - \frac{f}{\cos\theta} \quad (6)$$

**If the photon enters the active material from the well's wall and leaves from the outer wall:** If the photon enters from the well's wall, the total distance it covers inside the well cavity is $\dfrac{\pm c.\cos(\varphi) + \sqrt{r^2 - (c.\sin\varphi)^2}}{\sin\theta}$ (derived previously to be used in equation 5). While if it leaves from the detector's outer side, the distance it covers in the entire system (active material + well) is $\dfrac{\pm c.\cos(\varphi) + \sqrt{r^2 - (c.\sin\varphi)^2}}{\sin\theta}$ (derived previously to be used in equation 6). So the distance covered in the active material will be:

$$d_4 = \frac{\pm c.\cos(\varphi) + \sqrt{R^2 - (c.\sin\varphi)^2}}{\sin\theta} - \frac{\pm c.\cos(\varphi) + \sqrt{r^2 - (c.\sin\varphi)^2}}{\sin\theta} \quad (7)$$

**If the photon enters the active material from the inner side and leaves from the upper surface:** To derive the path length function $d_5$; Figure 5 shows three dimensional diagram for a photon entering from well's wall and leaving from detector's upper surface the total distance the photon covers in the entire system(well and active material) labeled as SF in the figure is $\dfrac{T - z}{\cos(\pi - \theta)}$ while the distance it covers in the well is $\dfrac{\overline{S'E}}{\sin(\pi - \theta)}$ S'E can be derived using the horizontal view in the same figure $\overline{S'E} = \overline{S'H} + \overline{H'E}$ So $\overline{S'E} = c.\cos(\varphi) + \sqrt{r^2 - (c.\sin\varphi)^2}$

$$= \frac{T}{\cos(\pi - \theta)} \frac{z}{} - \frac{\pm c.\cos(\varphi) + \sqrt{r^2 - (c.\sin\varphi)^2}}{\sin(\pi - \theta)} \quad (8)$$

If photon enters from the nearer arc          If photon enters from the farther arc

| OA: Radius of well (r) | $\varphi$: azimuth Angle | Nearer arc |
|---|---|---|
| OB: Radius of well (r) | S': projection of S (source) | Farther arc |
| $\overline{OX}$: $c.\sin(\varphi)$ | $\overline{S'X}$: $c.\cos(\varphi)$ | $\overline{S'Y}$: $c.\cos(\varphi)$ |
| $\overline{OY}$: $c.\sin(\varphi)$ | $\overline{XA} = \sqrt{r^2 - (c.\sin(\varphi))^2}$ | $\overline{YB} = \sqrt{r^2 - (c.\sin(\varphi))^2}$ |
| $\overline{S'A} = \sqrt{r^2 - (c.\sin(\varphi))^2} - c.\cos(\varphi)$ | | $\overline{S'B} = \sqrt{r^2 - (c.\sin(\varphi))^2} + c.\cos(\varphi)$ |

"c": distance between detector' axis and point source
"A": point where the photon enters the material from the nearer arc,
"B": point where the photon enters the material from the farther arc.
"O": Axial point belongs to the horizontal plane containing A or B.
"X": foot of perpendicular from O to S'A or S'B.

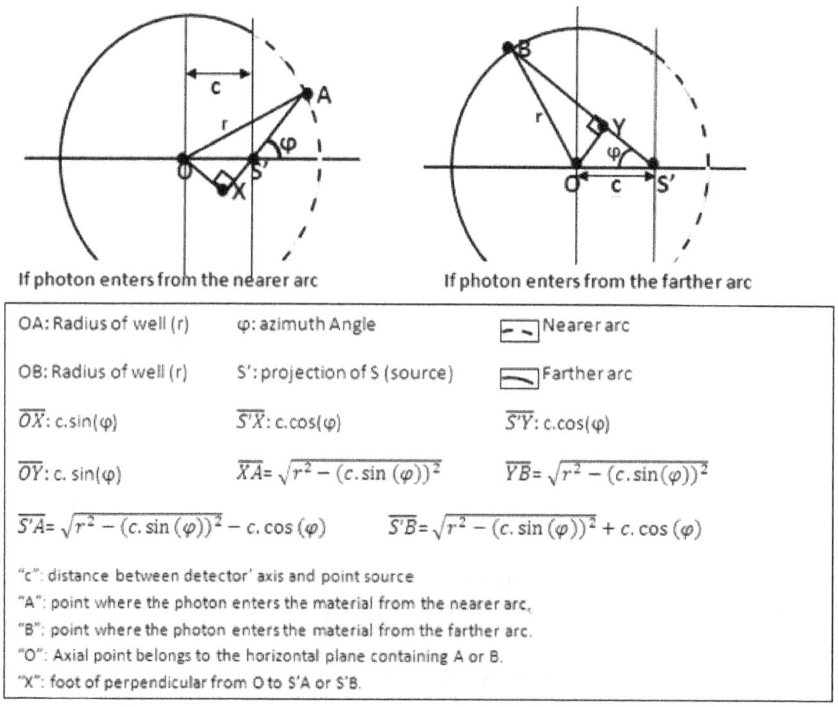

**Figure 3:** Horizontal View made to discuss the "$\pm$" sign in the equation of the path length: $d_2$.

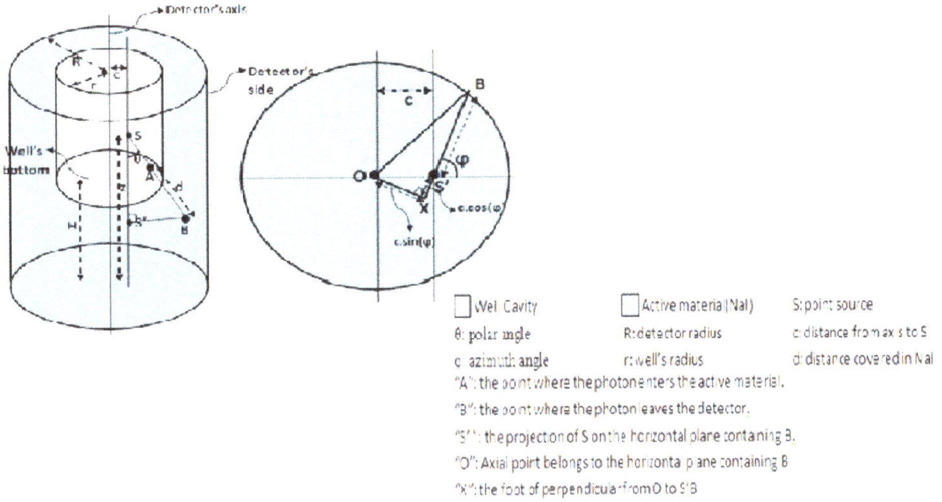


**Figure 4:** Three dimensional sketch and horizontal view made to derive the path length function: $d_3$.

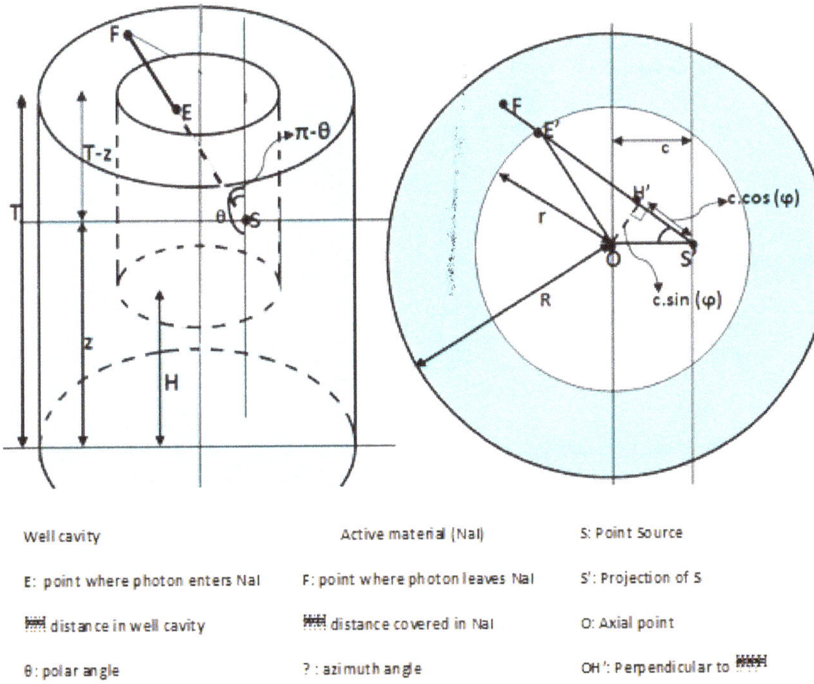


**Figure 5:** Three dimensional diagrams (left) and a projection to the plane containing the point E where the photon enters the active material (NaI).

Note: for an axial point source all items multiplied by "c" in the previous equations tend to zero.

## Total efficiency of an axial point source

By definition, the efficiency of a source - detector system is the number of photons detected divided by the number of photons emitted. So, taking the total solid angle to be 4π, and using equation (2) we obtain:

$$\varepsilon = \frac{1}{4\pi}\int_0^{\pi}\int_0^{2\pi} f_j \sin(\theta)\,d\varphi\,d\theta \qquad (9)$$

Where, $f_j = F_{att}\left(1-e^{-\mu.dj}\right)$ ($j \to 1-5$) $d_j$ can be one of the five functions described before in section 4.1. So to develop equation 9 we just want to replace $d_j$ ($j \to 1-5$) by its function and we integrate over the entire solid angle taking into account that it might depend on both φ and θ. But in the case of an axial point source, there is symmetry on the azimuth angle φ. We have to distinguish between two cases

**If the axial point source is located at a height,** $z > \frac{\mathbf{H}}{\mathbf{R-r}}.\mathbf{R}$ (Figure 6, right):

$$\varepsilon = \frac{1}{4\pi}\left(\int_0^{\theta_1}\int_0^{2\pi} f_1\,d\varphi\,d\theta + \int_{\theta_1}^{\theta_2}\int_0^{2\pi} f_3\,d\varphi\,d\theta + \int_{\theta_2}^{\theta_3}\int_0^{2\pi} f_4\,d\varphi\,d\theta + \int_{\theta_3}^{\theta_4}\int_0^{2\pi} f_5\,d\varphi\,d\theta\right) \qquad (10)$$

$$\theta_1 = \tan^{-1}\left(\frac{R}{z}\right);\quad \theta_2 = \tan^{-1}\left(\frac{r}{z-h}\right)\quad;\quad \theta_3 = \pi - \tan^{-1}\left(\frac{R}{H-z}\right);$$

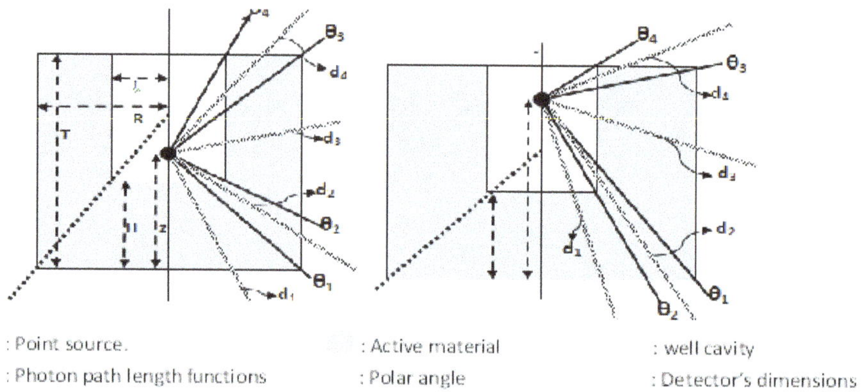


**Figure 6:** Vertical section view of the detector with an axial point source in two cases: Case 1 shows a lower point source below with z<H/(R-r).R (left). Case 2: higher point source with z>H/(R-r).R (to the right). Note: z, R, H and r were defined in Figure 1.

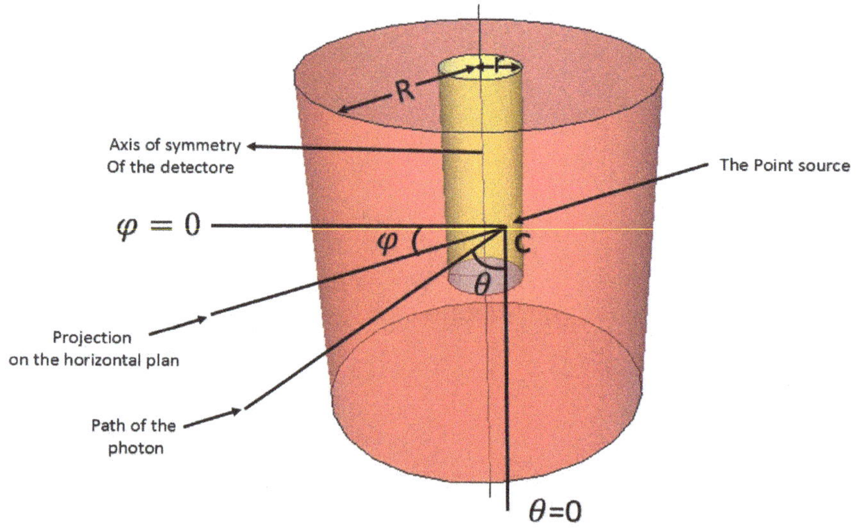


**Figure 7a:** Shows a 3D- drawing showing the references of the polar and azimuthal angles.

$$\theta_4 = \pi - tan^{-1}(\frac{r}{H-z}) \quad (11)$$

If the axial point source is located at a height, $z < \frac{\mathbf{H}}{\mathbf{R-r}}.\mathbf{R}$ (refer to Figure 6, left): applying the equation (2) gives:

$$\varepsilon = \frac{1}{4\pi}\left(\int_0^{\theta_2}\int_0^{2\pi} f_1 d\varphi d\theta + \int_{\theta_2}^{\theta_1}\int_0^{2\pi} f_2 d\varphi d\theta + \int_{\theta_1}^{\theta_3}\int_0^{2\pi} f_4 d\varphi d\theta + \int_{\theta_3}^{\theta_4}\int_0^{2\pi} f_5 d\varphi d\theta\right) \quad (12)$$

## Total efficiency of a non-axial point source

For a non-axial point source one can calculate the total efficiency by applying equation 9, but first we have to take references for the azimuth angle "φ" as well as for the polar angle "θ". Since each point should belong to certain diameter we take the diameter containing the point-source as the reference of azimuthal angle "φ=0" in order to obtain certain symmetry in the Figure 7a. As well as we take the line parallel to the detector axis passing by the point source is the reference of polar angle "θ=0". The plane passing by the point source and containing the axis of the detector as well as the horizontal line having φ=0; can be considered as plane of reference in our case because it cuts the detector into two identical parts. We divide the detector into four zones as indicated in Figure 7b. The aim of this technique is to make the equations easier. The two regions to the right are considered closer to the source, so the signs (±) in the equations (6 – 8) will turn to (–). Figure 7a shows a 3D- drawing showing the references of the polar and azimuthal angles.

**Zone A, the point is closer to the detector axis and $0 < \theta < \pi/2$:** To derive the mathematical equations, we will know first the photon path lengths allowed in this zone. Here the photon can enter either from the well cavity bottom or side and leave either from the outer side or bottom of detector. Thus, four path lengths will be allowed in this case which are: $d_1$, $d_2$, $d_3$ and $d_4$ (Tables 1 and 2). In this section we will study the different faiths when the photon enters the detector, as well as the different faiths when this photon leaves the detector, then we combine them to find different faiths of photon path length in the detector active volume. We will do the same strategy for the other zones. First we will study how the photon leaves the detector, we will make a projection on a horizontal plane containing the bottom of the detector so, as illustrated in Figure 8, we obtain:

- For $\theta < \theta_1$; the photon must leave from the bottom.
- For $\theta_1 < \theta < \theta_2$; the photon can leave from the side up to certain azimuth angle $\varphi_a$, while it leaves from the bottom when $\varphi > \varphi_a$.

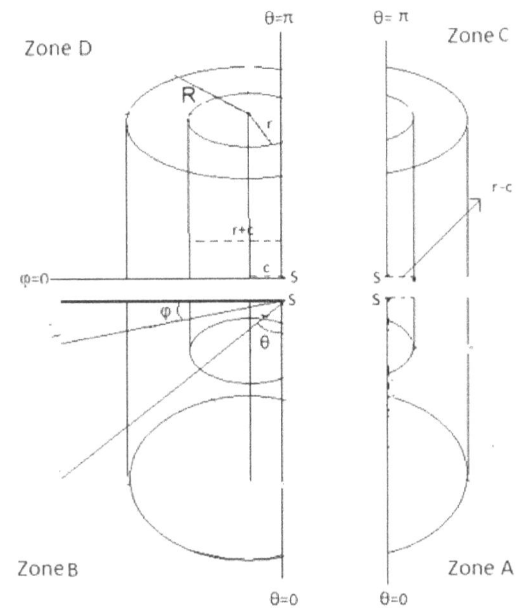


**Figure 7b:** Shows the decomposition of our system into four different zones.

| The proton leaves from / The proton enters from | $\theta < \theta_1$ bottom | $\theta_1 < \theta < \theta_2$ $\varphi < \varphi_a$ side | $\theta_1 < \theta < \theta_2$ $\varphi > \varphi_a$ bottom | $\theta > \theta_2$ side |
|---|---|---|---|---|
| $\theta < \theta_3$ the well cavity bottom | $d_1$ | $d_3$ | $d_1$ | $d_3$ |
| $\theta_3 < \theta < \theta_4$ and $\varphi < \varphi_a'$ the well cavity side | $d_2$ | $d_4$ | $d_2$ | $d_4$ |
| $\theta_3 < \theta < \theta_4$ and $\varphi > \varphi_a'$ the well cavity bottom | $d_1$ | $d_3$ | $d_1$ | $d_3$ |
| **$\theta > \theta_4$ the well cavity side** | **$d_2$** | **$d_4$** | **$d_2$** | **$d_4$** |

**Table 1:** The path length functions allowed for a photon striking part A of the detector. So, in general the photon path length can follow one of the path lengths d1, d2, d3 and d4. Table 1 summarizes the faiths that can be followed by the photon.

| The proton leaves from / The proton enters from | $\theta < \theta_2$ bottom | $\theta_2 < \theta < \theta_6$ $\varphi < \varphi_b$ bottom | $\theta_2 < \theta < \theta_6$ $\varphi > \varphi_b$ side | $\theta > \theta_2$ side |
|---|---|---|---|---|
| $\theta < \theta_4$ the well cavity bottom | $d_1$ | $d_1$ | $d_3$ | $d_3$ |
| $\theta_4 < \theta < \theta_5$ and $\varphi < \varphi_a'$ the well cavity bottom | $d_1$ | $d_1$ | $d_3$ | $d_3$ |
| $\theta_4 < \theta < \theta_5$ and $\varphi > \varphi_a'$ the well cavity side | $d_2$ | $d_2$ | $d_4$ | $d_4$ |
| $\theta_5 < \theta$ the well cavity side | $d_2$ | $d_2$ | $d_4$ | $d_4$ |

**Table 2:** The path length Functions allowed for a photon striking part B of the detector. So, in general the photon path length can follow one of the path lengths d1, d2, d3 and d4. Table 2 summarizes the faiths that can be followed by the photon.

- For $\theta > \theta_2$; the photon must leave from the side.

Also, Figure 8 allows us to derive the equations of $\theta_1$ and $\theta_2$:

$$\theta_1 = \tan^{-1}\left(\frac{R-c}{z}\right) \tag{13}$$

$$\theta_2 = \tan^{-1}\frac{\sqrt{R^2-c^2}}{z} \tag{14}$$

Also from Figure 8, it is easy to derive the equation of $\varphi_a$ by applying the cosine law in triangle SOC; we have: $(SO)^2 + (SC)^2 + 2\,(SO).(SC).\cos(\varphi_a) = (OC)^2$, which turns after substitution into: $c^2 + z^2.\tan^2(\theta) + 2.c.z.\tan(\theta).\cos(\varphi_a) = R^2$, thus $\varphi_a$ can be done by:

$$\varphi_a = \cos^{-1}\left(\frac{R^2 - c^2 - z^2.\tan\theta^2}{2.c.z.\tan\theta}\right) \tag{15}$$

Now to study how the photon can enter the detector we will make projection on the plane containing the bottom of the well, the projection of the polar angle "θ" on a horizontal plane is a circle with

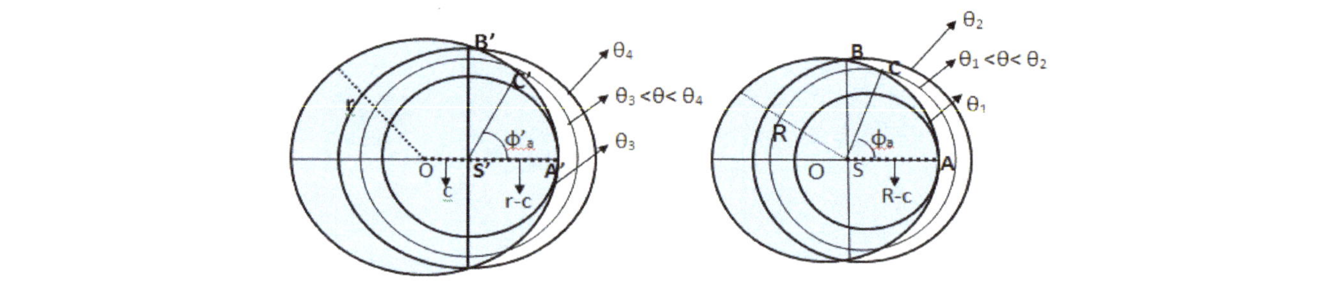


**Figure 8:** Projection on the horizontal plane containing the well's bottom (left) to show the projection of angles θ3 and θ4 and azimuth angle φa', and a projection on the detector's bottom (right) to show the projection of angles θ1 and θ2 and azimuth angle φa.

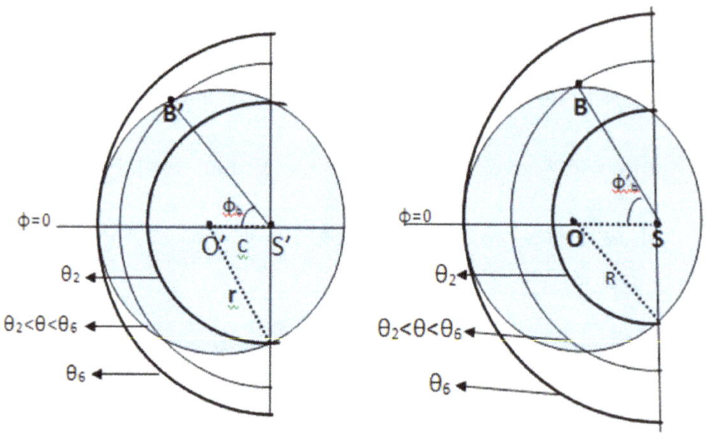


**Figure 9:** Projection on two horizontal planes (the plane containing the well's bottom to the left and the one containing the detector's bottom to the right) to discuss the path length functions followed by a photon crossing zone B of the detector.

radius [f x tan(θ)], where f designates the height of the source above the plane of projection. As shown in Figure 8; for $\theta < \theta_3$ the photon must enter from the bottom of well. But for $\theta_3 < \theta < \theta_4$, the photon can enter from the side if it has an azimuthal angle $\varphi < \varphi_b$ For $\theta > \theta_4$; the photon must enter from the side. Also from Figure 8 we can derive easily the functions of $\theta_3$ and $\theta_4$, it is clear: $\sin(\theta_4) = (\sqrt{(r^2-c^2)}/$ hypotenuse) while $\cos(\theta_4)$ = (f/ hypotenuse), so

$$\theta_4 = \tan^{-1}(\frac{\sqrt{r^2-c^2}}{f}) \quad (16)$$

and, in a similar manner:

$$\theta_3 = \tan^{-1}(\frac{r-c}{f}) \quad (17)$$

In addition we can use Figure 8 to derive the equation of $\varphi_a^{'}$ by applying the cosine law in triangle S'O'C'; we have: $(S'O')^2 + (O'C')^2 + 2.(S'O').(O'C').\cos(\varphi_a^{'}) = (S'C')^2$, which turns after substitution into: $c^2 + f^2.\tan^2(\theta) + 2.c.f.\tan(\theta).\cos(\varphi_a^{'}) = r^2$, thus $\varphi_a^{'}$ can be done by:

$$\varphi_a^{'} = \cos^{-1}(\frac{r^2-c^2-h^2.\tan(\theta)^2}{2.c.f.\tan(\theta)}) \quad (18)$$

**Zone B, the further zone with $0 < \theta < \pi/2$:** Just like we studied in zone A, the path length functions followed by the photon in the detector if it strikes zone B will depend on two things only: how it enters and how it leaves the detector material. The photon can enter from either the bottom or the side of the well cavity; we should study the different faiths depending on polar and azimuth angles. Starting by studying the faiths of entrance, we have to make projection on a plane containing the bottom of well cavity, thus according to Figure 9(left), we obtain:

For $\theta < \theta_4$; the photon must enter from the bottom, whatever the azimuth angle is $\theta_4 = \tan^{-1}\frac{\sqrt{r^2-c^2}}{h}$ where (as identified before in equation 16)

For $\theta_4 < \theta < \theta_5$; the photon enters the detector from the bottom if it has a small azimuthal angle $\varphi < \varphi_b$ or from the side if it has a large azimuthal angle $\varphi > \varphi_b^{'}$ where:

$$\theta_5 = \tan^{-1}\left(\frac{r+c}{f}\right) \quad (19)$$

For $\theta > \theta_5$; the photon must enter from well's side. While, if we study the faiths the photon will leave the detector according to the polar and azimuthal. For $\theta < \theta_2$; the photon must leave from the bottom whatever the azimuth angle is; where $\theta_2 = \tan^{-1}\frac{\sqrt{R^2-c^2}}{z}$ (as identified before in equation 14);

For $\theta_2 < \theta < \theta_6$; the photon might leave the detector from the bottom if it has a small azimuthal angle ($\varphi < \varphi_b$) or from the side if it has a large azimuthal angle ($\varphi > \varphi_b$), where

$$\theta_6 = \tan^{-1}(\frac{R+c}{z}) \quad (20)$$

- $\varphi_b$ is done in equation 21
- For $\theta > \theta_6$; the photon must leave from the side.

Here, because we are dealing with the bigger part of the detector, so the sign (±) in the equations (5 – 7) of $d_2$, $d_3$ and $d_4$ will turn to (+). By looking at Figure 9, we take the triangle *BSO* and applying the cosine law in simple mathematics. We have:

*(OB)*$^2$ = *(OS)*$^2$ + *(SB)*$^2$ +2.*(SO).(SB).cos*($\varphi_b$), but *(OS)* = *c*, *(OB)* = *r* and *(SB)* = *h tan*($\theta$), so we obtain:

$$\varphi_b = \cos^{-1}\left(\frac{c^2 + z^2.\tan\theta^2 - R_{,}^2}{2.c.z.\tan\theta}\right) \quad (21)$$

Now, it is easy to derive the equation of $\varphi_b'$, just by applying the same method and taking the triangle *S'O'B'*, we obtain

$$\varphi_b' \cos^{-1}\left(\frac{c^2 + f^2.\tan\theta^2 - r^2}{2.c.f.\tan\theta}\right) \quad (22)$$

**Zone C: The part of the detector close to the source with** $\frac{\pi}{2} < \theta < \pi$ **:** The particle must enter from the side, while it can leave either from the upper face or the outer side so two path lengths might take place here: $d_4$ and $d_5$. This depends on the polar and azimuth angles certainly, to study the path followed by the photon according to these angles we can make projection on a horizontal plane but now the horizontal plane will be the one containing the upper face of the detector. Here, unlike zones A and B, we need projection on one single plane. The projection is shown in Figure 10. By taking the triangle *OSY*; *OS* = *c*, *OY* = *r* and *SY* = *(T-z).tan(π-θ)*, we can derive the function of $\varphi_c$ and by taking the triangle OSX we can derive the equation of $\ddot{o}_c'$.

If $\theta < (\pi-\theta_7)$, the photon must leave from the side; thus it is going to follow the path length $d_4$, where:

$$\theta_7 = \tan^{-1}\left(\frac{\sqrt{R^2 - c^2}}{T - z}\right) \quad (23)$$

If $(\pi-\theta_7) < \theta < (\pi-\theta_8)$, the photon can leave either from upper face if its azimuthal angle is less then certain value we can call it $\varphi_c$ or from the side if it has an azimuthal angle $\varphi > \varphi_c$, where:

$$\varphi_8 = \tan^{-1}\left(\frac{R - c}{T - z}\right) \quad (24)$$

Also, $\varphi_c$ can be deduced from Figure 10, and in a way similar to $\varphi_a$:

$$\varphi_c = \cos^{-1}\left(\frac{R^2 - c^2 - (T-z)^2.(\tan(\pi-\theta))^2}{2.c.(T-z).\tan(\pi-\theta)}\right) \quad (25)$$

If $(\pi-\theta_8) < \theta < (\pi-\theta_9)$, the photon must leave from the upper face, where:

$$\theta_9 = \tan^{-1}\left(\frac{\sqrt{r^2 - c^2}}{T - z}\right) \quad (26)$$

If $(\pi-\theta_9) < \theta < (\pi-\theta_{10})$, the photon can either leave from the upper face if its azimuthal angle is less than certain value we call it $\varphi_c'$ or escape without hitting the active material of the detector for $\varphi > \varphi_c'$ where:

$$\theta_{10} = \tan^{-1}\left(\frac{r - c}{T - z}\right) \quad (27)$$

$$\varphi_c' = \cos^{-1}\left(\frac{r^2 - c^2 - (T-z)^2.(\tan(\pi-\theta))^2}{2.c.(T-z).\tan(\pi-\theta)}\right) \quad (28)$$

If $\theta > (\pi-\theta_{10})$, the photon leaves the detector without hitting its active material.

**Zone D: The farther part with θ > π/2:** Also for the farther part with polar angle $\theta > \frac{\pi}{2}$; the photon must enter from the side while it can leave either from side or upper face. We should study whether the photon will leave from the upper face or from the side by making a projection on the plane containing the upper face of detector just like we did in the previous part. The projection is shown in Figure 11.

- If $\theta < (\pi-\theta_{11})$, the photon must leave from the side and the only path way followed is $d_4$, where:

$$\theta_{11} = \tan^{-1}\left(\frac{R + c}{T - z}\right) \quad (29)$$

- If $(\pi-\theta_{11}) < \theta < (\pi-\theta_7)$, the photon can leave either from side or upper face depending on the azimuthal angle. It leaves from upper face for azimuthal angle $\varphi < \varphi_d$ and it leaves from side if $\varphi > \varphi_d$.

- If $(\pi-\theta_7) < \theta < (\pi-\theta_{12})$, the photon must leave from the upper face whatever the azimuthal angle was, where:

$$\theta_{12} = \tan^{-1}\left(\frac{r + c}{T - z}\right) \quad (30)$$

If $(\pi-\theta_{12}) < \theta < (\pi-\theta_9)$, the photon can escape without striking the material of detector for $\varphi < \varphi_d'$ or strike it entering from the side and leaving from the face if $\varphi = \varphi_d'$.

- For $\theta > (\pi-\theta_9)$, the photon cannot strike the detector.

Also, Figure 11 helps to derive the equations of the new azimuthal angles: $\varphi_d$ and $\phi_d$ in a manner very similar to $\phi_c$ and $\varphi_c'$.

$$\cos(\varphi_d') = \frac{(T-z)^2 .tan(\theta)^2 + c^2 - r^2}{(T-z).\tan\theta .c} \quad (31)$$

$$(\varphi_d) = \frac{(T-z)^2 .tan(\theta)^2 + c^2 - R^2}{(T-z).\tan\theta .c} \quad (32)$$

Thus the efficiency of the detector for a non-axial point source will be done by:

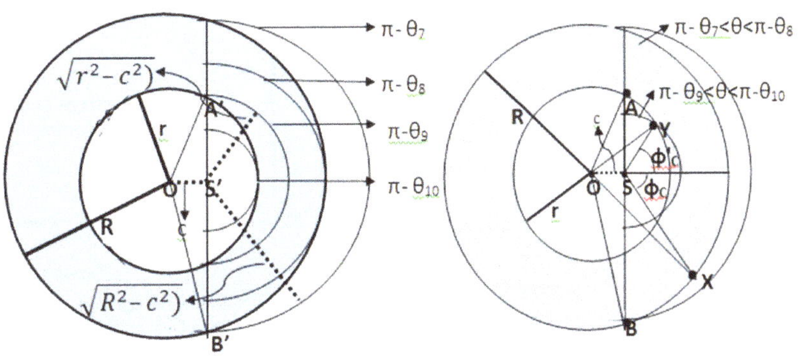


**Figure 10:** Projection to a horizontal plane containing the surface of detector the right figure can be used to determine the equations of $\varphi_c$ and $\varphi_c'$.

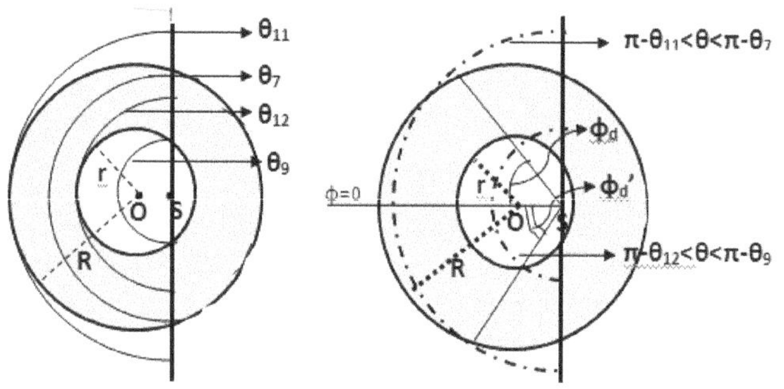


**Figure 11:** Projection on the upper plane of the detector.

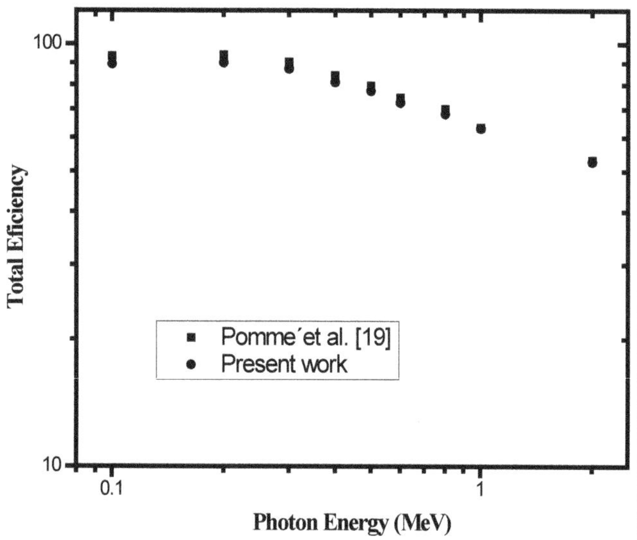


**Figure 12:** Shows the present work and Monte Carlo [19] values for an axial point source positioned at height z=8.55 cm, from the well bottom of detector IRMM, without source-walls.

$$\varepsilon_p = \frac{1}{2\pi}\Big\{\int_{\theta_0}^{\theta_1}\int_{\varphi_a'}^{\frac{\pi}{2}} f_1 d\varphi d\theta + \int_{\theta_1}^{\theta_3}\int_{0}^{\varphi_a} f_3 d\varphi d\theta + \int_{\theta_1}^{\theta_3}\int_{\varphi_a}^{\frac{\pi}{2}} f_1 d\varphi d\theta + \int_{\theta_3}^{\theta_4}\int_{0}^{\varphi_a'} f_4 d\varphi d\theta + \int_{\theta_3}^{\theta_2}\int_{\varphi_a'}^{\varphi_a} f_3 d\varphi d\theta + \int_{\theta_3}^{\theta_2}\int_{\varphi_a}^{\frac{\pi}{2}} f_1 d\varphi d\theta + \int_{\theta_2}^{\theta_4}\int_{0}^{\varphi_a'} f_4 d\varphi d\theta + \int^{\theta_4}\int^{\frac{\delta}{2}} f_3 d\varphi d +$$
$$\int_{\theta_4}^{\frac{\pi}{2}}\int_{\varphi_a'}^{\varphi_a} f_4 d\varphi d\theta + \int_{0}^{\theta_2}\int_{0}^{\frac{\pi}{2}} f_1 d\varphi d\theta + \int_{\theta_2}^{\theta_6}\int_{0}^{\varphi_b} f_1 d\varphi d\theta + \int_{\theta_2}^{\theta_6}\int_{\varphi_b}^{\frac{\pi}{2}} f_3 d\varphi d\theta + \int_{\theta_6}^{\theta_4}\int_{0}^{\frac{\pi}{2}} f_3 d\varphi d\theta + \int_{\theta_4}^{\theta_5}\int_{0}^{\varphi_b'} f_3 d\varphi d\theta + \int_{\theta_4}^{\theta_5}\int_{\varphi_b'}^{\frac{\pi}{2}} f_4 d\varphi d\theta + \int_{\theta_5}^{\frac{\pi}{2}}\int_{0}^{\frac{\pi}{2}} f_4 d\varphi d\theta$$
$$\int_{\frac{\pi}{2}}^{\pi-\theta_7}\int_{0}^{\frac{\pi}{2}} f_4 d\varphi d\theta + \int_{\pi-\theta_7}^{\pi-\theta_8}\int_{0}^{\varphi_c} f_4 d\varphi d\theta + \int_{\pi-\theta_7}^{\pi-\theta_8}\int_{\varphi_c}^{\frac{\pi}{2}} f_5 d\varphi d\theta + \int_{\pi-\theta_8}^{\pi-\theta_9}\int_{0}^{\frac{\pi}{2}} f_5 d\varphi d\theta + \int_{\pi-\theta_9}^{\pi-\theta_{10}}\int_{0}^{\varphi_c'} f_5 d\varphi d\theta + \int_{\theta_4}^{\frac{\delta}{2}}\int_{\varphi_b'}^{\varphi_b'} f_4 d\varphi d\theta + \quad (33)$$
$$\int_{\frac{\pi}{2}}^{\pi-\theta_{11}}\int_{0}^{\frac{\pi}{2}} f_4 d\varphi d\theta + \int_{\pi-\theta_{11}}^{\pi-\theta_7}\int_{0}^{\varphi_d} f_5 d\varphi d\theta + \int_{\pi-\theta_{11}}^{\pi-\theta_7}\int_{\varphi_d}^{\frac{\pi}{2}} f_4 d\varphi d\theta + \int_{\pi-\theta_7}^{\pi-\theta_{12}}\int_{0}^{\frac{\pi}{2}} f_5 d\varphi d\theta + \int_{\pi-\theta_{12}}^{\pi-\theta_9}\int_{\varphi_d'}^{\frac{\pi}{2}} f_5 d\varphi d\theta$$

Where, $f$ is as identified before in equation 2. The previous equation can be used for $\theta_3 < \theta_2$ and $\theta_6 < \theta_4$.

## Absolute Efficiency for Disk and Cylindrical Sources

The efficiency of a circular disk source of radius $S$ can be done by:

$$\varepsilon_d = \frac{2}{S^2}\int_0^S \varepsilon_P . c\, dc \quad (34)$$

Where $\varepsilon_p$ is the efficiency of a non-axial point source, as identified before in equation (33).In the case of cylindrical sources, we can calculate the efficiency by making an integration of $\varepsilon_p$ over $dc$ and $d_h$ and taking into consideration the efficiencies of many point sources on the same shell are equals. So for a cylinder of radius $S$ and height $H$ we have:

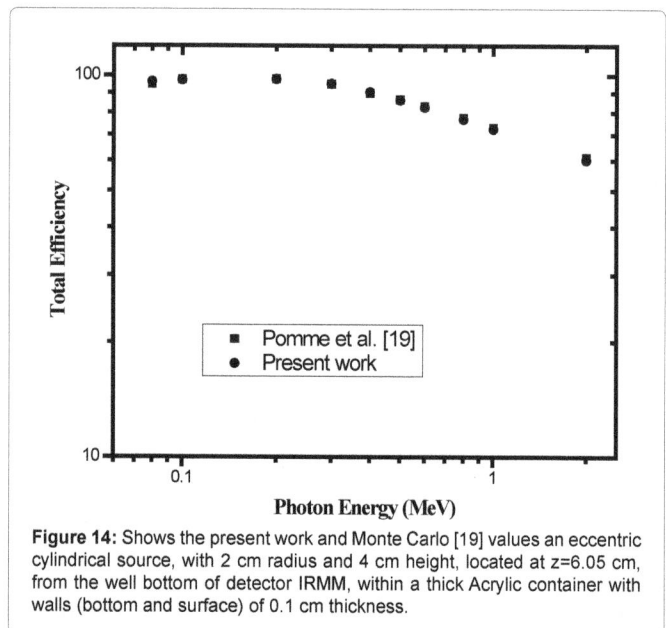

$$\varepsilon_{cyl} = \frac{2}{S^2 \times H}\int_0^H\int_0^S \varepsilon_P . c\, dc\, dh \quad (35)$$

## Results and Comparisons

In this study we used two well type detectors [15], (1) IRMM is used in Figures 12-17 and (2) CIEMAT is used in Figure 18. The dimensions of the two detectors are given in Table 3. The following figures illustrate the comparison between the theoretical calculated values (present work) and the published Monte Carlo values for different sources. The percentage deviation is given by:

$$\Delta = \frac{\varepsilon_{expt.} - \varepsilon_{theo.}}{\varepsilon_{expt.}} \times 100\% \quad (36)$$

where, $\varepsilon_{theo.}$ and $\varepsilon_{expt.}$ are the theoretical (present work) and the Monte Carlo efficiencies, respectively [17-21]

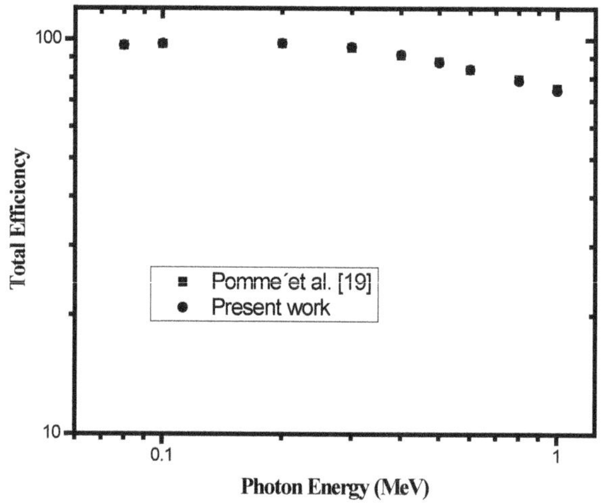


**Figure 13:** Shows the present work and Monte Carlo [19] values for a non-axial point source positioned at height z=5.66 cm, from the well bottom of detector IRMM, without source-walls.

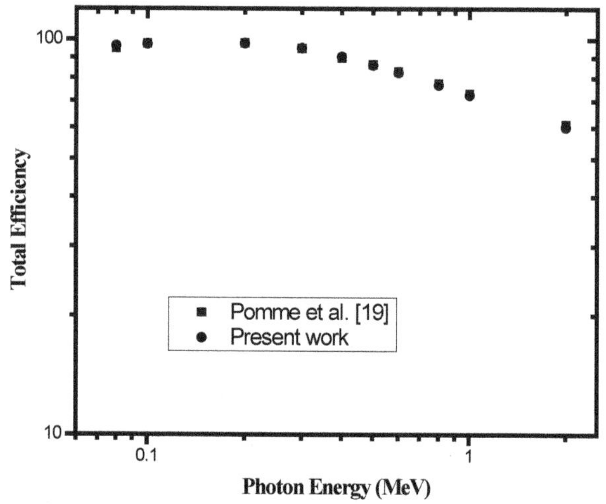


**Figure 14:** Shows the present work and Monte Carlo [19] values an eccentric cylindrical source, with 2 cm radius and 4 cm height, located at z=6.05 cm, from the well bottom of detector IRMM, within a thick Acrylic container with walls (bottom and surface) of 0.1 cm thickness.

## Conclusions

The present approach offers straightforward mathematical equations to calibrate well-type NaI(Tl) scintillation detector over a wide energy range. This approach is based on simple mathematical integrals and computer programs to calculate the detection efficiency. It does not need neither optimization of the detector parameters or previous knowledge of a standard source values. The comparison between calculated and reference values shows that the difference is very small (never escaped 8% and reduced to less than 2% in most cases).

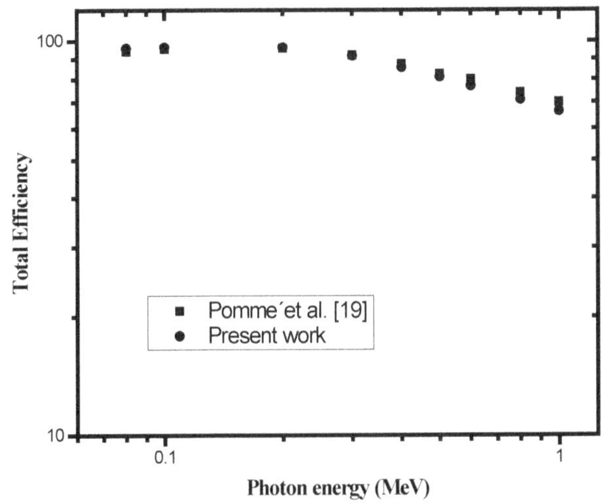


**Figure 15:** Shows the present work and Monte Carlo [19] values for a co-axial cylindrical source, with 2 cm radius and 4 cm height, and its bottom is located at a height z=6.05 cm, from the well bottom of detector IRMM, within a thick Acrylic container with walls (bottom and surface) of 0.1 cm thickness.

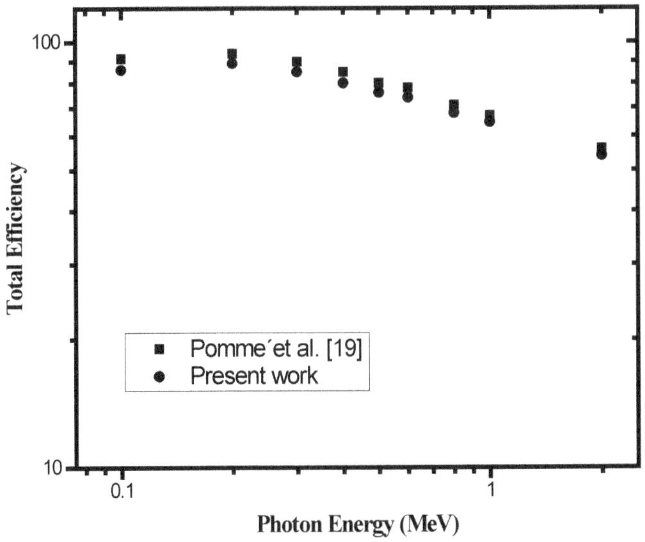


**Figure 16:** Shows the present work and Monte Carlo [19] values for an axial point source located at a height z=5.55 cm, from the well bottom of detector IRMM, within a thick Acrylic container with walls (bottom and surface) of 2.5 cm thickness.

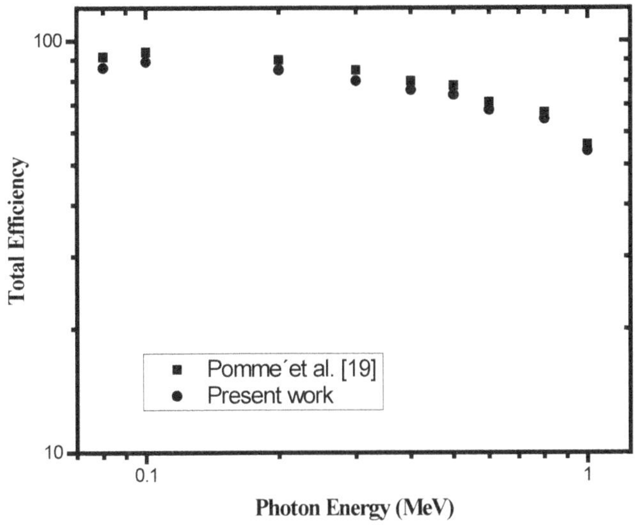


**Figure 17:** Shows the present work and Monte Carlo [19] values for an axial point source located at a height z=7.85 cm, from the well bottom of detector IRMM, within an Aluminum container with walls of 1 cm thickness and its bottom and surface of thickness 0.2 cm.

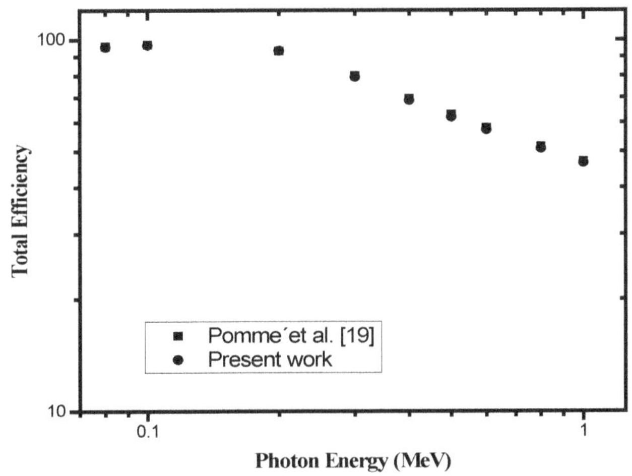


**Figure 18:** Shows the present work and Monte Carlo [19] values for a co-axial disk source, with radius=0.15 cm, located at a height z=2.61 cm, from the well bottom of detector CIEMAT, within a very thin Acrylic container.

| | IRMM | CEIMAT |
|---|---|---|
| The detector radius (R) | 7.62 | 3.81 |
| The well radius (r) | 2.62 | 1.45 |
| The entire height (T) | 15.23 | 7.62 |
| Bottom to well distance (H) | 5.43 | 2.54 |
| Well's cover thickness Made of Aluminum and Aluminum Oxide | 0.12 | 0.06 |
| Well's bottom | 0.12 | 0.06 |

**Table 3**: Shows the dimensions (are in cm) of the two well-type detectors.

## References

1. Selim YS, Abbas MI (1996) Direct Calculation of the total efficiency of cylindrical scintillation detectors for extended circular sources. Radiation physics and chemistry 48: 23-27.
2. Selim YS, Abbas MI, Fawzy MA (1998) Analytical calculation of the efficiencies of gamma scintillators efficiencies. I: total efficiency for coaxial disk sources. Radiat Phys Chem 53: 589-592.
3. Selim YS, Abbas MI (2000) Analytical calculations of gamma scintillators efficiencies. II: total efficiency for wide co-axial disk sources. Radiat Phys Chem 53: 15-19.
4. Abbas MI (2001) HPGe detector photopeak efficiency calculation including self-absorption and coincidence corrections for Marinilli beaker sources using compact analytical expressions. Appl Radiat Isoto 54: 761-768.
5. Abbas MI (2001) A direct mathematical method to calculate the efficiencies of a parallelepiped detector for an arbitrarily positioned point source. Radiat Phys Chem 60: 03-09.
6. Abbas MI (2001)Analytical formulae for well-type NaI(Tl) and HPGe detectors efficiency computation. Appl Radiat Isot 55: 245-252.
7. Abbas MI Selim YS, Bassiouni M (2001) HPGe Detector Photopeak Efficiency Calculation including Self-Absorption and Coincidence Corrections for Cylindrical Sources using Compact Analytical Expressions. Radiat Phys Chem 61: 429-431.
8. Abbas MI Selim, YS (2002) Calculation of relative full-energy peak efficiencies of well-type detectors. Nucl Instrum Meth A 480: 651-657.
9. Abbas MI (2006) Analytical calculations of the solid angles subtended by a well-type detector at a point and extended circular source. Appl Radiat Isot 64: 1048-1056.
10. Abbas MI, Nafee S, Selim YS (2006) Calibration of cylindrical detectors using a simplified theoretical approach. Appl Radiat Isot 64: 1057-1064.
11. Abbas MI (2011) A new analytical method to calibrate cylindrical phoswich and $LaBr_3$(Ce) scintillation detectors. Nucl Instrum Meth A 621: 413-418.
12. Abbas MI (2010) Analytical formulae for borehole scintillation detectors efficiency calibration. Nucl Instrum Meth A 622: 171-175.
13. Grosswendt B, Waibel E (1976) Monte Carlo calculation of the intrinsic gamma ray efficiencies of cylindrical NaI(Tl) detectors. Nucl Instrum Methods 133: 25-28.
14. Hernandez F, El-Daoushy F (2003) Accounting for incomplete charge collection in Monte Carlo simulations of the efficiency of well-type Ge-detectors. Nucl Instrum Meth A 498: 340-351.
15. Pomme´ S, Sibbens G, Vidmar T, Camps J, Peyres V (2009) Detection efficiency calculation for photons, electrons and positrons in a well detector. Part II: Analytical model versus simulations. Nuclear Instruments and Methods in Physics Research A 606: 501-507.
16. Ahmed AH (2003) Accurate Direct Mathmatical Determination of The Efficiencies Of Gamma Detectors Arising From Radioactive Sources Of Different Shapes. Alexandria University.
17. Blaauw M (1993) To use of Sources Emitting Coincident γ-rays for Determination of Absolute Efficiency Curves of Highly Efficient Ge Detectors. Nucl Instrum Meth A 332: 493-500.
18. Sima O (2000) Accurate calculation of total efficiency of Ge well-type detectors suitable for efficiency calibration using common standard sources. Nucl Instrum Meth A 450: 98-108.
19. Badawi MS, El-Khatib AM, Krar ME (2013) New Numerical Simulation Approach to Calibrate the NaI(Tl) Detectors Array Using Non-Axial Extended Spherical Sources. Journal of Instrumentation 8: 11-20.
20. Badawi MS, Elzaher MA, Thabet AA, El-khatib, AM (2013) An empirical formula to calculate the full energy peak efficiency of scintillation detectors. Appl Radiat Isot 74: 46-49.
21. L'Annunziata MF (2011) Handbook of Radioactivity Analysis. (2ndedn) Academic press.

# Identification of Trends and Wave Regularities according to Statistical Data of Fluctuations of Mass of Glaciers

**Mazurkin PM***

*Doctor of Engineering, Academician of Russian Academy of Natural History and Russian Academy of Natural Sciences, Volga State University of Technology, Russia*

**Abstract**

The method of identification is shown on the example of tabular data of measurements of six parameters of subgroup of EEE on fluctuations in 2013 at balance of mass of 2528 glaciers of Earth. The equations of a trend and oscillatory indignations on the basis of steady laws on the generalized wave function in the form of an asymmetric wavelet signal with variables of amplitude and the period of fluctuation are received. Schedules of components of the generalized model of a wavelet signal allow to see visually a picture of mutual influence of all six parameters of subgroup of EEE at fluctuations of mass of glaciers. On the revealed equations it is possible to carry out the amplitude-frequency analysis.

**Keywords**: Glaciers; Fluctuation; Balance of weight; Factors; Regularities

## Introduction

Unlike deductive approach to wavelet analysis proceeding from the equations of classical mathematics inductive approach when statistical selection is primary is offered and concerning it the structure and values of parameters of the *generalized wave function* is identified.

Any phenomenon (time cut) or process (change in time) according to sound tabular statistical quantitative data (a numerical field) inductively can be identified the sum of asymmetric wavelet signals of a look

$$y = \sum_{i=1}^{m} y_i, y_i = A_i \cos(\pi x / pi - a_{8i}) \quad (1)$$

$$A_i = a_{1i} x^{a_{2i}} \exp(-a_{3i} x^{a_{4i}}), \quad p_i = a_{5i} + a_{6i} x^{a_{7i}}$$

where $y$ - indicator (dependent variable), $i$ - number of the making statistical model (1), $m$ - the number of members of model depending on achievement of the remains from (1) error of measurements, $x$ - explanatory variable, $Ai$-amplitude (half) of fluctuation (ordinate), $pi$ - half-cycle of fluctuation (abscissa), $a_1 ... a_8$.

-The parameters of model (1) determined in the program environment CurveExpert.

On a formula (1) with two fundamental physical constants e (Napier's number or number of time) and π (Archimedes's number or number of space) the quantized wavelet signal is formed from within the studied phenomenon and/or process.

## Basic Data for Statistical Modeling

In geoecology different types of natural objects [1], and among them, undoubtedly are considered by us, the important place is taken by glaciers as climate regulators. In article of regularity of distribution of glaciers for 2013 are given for the first time. On basic data [1] the fragment of the matrix including six factors is shown in Table 1 (identifiers are accepted according to the original). Data are located asymmetrically in Table 2.

Each of factors can have two states: first, a factor as the influencing variable $x$; secondly, the same factor is accepted as a dependent indicator of $y$. On them it is possible to assume that amplitude and the period of fluctuations on the general model. (1) submit to the biotechnical law [2-5]. Because of uncertainty of the direction of a vector "better → worse" at values of each factor we don't reveal regularity the rank distributions. Therefore the rating of glaciers is also not carried out.

In this regard the correlation coefficient the rank distributions is equal 1, and the factorial analysis we will carry out at all $6^2 - 6=30$ binary relations.

## A rating of Factors as the Influencing Variables and Indicators

The method of the factorial analysis offered by us allows not to think a priori of ratios between separate parameters of the studied system (system contains 2528 glaciers in our example from the Table 1). As a result the psychological barrier at researchers gets off: many binary relations for them will be unexpected. Therefore as our practice [5] showed, the factorial analysis the unique equation of type (1) allows to find unexpected scientific solutions in the field of research [6-11]. If some factorial communications are unusual and thus are highly adequate, new technical solutions are theoretically shown here. And often at the level of inventions of world novelty [11]. Thus the repeating process identification of the law (1) on one binary relation we designated the wavelet analysis [8,10].

Without waves, that is changes only on amplitude at very long wave, incommensurably bigger on the fluctuation period to an interval of measurements, are formed the so-called determined binary relations. All of them are special cases of a formula (1).

The square correlation matrix received after the analysis of the binary relations between all six variables accepted on basic data from Table 1 is given in Table 3. Here the rating of factors is given.

The coefficient of a correlative variation for 2528 glaciers of Earth is equal 22.0009/62=0.6111. This criterion is applied when comparing various systems, for example, of different groups of glaciers with each

***Corresponding author:** Mazurkin PM, Doctor of Engineering, professor, Academician of Russian Academy of Natural History and Russian Academy of Natural Sciences, Volga State University of Technology, Russia
E-mail: kaf_po@mail.ru

| Number | POINT_ LATITUDE | POINT_ LONGITUDE | POINT_ ELEVATION | POINT_WINTER_ BALANCE | POINT_SUMMER_ BALANCE | POINT_ANNUAL_ BALANCE |
|---|---|---|---|---|---|---|
| | EEE6 | EEE7 | EEE8 | EEE9 | EEE10 | EEE11 |
| 1 | | | 100 | | | 100 |
| 2 | | | 398 | | | 500 |
| ... | ... | ... | ... | ... | ... | ... |
| 71 | | | 1073 | | | -63 |
| 72 | -54.78136 | -68.40169 | 1038 | 1015 | -644 | 371 |
| 73 | -54.78182 | -68.40249 | 1036 | 947 | -540 | 406 |
| ... | ... | ... | ... | ... | ... | ... |
| 101 | -54.78048 | -68.40171 | 1073 | 743 | -984 | -241 |
| 102 | -16.3032 | -68.108 | 5053 | | | -958 |
| 103 | -16.3025 | -68.1083 | 5056 | | | -1385 |
| ... | ... | ... | ... | ... | ... | ... |
| 620 | | | 4058 | | | 120 |
| 621 | | | 3796 | 20 | -3719 | -3699 |
| 622 | | | 3828 | 99 | -2565 | -2466 |
| ... | ... | ... | ... | ... | ... | ... |
| 2526 | 60.40417 | -148.9067 | 1053 | | | -2090 |
| 2527 | 60.41974 | -148.9207 | 1283 | | | 0 |
| 2528 | 60.42495 | -148.9371 | 1367 | | | 960 |

**Table 1:** Matrix of basic data for statistical modelling.

| Influencing factors *x* | Dependent factors (indicatorsy) | | | | | |
|---|---|---|---|---|---|---|
| | EEE6 | EEE7 | EEE8 | EEE9 | EEE10 | EEE11 |
| EEE6 - POINT LATITUDE *[decimal degree]* | 1623 | 1623 | 1623 | 388 | 388 | 1591 |
| EEE7 - POINT LONGITUDE *[decimal degree]* | 1623 | 1623 | 1623 | 388 | 388 | 1591 |
| EEE8 - POINT ELEVATION *[m a.s.l.]* | 1623 | 1623 | 2467 | 531 | 531 | 2467 |
| EEE9 - POINT WINTER BALANCE *[mm w.e.]* | 388 | 388 | 531 | 531 | 531 | 531 |
| EEE10 - POINT SUMMER BALANCE *[mm w.e.]* | 388 | 388 | 531 | 531 | 531 | 531 |
| EEE11 - POINT ANNUAL BALANCE *[mm w.e.]* | 1591 | 1591 | 2467 | 531 | 531 | 2467 |

**Table 2:** Quantity of the measured points of balance of mass of glaciers, piece.

| Influencing factors *x* | Dependent factors Indicators *y* | | | | | | Sum | Place |
|---|---|---|---|---|---|---|---|---|
| | EEE6 | EEE7 | EEE8 | EEE9 | EEE10 | EEE11 | $\Sigma r$ | $I_x$ |
| EEE6 | 1 | 0.9732 | 0.9491 | 0.4425 | 0.3985 | 0.4551 | 4.2184 | 2 |
| EEE7 | 0.9688 | 1 | 0.4305 | 0.4154 | 0.4629 | 0.4156 | 3.6932 | 3 |
| EEE8 | 0.8178 | 0.8611 | 1 | 0.7201 | 0.6835 | 0.4421 | 4.5246 | 1 |
| EEE9 | 0.2237 | 0.4492 | 0.5431 | 1 | 0.2265 | 0.4721 | 2.9146 | 6 |
| EEE10 | 0.6408 | 0.5571 | 0.1557 | 0.2301 | 1 | 0.7309 | 3.3146 | 5 |
| EEE11 | 0.3703 | 0.4978 | 0.1719 | 0.5363 | 0.7592 | 1 | 3.3355 | 4 |
| Sum $\Sigma r$ | 4.0214 | 4.3384 | 3.2503 | 3.3444 | 3.5306 | 3.5158 | 22.0009 | - |
| Place $I_y$ | 2 | 1 | 6 | 5 | 3 | 4 | - | 0.6111 |

**Table 3:** Correlation matrix and rating of factors on the binary relations.

other. Perhaps also comparison of all glaciers with other objects sew planets, for example, with deserts or with agroecological classes of a soil cover.

From six influencing variables on the first place there was EEE8 factor. On the second place EEE6 factor, and was located on the third place – EEE7. Among dependent indicators on the first place there is EEE7 factor. In second place is the factor EEE6, and the third -EEE10.

## Strong Binary Relations

At correlation coefficient more than 0.7 binary relations between factors become strong (tab. 4As a rule, the accounting of wave indignation, additional to a trend, gives significant increase in adequacy to the revealed regularity on a formula (1). But glaciers do not yet have good data, because many of the values of the six factors have empty cells: the best matrix is illustrated in Table 4 completely filled with cells.

Because of low completeness of a matrix from 30 formulas received only eight strong communications that makes 22.2%. Thus as the influencing variable EEE9 factor -POINT WINTER BALANCE was excluded. But as dependent indicators remained all six factors. From eight strong communications three (37.50%) treat trends.

Hierarchy of strong communications following (Table 5): 1) 0.9732 – EEE7=f(EEE6); 2) 0.9688 – EEE6=f(EEE7); 3) 0.9491 – EEE8=f(EEE6); 4) 0.8611 – EEE7=f(EEE8); 5) 0.8178 – EEE6=f(EEE8); 6) 0.7592 – EEE10=f(EEE11); 7) 0.7309 – EEE11=f(EEE10); 8) 0.7201 – EEE9=f(EEE8).

In Table 5 parameters of statistical models on the general formula

(1) which values are written down in a compact matrix form with significant Figures 1-7 are given

From data of Table 5 it is visible that some regularities required replacement of variables because of negative values of the influencing variables.

From schedules in Figure 1 it is visible that the function EEE7=f(EEE6) supports three members from whom the first two are a difference of two laws: first, law of exponential growth; secondly, indicative law. The third component is wavelet on a formula (1) with very high coefficient of correlation 0.9362. Therefore the main contribution to formation of the binary relation is carried out by wave function with a variable amplitude under the law of exponential death and the variable period of fluctuation. Thus the initial stage of fluctuation is equal 2×4.87479 ≈ 9.75 degrees. In process of increase in width fluctuation calms down because of growth of a half-cycle.

Remains EEE7 after three members of model have significant values at the latitudes of 40-65 degrees. Therefore if it is necessary, the wavelet analysis of microwaves on the remains for this width is possible further. The EEE8 function=f(EEE6) it consists of only the binomial trend. According to Table 5 it contains the sum from two laws: first, law of exponential death; secondly, biotechnical law [2-5]. The remains show small fluctuation which can be identified a formula (1), however adequacy of a wave component will be very small.

Other schedules have similar explanations. Therefore glaciers, as well as other types of natural objects [5], have natural distributions.

The maximum number of members of the statistical model EEE6=f(EEE7) is equal to four that corresponds to computing opportunities of the program CurveExpert-1.40 environment.

| Influencing factors *x* | Dependent factors indicators *y* ) | | | | | |
|---|---|---|---|---|---|---|
| | **EEE6** | **EEE7** | **EEE8** | **EEE9** | **EEE10** | **EEE11** |
| EEE6 - POINT LATITUDE | | **0.9732** | **0.9491** | | | |
| EEE7 - POINT LONGITUDE | **0.9688** | | | | | |
| EEE8 - POINT ELEVATION | **0.8178** | **0.8611** | | **0.7201** | | |
| EEE10 - POINT SUMMER BALANCE | | | | | | **0.7309** |
| EEE11 - POINT ANNUAL BALANCE | | | | | **0.7592** | |

**Table 4:** Correlation matrix of the strong binary relations at r>0.7.

| Indi- cators *y* | Asymmetric wavelet $y_i = a_{1i} x^{a_{2i}} \exp(-a_{3i} x^{a_{4i}}) \cos(\pi x/(a_{5i} + a_{6i} x^{a_{7i}}) - a_{8i})$ | | | | | | | | Correl. coeffic. *r* |
|---|---|---|---|---|---|---|---|---|---|
| | Amplitude (half) of fluctuation | | | | Fluctuation half-cycle | | | Shift | |
| | *a1i* | *a2i* | *a3i* | *a4i* | *a5i* | *a6i* | *a7i* | *a8i* | |
| The influence EEE6 *[decimal degree]* | | | | | | | | | |
| EEE71 Figure 1 | 4.23078 | 0 | -0.16452 | 0.54045 | 0 | 0 | 0 | 0 | 0.9732 |
| | -0.31020 | 1.48076 | 0 | 0 | 0 | 0 | 0 | 0 | |
| | 132.72384 | 0 | 8.79872e-5 | 1.76602 | 4.87479 | 0.29176 | 0.56033 | 0.24849 | |
| EEE81 Figure 1 | 1.66509e6 | 0 | 0.31234 | 0.94742 | 0 | 0 | 0 | 0 | 0.9491 |
| | 7.82146e-5 | 4.91282 | 0.018337 | 1.20027 | 0 | 0 | 0 | 0 | |
| The influence EEE7 *[decimal degree]* | | | | | | | | | |
| EEE62 Figure 2 | 64.56655 | 0 | 0.00076040 | 1.01838 | 0 | 0 | 0 | 0 | 0.9688 |
| | -1.24710e-15 | 6.59346 | 0 | 0 | 0 | 0 | 0 | 0 | |
| | -2.35962e-7 | 5.20107 | 0.042456 | 1 | 12.92183 | 0.0036148 | 1.03167 | 2.65582 | |
| | -0.00052866 | 3.02216 | 0.025670 | 1 | 3.80507 | 9.27492e-5 | 1 | 4.71131 | |
| The influence EEE8 *[m a.s.l.]* | | | | | | | | | |
| EEE7 Figure 3 | -87.70507 | 0 | -8.09363e-5 | 1.07588 | 0 | 0 | 0 | 0 | 0.8611 |
| | 2.50470e-6 | 2.55776 | 0.00026769 | 1.12617 | 0 | 0 | 0 | 0 | |
| | -9.59526e-8 | 3.19687 | 0.0017158 | 0.99993 | 157.16690 | 0.28654 | 0.88567 | 6.01043 | |
| EEE6 Figure 4 | 7.25053 | 0 | -0.00032396 | 1.09546 | 0 | 0 | 0 | 0 | 0.8178 |
| | -6.85611e-17 | 5.03862 | 0 | 0 | 0 | 0 | 0 | 0 | |
| | -0.00012515 | 2.24860 | 0.0020534 | 1 | 2922.5439 | -0.48413 | 0.99988 | -2.92162 | |
| EEE9 Figure 5 | 59.61396 | 0 | -0.14457 | 0.46269 | 0 | 0 | 0 | 0 | 0.7201 |
| | -2.90947e-7 | 3.11597 | 0 | 0 | 0 | 0 | 0 | 0 | |
| | -0.00062264 | 2.34099 | 0.0018640 | 1.00057 | 846.23612 | -0.042840 | 0.99814 | -0.80809 | |
| The influence EEE11 *[mm w.e.]* | | | | | | | | | |
| EEE103 Figure 6 | -28940.548 | 0 | 0 | 0 | 0 | 0 | 0 | 0 | 0.7592 |
| | 270.01817 | 0.52442 | 2.61663e-5 | 0.99761 | 0 | 0 | 0 | 0 | |
| The influence EEE10 *[mm w.e.]* | | | | | | | | | |
| EEE113 Figure 7 | -11303.0999 | 0 | 0 | 0 | 0 | 0 | 0 | 0 | 0.7308 |

**Table 5:** Parameters of the strong binary relations at correlation coefficient r≥0.7.

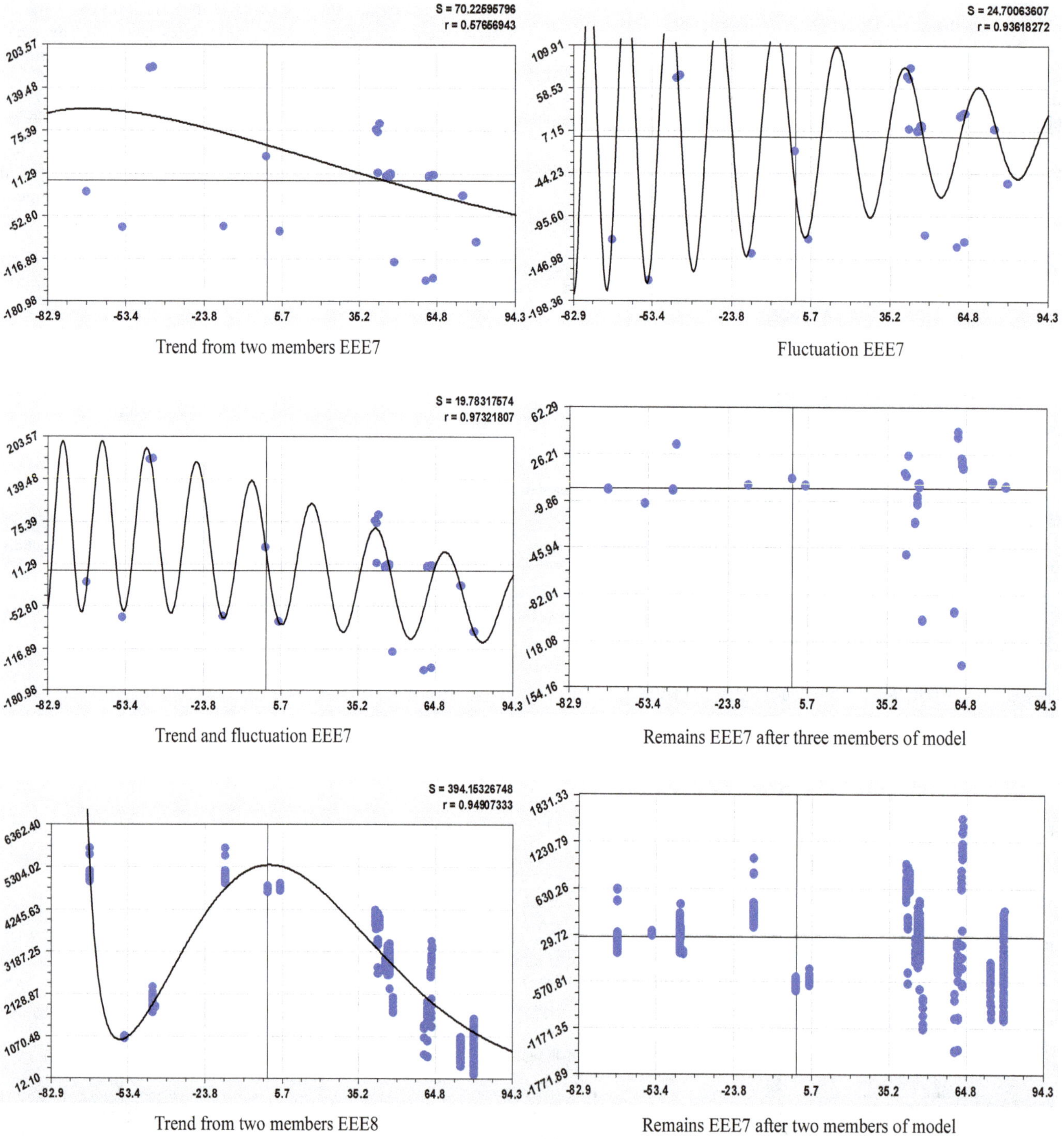


**Figure 1:** Schedules of models of influence of a factor of EEE6 on change of factors of EEE7 and EEE8.

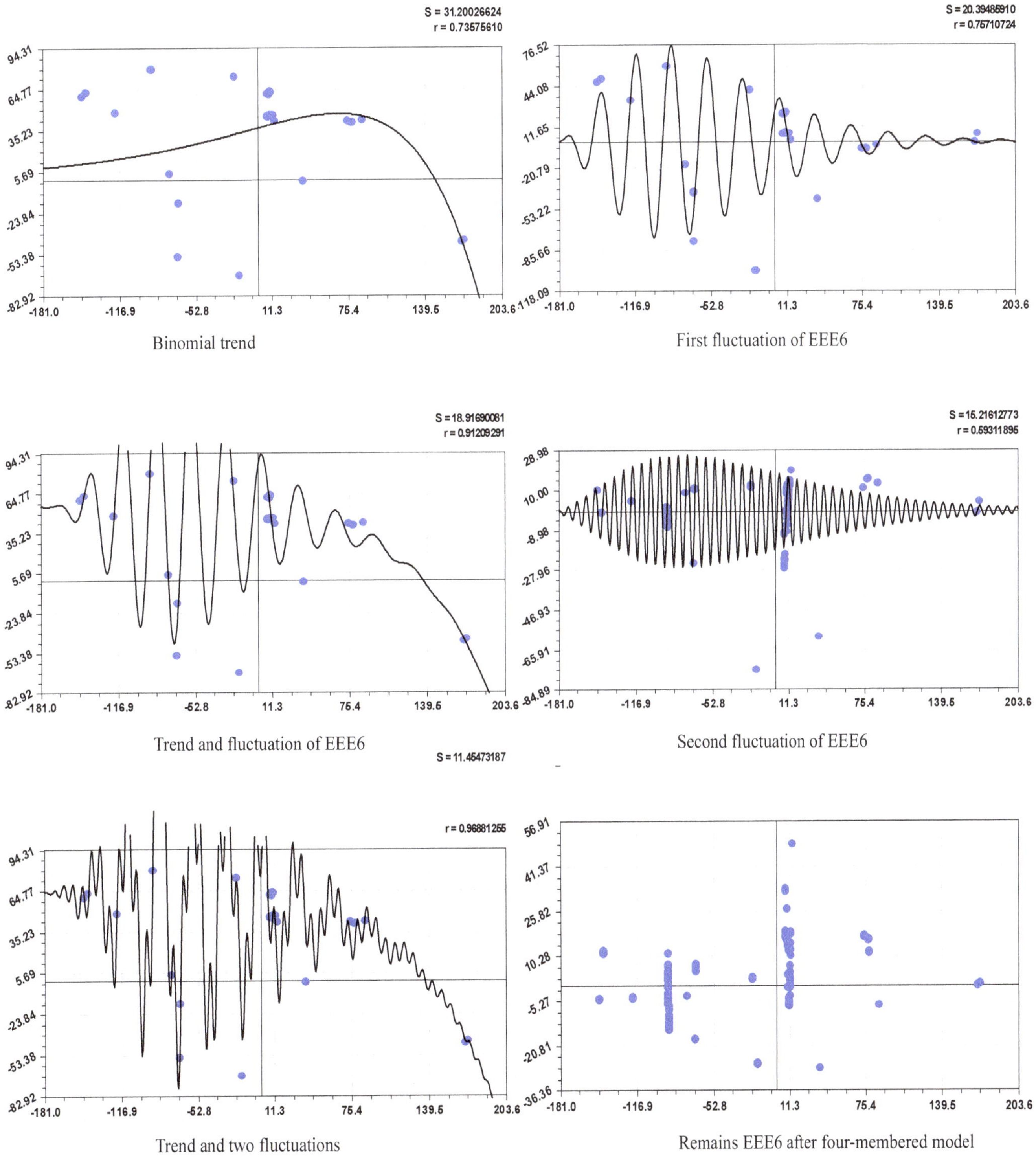


**Figure 2**. Schedules of models of influence of EEE7 on change of EEE6.

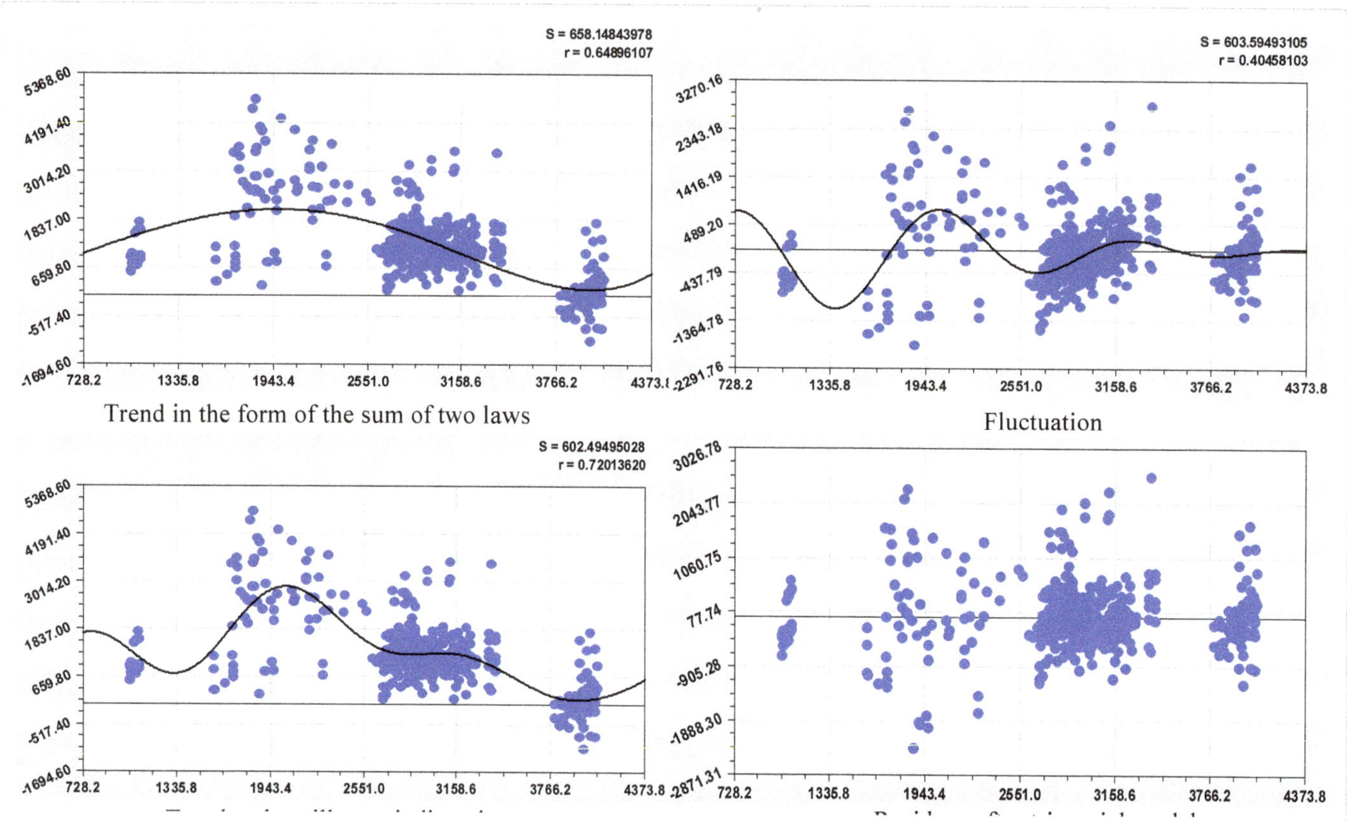


**Figure 3**. Schedules of models of influence of EEE8 on change of EEE7.

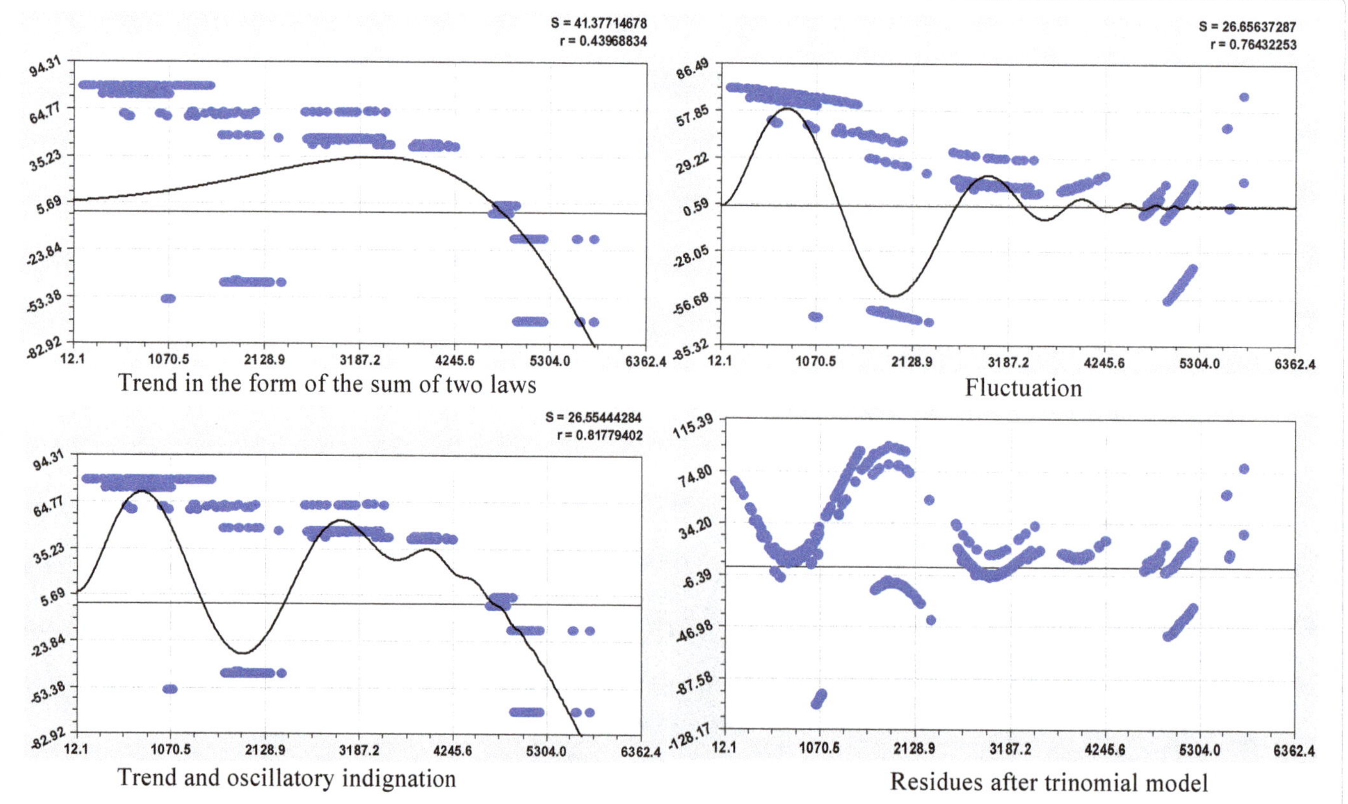


**Figure 4**. Schedules of models of influence of EEE8 on change of EEE6.

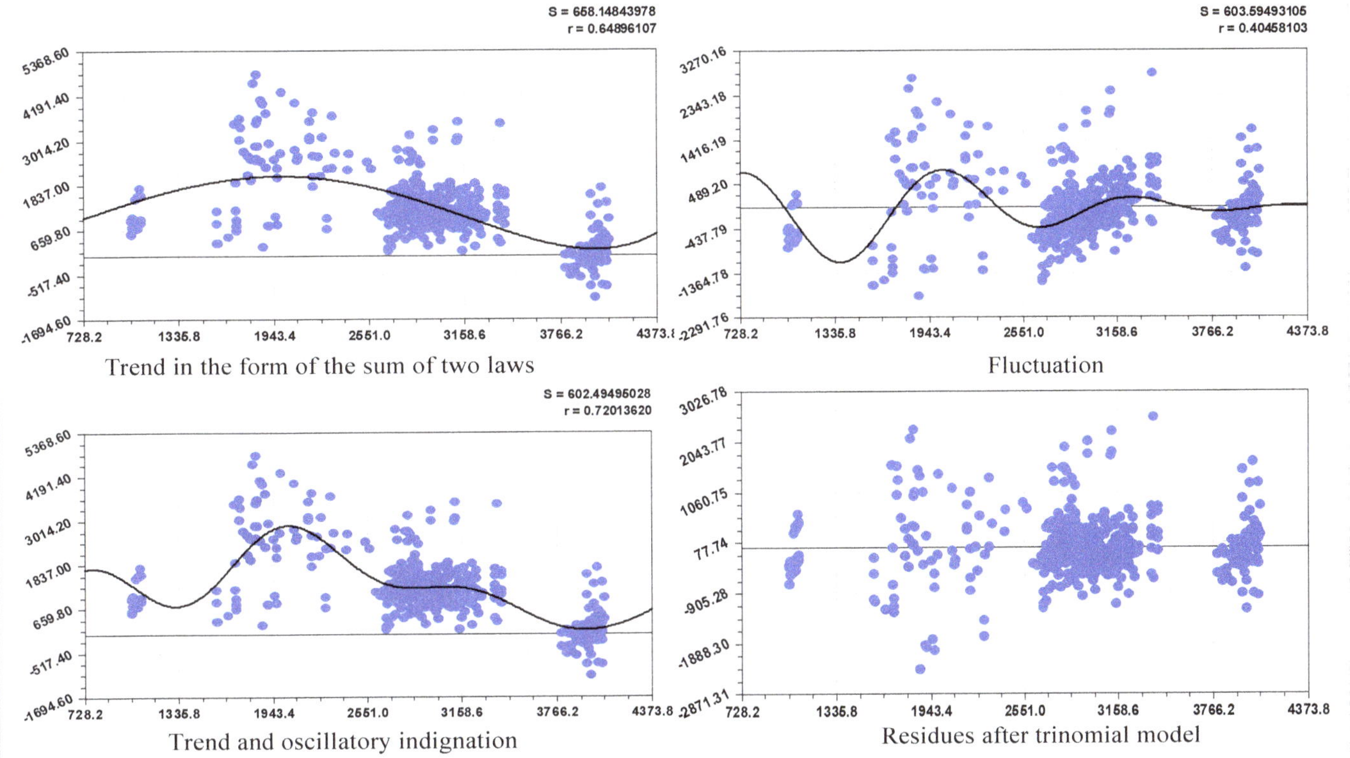


**Figure 5**. Schedules of models of influence of EEE8 on change of EEE9.

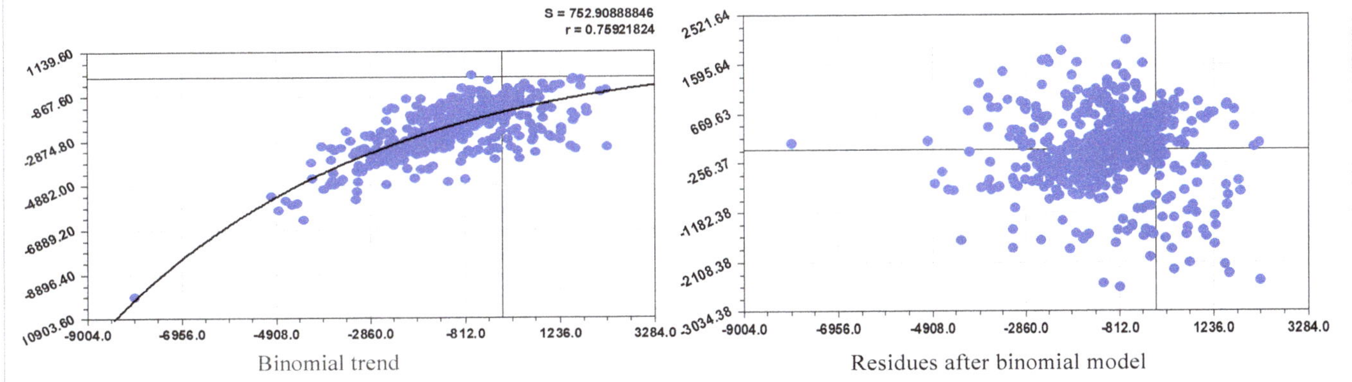


**Figure 6**. Schedules of models of influence of EEE11 on change of EEE10.

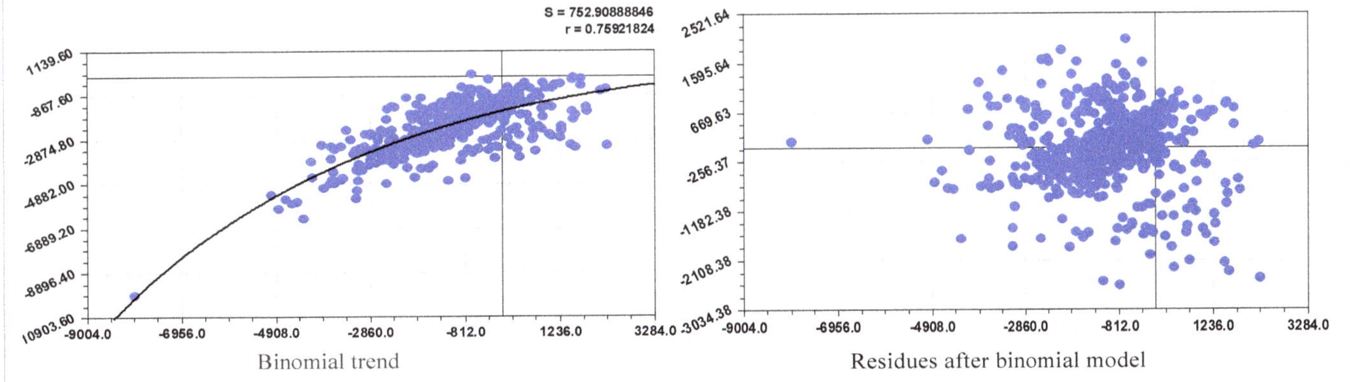


**Figure 7**. Schedules of models of influence of EEE10 on change of EEE11

For the full wavelet analysis it is necessary to develop the special program environment according to our scenarios of statistical modeling for a supercomputer of a petaflop class. Thus the new program environment for large volumes of the Table of basic data will be universal for science.

The analysis of schedules according to amplitude-frequency characteristics shows that the system of glaciers possesses a certain property of wave adaptation to living conditions on the planet Earth.

For some binary relations the number of members in the general statistical model can exceed 100-120 pieces. In this case there is a possibility of carrying out the fractal analysis for group of wavelets on mega, macro, meso and to micro fluctuations.

## Conclusion

Applicability of statistical model (1) to parameters of points of fluctuations of balance at the mass of glaciers of Earth is proved. As a result each binary relation contains a trend and wavelet signals. And the trend is a special case of fluctuation of a wavelet, superlong on the period. As a result the general statistical model represents the plait consisting of a set of lonely waves with variables amplitude and the period of fluctuations. After statistical modeling the factorial analysis allowing making ratings of factors as the influencing parameters and as dependent indicators is carried out.

The offered methodology of identification allows allocating waves of the binary relations between the measured factors at glaciers. Thus for 2528 glaciers of subgroup of EEE finite-dimensional wavelets which can be compared further with heuristic representations of experts are characteristic. The method of identification allows to allocate significant parameters of glaciers and the binary relations between them in which it will be necessary to increase the accuracy of future measurements. Thus from different subgroups we allocated 26 factors, but their joint analysis is disturbed by a disagreement between Tables of data.

### References

1. Mazurkin PM(2006) Geoecology: Patterns of Modern Natural Science.
2. http://wgms.ch/
3. Mazurkin PM (2009) Biotechnical law and designing of adequate models. Achievements of modern natural sciences 9: 125-129.
4. Mazurkin PM (2009) The biotechnical law, algorithm in intuitive sense and algorithm of search of parameters .Achievements of modern natural sciences 9: 88-92.
5. Mazurkin PM (2009) The biotechnical principle in statistical modeling. Achievements of modern natural sciences. 9: 107-111.
6. Mazurkin PM (2013) Identification of statistical steady regularities. Science and world international scientific magazine 3: 28-33.
7. Mazurkin PM (2015) Invariants of the Hilbert Transform for 23-Hilbert Problem. Advances in Sciences and Humanities. 1: 1-12.
8. Mazurkin PM (2014) Method of identification. 14th international multidisciplinary scientific geoconference SGEM2014, Russia.
9. Mazurkin PM (2014) The decision 23-oh Gilbert's problems. Interdisciplinary researches in the field of mathematical modelling and informatics. Materials of the 3rd scientific and practical Internet conference.
10. Mazurkin PM (2014) Statistical modelling of entire prime numbers. International Journal of Engineering and Technical Research 2: 148-158.
11. Mazurkin PM, Filonov AS (2006) Mathematical modelling. Identification of one-factorial statistical regularities: manual.

# Pisot-K Elements in the Field of Formal Power Series over Finite Field

**Kthiri H***

*Department of Mathematics, University of Sfax, Tunisa*

**Abstract**

In this paper, we will give a criteria of irreducible polynomials over $\mathbb{F}_q[X]$ where $\mathbb{F}_q$ is a finite field. We will present an estimation for the number of the Pisot-k power formal series, precisely we will give their degrees and their logarithmic heights.

**Keywords:** Formal power series; Irreducible polynomials; Pisot-2 series. |z|<1

## Introduction

In 2007, Gabriel Ponce, Melanie Ruiz, Emily McLeod Schnitger and Noah Simon have discuss certain sets of algebraic integers related to Pisot numbers (An integer algebraic number $\alpha>1$ is called a Pisot number if all its conjugates, different from $\alpha$ lie in the open disc $|z|<1$ on the complex plane ) and Salem number (An integer algebraic number $\alpha>1$ is called a Salem number if all its conjugates, different from $\alpha$ lie in the disc $|z|<1$ and it has at least one conjugate having modulus equal to 1) Also their properties.

Much is known about Pisot numbers, for example, if the integer coefficients of the minimal polynomial of $\alpha$ $P(z)=z^d+q_{d-1}z^{d-1}+\cdots+q_0$, satisfy $1+\sum_{i=0}^{d-1}q_i<0$ and $P(x)=x^d-a_{d-1}x^{d-1}-\cdots-a_0$. then a is a Pisot number. Also, for $\alpha\in s$ it is known that $\alpha^n\in s$ One important theorem about Pisot numbers is that the set $S$ is closed and there are many known ways to construct them.

Let $\alpha$ be a root dominate of polynomial $P(x)=x^d-a_{d-1}x^{d-1}-\cdots-a_0$. [1].

$a_{d-1}\geq a_{d-2}\geq\cdots\geq a_0>0$ where $a_i\in\mathbb{N}^*$ then $\alpha$ is a Pisot number. If $a_{d-1}>\sum_{i=0}^{d-2}|a_i|+1$ where $a_i\in\mathbb{Z}$ and $a_0\neq0$ then $\alpha$ is a Pisot number.

If $a_{d-1}>\sum_{i=2}^{d}a_i$ ,$a_i\geq0$ and $a_d\neq0$, then $\alpha$ is a Pisot number [2].

In comparison, little is known about Salem numbers, and their construction is difficult. There are still many open questions about Salem numbers, including determining the infest of the set.

**Definition 1.1:** *The reciprocal polynomial of a polynomial P is defined by* $P^*(z)=z^{\deg P}P(\frac{1}{z})$. *A polynomial is called reciprocal if P=P* In general, we will denote the reciprocal polynomial of P by Q.*

The minimal polynomial of a Salem number will always be reciprocal, and thus, there are no Salem numbers with odd degree or degree less than 4. Constructing Salem numbers is much more difficult than Pisot numbers. For instance, graph theory can be used to construct some but not nearly all. The smallest Salem number is still not known, though it is conjectured to be largest root of $1+z-z^3-z^4-z^6-z^7+z^9+z^{10}$.

They have defined new sets of generalized Pisot numbers and they have concerned with the arithmetic properties and limit points of one of these sets, Pisot-2 pairs (A pair of real distinct algebraic conjugates $(\alpha_1,\alpha_2,\ldots,\alpha_k)$ is called a Pisot-k uplet or o(k)-Pisot number if $\alpha_1,\alpha_2,\ldots,\alpha_k$ are greater than 1 and all remaining conjugates have modulus strictly less than 1) We denote the set of Pisot-k pairs by $S_k$ For this set they have obtained some results analogous to those known about Pisot numbers. $(\alpha_1,\alpha_2)\in S_2$

**Proposition 1.1:** *If* $(\alpha_1,\alpha_2)\in S_2$ *then* $(\alpha_1^n,\alpha_2^n)\in S_2$ *for all* $n\in\mathbb{N}$.

**Proposition 1.2:** *If* $(\alpha_1,\alpha_2)\in S_2$ *then* $\{\alpha_1^n+\alpha_2^n\}\to0$ *for all* $n\in\mathbb{N}$

**Proposition 1.3:** *If* $(\alpha_1,\alpha_2)\in S_2$ *with minimal polynomial,* $P\in\mathbb{Q}[X]$ *of degree 3 and P(0)=1 then* $\alpha_1,\alpha_2\in S_2$

**Theorem 1.4:** *The limit points of* $S_2$ *lie either in* $S_2$ S× or $\mathbb{R}$× S

A great deal is known about this set. Then, they have discussed another set, Pisot (o)-2 numbers, and its connections to Salem numbers, including a relationship with the infemum of Salem numbers. Finally, They have giving arithmetic properties of these Pisot (o)-2 numbers. In this paper, we consider an analogue of this concept in algebraic function over finite fields.

Recall that in 1962 Batemen and Duquettes introduced and characterized the elements of Pisot and Salem in the field of formal power series [3].

**Theorem 1.5:** *[4] Let.* $w\in\mathbb{F}_q((X^{-1}))$ *an algebraic integer over* $\mathbb{F}_q[X]$ *and its minimal polynomial be*

$$P(y)=y^n+A_{n-1}y^{n-1}+\cdots+A_0,A_i\in\mathbb{F}_q[X]$$

Then $w$ is a Pisot (respectively Salem) series if and only if $|A_{n-1}|>\max_{i\neq n-1}|A_i|$( respectively $|A_1|=\max_{i\neq n-1}|A_i|$).

Chandoul, Jellali and Mkaouar have improve the Theorem of Bateman and Duquette [4] on $\mathbb{F}_q((X^{-1}))$ and this while establishing the same result with weaker hypotheses [4].

**Theorem 1.6:** *[2] Let* $w\in\mathbb{F}_q((X^{-1}))$ *be Pisot (respectively Salem) series if and only if w is a root of polynomial*

$$P(y)=y^n+A_{n-1}y^{n-1}+\cdots+A_0,\ A_i\in\mathbb{F}_q[X]$$

where $|A_{n-1}|>\max_{i\neq n-1}|A_i|$( respectively $|A_{n-1}|=\max_{i\neq n-1}|A_i|$).

In the same setting and on the field of the real, Brauer, gave a criteria of irreducibility on $\mathbb{Q}$ [5]

**Theorem 1.7:** *If* $a_1,a_2,\cdots,a_{n-1}$ *are of the integer such that* $a_{n-1}\geq a_{n-2}\geq\cdots\geq a_0>0$, *thenthepolynomial* $x^n-a_{n-1}x^{n-1}-a_{n-2}x^{n-2}-\cdots-a_0$ *is irreducible on* $\mathbb{Q}$

***Corresponding author:** Kthiri H, Department of Mathematics, University of Sfax, Tunisa, E-mail: hassenkthiri@gmail.com

Chandoul, Jellali and Mkaouar are constructs the analog of the theorem of Brauer in the case of the polynomials to coefficients in $_q$ [X] [4].

**Theorem 1.8:** *Let* $\Lambda(Y)=Y^d+\lambda_{d-1}Y^{d-1}+\cdots+\lambda_0$ *where* $\lambda_i\in\mathbb{F}_q[X],\lambda_0\neq 0$ *and* $\deg\lambda_{d-1}>\deg\lambda_i$, *for all i≠d-1 then* $\Lambda$ *is irreducible on* $\mathbb{F}_q$ *[X].*

This last theorem permits to give a new evaluation to calculate the number of elements of Pisot on $\mathbb{F}_q((X^{-1}))$ being given their $n$ degrees and their logarithmic heights $h$ This evaluation is illustrated in the following result:

$$\mathcal{P}(n,h)=(q-1)(q^h-1)q^{(n-1)h}.$$

The smallest Pisot number, in the real case, is the only real root of the polynomial $x^3$-$x$-1 known as the plastic number or number of money (approximately 1,324718 But, as we have said, the smallest accumulation point of all Pisot numbers is the golden number. Chandoul, Jellali and Mkaouar are prove in [5] that the minimal polynomial of the smallest Pisot element (SPE) of degree $n$ is $P(Y) = Y^n-\alpha XY^{n-1}-\alpha^n$ where $\alpha$ is the least element of $F_q\backslash\{0\}$ Moreover, the sequence $(w_n)_{n1}$ is decreasing one and converge to $\alpha X$ In the present paper we give generalized Pisot and Salem series in the field of formal power series over finite field. $P(Y) = Y^n-\alpha XY^{n-1}-\alpha^n$

The paper is organized as follows. In this section, we give some preliminary definitions, we define Pisot numbers and Salem numbers. We give the well known properties of its. In section 2 we introduce the field of formal power series over finite field, we define a totally order, the lexicographic order, over a field of formal power series with coefficients in a finite field. In section 3 We give the arithmetic Properties of Pisot (o)-k series and the criteria of irreducible of this series.

## of Formal Series $\mathbb{F}_q((X^{-1}))$

Let be $\mathbb{F}_q$ a finite field of $q$ elements, $\mathbb{F}_q$ [X] the ring of polynomials with coefficient I $_q$, $\mathbb{F}_q(X)$ the field of rational functions, $\mathbb{F}_q(X,\beta)$ the minimal extension of $\mathbb{F}_q$ containing $X$ and $\beta$ and $\mathbb{F}_q(x,\beta)$ the minimal ring containing $X$ and $\beta$. Let $\mathbb{F}_q((X^{-1}))$ be the field of formal power series of the form :

$$w=\sum_{k=-\infty}^{l}w_kX^k,\quad w_k\in\mathbb{F}_q$$

where

$$l=\deg w:=\begin{cases}\max\{k:w_k\neq 0\} & for\ w\neq 0\\ -\infty & for\ w=0.\end{cases}$$

Define the absolute value by

$$|w|=\begin{cases}q^{\deg w} & for\ w\neq 0\\ 0 & for\ w=0.\end{cases}$$

Then |.| is not archimedean. It fulfills the strict triangular inequality

$$|w+v|\leq\max(|w|,|v|)\qquad and$$

$$|w+v|=\max(|w|,|v|)\qquad if\ |w|\neq|v|.$$

For $f\in\mathbb{F}_q((X^{-1}))$ define the integer (polynomial) part $[w]=\sum_{k=0}^{l}w_kX^k$ where the empty sum, as usual, is defined to be zero. Therefore $f\in\mathbb{F}_q$ [X] and ($w$-[$w$]) is in the unit disk $D(0,1)$ for all $w\in\mathbb{F}_q((X^{-1}))$. As explained by Sprindzuk a non archimedean absolute value on $\mathbb{F}_q((X^{-1}))$ is defined by $|w|=q^{-s}$ It is clear that, for all $P\in\mathbb{F}_q[X],|P|=q^{degP}$ and, for all $Q\in\mathbb{F}_q$ [X];such that Q≠0 $|\frac{P}{Q}|=q^{degP-degQ}$ [6].

We know that $\mathbb{F}_q((X^{-1}))$ is complete and locally compact with respect to the metric defined by this absolute value. We denote by $\overline{\mathbb{F}}_q((X^{-1}))$ an algebraic closure of $\mathbb{F}_q((X^{-1}))$ We note that the absolute value has a unique extension to $\overline{\mathbb{F}}_q((X^{-1}))$. Abusing a little the notations, we will use the same symbol |.| for the two absolute values. For all polynomial ($P$≠0)

$$P(Y)=Y^n+A_1Y^{n-1}+\cdots+A_n,A_i\in\mathbb{F}_q[X]$$

We define the logarithmic hauteur $\mathbb{H}(P)$ of $P$ by

$$\mathcal{H}(P)=\log_q\max_{0\leq i\leq n}|A_i|=\max_{0\leq i\leq n}\deg(A_i)$$

Where $\log_q x$ designate the logarithmic function in the $q$ basis. For all element algebraic $w\in_q((X^{-1}))$ one will note by $\mathbb{H}(w)$ the logarithmic hauteur of his minimal polynomial.

**Theorem 2.1:** *Let* $w\in\mathbb{F}_q((X^{-1}))$ *is of Pisot number if and only if it exist* $\lambda\in_q((X^{-1}))\backslash\{0\}$ *such that* $\lim_{n\to+\infty}\{\lambda w^n\}=0$; *Moreover* $\lambda$ *can be chosen to belong to* $\mathbb{F}_q(X)(w)$.

Recall that $\mathbb{F}_q((X^{-1}))$ contains Pisot elements of any degree over $\mathbb{F}_q(X)$ Indeed, consider the polynomial $Y^n-aY^{n-1}-b$ where $a,b\in\mathbb{F}_q\backslash\{0\}$ it can be seen easily, considering its Newton polygon, that the polynomial, which is irreducible over $_q(X)$ has a root $w\in\mathbb{F}_q((X^{-1}))$ such that $|w|>1$ and all of its conjugates in $\overline{\mathbb{F}}_q((X^{-1}))$ have an absolute value strictly smaller than 1.

**Definition 2.1:** *An uplet of series algebraic conjugates*$(w_1,\ldots w_k)\mathbb{F}_q((X^{-1})))^k$ *is called a series of Pisot-k uplet (o(k)-Pisot series) if* $w_1$ *integer algebraic such that* $|w_1|,\ldots,|w_k|$ *are greater than 1 and all remaining conjugates have modulus strictly less than 1 We denote the set of Pisot-k uplet by* $S_k'$.

**Example 2.2**

1) Series of Pisot-2 of degree 2 on $\mathbb{F}_2((X^{-1}))$ $(w_1\ldots w_k)\mathbb{F}_q((X^{-1})))^k$

Let

$$P(Y)=Y^2+Y+X^2+X+1.$$

Then P is irreducible over $\mathbb{F}_2[X][Y]$ Now we show that $P$ has two roots $|Y_1|>1|Y_2|>1$ such that $(Y_1, Y_2)\in\mathbb{F}_2((X^{-1}))^2$ Let $Y=\sum_{i=-1}^{\infty}Y_iX^{-i}\in\mathbb{F}_2((X^{-1}))$ one root of $P$.Then

$$\begin{cases}Y_1=X+1+\frac{1}{X^2}+\cdots=X+1+\frac{1}{Z_1} & such\ that\ |Z_1|>2;\\ Y_2=X+\frac{1}{X}+\cdots=X+\frac{1}{Z_2} & such\ that\ |Z_2|>1.\end{cases}$$

For $Y_1$ we have

$$Y_1^2+Y_1+X^2+X+1=0.$$

So

$$Z_1^2+Z_1+1=0.$$

Therefore $Z_1$ is root of Polynomial irreducible and $Z_1\in\mathbb{F}_2((X^{-1}))$ So $Y_1\in\mathbb{F}_2((X^{-1}))$

For $Y_2$ we have

$$Y_2^3+Y_2+X^2+X+1=0.$$

Then

$Z_2^2 + Z_2 + 1 = 0.$

Therefore $Z_1$ is root of Polynomial irreducible and $Z_2 \in \mathbb{F}_2((X^{-1}))$ So $Y_1 \in \mathbb{F}_2((X^{-1}))$

Series of Pisot-2 of degree 3 on $\mathbb{F}_2((X^{-1}))$

Let $P(Y) = Y^3 + (X^2+1)Y^2 + (X^3+X)Y + 1.$

Then $P$ is irreducible over $\mathbb{F}_2[X][Y]$ because $P$ and his reciprocal polynomial $Q(Y) = Y^3 + (X^3+X)Y^2 + (X^2+1)Y + 1$ are the same natures. However $Q$ is of type $(I)$ then it is irreducible. Therefore $P$ is also.

Now we show that $P$ has two roots $|Y_1|>1$ and $|Y_2|>1$ such that $(Y_1, Y_2) \in \mathbb{F}_2((X^{-1}))^2$ and $|Y_3|<1$ Let $Y = \sum_{i=-2}^{\infty} Y_i X^{-i} \in \mathbb{F}_2((X^{-1}))$ one root of $P$ Then $(Y_1, Y_2) \in \mathbb{F}_2((X^{-1}))^2$

$$\begin{cases} Y_1 = X^2 + X + \dfrac{1}{X^2} + \cdots = X^2 + X + \dfrac{1}{Z_1} & \text{such that } |Z_1| > 2; \\ Y_2 = X + 1 + \dfrac{1}{X} + \cdots = X + 1 + \dfrac{1}{Z_2} & \text{such that } |Z_2| > 1; \\ Y_3 = \dfrac{1}{X} + \cdots = \dfrac{1}{Z_3} & \text{such that } |Z_3| > 1. \end{cases}$$

For $Y_1$: we have

$Y_1^3 + (X^2+1)Y_1^2 + (X^3+X)Y_1 + 1 = 0.$

Then

$Z_1^3 + (X^4+X^3+X^2+X)Z_1^2 + (X+1)Z_1 + 1 = 0.$

Therefore $Z_1$ is of type $(I)$ and $Z_1 \in \mathbb{F}_2((X^{-1}))$ So $Y_1 \in \mathbb{F}_2((X^{-1}))$

For $Y_2$: we have

$Y_2^3 + (X^2+1)Y_2^2 + (X^3+X)Y_2 + 1 = 0.$

Then

$Z_1^3 + (X^3+X)Z_1^2 + (X^2+1)Z_1 + 1 = 0.$

So $Z_2$ is of type $^{(I)}$ and $Z_2 \in \mathbb{F}_2((X^{-1}))$ So $Y_2 \mathbb{F}_2((X^{-1}))$

For $Y_3$:we have $Y_3 = \dfrac{1}{Z_3}$ such that $|Z_3|>1$ Then

$Y_3^3 + (X^2+1)Y_3^2 + (X^3+X)Y_3 + 1 = 0.$

$Z_3^3 + (X^3+X)Z_3^2 + (X^2+1)Z_3 + 1 = 0.$

Thus $Z_3$ is of type $(I)$ and $Z_3 \mathbb{F}_2((X^{-1}))$ So $Y_3 \in \mathbb{F}_2((X^{-1}))$

## Results

### Arithmetic properties of Pisot-k uplet (respectively Salem) series

In this section we discuss some basic arithmetic properties of $S_k'$. It is known that $(\alpha,\beta) \in S_2$ if $(\alpha,\beta) \in S_2$ then $(\alpha^n,\beta^n) \in S_2$ $\forall n \in \mathbb{N}$ Also, we have that $\lim_{n\to\infty}\{\alpha^n,\beta^n\} \to \{0\}$, where $\{x\}$ denotes the fractional part of $x$. Our last proposition relates a specific subset of $S_k'$ to $S'$ The algebraic closure of $\mathbb{F}_q((X^{-1}))$ will be denoted by $\overline{\mathbb{F}}_q((X^{-1}))$.

**Proposition 3.1:** *Let* $(w_1,\cdots,w_k) \in S_k'$ *(respectively.* $\in T_k'$*), then* $(w_1^n,\cdots,w_k^n) \in S_k'$ *(respectively.* $\in T_k'$ *) for all* $n\mathbb{N}^*$.

*Proof.* Let $(w_1,\cdots,w_k) \in S_k'$ (respectively. $\in T_k'$) and $M \in \mathbb{F}_q[X][Y]$ the minimal polynomial of $w$ and $w = w_1, \ldots w_d$ the conjugates of $w$. Then there exist exactly $k$ conjugates $w = w_1, \ldots w_k$ of $w$ that lie outside the unit disc. Let $w_{k+1} \ldots w_d$ denote the other roots of $M$.

We know that the product of any two algebraic is, itself, an algebraic. Since $w_1$ is an algebraic, then$\forall n \in \mathbb{N}$ $\forall n \in \mathbb{N}, w_1^n$ a is also an algebraic. Let $P \in \mathbb{F}_q[X][Y]$ be the minimal polynomial of $w_1^n$. Now consider the embedding $_i$ of $\mathbb{F}_q(X)$ $w_1$ into $\overline{\mathbb{F}}_q((X^{-1}))$, which fixes $\mathbb{F}_q(X)$ and maps $w_1$ to $w_i$

$$P(w_i^n) = P((\sigma_i(w_1)^n)) = P(\sigma_i(w_1^n)) = \sigma_i(P(w_1^n)) = \sigma_i(0) = 0$$

So for all $i \le d$ $w_i^n$ satisfies $P(Y)=0$ We have, $[\mathbb{F}_q(X)(w_1^n) : \mathbb{F}_q(X)] \le [\mathbb{F}_q(X)(w_1) : \mathbb{F}_q(X)]$. This shows that $\deg(P) \le \deg(M)$ So $w_1^n, w_2^n, \cdots, w_d^n$ are all the roots of $P$

If $k+1 \le i \le d$ then $|w_i^n| = |w_i|^n < 1$ (respectively. there are at least $k+1 \le j \le d$ such that $|w_j^n| = |w_j|^n = 1$) and $|w_i^n| = |w_i|^n > 1$ for $I = 1 \ldots k$

Therefore $(w_1^n,\cdots,w_k^n) \in S_k'$( respectively. $\in T_k'$ ) for all $n\mathbb{N}^*$

**Proposition 3.2:** *Let* $(w_1,\cdots,w_k) \in S_k'$ *then* $\{w_1^n + \cdots, w_k^n\} \to 0$ *as* $n \to \infty$

*Proof.* Let $w_1$ be an Pisot series and $w_1 .. w_d$ its conjugates. By the proof of theorem 3.1, for all $n \in \mathbb{N}$ a $w_1^n, \cdots, w_k^n$ are the roots of some degree $d$ irreducible polynomial, $P_n$ in $\mathbb{F}_q[X]$ Also,

$$tr(P_n) = \sum_{i=1}^{d} w_i^n \in \mathbb{F}_q[X]$$

So $\{tr(P_n)\}=0$ The above can be rewritten as

$$\{tr(P_n) = \sum_{i=1}^{d} w_i^n\} = \{w_1^n + \cdots + w_k^n + \sum_{i=k+1}^{d} w_i^n\}$$

Since, for $k+1 \le i \le d$ by definition $|w_i|<1$ therefore $w_i^n \to 0$. Thus $\{\sum_{i=k+1}^{d} w_i^n\} \to 0$. Therefore $\{w_1^n + \cdots w_k^n\} \to 0$.

**Proposition 3.3:** *Let* $(w_1,\cdots,w_k) \in S_k'$ *with minimal polynomial* $p \in \mathbb{F}_q[X][Y]$ *of degree* $k+1$ *and* $w = w_1, \ldots w_k \ldots w_{k+1}$ *the conjugates of* $w$ *If* $w$ *is unit then* $\dfrac{(-1)^{k-1}}{c} w_1 \cdots w_k \in S'$.

*Proof.* Let $(w_1,\cdots,w_k) \in S_k'$ with minimal polynomial $P$ has degree $k+1$ and $P(0)=C$

Let $w_{k+1}$ be the $k+1$ root of $P$ Since

$P(Y) = (Y - w_1)(Y - w_2) \cdots (Y - w_{k+1}).$

Consider

$$\frac{1}{w_1}, \frac{1}{w_2}, \cdots, \frac{1}{w_{k+1}} = \frac{(-1)^{k+1}}{c} w_1 \cdots w_k.$$

Clearly $Q$ is unit, irreducible over $\mathbb{F}_q[X][Y]$. and has roots $\frac{1}{w_1}, \frac{1}{w_2}, \cdots, \frac{1}{w_{k+1}} = \frac{(-1)^{k+1}}{c} w_1 \cdots w_k$. We have $|\dfrac{1}{w_{k-1}}| = |\dfrac{(-1)^{k+1} w_1 \cdots w_k}{c}| = |w_1 \cdots w_k| > 1$ and $|\dfrac{1}{w_i}| < 1$ for $i=1, 2, \ldots k$ Therefore $(-1)^{k+1} w_1 \cdots w_k$ is a Pisot series.

### Formal Pisot-k pairs (respectively Salem) Series

**Theorem 3.4:** *Let the polynomial*

$$P(Y) = A_n Y^n + A_{n-1} Y^{n-1} + A_{n-2} Y^{n-2} + \cdots + A_1 Y + A_0,\ A_0 \ne 0,\ A_i \in \mathbb{F}_q[X].$$

$P$ has exactly $k$ roots that lie outside the unit disc sauch that one the these roots are the biggest and all remaining roots have modulus strictly less than 1(respectively the other roots are have modulus inferior or equal to 1 and at least exist a root of module equals to 1) if and only if $|A_{n-k}| > \sup_{i \ne n-k} |A_i|$. respectively $|A_{n-k}| > \sup_{i \ne n-k} |A_i|$ and $|A_{n-k}| = \sup_{i < n-k} |A_i|$.)

*Proof.* Let $w = w_1, \ldots w_2, \ldots w_n$ be the roots of $P(Y)$ such that

$$|w_1|>|w_2|\geq\cdots\geq|w_k|>1>|w_{k+1}|\geq\cdots\geq|w_n|.$$

We have

$$\begin{cases}
\left|\sum_{1\leq i\leq n} w_i\right| = \left|\frac{A_{n-1}}{A_n}\right| \leq |w_1| & <|w_1\cdots w_k| = \left|\frac{A_{n-k}}{A_n}\right| \\
\left|\sum_{1\leq i_1<i_2\leq n} w_{i_1}w_{i_2}\right| = \left|\frac{A_{n-2}}{A_n}\right| \leq |w_1w_2| & <|w_1\cdots w_k| = \left|\frac{A_{n-k}}{A_n}\right| \\
\vdots & \vdots \\
\left|\sum_{1\leq i_1<i_2\leq\cdots\leq i_{k-1}\leq n} w_{i_1}w_{i_2}\cdots w_{i_{k-1}}\right| = \left|\frac{A_{n-k+1}}{A_n}\right| \leq |w_1\cdots w_{k-1}| & <|w_1\cdots w_k| = \left|\frac{A_{n-k}}{A_n}\right| \\
\left|\sum_{1\leq i_1<i_2\leq\cdots\leq i_{k+1}\leq n} w_{i_1}w_{i_2}\cdots w_{i_{k+1}}\right| = \left|\frac{A_{n-k-1}}{A_n}\right| \leq |w_1\cdots w_{k+1}| & <|w_1\cdots w_k| = \left|\frac{A_{n-k}}{A_n}\right| \\
\vdots & \vdots \\
\left|\prod_{1\leq i\leq n} w_i\right| = \left|\frac{A_0}{A_n}\right| = |w\cdots w_n| & <|w\cdots w_k| = \left|\frac{A_{n-k}}{A_n}\right|
\end{cases}$$

Then

$$|A_{n-k}|>\sup_{i\neq n-k}|A_i|.$$

Second, Prove the sufficiency by the symmetrical relations of the roots of a polynomial. Let $w = w_1,\ldots w_2,\ldots w_n$ be the roots of $P(Y)$ such that $|w_1|>|w_2|\geq|w_3|\geq\cdots\geq|w_n|$. Then

$$\left|\frac{A_{n-k+1}}{A_n}\right| = \left|\sum_{1\leq i_1<i_2\leq\cdots\leq i_{k-1}\leq n} w_{i_1}w_{i_2}\cdots w_{i_{k-1}}\right| = |w_1\cdots w_{k-1}| < \left|\frac{A_{n-k}}{A_n}\right| = |w_1\cdots w_k|.$$

So

$$|w_k|>1 \; and \; |w_1|>|w_2|\geq|w_3|\geq\cdots\geq|w_k|>1.$$

on the other hand

$$\left|\frac{A_{n-k-1}}{A_n}\right| = \left|\sum_{1\leq i_1<i_2\leq\cdots\leq i_{k+1}\leq n} w_{i_1}w_{i_2}\cdots w_{i_{k+1}}\right| = |w_1\cdots w_{k+1}| < \left|\frac{A_{n-k}}{A_n}\right| = |w_1\cdots w_k|.$$

Then

$$|w_{k+1}|<1 \; and \; 1>|w_{k+1}|>|w_{k+2}|\geq\cdots\geq|w_n|.$$

Now, if $|A_{n-k}| = \sup_{i<n-k}|A_i|$.

Let $w = w_1,\ldots w_2,\ldots w_k$ be the roots of $P(Y)$ such that

$$|w_1|>|w_2|\geq\cdots\geq|w_k|>1>|w_{k+1}|\geq\cdots\geq|w_n| \; and \; \exists\, k+1\leq j\leq n \; such\; that \; |w_j|=1.$$

We have

$$\begin{cases}
\left|\sum_{1\leq i\leq n} w_i\right| = \left|\frac{A_{n-1}}{A_n}\right| \leq |w_1| & <|w_1\cdots w_k| = \left|\frac{A_{n-k}}{A_n}\right| \\
\left|\sum_{1\leq i_1<i_2\leq n} w_{i_1}w_{i_2}\right| = \left|\frac{A_{n-2}}{A_n}\right| \leq |w_1w_2| & <|w_1\cdots w_k| = \left|\frac{A_{n-k}}{A_n}\right| \\
\vdots & \vdots \\
\left|\sum_{1\leq i_1<i_2\leq\cdots\leq i_{k-1}\leq n} w_{i_1}w_{i_2}\cdots w_{i_{k-1}}\right| = \left|\frac{A_{n-k+1}}{A_n}\right| \leq |w_1\cdots w_{k-1}| & <|w_1\cdots w_k| = \left|\frac{A_{n-k}}{A_n}\right| \\
\left|\sum_{1\leq i_1<i_2\leq\cdots\leq i_{k+1}\leq n} w_{i_1}w_{i_2}\cdots w_{i_{k+1}}\right| = \left|\frac{A_{n-k-1}}{A_n}\right| \leq |w_1\cdots w_{k+1}| & \leq|w_1\cdots w_k| = \left|\frac{A_{n-k}}{A_n}\right| \\
\vdots & \vdots \\
\left|\sum_{1\leq i_1<i_2\leq\cdots\leq i_j\leq n} w_{i_1}w_{i_2}\cdots w_{i_j}\right| = \left|\frac{A_{n-(k+j)}}{A_n}\right| \leq |w_1\cdots w_{k-j}| & \leq|w_1\cdots w_k| = \left|\frac{A_{n-k}}{A_n}\right| \\
\vdots & \vdots \\
\left|\prod_{1\leq i\leq n} w_i\right| = \left|\frac{A_0}{A_n}\right| = |w\cdots w_n| & \leq|w\cdots w_k| = \left|\frac{A_{n-k}}{A_n}\right|
\end{cases}$$

Then

$$|A_{n-k}|>\sup_{i>n-k}|A_i| \; and \; |A_{n-k}| = \sup_{i<n-k}|A_i|.$$

**Consequence 3.5:** *Let the polynomial*

$$P(Y) = A_nY^n + A_{n-1}Y^{n-1} + A_{n-2}Y^{n-2} + \cdots + A_1Y + A_0,\; A_0\neq 0,\; A_i\in\mathbb{F}_q[X].$$

*P* has exactly *k* roots that lie outside the unit disc sauch that one the these roots are the biggest and all remaining roots have modulus strictly less than 1(respectively the other roots are have modulus inferior or equal to 1 and at least exist a root of module equals to 1) if and only if $|A_{n-k}|>|A_{n-k+1}|>\cdots>|A_{n-1}|>\sup_{i<n-k}|A_i|$. respectively $|A_{n-k}|>\sup_{i\neq n-k}|A_i|$ and $|A_{n-k}| = \sup_{i<n-k}|A_i|$.)

**Corollary 3.6:** *Let P be the polynomial*

$$\Lambda(Y) = \lambda_nY^n + \lambda_{n-1}Y^{n-1} + \lambda_{n-2}Y^{n-2} + \cdots + \lambda_1Y + \lambda_0,\; \lambda_0\neq 0,\; \lambda_i\in\mathbb{F}_q[X]$$

such that $|\lambda_{n-k}|>\sup_{i\neq n-k}|\lambda_i|$. If $|\lambda_{n-k}|\geq|\lambda_{n-k+1}||\lambda_{n-1}|$, then $\Lambda$ has no roots in $\mathbb{F}_q((X^{-1}))$

*Proof.* By the previous Theorem Λ has *k* roots of modules >1 with a different value and the other roots are of modules <1 Let $w = w_1,\ldots w_2,\ldots w_n$ be the roots of $P(Y)$ such that

$$|w_1|>|w_2|\geq\cdots\geq|w_k|>1>|w_{k+1}|\geq\cdots\geq|w_n|.$$

We have

$$|\lambda_{n-k}|\geq|\lambda_{n-k+1}||\lambda_{n-1}|.$$

And so

$$|w_1\cdots w_k|\geq|w_1\cdots w_{k-1}||w_1|.$$

This gives us that

$$|w_k|\geq|w_1|.$$

witch is absurd. So Λ is irreducible on $\mathbb{F}_q[X][Y]$

**Theorem 3.7:** *Let* $(w_i)\in\mathbb{F}_q((X^{-1}))$ *for* $i=1,\cdots,k$ *witch* $|w_1|>|w_2|\geq\cdots|w_k|>1$ *be the are roots of the polynomial*

$$\Lambda(Y) = Y^n + \lambda_{n-1}Y^{n-1} + A_{n-2}Y^{n-2} + \cdots + \lambda_1Y + \lambda_0,\; \lambda_i\in\mathbb{F}_q[X],\; \lambda_0\neq 0 \; and \; |\lambda_{n-2}|>\sup_{i\neq n-2}|\lambda_i|$$

If $\deg\lambda_{n-k} = \deg\lambda_{n-k+1} + \deg\lambda_{n-1}$, then Λ is the minimal polynomial of *w* and $(w = w_1,\ldots w_k)\in S_k$

*Proof.* By the Theorem 3.4, *P* has exactly *k* conjugates of $w = w_1,\ldots w_k$ that lie outside the unit disc and all remaining conjugates have modulus strictly less than <1 Let's $w_i$ for *i*=*k*+1,…*n* witch $|w_i|<1$ the authors roots of *P*. Show that Λ(*Y*) is irreducible.

By the condition of the Theorem, $\Lambda(0)=\lambda_0\neq 0$ hence, all roots of the polynomial Λ(*Y*). are not equal to 0

Let $\Lambda(Y) = \Lambda_1(Y).\,\Lambda_2(Y)$ where $\Lambda_i(Y)$, *i* =1,2, has of the coefficients in $\mathbb{F}_q[X]$.

Suppose in the first $w_1,\ldots w_k$ are the roots of $\Lambda_1$ and the other roots are of $\Lambda_2$ Clearly, the absolute value of leading coefficient of the polynomial $\Lambda_2$ superior or equal 1 with is absurd because $\Lambda_2$ has only roots $w_i$ such that

$0<|w_i|<1$

Suppose in the second that $\Lambda_1$ is the polynomial of the series $w_1,\ldots w_{k'}$ and $\Lambda_2$ the polynomial of the series $w_{k'+1}\ldots w_{k''}$ such that $k'+k''=k$

$$\begin{aligned}
\Lambda(Y) &= \Lambda_1(Y).\Lambda_2(Y) \\
&= (Y^s + A_{s-1}Y^{s-1} + \cdots + A_{s-k'}Y^{s-k'} + \cdots + A_1Y + A_0)(Y^m + B_{m-1}Y^{m-1} + \cdots \\
&\quad + B_{m-k''}Y^{m-k''} + \cdots + B_1Y + B_0)
\end{aligned}$$

Then we have

$$\begin{cases} \lambda_{n-1} = & A_{s-1} + B_{m-1} \\ \lambda_{n-k+1} = & \cdots + A_{s-k'+1} B_{m-k''} + A_{s-k'} B_{m-k''+1} + \cdots \\ \lambda_{n-k} = & \cdots + A_{s-k'} B_{m-k''} + \cdots \end{cases}$$

This gives us that

$$\begin{cases} \deg \lambda_{n-1} = & \sup(\deg A_{s-1}; \deg B_{m-1}) = \deg A_{s-1} \ because \ |A_{s-1}| = |w_1| > |w_i| \ for \ i \neq 1 \\ \deg \lambda_{n-k} = & \deg A_{s-k'} + \deg B_{m-k''} \\ \deg \lambda_{n-k+1} = & \sup(\deg A_{s-k'} + \deg B_{m-k''+1}; \deg A_{s-k'+1} + \deg B_{m-k''}). \end{cases}$$

If $\deg \lambda_{n-k+1} = \deg A_{s-k'} + \deg B_{m-k''+1}$, then

$$\deg \lambda_{n-k} = \deg A_{s-k'} + \deg B_{m-k''}.$$

And

$$\deg \lambda_{n-k} = \deg \lambda_{n-k+1} + \deg \lambda_{n-1} + \deg A_{n-1}.$$

$$\deg A_{s-k'} + \deg B_{m-k''} = \deg A_{s-k'} + \deg B_{m-k''+1} + \deg A_{s-1}.$$

Also

$$\deg B_{m-k''} = \deg B_{m-k''+1} + \deg A_{s-1}.$$

$$|B_{m-k''}| = |B_{m-k''+1}| + |\deg A_{s-1}|.$$

This gives us that

$$|w_{k'+1} \cdots w_k''| = |w_{k'+1} \cdots w_{m-k''+1}| |w_1|.$$

Therefore

$$|w_k''| = |w_1|.$$

With is absurd.

Now if $\deg \lambda_{n-k+1} = \deg A_{s-k'+1} + \deg B_{m-k''}$, then

$$\deg \lambda_{n-k} = \deg A_{s-k'} + \deg B_{m-k''}.$$

And

$$\deg \lambda_{n-k} = \deg \lambda_{n-k+1} + \deg \lambda_{n-1} + \deg A_{n-1}.$$

$$\deg A_{s-k'} + \deg B_{m-k''} = \deg A_{s-k'+1} + \deg B_{m-k''} + \deg A_{s-1}.$$

This gives us that

$$|A_{s-k'}| = |A_{s-k'+1}| . |\deg A_{s-1}|.$$

$$|A_{s-k'}| = |A_{s-k'+1}| . |\deg A_{s-1}|.$$

We obtain

$$|w_1 \cdots w_{k'}| = |w_1 \cdots w_{k'-1}| |w_1|.$$

Therefore

$$|w_{k'}| = |w_1|.$$

With is absurd. $w_1, \ldots w_{k'+1} \cdots w_k''$

Let $\Lambda(Y)$=where $\Lambda_i(Y), i=1,2,3$ has of the coefficients in $\mathbb{F}_q[X]$. Suppose in the first $w_1, \ldots w_k$ are the roots of $\Lambda_1$ $w_{k'+1} \cdots w_k''$ are the roots of $\Lambda_2$, such that $k'+k''=k$ and the other roots are of $\Lambda_3$. Clearly, the absolute value of leading coefficient of the polynomial $\Lambda_3$ superior or equal 1 with is absurd because $\Lambda_2$ has only roots $w_i$ such that $0<|w_i|<1$.

We conclude that $\Lambda$ is the minimal polynomial of and $(w_1 \ldots w_k) \in S_k$

**Corollary 3.8:** *Let $\mathbb{P}_k(n,h)$ the set of the element of Pisot-k series in $_q((X^{-1}))$ of degree n and of height logarithmic h Then the numbers of the elements of $\mathbb{P}_k(n,h)$ are*

$$|\mathcal{P}_k(n,h)| \leq (q-1)(q^{h-1}-1)q^{n(h-1)-1}.$$

*Proof.* The set of the minimal polynomials of the elements of $\mathbb{P}_k(n,h)$ is

$$\mathcal{L}_k(n,h) = \{P(Y) = Y^n + A_{n-1}Y^{n-1} + \cdots + A_{n-k}Y^{n-k} + \cdots + A_0 :$$

$$h = \deg A_{n-k} > \sup \deg_{i \neq n-k} A_i \ and \ \deg A_{n-k} = \deg A_{n-k+1} + \deg A_{n-1}.\}$$

By the consequence 3.5, we have $h = |A_{n-k}| > |A_{n-k+1}| > \cdots > |A_{n-1}| > \sup_{i<n-k} |A_i|$. we obtain

$$\begin{cases} \deg A_{n-k} = h, \ then & \# A_{n-k} = (q-1)q^h \\ \deg A_{n-k+1} \leq h-1, \ then & \# A_{n-1} \leq q^{h-1} \\ \vdots & \vdots \\ \deg A_{n-1} \leq h-k+1, \ then & \# A_{n-1} \leq q^{h-k+1} \\ \deg A_{n-k-1} \leq h-k, \ then & \# A_{n-1} \leq q^{h-k} \\ \vdots & \vdots \\ \deg A_0 \leq h-k-1, \ and \ A_0 \neq 0, \ then & \# A_0 \leq q^{h-k} - 1 \end{cases}$$

Therefore

$$|\mathcal{L}_k(n,h)|$$

$$\leq (q-1)q^h.q^{h-1}.\cdots q^{h-k+1} q^{(h-k)(n-k+1)}(q^{h-k}-1)$$

$$\leq (q-1)(q^{h-1}-1)q^{(h-k)(n-k-1)}(q^{h-k}-1)q^{[h+(h-1)+\cdots(h-k+1)]}$$

$$\leq (q-1)(q^{h-1}-1)q^{(h-k)(n-k+1)}(q^{h-k}-1)q^{\frac{k}{2}(2h-k+1)}.$$

## Conclusion

The presentation and estimation for the number of the Pisot-k power formal series, precisely we will give their degrees and their logarithmic heights.


### Acknowledgements

I would like to thank Dr. Mabrouk ben ammar and Amara chandoul for all of his guidance and instruction, and for introducing me to the study of Pisot numbers.


### References


1. Brauer A (1951) On algebraic equations with all but one root in the interior of the unit circle. Math Nachr 4: 250-257.
2. Akiyama S (2006) Positive finiteness of number systems. Number theory , Tradition and Modernization.
3. Bateman P, Duquette AL (1962) The analogue of Pisot-Vijayaraghavan numbers in fields of power series. Ill J Math 6: 594-406.
4. Chandoul A, Jellali M, Mkaouar M (2011) Irreducibility criterion over finite fields. Communication in Algebra 39: 3133-3137.
5. Chandoul A, Jellali, Mkaouar M (2013) The smallist Pisot element over $\mathbb{F}_q((X^{-1}))$, Communication in Algebra 56: 258-264.
6. Sprindzuk VG (1963) Mahler's problem in metric number theory, Translaion of Mathematical monographs. Amer math Soc.

# Identification of Wave Regularities According to Statistical Data of Parameters of 24 Pulsars

**Mazurkin PM***

*Doctor of Engineering, Academician of Russian Academy of Natural History and Russian Academy of Natural Sciences, Volga State University of Technology, Russia*

**Abstract**

The method of identification is shown on the example of tabular these measurements of six parameters at 24 pulsars. The equations of a trend and oscillatory indignations on the basis of steady laws on the generalized wave function in the form of an asymmetric wavelet signal with variables of amplitude and the period of fluctuation are received. On the remains it is possible to receive a set of microfluctuations and to bring identification to an error of measurements. Schedules of components of the generalized model of a wavelet signal allow to see visually a picture of mutual influence of all six parameters of pulsars. On the revealed equations it is possible to carry out the amplitude-frequency analysis. Quality of basic data is estimated by rank distributions of values of parameters of pulsars. Thus ranging of values of parameters on a preference preorder vector "better→worse" is carried out in the beginning, the rating of pulsars is formed further.

**Keywords**: Wavelet; Identification; Pulsars; Parameters; The relations; Regularities

## Introduction

Unlike deductive approach to wavelet analysis proceeding from the equations of classical mathematics inductive approach when statistical selection is primary is offered and concerning it the structure and values of parameters of the generalized wave function [1-17] is identified. Any phenomenon (time cut) or process (change in time) according to sound tabular statistical quantitative data (a numerical field) inductively can be identified the sum of asymmetric wavelet signals of a look:

$$y = \sum_{i=1}^{m} y_i, y_i = A_i \cos(\pi x / pi - a_{8i}) \qquad (1)$$

$$A_i = a_{1i} x^{a_{2i}} \exp(-a_{3i} x^{a_{4i}}), \quad p_i = a_{5i} + a_{6i} x^{a_{7i}}$$

Where $y$ - indicator (dependent variable), $i$ - number of the making statistical model (1), $m$ - the number of members of model depending on achievement of the remains from (1) error of measurements, $x$ - explanatory variable, $Ai$-amplitude (half) of fluctuation (ordinate), $pi$ - half-cycle of fluctuation (abscissa), $a1...a8$ - the parameters of model (1) determined in the program environment On a formula (1) with two fundamental physical constants e (Napier's number or number of time) and π (Archimedes's number or number of space) the quantized wavelet signal is formed from within the studied phenomenon and/or process.

## Basic Data for Statistical Modeling

The pulsars found in the Einstein @Home project are given in article [18]. The data selected for statistical modeling are provided in Table 1. In Table 1 symbols with preference vectors are accepted: P: spin periods; P Epochepochal period of spin; DM: Dispersion Measure; S1400: Flux Densities; D: Estimated Distance; S: Significance are given for reproducibility reasons.

In total from data [18] it was succeeded to allocate six factors having quantitative values. On them it is possible to assume that amplitude and the period of fluctuations on the general model (1) submit to the biotechnical law [2-5]. In the beginning we will consider rank distributions of values of each factor on a vector of preference and we will determine a rating of pulsars by the sum of ranks, and then we will carry out the factorial analysis of all $6^2 - 6=30$ binary relations.

## Rank Distributions of Values of Factors

Any factors have an accurate vector orientation. The person understands an orientation of changes therefore only two options of a vector of preference are possible:

| PSR | *P (s)* | P Epoch (MJD) | DM(pc $cm^{-3}$) | *S1400 (mJy)* | *D (kpc)* | S |
|---|---|---|---|---|---|---|
| J0811-38 | 0.482594 | 50824.5 | 336.2 | 0.3 | 6.2 | 15.6 |
| J1227-6208b | 0.03453 | 51034.1 | 363.2 | 0.8 | 8.4 | 17.9 |
| J1305-66 | 0.197276 | 51559.7 | 316.1 | 0.2 | 7.5 | 15.5 |
| J1322-62 | 1.044851 | 50591.6 | 733.6 | 0.3 | 13.2 | 23.1 |
| J1637-46 | 0.493091 | 50842.9 | 660.4 | 0.7 | 7 | 17.2 |
| J1644-44 | 0.173911 | 51030.2 | 535.1 | 0.4 | 6.2 | 14.1 |
| J1644-46 | 0.250941 | 50839 | 405.8 | 0.8 | 4.8 | 13.2 |
| J1652-48b | 0.003785 | 51373.3 | 187.8 | 2.7 | 3.3 | 22.3 |
| J1726-31b | 0.12347 | 51026.4 | 264.4 | 0.4 | 4.1 | 15.9 |
| J1748-3009b | 0.009684 | 51495.1 | 420.2 | 1.4 | 5 | 18 |
| J1750-2536b | 0.034749 | 50593.8 | 178.4 | 0.4 | 3.2 | 15.9 |
| J1755-33 | 0.959466 | 52080.6 | 266.5 | 0.2 | 5.7 | 21.2 |
| J1804-28 | 1.273011 | 51973.7 | 203.5 | 0.4 | 4.2 | 13.2 |
| J1811-1049+ | 2.623859 | 55983.5 | 253.3 | 0.3 | 5.5 | 29.2 |
| J1817-1938+ | 2.046838 | 55991.8 | 519.6 | 0.1 | 8.6 | 16.9 |
| J1821-0331+ | 0.902316 | 55980.9 | 171.5 | 0.2 | 4.3 | 28.3 |
| J1838-01 | 0.183295 | 51869.1 | 320.4 | 0.3 | 6.9 | 16.7 |
| J1838-1849+ | 0.488242 | 55991.9 | 169.9 | 0.4 | 4.5 | 31.7 |
| J1840-0643+ | 0.035578 | 55930 | 500 | 1.2 | 6.8 | 18.2 |
| J1858-0736 | 0.551059 | 56108.5 | 194 | 0.3 | 5 | 16.7 |

**Table 1:** Parameters of 24 pulsars.

***Corresponding author:** Mazurkin PM, Doctor of Engineering, professor, Academician of Russian Academy of Natural History and Russian Academy of Natural Sciences, Volga State University of Technology, Russia
E-mail: kaf_po@mail.ru

a) Better it is less (yes better, the symbol ↓on a vector "better→worse");

б) Better it is more (and it is good, therefore in the Table 2 a symbol ↑).

In function=RANG(P1;P$1:P$24;1) for the first indicator P in the program Excel environment the following symbols are accepted: P1: identifier of a column and the first line; P$1: the first line of the ranged column; P$24: the last line of the ranged column according to the Table 1; 0∨1: ranging on decrease (0) or to increase (1).

Ranks change from zero therefore it is necessary from results of ranging in the program Excel environment to subtract unit can be clearly understood from Table 2 and Figure 1.

The interrelation of a factor from most on rank distribution proves to be good quality or quality of basic data and it serves for check of their reliability.

The analysis of good quality of basic data is made on coefficient of correlation r of the equation

*y=f* (*R*=0,1,2,3,...) of rank distribution of a factor on the general formula

$Y=Y0\ \exp(\pm aR^{b})$ ()

where Y - the ranged parameter, Y0- initial value of parameter, a - activity of exponential growth or death of values of parameter; b - intensity of growth or recession.

Parametrical identification [8-10,12,18] of formulas (1) received the equations:

$$S1400 = 2.70997\exp\left(-0.49867R_{S1400}^{0.59345}\right), r = 0.9918 \quad (2)$$

$$D = 13.00645\exp\left(-0.32575R_{D}^{0.43075}\right), r = 0.9887 \quad (3)$$

$$DM = 134.86273\exp(0.052517R_{DM}^{1.09651}), r = 0.9875 \quad (4)$$

$$S = 36.27767\exp(-0.14785R_{P}^{1.53169}), r = 0.9829 \quad (5)$$

$$P = 0.10117\exp(0.026123R_{P}^{1.53169}), r = 0.9804 \quad (6)$$

$$P_{Epoch} = 49452.9768\exp(0.0050522R_{PEpoch}), r = 0.8697 \quad (7)$$

On decrease of coefficient of correlation a rating of factors on quality of measurements the following: 1: S1400; 2: D; 3: DM; 4: S; 5: P and 6: P Spoch. Thus it appeared that it is convenient to use ranks instead of factors as remove a mathematical problem of "curse of dimensionality", for example, at a rating on a set of diverse indicators.

## Rating of Pulsars

The general vector of heuristic preference "better→worse" leads all considered factors to one "denominator" that allows to estimate the sum of ranks (Table 3), even without mathematical justification, ratings of subjects and objects (in our example among 24 pulsars).

From data of Table 3 it is visible that on the first place on an indicator I there is J1227- 6208b pulsar, on the second - J1652-48b and on the third J1748-3009b. And among the factors for the indicator IF on the first place is a factor in S1400, the second S and in third place D. And the three factors P, P Epoch and DM took fourth place. The rating of pulsars (Figure 2) changes under the law of exponential growth

$$\Sigma R=44.63386\ exp(0.0405641 I^{0.90632}) \quad (8)$$

From the remains in Figure 2 it is visible that at bigger quantity of pulsars in addition to the equation (8) also wave components on model are possible (1).

## Binary Relations between Factors

From six factors of all are possible $6^2$ – 6=30 binary relations. Six distributions are rank. Correlation matrix, including and rank distributions, it is given in Table 4.

| PSR | RP | P (s) | $R_{PE}$ | P Epoch (MJD) | $R_{DM}$ | DM (pc $cm^{-3}$ | $R_{S1400}$ | S1400 (mJy | $R_D$ | D (kpc) | $R_S$ | S |
|---|---|---|---|---|---|---|---|---|---|---|---|---|
| J0811-38 | 12 | 0.482594 | 2 | 50824.5 | 14 | 336.2 | 14 | 0.3 | 8 | 6.2 | 17 | 15.6 |
| J1227-6208b | 2 | 0.03453 | 9 | 51034.1 | 15 | 363.2 | 4 | 0.8 | 2 | 8.4 | 10 | 17.9 |
| J1305-66 | 9 | 0.197276 | 12 | 51559.7 | 12 | 316.1 | 20 | 0.2 | 3 | 7.5 | 18 | 15.5 |
| J1322-62 | 20 | 1.044851 | 0 | 50591.6 | 23 | 733.6 | 14 | 0.3 | 0 | 13.2 | 5 | 23.1 |
| J1455-59 | 7 | 0.176191 | 4 | 50841.7 | 18 | 498 | 1 | 1.6 | 4 | 7 | 20 | 14 |
| J1601-50 | 16 | 0.860777 | 6 | 50993.6 | 0 | 59 | 8 | 0.4 | 21 | 3.6 | 3 | 29.1 |
| J I 619-42 | 19 | 1.023152 | 16 | 51975.6 | 3 | 172 | 7 | 0.6 | 20 | 3.7 | 0 | 35.4 |
| J1626-44 | 11 | 0.308354 | 13 | 51718.6 | 11 | 269.2 | 14 | 0.3 | 14 | 4.8 | 21 | 13.2 |
| J1637-46 | 14 | 0.493091 | 5 | 50842.9 | 22 | 660.4 | 6 | 0.7 | 4 | 7 | 11 | 17.2 |
| J1644-44 | 6 | 0.173911 | 8 | 51030.2 | 21 | 535.1 | 8 | 0.4 | 8 | 6.2 | 19 | 14.1 |
| J1644-46 | 10 | 0.250941 | 3 | 50839 | 16 | 405.8 | 4 | 0.8 | 14 | 4.8 | 21 | 13.2 |
| J1652-48b | 0 | 0.003785 | 10 | 51373.3 | 5 | 187.8 | 0 | 2.7 | 22 | 3.3 | 6 | 22.3 |
| J1726-31b | 5 | 0.12347 | 7 | 51026.4 | 9 | 264.4 | 8 | 0.4 | 19 | 4.1 | 15 | 15.9 |
| J1748-3009b | 1 | 0.009684 | 11 | 51495.1 | 17 | 420.2 | 2 | 1.4 | 12 | 5 | 9 | 18 |
| J1750-2536b | 3 | 0.034749 | 1 | 50593.8 | 4 | 178.4 | 8 | 0.4 | 23 | 3.2 | 15 | 15.9 |
| J1755-33 | 18 | 0.959466 | 17 | 52080.6 | 10 | 266.5 | 20 | 0.2 | 10 | 5.7 | 7 | 21.2 |
| J1804-28 | 21 | 1.273011 | 15 | 51973.7 | 7 | 203.5 | 8 | 0.4 | 18 | 4.2 | 21 | 13.2 |
| J1811-1049+ | 23 | 2.623859 | 20 | 55983.5 | 8 | 253.3 | 14 | 0.3 | 11 | 5.5 | 2 | 29.2 |
| J I 817-1938+ | 22 | 2.046838 | 21 | 55991.8 | 20 | 519.6 | 23 | 0.1 | 1 | 8.6 | 12 | 16.9 |
| J1821-0331+ | 17 | 0.902316 | 19 | 55980.9 | 2 | 171.5 | 20 | 0.2 | 17 | 4.3 | 4 | 28.3 |
| J1838-01 | 8 | 0.183295 | 14 | 51869.1 | 13 | 320.4 | 14 | 0.3 | 6 | 6.9 | 13 | 16.7 |
| J1838-1849+ | 13 | 0.488242 | 22 | 55991.9 | 1 | 169.9 | 8 | 0.4 | 16 | 4.5 | 1 | 31.7 |
| J1840-0643+ | 4 | 0.035578 | 18 | 55930 | 19 | 500 | 3 | 1.2 | 7 | 6.8 | 8 | 18.2 |
| J1858-0736 | 15 | 0.551059 | 23 | 56108.5 | 6 | 194 | 14 | 0.3 | 12 | 5 | 13 | 16.7 |

**Table 2**: Rank distributions of six parameters of 24 pulsars.

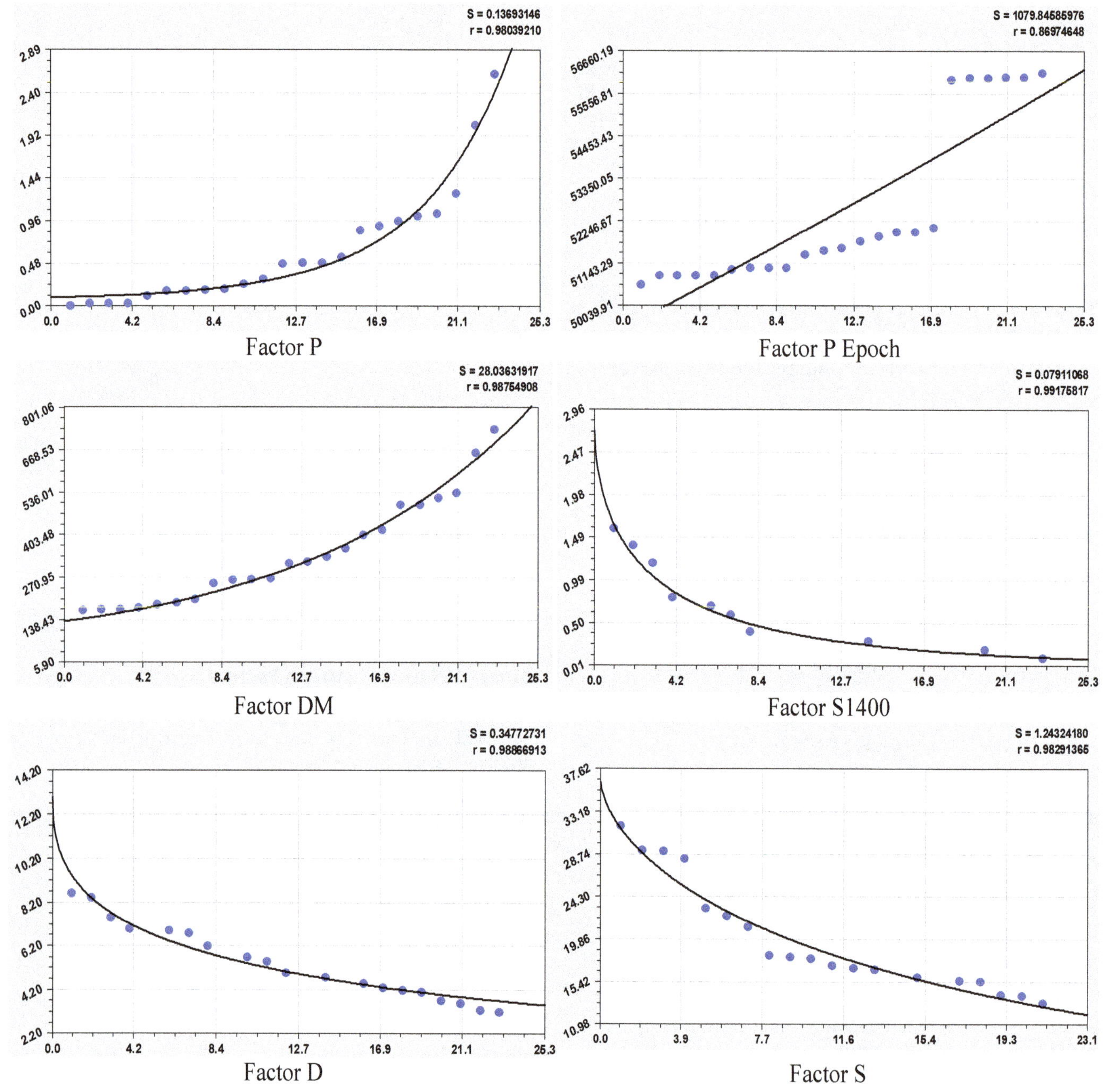


**Figure 1:** Schedules the rank of distributions of values at six parameters of 24 pulsars.

The minimum narrowness of factorial communication is observed at the mathematical DM=f(P) function, and the maximum coefficient of correlation at S1400=f(DM) ratio with the accounting of wave indignations of parameters of pulsars.

## A Rating of Factors as the Influencing Variables and Indicators

From data of Table 4 it is visible that each measured factor can be considered in two roles:

First, as the influencing variable; secondly, as dependent indicator. Thus the method of the factorial analysis offered by us allows not thinking a priori of ratios between separate parameters of the studied system (in our example system from 24 pulsars). As a result the psychological barrier at re- searchers gets off. Many binary relations for the researcher will be unexpected. Therefore as our practice showed, the factorial analysis the unique equation of type (1) allows finding unexpected solutions in the field of research. If some factorial communications are unusual and thus are highly adequate, new technical solutions, and often at the level of inventions of world novelty are shown [18]. Thus repeated identification (1) on a single binary relation we called wavelet analysis [1,6,7,11,13-17].

Without waves, that is changes only on amplitude at very long wave, incommensurably bigger on the fluctuation period to an

| PSR | Ranks of values of factors | | | | | | ∑R | Place I |
|---|---|---|---|---|---|---|---|---|
| | $R_P$ | $R_{PE}$ | $R_{DM}$ | $R_{S1400}$ | $R_D$ | $R_S$ | | |
| J0811-38 | 12 | 2 | 14 | 14 | 8 | 17 | 67 | 13 |
| J1227-6208[b] | 2 | 9 | 15 | 4 | 2 | 10 | 42 | **1** |
| J1305-66 | 9 | 12 | 12 | 20 | 3 | 18 | 74 | 17 |
| J1322-62 | 20 | 0 | 23 | 14 | 0 | 5 | 62 | 9 |
| J1455-59 | 7 | 4 | 18 | 1 | 4 | 20 | 54 | 4 |
| J1601-50 | 16 | 6 | 0 | 8 | 21 | 3 | 54 | 4 |
| J I 619-42 | 19 | 16 | 3 | 7 | 20 | 0 | 65 | 12 |
| J1626-44 | 11 | 13 | 11 | 14 | 14 | 21 | 84 | 22 |
| J1637-46 | 14 | 5 | 22 | 6 | 4 | 11 | 62 | 9 |
| J1644-44 | 6 | 8 | 21 | 8 | 8 | 19 | 70 | 16 |
| J1644-46 | 10 | 3 | 16 | 4 | 14 | 21 | 68 | 14 |
| J1652-48[b] | 0 | 10 | 5 | 0 | 22 | 6 | 43 | **2** |
| J1726-31[b] | 5 | 7 | 9 | 8 | 19 | 15 | 63 | 11 |
| J1748-3009[b] | 1 | 11 | 17 | 2 | 12 | 9 | 52 | **3** |
| J1750-2536[b] | 3 | 1 | 4 | 8 | 23 | 15 | 54 | 4 |
| J1755-33 | 18 | 17 | 10 | 20 | 10 | 7 | 82 | 20 |
| J1804-28 | 21 | 15 | 7 | 8 | 18 | 21 | 90 | 23 |
| J1811-1049[+] | 23 | 20 | 8 | 14 | 11 | 2 | 78 | 18 |
| J I 817-1938[+] | 22 | 21 | 20 | 23 | 1 | 12 | 99 | 24 |
| J1821-0331[+] | 17 | 19 | 2 | 20 | 17 | 4 | 79 | 19 |
| J1838-01 | 8 | 14 | 13 | 14 | 6 | 13 | 68 | 14 |
| J1838-1849[+] | 13 | 22 | 1 | 8 | 16 | 1 | 61 | 8 |
| J1840-0643[+] | 4 | 18 | 19 | 3 | 7 | 8 | 59 | 7 |
| J1858-0736 | 15 | 23 | 6 | 14 | 12 | 13 | 83 | 21 |
| ∑R | 276 | 276 | 276 | 242 | 272 | 271 | 1613 | - |
| Place I / F | 4 | 4 | 4 | 1 | 3 | 2 | - | - |

**Table 3:** Ranks of values of six parameters and rating among 24 pulsars.

| Influencing factors *x* | Dependent factors (indicators *y*) | | | | | |
|---|---|---|---|---|---|---|
| | *P* (s) | *P* Epoch (MJD) | DM (pc cm$^{-3}$) | *S*1400 (mJy) | *D* (kpc) | *S* |
| Spin periods P (s) | | 0.4978 | ***0.0608*** | 0.8640 | 0.2688 | 0.4270/0.8024 |
| *P* Epoch (MJD) | 0.4741 | | 0.1904 | 0.2097 | 0.0652 | 0.3687 |
| Dispersion measure DM (pc cm3) | 0.1039/0.6948 | 0.1904 | | 0.0987/**0.9846** | 0.8103/0.9169 | 0.6445/0.8734 |
| Flux densities S1400 (mJy) | 0.6069 | 0.3223 | 0.4914 | | 0.2089 | 0.2462 |
| Estimated distance D (kpc) | 0.1602/0.7742 | 0.0648 | 0.8086 | 0.2188 | | 0.4270/0.6403 |
| Significance are given for reproducibility reasons S | 0.4335 | 0.3951/0.9145 | 0.4800/0.8602 | 0.2455/0.8576 | 0.4287/0.9163 | |

**Table 4:** Correlation matrix of the binary relations between six factors of 24 pulsars.

| Influencing factors *x* | Dependent factors (indicators *y*) | | | | | | Sum Σr | Place $I_x$ |
|---|---|---|---|---|---|---|---|---|
| | *P* (s) | *P* Epoch (MJD) | DM (pc cm$^{-3}$) | *S*1400 (mJy) | *D* (kpc) | *S* | | |
| *P* (s) | 0.9804 | 0.4978 | 0.0608 | 0.8640 | 0.2688 | 0.4270 | 3.0988 | 1 |
| *P* Epoch (MJD) | 0.4741 | 0.8697 | 0.1904 | 0.2097 | 0.0652 | 0.3687 | 2.1778 | 6 |
| DM (pc cm$^{-3}$) | 0.1039 | 0.1904 | 0.9875 | 0.0987 | 0.8103 | 0.6445 | 2.8353 | 4 |
| *S*1400 (mJy) | 0.6069 | 0.3223 | 0.4914 | 0.9918 | 0.2089 | 0.2462 | 2.8675 | 3 |
| *D* (kpc) | 0.1602 | 0.0648 | 0.8086 | 0.2188 | 0.9887 | 0.4270 | 2.6681 | 5 |
| *S* | 0.4335 | 0.3951 | 0.4800 | 0.2455 | 0.4287 | 0.9829 | 2.9657 | 2 |
| Sum Σr | 2,7590 | 2,3401 | 3,0187 | 2,6285 | 2,7706 | 3,0963 | 16,6132 | - |
| Place of Indicator $I_y$ | 4 | 6 | 2 | 5 | 3 | 1 | - | 0.4615 |

**Table 5:** Rating of factors on the determined relations (trends).

interval of measurements, are formed the so-called determined binary relations. Rank distributions, as a rule, are accepted in the form of not wave steady laws [2-5,8-10,12].

The square correlation matrix received after the analysis the rank distributions and the bi- nary relations between all six variables accepted on basic data from Table 1 is given in Table 5.

Here all 36 relations are considered. Besides, summation in the lines and columns two ratings on decrease of this sum turned out: first, a rating of factors as the influencing variables; secondly, rating of factors as dependent indicators. We will define the second rating (Table 6) taking into account identification of wave indignations (1).

The coefficient of a correlative variation for 24 pulsars is equal

| Indicators *y* | Asymmetric wavelet $y_i=a_{1i}x^{a_{2i}}\exp(-a_{3i}x^{a_{4i}})\cos(\pi x/a_{5i}+a_{6i}x^{a_{7i}})-a_{8i}$ | | | | | | | | Correl. coeffic. *r* |
|---|---|---|---|---|---|---|---|---|---|
| | **Amplitude (half) of fluctuation** | | | | **Fluctuation half-cycle** | | | **Shift** | |
| | $a_{1i}$ | $a_{2i}$ | $a_{3i}$ | $a_{4i}$ | $a_{5i}$ | $a_{6i}$ | $a_{7i}$ | $a_{8i}$ | |
| | | | | The influence *P* (s) | | | | | |
| *S*1400 (Figure 3) | 2050.7218 | 0 | 7.92517 | 0.096413 | 0 | 0 | 0 | 0 | 0.8640 |
| | -1598.1935 | 0.13364 | 8.27987 | 0.14063 | 0 | 0 | 0 | 0 | |
| *S* (Figure 4) | 1.53821e6 | 0 | 11.67158 | 0.0081309 | 0 | 0 | 0 | 0 | 0.8024 |
| | -15.39225 | 0.42954 | 0 | 0 | 3.97475 | -2.58176 | 0.34198 | -1.26394 | |
| | | | | The influence DM (pc cm$^{-3}$) | | | | | |
| *S*1400 (Figure 5) | 2.04947e-9 | 3.72097 | 0.00040479 | 1.41209 | 0 | 0 | 0 | 0 | 0.9846 |
| | 0.63017 | 0 | -0.0010119 | 1 | 0.091922 | 0.060427 | 0.92986 | -1.25936 | |
| | -1.98879e-99 | 48.97221 | 0.011710 | 1.48994 | 11.11160 | 0.0027704 | 0.99620 | -3.00858 | |
| | 1.79820e-71 | 30.68609 | 0.051046 | 1.00882 | 16.28551 | 0.020073 | 1.07326 | 2.77439 | |
| *D* (Figure 6) | 3.34608 | 0 | -0.0015439 | 1 | 0 | 0 | 0 | 0 | 0.9169 |
| | -2.34611e-6 | 2.71288 | 0.080460 | 0.59551 | 196.36141 | -0.018607 | 1.24995 | -3.39603 | |
| *S* (Figure 7) | 29.79584 | 0 | -2.61436e-5 | 1.43411 | 0 | 0 | 0 | 0 | 0.8734 |
| | -5.16004e-6 | 2.85167 | 0.0050163 | 0.99972 | 0 | 0 | 0 | 0 | |
| | 1.19274e-19 | 10.57480 | 0.050357 | 0.99990 | 91.63358 | -0.13742 | 1.00030 | 0.49007 | |
| | | | | The influence *D* (kpc) | | | | | |
| *P* (Figure 8) | 7.26123e-6 | 0 | -10.71388 | 0.045383 | 0 | 0 | 0 | 0 | 0.7743 |
| | 1.59048e-12 | 82.10764 | 34.41670 | 0.69352 | 1.15158 | -0.0050928 | 2.22537 | -0.54182 | |
| DM | 7.64387 | 5.36583 | 15.53911 | 0.19098 | 0 | 0 | 0 | 0 | 0.8086 |
| | | | | The influence *S* | | | | | |
| *P* Epoch (Figure 9) | 71154.646 | 0 | 0.041508 | 1 | 0 | 0 | 0 | 0 | 0.9145 |
| | 16.95545 | 2.85827 | 0.074222 | 1 | 0 | 0 | 0 | 0 | |
| | 4.36354e-97 | 118.48896 | 5.41816 | 1.04780 | 0.029084 | 0.020098 | 0.79277 | 4.45296 | |
| | 1.15086e-81 | 74.94389 | 1.06061 | 1.18174 | 1.62870 | -0.012183 | 1.13631 | -0.72450 | |
| DM (Figure 10) | 2.95376 | 2.76075 | 0.17533 | 1 | 0 | 0 | 0 | 0 | 0.8602 |
| | -1.19512e-13 | 17.08714 | 0.77904 | 1 | 3.16941 | -0.19665 | 0.61241 | -1.19205 | |
| *S*1400 (Figure 11) | 7.88199e-6 | 4.31945 | 0.00024563 | 2.84411 | 0 | 0 | 0 | 0 | 0.8576 |
| | -5.87618e-11 | 8.68372 | 0.0018541 | 2.37349 | 4.86794 | -0.098820 | 0.97998 | -3.36086 | |
| *D* (Figure 12) | 0.014339 | 2.62626 | 0.018621 | 1.49231 | 0 | 0 | 0 | 0 | 0.9163 |
| | 2.66617e-34 | 35.99771 | 1.08513 | 1.09501 | 1.19527 | 0.00025715 | 1.94603 | -0.63065 | |
| | -3.92413e8 | 0 | 10.53668 | 0.22211 | 0.056832 | 0.015024 | 1.05854 | -1.52698 | |
| | 0.023354 | 4.22841 | 1.20530 | 0.67921 | 1.19570 | -0.081320 | 0.30418 | -2.16059 | |

**Table 6:** Parameters of the strong binary relations at correlation coefficient $r \geq 0.7$

| Influencing factors *x* | Dependent factors (indicator *y*) | | | | | | Sum $\sum r$ | Place *I x* |
|---|---|---|---|---|---|---|---|---|
| | *P* (s) | *P* Epoch (MJD) | DM (pc cm$^{-3}$) | *S*1400 (mJy) | *D* (kpc) | *S* | | |
| *P* (s) | 0.9804 | 0.4978 | 0.0608 | 0.8640 | 0.2688 | **0.8024** | 3.4742 | 4 |
| *P* Epoch (MJD) | 0.4741 | 0.8697 | 0.1904 | 0.2097 | 0.0652 | 0.3687 | 2.1778 | 6 |
| DM (pc cm$^{-3}$) | **0.6948** | 0.1904 | 0.9875 | **0.9846** | **0.9169** | **0.8734** | 4.6476 | 2 |
| *S*1400 (mJy) | 0.6069 | 0.3223 | 0.4914 | 0.9918 | 0.2089 | 0.2462 | 2.8675 | 5 |
| *D* (kpc) | **0.7743** | 0.0648 | 0.8086 | 0.2188 | 0.9887 | **0.6403** | 3.4955 | 3 |
| *S* | 0.4335 | **0.9145** | **0.8602** | **0.8576** | **0.9163** | 0.9829 | 4.9650 | 1 |
| Sum $\sum r$ | 3.9640 | 2.8595 | 3.3989 | 4.1265 | 3.3648 | 3.9139 | 21.6276 | - |
| Place of Indicator *I y* | 2 | 6 | 4 | 1 | 5 | 3 | - | 0.6008 |

**Table 7:** Rating of factors on the determined (trends) and the wave relations.

21.6276/62=0.6008. This criterion is applied when comparing various systems, for example, of different groups of pulsars with each other.

From six influencing variables on the first place there was S factor (and according to the Table 5 a factor P). In second place fit factor DM, and in third place - D. Among dependent indicators on the first place there is S1400 factor. On the second place is occupied by the factor P, and the third - S.

## Strong Binary Relations

At correlation coefficient more than 0.7 binary relations between factors become strong (Table 7). As a rule, the concept of wave indignation of the Universe gives significant increase in adequacy of the revealed regularities on a formula (1). From 11 strong communications only two treat the determined relations.

In Tables 6-9 and in Figures 3-12 parameters of models which

| Influencing factors *x* | Dependent factors (indicator *y*) | | | | | |
|---|---|---|---|---|---|---|
| | *P* (s) | *P* Epoch (MJD) | DM (pc cm$^{-3}$) | *S*1400 (mJy) | *D* (kpc) | *S* |
| *P* (s) | | | | 0.8640 | | **0.8024** |
| DM (pc cm$^{-3}$) | | | | **0.9846** | **0.9169** | **0.8734** |
| *D* (kpc) | **0.7743** | | 0.8086 | | | |
| *S* | | **0.9145** | **0.8602** | **0.8576** | **0.9163** | |

**Table 8:** Rating of factors on the determined (trends) and the wave relations.

| Number i | Asymmetric wavelet $y_i = a_{1i} x a^{2i} exp(-a_{3i} x^{a4i}) cos(\pi x / a_{5i} + a_{6i} x^{a7i}) - a^{8i}$ | | | | | | | | Correl. coeffic. *r* |
|---|---|---|---|---|---|---|---|---|---|
| | Amplitude (half) of fluctuation | | | | Fluctuation half-cycle | | | shift | |
| | $a_{1i}$ | $a_{2i}$ | $a_{3i}$ | $a_{4i}$ | $a_{5i}$ | $a_{6i}$ | $a_{7i}$ | $a_{8i}$ | |
| 1 | 2.04947e-9 | 3.72097 | 0.00040479 | 1.41209 | 0 | 0 | 0 | 0 | 0.9846 |
| 2 | 0.63017 | 0 | -0.0010119 | 1 | 0.091922 | 0.060427 | 0.92986 | -1.25936 | |
| 3 | -1.98879e-99 | 48.97221 | 0.011710 | 1.48994 | 11.11160 | 0.0027704 | 0.99620 | -3.00858 | |
| 4 | 1.79820e-71 | 30.68609 | 0.051046 | 1.00882 | 16.28551 | 0.020073 | 1.07326 | 2.77439 | |
| 5 | 2.24333e-50 | 38.03372 | 0.45820 | 1.00739 | 0.49970 | 0 | 0 | 1.79228 | 0.7520 |
| 6 | 1.74209e-12 | 6.20719 | 0.042832 | 1 | 0.99751 | -4.47582e-8 | 1 | 0.23907 | 0.4587 |
| 7 | -1.89081e-26 | 17.94744 | 0.20496 | 1.00222 | 3.99532 | -9.33281e-7 | 1 | 0.83797 | 0.3678 |
| 8 | 8.50896e-54 | 24.31598 | 0.017622 | 1.22135 | 13.86875 | -0.0083353 | 0.81558 | -3.78088 | 0.6264 |
| 9 | 8.15332e-52 | 22.01510 | 0.043386 | 1 | 2.01821 | 5.58641e-9 | 1 | -3.83885 | 0.6344 |
| 10 | 2.80946e-7 | 2.05017 | 0.0036490 | 1 | 0.99921 | 6.42334e-8 | 1 | -6.70016e-6 | 0.1153 |
| 11 | -2.38324e-28 | 13.09090 | 0.039405 | 1.03089 | 1.09087 | 0 | 0 | 0.60793 | 0.6013 |
| 12 | 6.00812e-8 | 2.65338 | 0.015604 | 0.86794 | 2.08923 | 0 | 0 | 2.00084 | 0.5683 |

**Table 9:** Full wavelet analysis of the binary relation of S1400=f(DM).

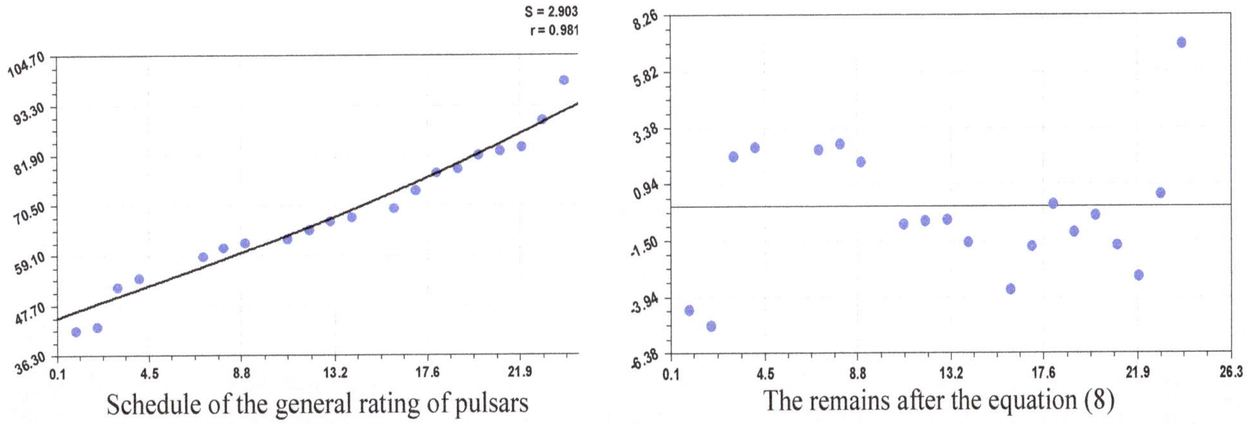


**Figure 2:** Schedules of rating distribution of 24 pulsars.

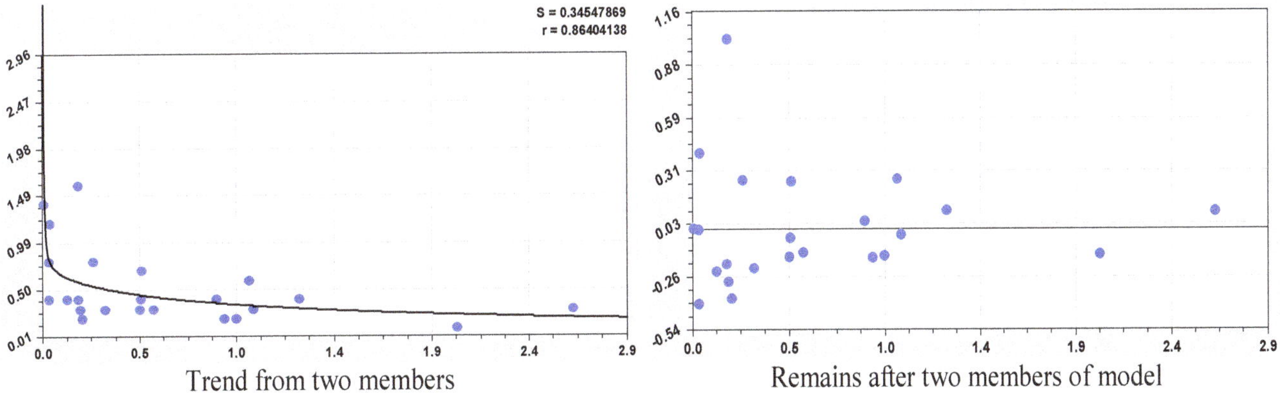


**Figure 3:** Schedules of models of influence of *P* (s) on change of *S*1400 (mJy).

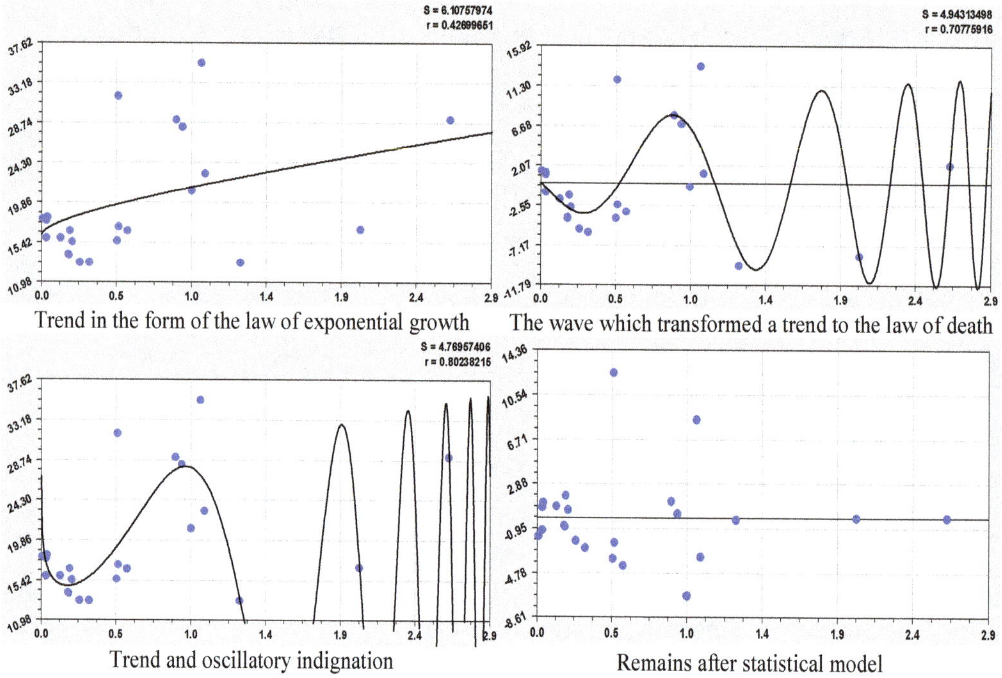


**Figure 4:** Schedules of models of influence of *P* (s) on change of *S*.

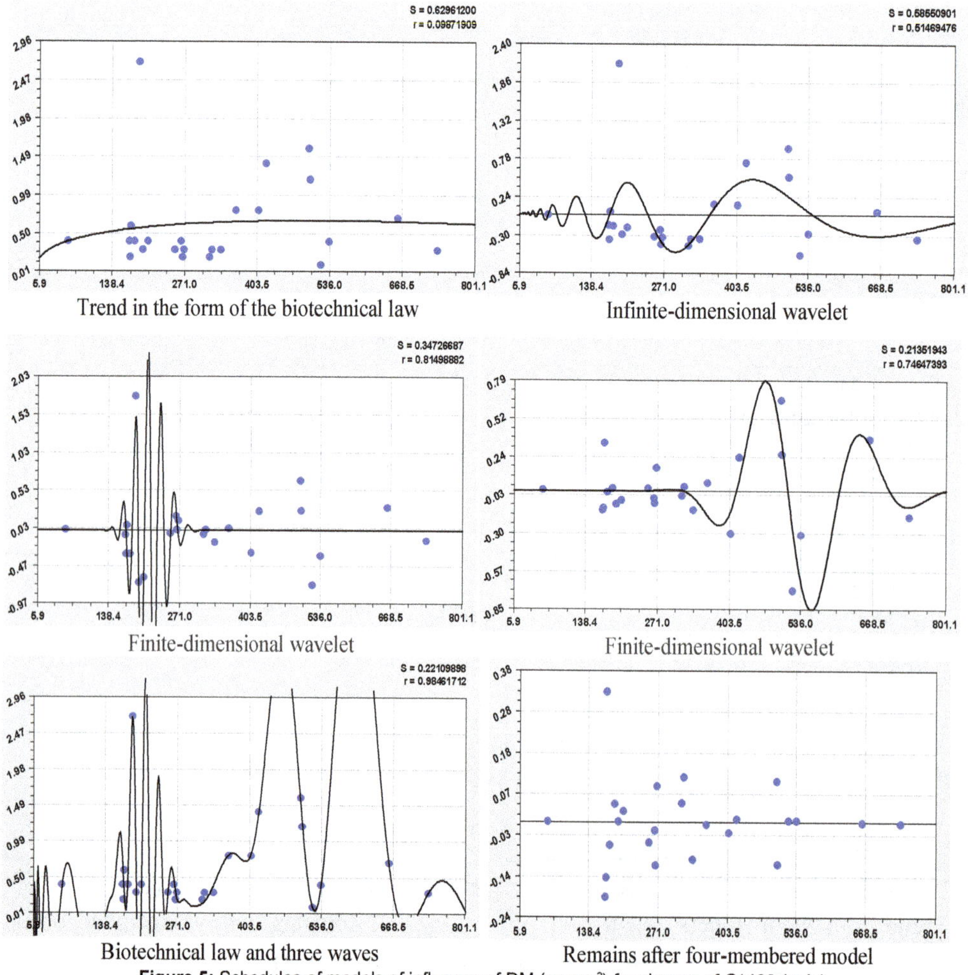


**Figure 5:** Schedules of models of influence of DM (pc cm$^{-3}$) for change of *S*1400 (mJy).

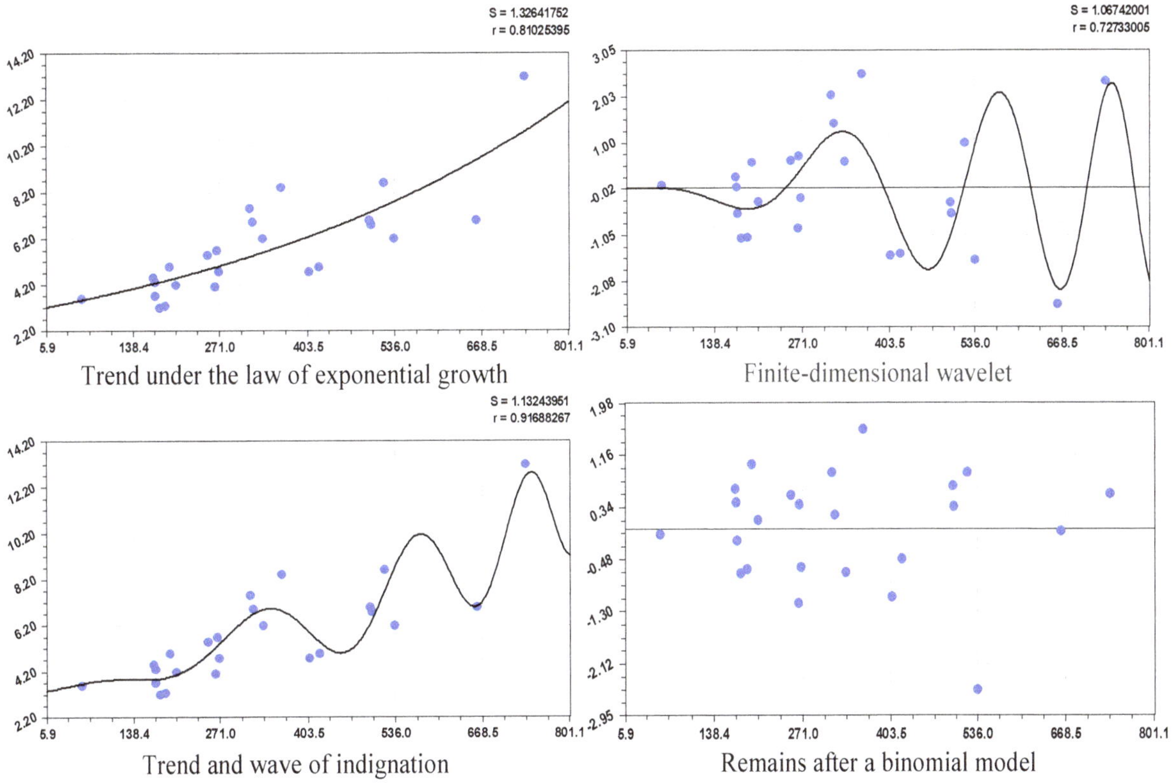


**Figure 6:** Schedules of models of influence of DM (pc cm$^{-3}$) for change of $D$ (kpc).

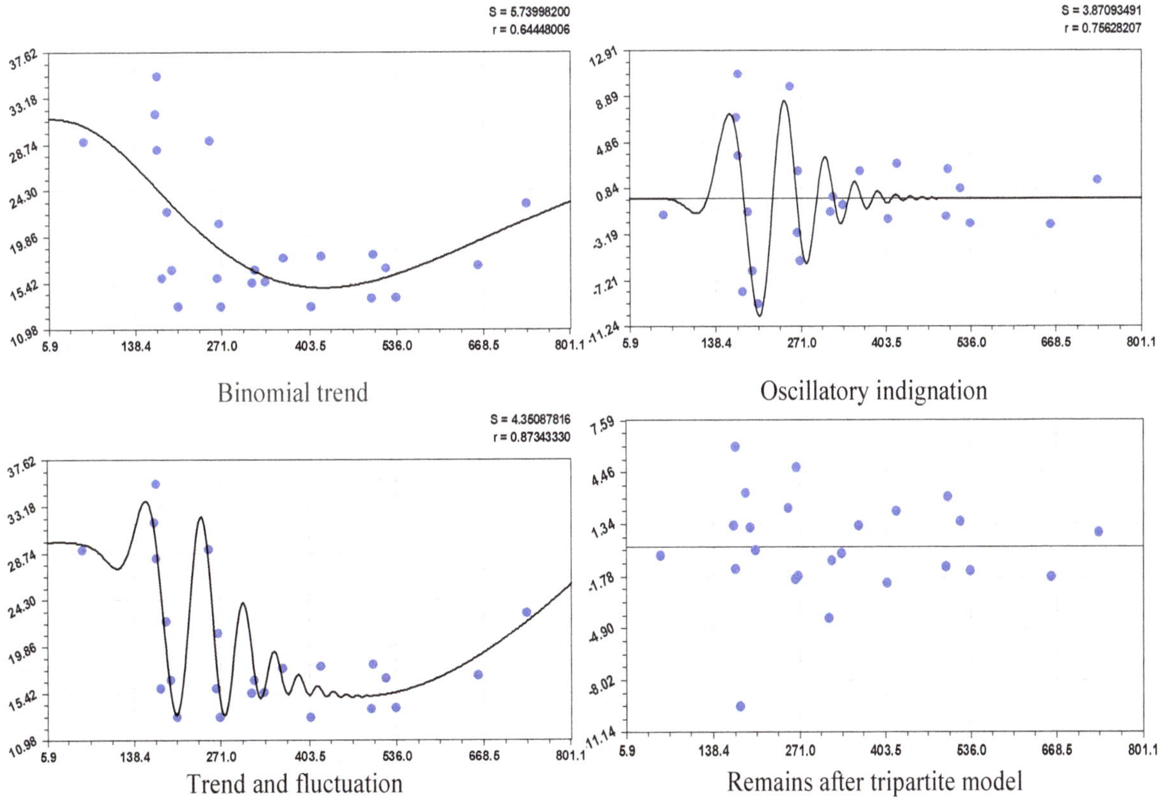


**Figure 7:** Schedules of models of influence of DM (pc cm$^{-3}$) for change of $S$.

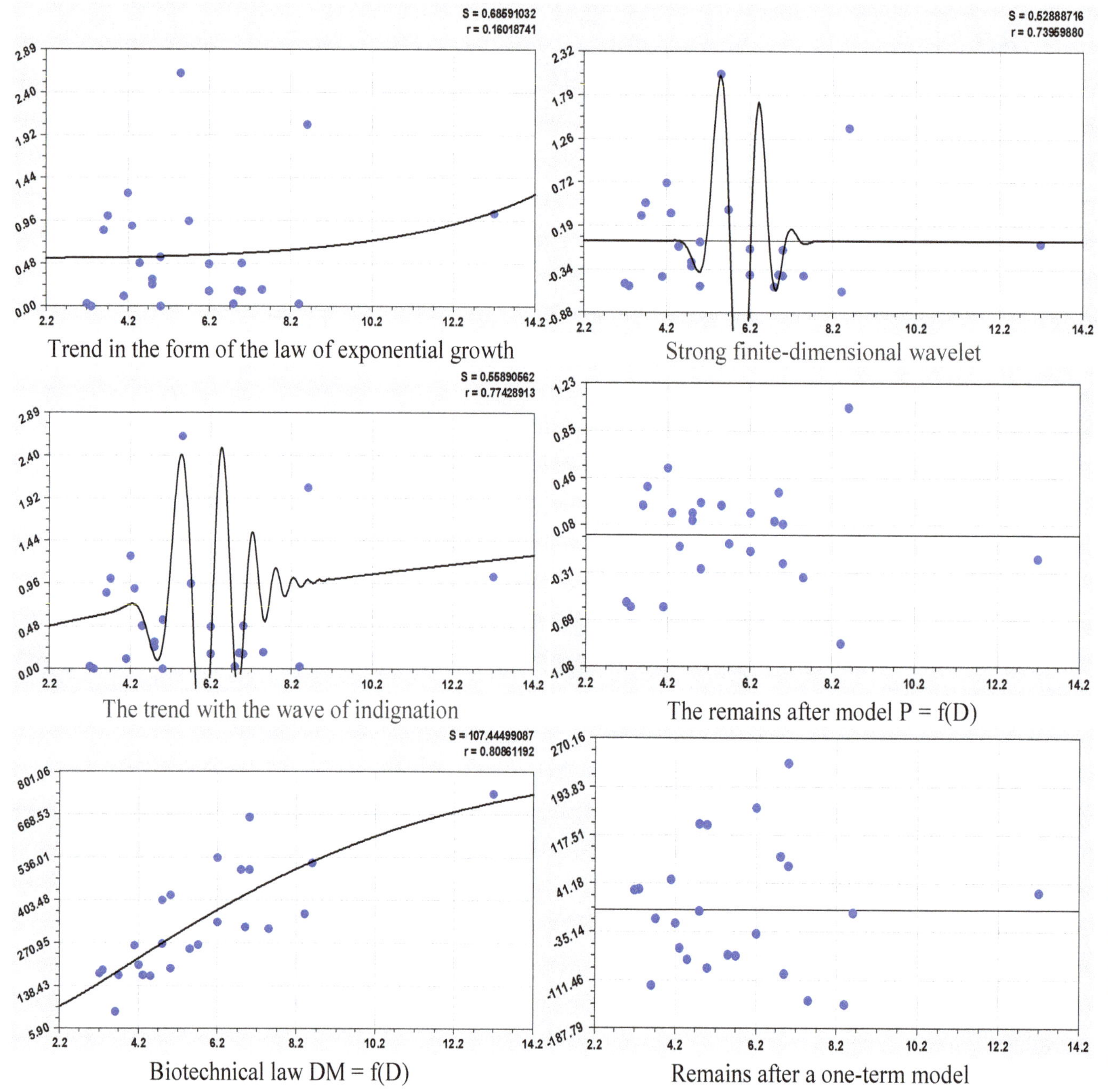


**Figure 8:** Schedules of models of influence of *D* (kpc) on change of *P* (s) and DM (pc $cm^{-3}$).

values are written down in a compact matrix form with five significant figures are given.

The maximum number of members of statistical model is equal to four that corresponds to computing opportunities of the program CurveExpert-1.40 environment. For the full wavelet analysis it is necessary to develop the special program environment according to our scenarios of statistical modeling for a supercomputer of a petaflop class. Thus the new program environment for large volumes of the table of basic data will be universal for all branches of science.

In Figure 8 influence of D (kpc) on change of two factors of P (s) and DM (pc $cm^{-3}$) is shown. Thus change of P (s) received fluctuation in the form of a finite-dimensional wavelet.

The analysis of schedules according to amplitude-frequency characteristics shows that the system of pulsars possesses a certain property of wave adaptation.

## Wavelet Analysis of Factor Relations

Consider the possibility of further identification model S1400=f(DM) with the highest adequacy 0.9846 (Table 9 and Figure 13). The price of division of a factor of S1400 according to Table 1 is

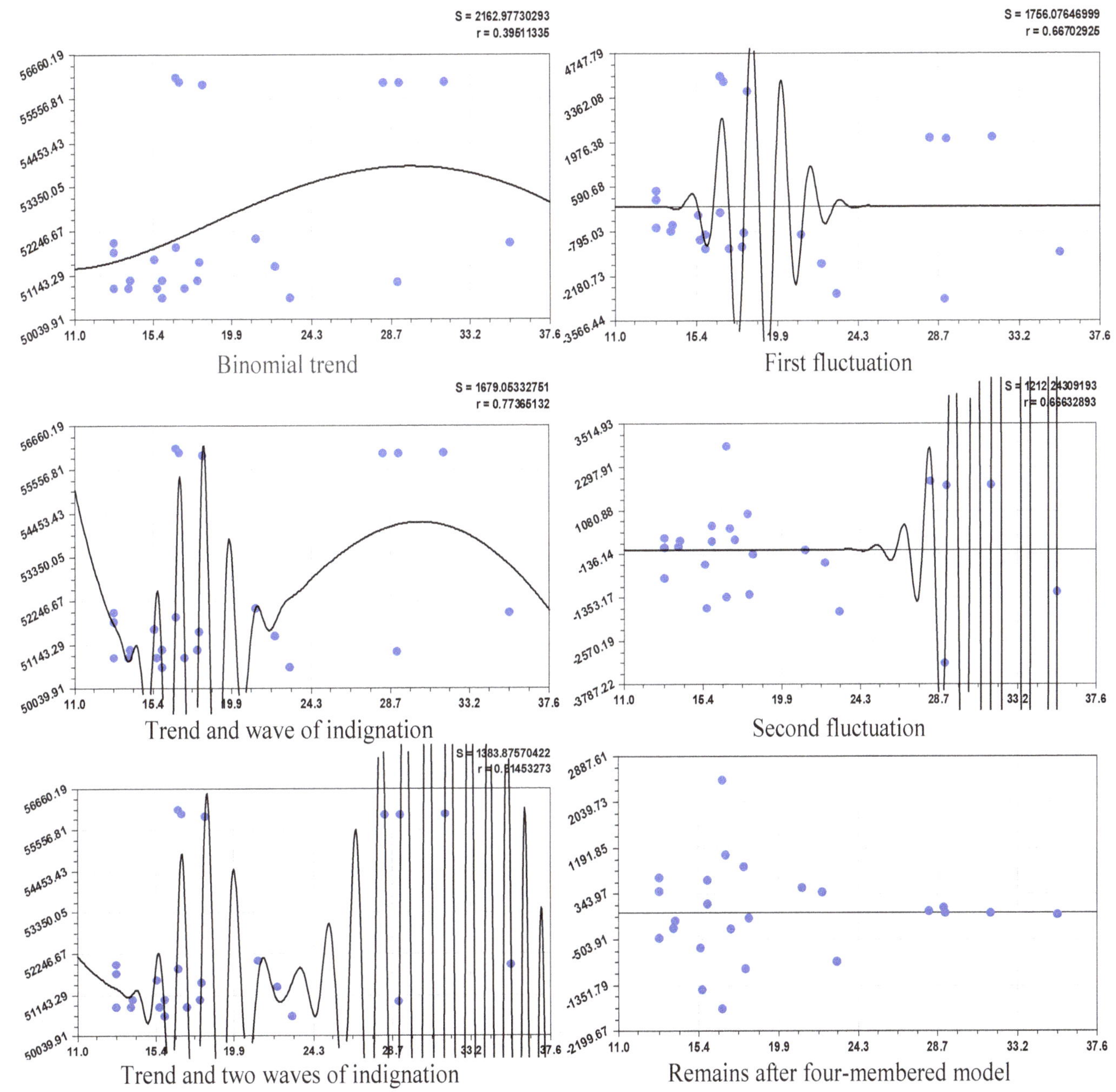


**Figure 9:** Schedules of models of influence of *S* on change of *P* Epoch (MJD).

equal 0.1 (mJy).Then the measurement error equal to ±0.05 (mJy), and the remainder by point graphics in Figure 13 become smaller this error. The identification process is stopped, wherein the wavelet analysis is completed. Apparently from the schedule of four-membered model in Figure 13, influence of a factor of DM for a factor of S1400 gives three clusters (a plot on abscissa axis):

1) Initial site of DM=0-130 (pc $cm^{-3}$)

2) Average site of DM=130-300 (pc $cm^{-3}$)

3) Extreme site of DM=300-800 (pc $cm^{-3}$).

At each of the sites on the abscissa is finite wavelet signal.

For some binary relations the number of members in the general statistical model can exceed 100-120 pieces. In this case there is a possibility of carrying out the fractal analysis for group of wavelets on mega, macro, meso and to micro fluctuations.

## Conclusion

Applicability of statistical model (1) to parameters of pulsars is proved. As a result each binary relation contains a trend and wavelet signals. Moreover, the trend is a special case of very long period oscillations of the wavelet. As a result the general statistical model represents the plait consisting of a set of lonely waves with variables amplitude and the period of fluctuations.

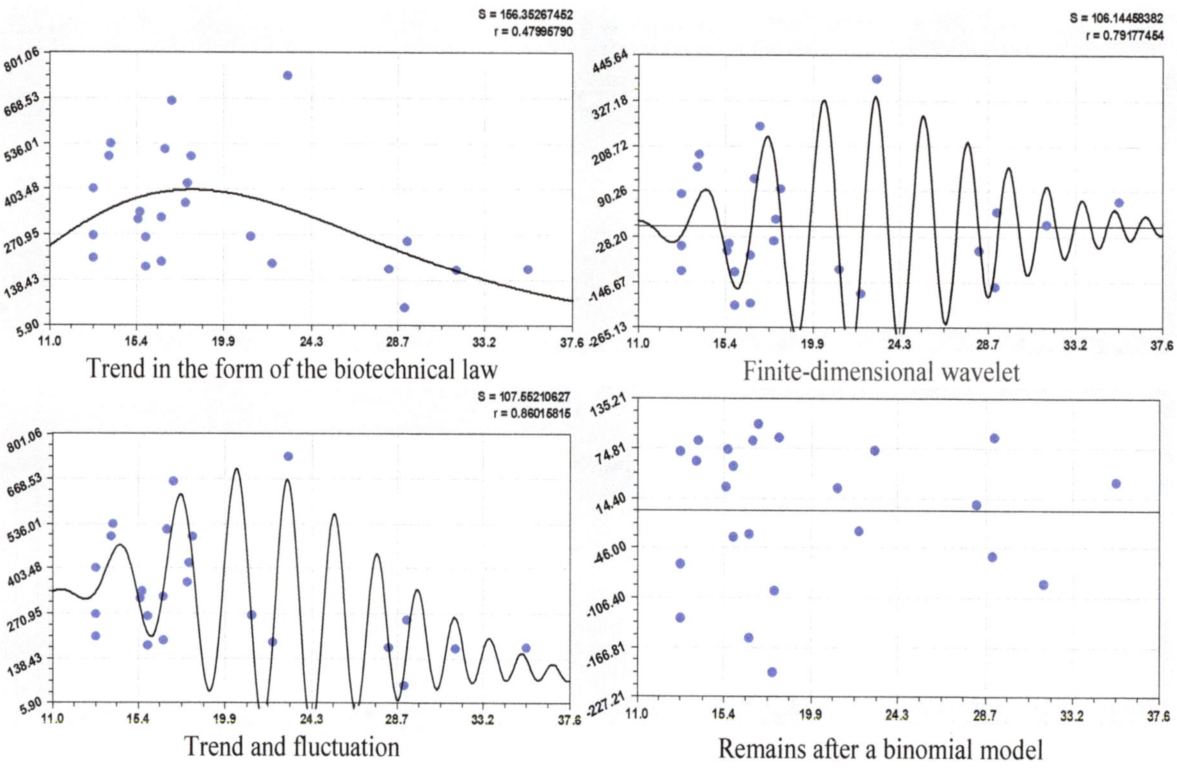


**Figure 10:** Schedules of models of influence of *S* on change of DM (pc cm$^{-3}$).

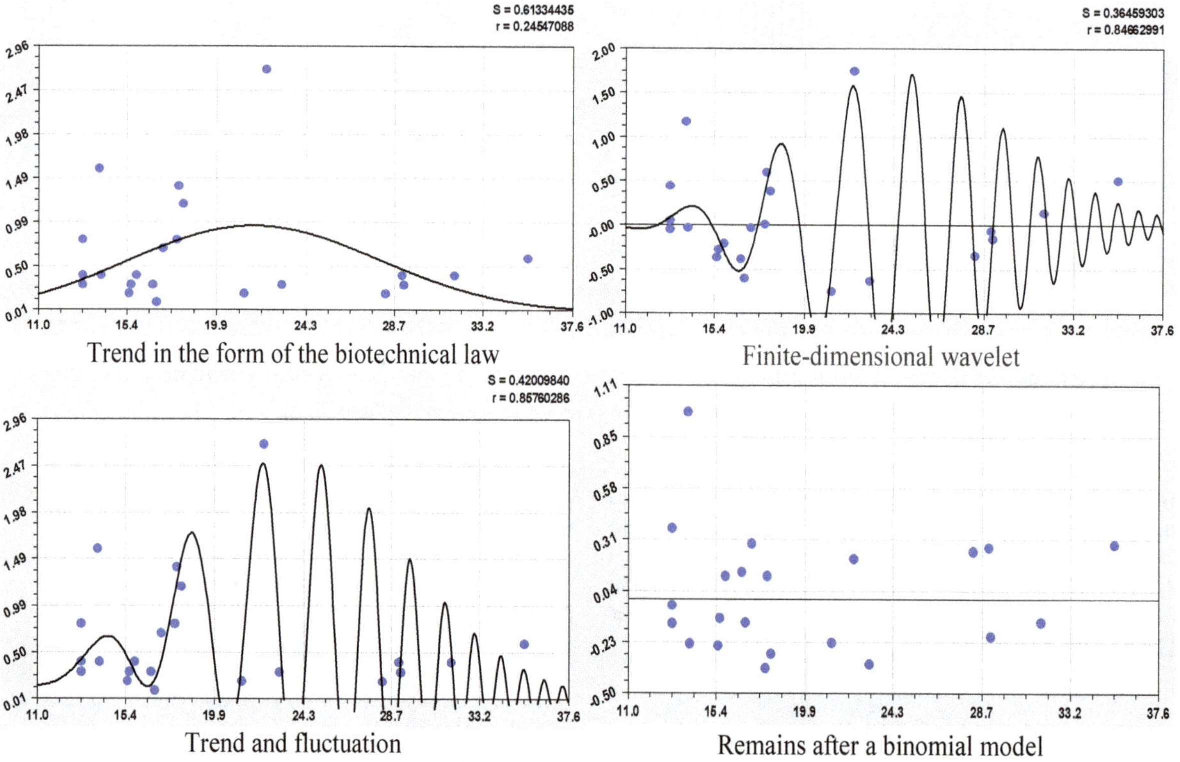


**Figure 11:** Schedules of models of influence of *S* on change of *S*1400 (mJy).

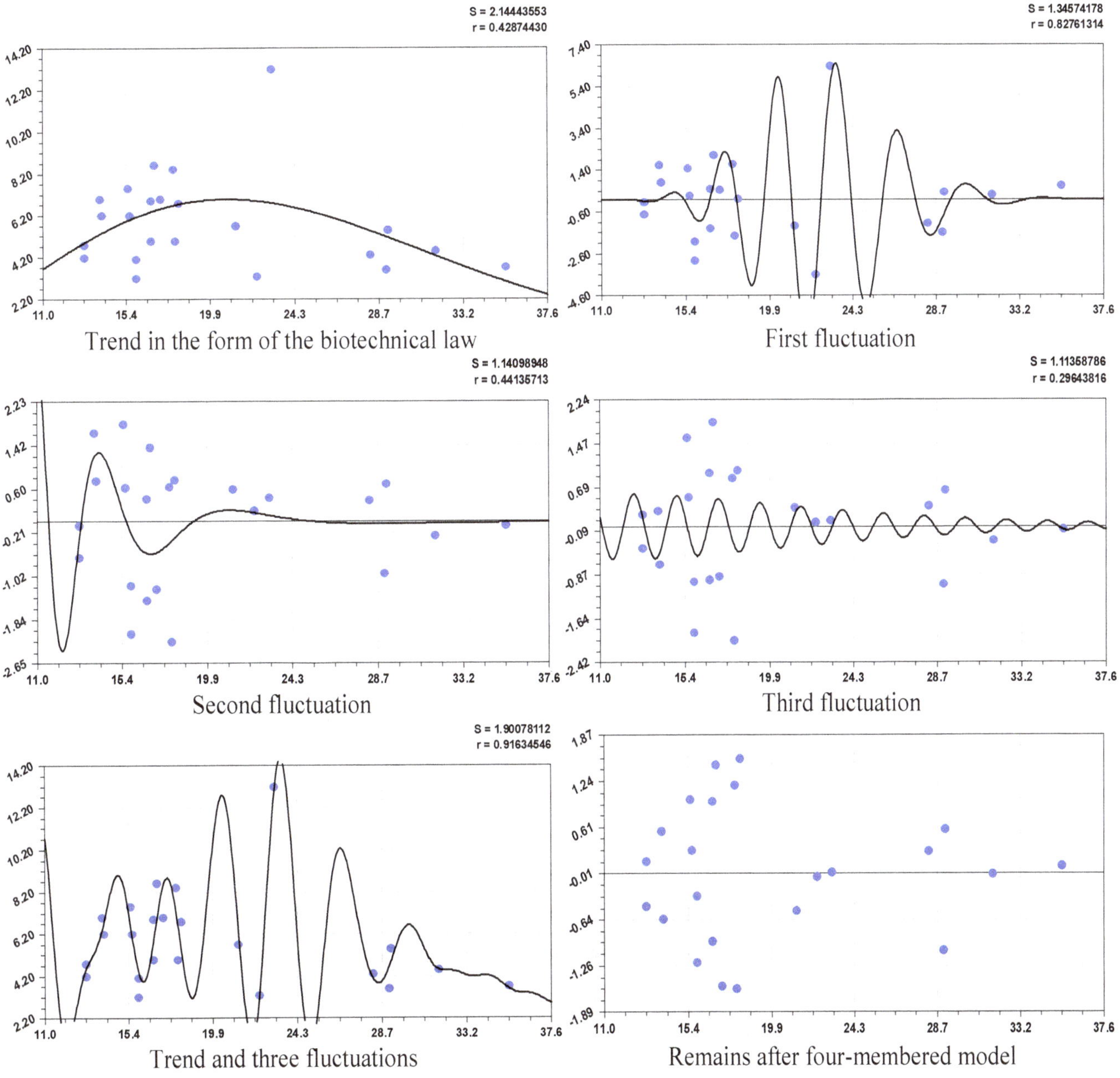


**Figure 12:** Schedules of models of influence of *S* on change of *D* (kpc).

Quality control input data can be estimated rank distribution of values of parameters of pulsars and the ability to detect the wave patterns of reporting to the design of the same wavelet signal (1). Thus without modeling, only due to ordering of values of parameters on a preference preorder vector "better→worse", it is possible to make a rating of pulsars.

After statistical modeling carried out factor analysis which allows to make the ratings of factors as influencing parameters and how dependent indicators. For strong factorial relations additionally conducted a wavelet analysis, in which re- identification patterns (1) to ascertain residues below the error of measurement of parameters of pulsars. This set of wavelet signals can then be subjected to fractal analysis.

The offered methodology of identification allows to allocate waves of the binary relations between the measured factors. Thus for 24 pulsars characterized by nit dimensional wavelets, which can then be compared with the heuristic views of specialists. The method of identification allows to allocate the most significant parameters and the binary relations between them at pulsars on which it will be necessary to increase the accuracy of future measurements.

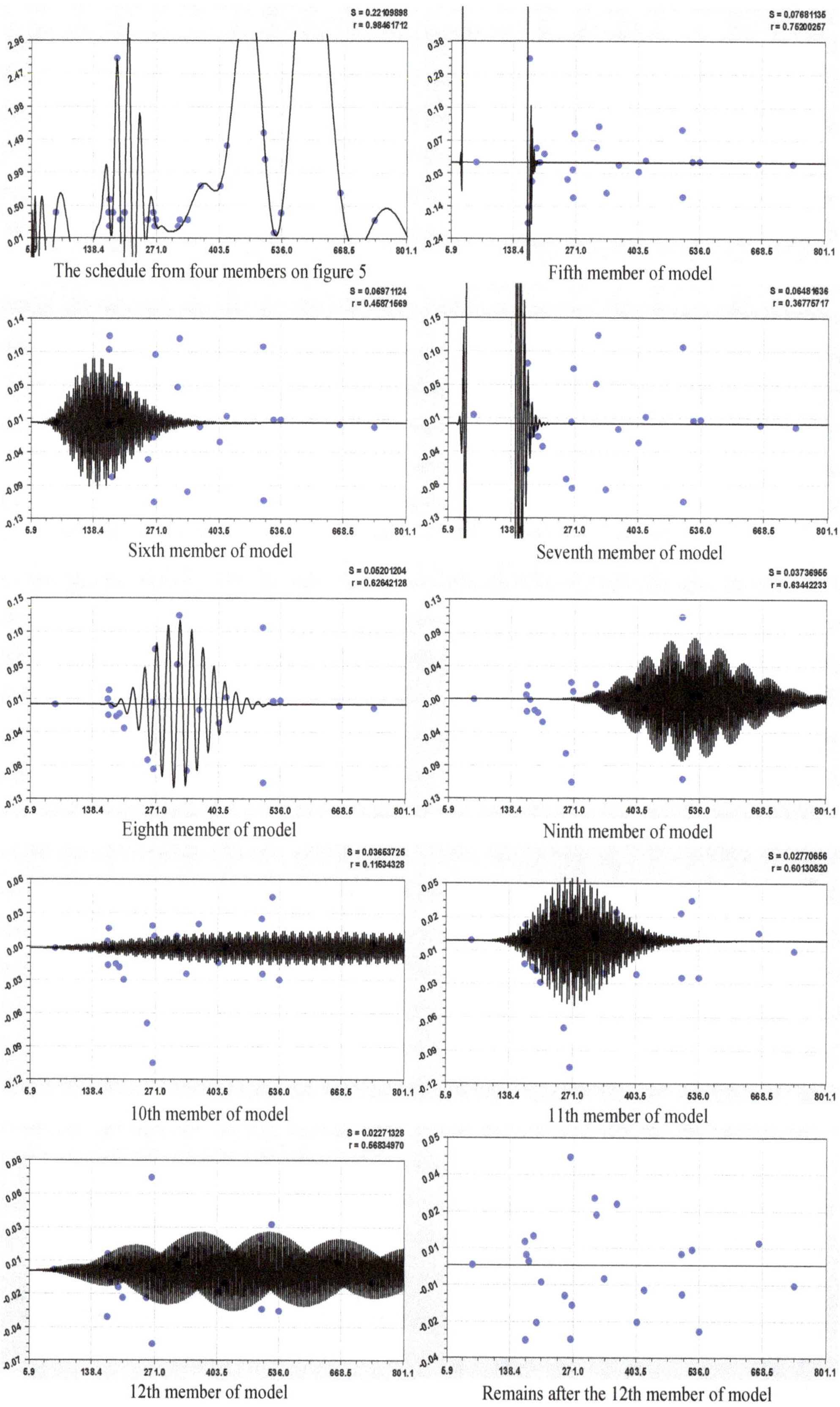


**Figure 13:** Schedules of influence of DM (pc cm$^{-3}$) for change of *S*1400 (mJy), additional to Figure 5.

## References


1. Mazurkin PM (2014) Asymmetric Wavelet Signal of Gravitational Waves. Applied Mathematics and Physics 2: 128-134.
2. Mazurkin PM (2009) Biotechnical law and designing of adequate models. Achievements of modern natural sciences 9: 125-129.
3. Mazurkin PM (2009) The biotechnical law, algorithm in intuitive sense and algorithm of search of parameters. Achievements of modern natural sciences 9: 88-92.
4. Mazurkin PM (2009) The biotechnical principle in statistical modelling. Achievements of modern natural sciences 9: 107-111.
5. Mazurkin PM (2009) Biotechnical principle and steady laws of distribution. Achievements of modern natural sciences 9: 93-97.
6. Mazurkin PM (2013) Dynamics of alpha activity of pattern $^{239}PU$ in different time scales. Science and World: International Scientific Journal 2: 20-26.
7. Mazurkin PM (2013) Dynamics of alpha activity $^{239}PU$ in stages of solar eclipse. Science and World: International Scientific Journal 4: 20-26.
8. Mazurkin PM (2013) Identification of statistical steady regularities. Science and world: International scientific magazine 3: 28-33.
9. Mazurkin PM (2014) Method of identification $14^{th}$ International multidisciplinary scientific geo-conferenct & SGEM, Russia.
10. Mazurkin PM (2014) The decision 23-oh Gilbert's problems. Interdisciplinary researches in the field of mathematical modeling and informatics. Materials of the 3rd scientific and practical Internet conference.
11. Mazurkin PM (2014) The wavelet analysis of alpha activity $^{239}PU$ of the solar eclipse. Science and World International Scientific Journal 1: 94-104.
12. Mazurkin PM (2014) Statistical modeling of entire prime numbers. International Journal of Engineering and Technical Research 2: 148-158.
13. Mazurkin PM (2013) Wavelet analysis of alpha activity of a sample $^{239}Pu$ solar eclipse. Science and world: international scientific magazine 1: 94-104.
14. Mazurkin PM (2014) Wavelet analysis of hour increments of alpha activity $^{239}PU$ at the maximum of the solar eclipse. SCIENCE AND WORLD: International scientific journal 2: 46-55.
15. Mazurkin PM (2014) Wavelet analysis of hour increments of alpha activity $^{239}PU$ after a solar eclipse. Science and the world: international scientific magazine 1: 31-40.
16. Mazurkin PM (2014) Wavelet signals of gravitational waves from pulsars. Researches of the main directions of technical and physical and mathematical science: The collection of scientific works on materials II of the International scientific conference.
17. Mazurkin PM, Filonov AS (2006) Mathematical modeling Identification of one-factorial statistical regularities: manual. Yoshkar-Ola: MarSTU 292.
18. Knispel B, Eatough RP, Kim H, Keane EF, Allen B, et al. (2013) Einstein@ home discovery of 24 pulsars in the parkes multi-beam pulsar survey.The Astrophysical Journal 774: 93.

# Relation between the Gravitational and Magnetic Fields

**Baixauli JG***

*Independent Researcher, Spain*

**Abstract**

Quantum and relativistic phenomena can be explained by the hypothesis that both the universe and the elementary particles are formed by atoms' four spatial dimensions. It is also possible to deduce and calculate properties of elementary particles, such as mass, electric charge, spin, radio, etc. Based on this hypothesis, the fundamental constants can be calculated depending only on the speed of light c. Moreover, the Schrodinger equation can be deduced and the equality of inertial and gravitational mass can be demonstrated. The aim of this article is to establish a relationship between electron mass and electron charge, and thereby calculate the electrical charge depending on the mass of the electron, Planck's constant h and the speed of light c. Through this relationship, extended to the gravitational and magnetic fields, it is possible to establish a relationship between the constants of both fields.

**Keywords:** Gravitational and magnetic fields; Quantum gravity

## Introduction

Magnetic and electrical phenomena were already known in Ancient Greece. In the 17th and 18th Centuries, electromagnetic phenomena were studied separately. James Clerk Maxwell described the electric and magnetic fields using a set of equations in 1861, unifying the two fields into one: the electromagnetic field. In Newtonian physics, the gravitational field is defined as the force per unit mass that experiences a point particle in the presence of a mass. In general relativity, gravity is due to the curvature of space time. The presence of a mass curves space time, causing bodies to move along those lines curves denominated geodesic. General relativity assumes that space time is continuous. However, there is no experimental evidence for it. Are space and time continuous? Or are we only convinced of that continuity as a result of education? In recent years, both physicists and mathematicians have asked whether it is possible that space and time are discrete. If we could probe to size scales that were small enough, would we see "atoms" of space, irreducible pieces of volume that cannot be broken into anything smaller? [1]. Quantification of space time allows us to distinguish elementary particles from each other in a simple and natural way [2,3].

Minimum volume, length or area are measured in the units of Planck [1]. Theories related to quantum gravity, such as string theory and doubly special relativity, as well as black hole physics, predict the existence of a minimum length [4,5]. The familiar concept of a space-time continuum implies that it should be possible to measure always smaller and smaller distances without any finite limit [2]. Heisenberg, who insisted on expressing quantum mechanical laws in terms of measurable observables, questioned already the validity of this postulate [6]. Heisenberg said that physics must have a fundamental length scale, and with Planck's constant h and the speed of light c, allow the derivation of the masses of the particles [7,8]. Planck's length has been considered as the shortest distance having any physical meaning. A fundamental (minimal) length scale naturally emerges in any quantum theory in the presence of gravitational effects that accounts for a limited resolution of space-time. As there is only one natural length scale we can obtain by combining gravity (G), quantum mechanics (ħ) and special relativity (c), this minimal length is expected to appear at the Planck scale [9]. Padmanabhan shows that the Planck length provides a lower limit of length in any suitable physical space-time [10,11]. It is impossible to construct an apparatus which will measure length scales smaller than Planck length. These effects exist even in flat space-time because of vacuum fluctuations of gravity [11]. Quantum particles in discrete spacetime are studied in relation to relativistic dynamics [12,13]. Farrelly and Short studied the causal evolution of a single particle in discrete spacetime [14].

There is evidence of discrete structures on the largest scales, for example superclusters and the redshift [15]. Cowan already said in 1969 that redshift can only occur with discrete values [16]. This was subsequently confirmed by Karlsson [17].

## Relations between Rest Mass and Electric Charge

The hypothesis is that the universe is composed of four spatial dimensions Planck atoms, with two possible states: rest state and rotational movement. Rest atoms are empty space, and the rotational motion of the atoms gives rise to different properties of the particles. Of the four dimensions, three are seen as space (x, y and z) and the fourth (u=ct) is observed as time. The 4D Planck atoms are atoms of space and time that Smolin says. To simplify the drawing (Figure 1), only three dimensions are considered: *r(x,y)* and *u*. If space is made up of Planck-size atoms, each Planck atom can only be resting or turning on itself. The rotations can be in three-dimensional space or in the fourth dimension. The rotational energy in the fourth dimension gives rise to the rest mass and period originating the electric charge.

The energy of the particle can be expressed as:

$$E = mc^2 = \hbar\omega_u = \frac{1}{2}\hbar\omega_e \tag{1}$$

Where m is the rest mass of the particle, c the speed of light, ħ the reduced Planck constant (*h/2π*) and $\omega u$ rotation in the fourth dimension (Figure 2). A 4D Planck atom may rotate both in three dimensional space and in the fourth dimension (u, Figure 2), which results in the following combinations:

***Corresponding author:** Baixauli JG, Independent Researcher, Spain
E-mail: jgarrigu@eln.upv.es

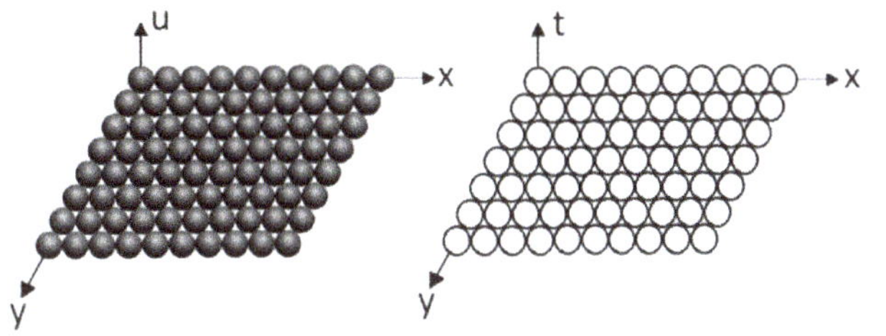


**Figure 1:** A 3D flat universe is seen as 2D and t.

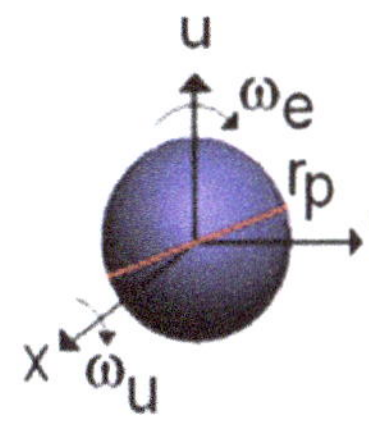


**Figure 2:** Particle rotations or 4D Planck atom.

- Zero rotations. The vacuum space.
- One spatial rotation $\omega_e$. Neutrinos.
- One rotation in the fourth dimension $\omega_u$ Photons.
- Two rotations, one spatial rotation ($\omega_e$) and one rotation in the fourth dimension $\omega_u$. Electrons and quarks of the first generation.

Equation (1) can be expressed in terms of period (wu=2 $\pi/T_u$), resulting in:

$$E = mc^2 = \frac{h}{T_u} \tag{2}$$

The electric charge will be due to the rotation (one of three possible rotations) or therefore, the electric charge can be expressed as:

$$q = \frac{1}{c^2}\frac{\delta V_{4D}}{\delta u} = 2\pi^2 T_u \tag{3}$$

Substituting (Tu) in Equation (2), the following is obtained:

$$E = mc^2 = 2\pi^2 \frac{h}{q} \tag{4}$$

Where m is the electron rest mass, and electric charge is in seconds.

$$q = 2\pi^2 \frac{h}{mc^2} = 1,59710^{-19} s \tag{5}$$

To preserve the units, it is only necessary to multiply Equation (2) by the current unit (I=1 A). A coulomb is an arbitrary unit of electrical charge. Current theory allows one to measure the electrical charge but not to explain it. Simply, the electric charge is defined as an intrinsic or fundamental property of matter.

$$E = mc^2 = 2\pi^2 \frac{h}{q} I \tag{6}$$

The rest mass and charge of the electron were calculated above, according to Planck's constant h and the speed of light c [18].

## Corrections

When measurements are made, three factors should be taken into account to avoid errors in the value of the parameter measured in the laboratory or mathematically calculated.

These factors are:

- *Light refraction index (n)* some parameters are measured in a vacuum and others in the laboratory. Everything depends on the speed of light, so it will be necessary to take into account the light refraction index in the atmosphere.
- *Special Relativity* (SR) Assuming that the Earth moves to 21000 km/s, due to the movement of the Sun around the galaxy, the movement of the galaxy in the direction of the Virgo cluster and other possible movements due to the gravitational attraction, should be taken into account.
- *Gravitational system* General relativity establishes that time elapses more slowly in a gravitational system than far from the gravitational potential.

This difference over time relative to gravitational potential is:

$$t' = t\sqrt{1 - \frac{2GM}{Rc^2}} \tag{7}$$

Where t is the time measured away from the gravitational system.

Taking into account n and SR, and I0 current of 1 A becomes:

$$I = I_0 \frac{n}{\sqrt{1 - \frac{v^2}{c^2}}} = 1,0028A \tag{8}$$

Measures should note the time and the speed of light in the laboratory, because the intensity is defined as coulombs per unit time (I=q/t). Then Equation (5) becomes:

$$E = mc^2 = 2\pi^2 \frac{h}{mc^2} I_0 \frac{n}{\sqrt{1 - \frac{v^2}{c^2}}} = 1,602010^{-19} C \tag{9}$$

Where h=6.6260700410-34 Js, m=9.109383560 10-31 kg, c=299792458 m s-1, n=1.00029260 and I0=1 A.

Although both electromagnetic field and the gravitational field are well known, it may be interesting to review some basic concepts of both fields. Specifically, referring to the orbits.

## Electromagnetic Field

In classical physics the electric field is produced by the accumulation of electrical charge, while the magnetic field is due to electric current. In relativistic physics, an observer at rest with respect to a reference system will measure a different value than that an observer moving relative to the reference system mentioned above would measure.It has been shown that the electric charge is the time it takes the Planck atom to spin in the fourth dimension, and that the mass is the energy of this rotation or the magnetic field energy. Therefore, the electromagnetic and gravitational fields must be related, as are the mass and electric charge.

## Magnetic Orbits

When a negative charge q moving with velocity **v** penetrates a magnetic field ***B***, the charge is subject to a force. If the field is perpendicular to the velocity, the modulus of the magnetic force can be calculated with Equation (10) (Figure 3):

$$F_m = qvB \tag{10}$$

The radius of the circular orbit is determined by the condition that the centripetal force of magnetic origin is compensated by the centrifugal force, then:

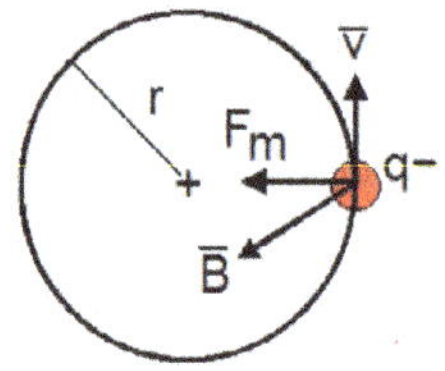

**Figure 3:** Movement of negative charges in a magnetic field.

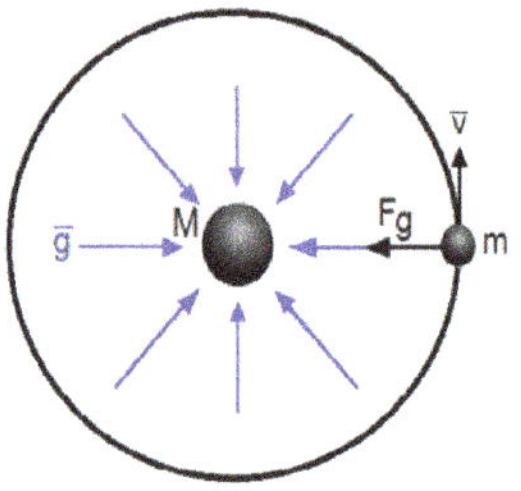

**Figure 4:** Movement of a mass m in a gravitational field.

$$q v B = \frac{mv^2}{r} \quad (11)$$

## Gravitational Orbits

If a body of mass m, moving with velocity v, penetrates a gravitational field g, it will be subject to an attractive force. If the kinetic and potential energies remain constant over time, the body will describe a circular orbit (Figure 4).

The gravitational force will go in the direction of the field and the modulus will be:

$$F_g = m g \quad (12)$$

To calculate the radius of the orbit, the dynamics of uniform circular motion is applied (i.e. Newton's Second Law):

$$m g = \frac{mv^2}{r} \quad (13)$$

## Relationship between Fields

The units of the gravitational and magnetic field are $m.s^{-2}$ and $kg{\cdot}s^{-2}.A^{-1}$ respectively. If both fields produce the same phenomenon, why do they not have the same units? We raise the issue conversely: suppose that a particle of mass m and charge q+ describes a circular orbit with speed v, about X field; find the term and nature of the field X. Considering the mass-dependent phenomenon, it is a gravitational field, but if we consider that the attraction depends on the electric charge, a magnetic field is obtained. From Equations (11) and (13):

$$m g = \frac{mv^2}{r} = q v B \quad (14)$$

And

$$\frac{g}{B} = \frac{qv}{m} \quad (15)$$

If g and B are equal to 1 ($g=B=1$):

$$m = qv \quad (16)$$

In a universe formed by four spatial dimensions (Planck spheres) that expands at the speed of light, it is clear that the previous Equation (16) must hold, because there are only space and movement. The movement can be observed in different ways, such as time, frequency, electric charge, rest mass, etc. If the electric charge is the time that a Planck atom takes to turn one spin in the fourth dimension, the rest mass must be related to the conditions of Planck.

## Conditions of Planck

If Equation (15) is applied under the conditions of Planck, it is:

$$\frac{g_p}{B_p} = \frac{q_p C}{m_p} \quad (17)$$

Where qp is the Planck charge and mp the Planck mass. Gravitational and Coulomb forces are equal under Planck conditions, then:

$$G\frac{m_p m_p}{r_p^2} = k\frac{q_p q_p}{r_p^2} \quad (18)$$

Where k is Coulomb constant and rp is the Planck radius, which is the Planck atom diameter. Then:

$$m_p = \sqrt{\frac{k}{G}} q_p \quad (19)$$

Multiplying and dividing by the speed of light:

$$m_p = \sqrt{\frac{k}{Gc^2}} q_p c = \sqrt{4\pi}\sqrt{\frac{G}{\mu_0}} q_p c = q_p v \quad (20)$$

As the rotation of the Planck atom decreases, the speed decreases until the minimum energy state. This state matches the conditions of the electron in the free state, and its speed:

$$v = \alpha c \quad (21)$$

Considering both Equations (20), (21):

$$G = \alpha^2 \mu_0 \quad (22)$$

$6.67408{\cdot}10^{-11} = (7.2973525664\ 10^{-3})^2 \times 1.2566370614{\cdot}10^{-6} = 6.69176{\cdot}10^{-11}$

Obviously, in this equation the units do not match, because neither quantum mechanics nor general relativity explain rest mass or electric charge. These theories only assign arbitrary units to them and consider them independent intrinsic properties.

If it is intended to recover the typical units of these parameters, Equation (22) can be simply multiplied as follows:

$$G J^2 = \alpha^2 \mu_0 I^2 \quad (23)$$

Where J is the linear mass density ($J=1$ kg $m^{-1}$) and current I ($I=1$ A).

We must consider that we are in a gravitational system, which moves in space and also some physical quantities are measured in the laboratory.

$$J = J_0 \frac{n}{1 - \frac{v2}{c2}} = 1.0052\text{kg}m^{-1} \quad (24)$$

$$G\frac{J^2}{I^2} = 6.67408.10^{-11}\ 1.0049 = 6.707010^{-11};\ \alpha^2\mu_0 = 6.69176.10^{-11}$$

Now the error is greater than that between rest mass and electric charge, due to the fact that in the previous Equation (24) the gravitational constant is directly involved. This gravitational constant is the most difficult to be measured due to its small magnitude, despite being well known. The gravitational constant $G$ has been previously calculated on the basis of Planck's constant $h$ and the speed of light c [18]. The permittivity **0**, Planck's constant h and the fine structure constant have been previously calculated according to the speed of light c [18].

## Conclusion

Rotation of Planck atoms in the fourth dimension gives rise to the electron charge; the energy of this rotation or electromagnetic field energy gives rise to the mass. The charge is the time it takes in returning in the fourth dimension.

Therefore, in the electron:

- Rotation in the fourth dimension, which originates the electric charge and so, the electric field is caused.
- Rotation in three-dimensional space generates the moving charge and therefore the magnetic field.
- The energy of the rotation, which we call mass, leads to the gravitational field.

It is evident that by changing the rotation direction, the polarity of the electric charge is changed. Clockwise rotation in the fourth dimension can only be cancelled with another anti-clockwise rotation, resulting from the Principle of Conservation of Charge.The mass of a Planck atom decreases with decreasing rotation, just as the electric charge increases and the wavelength or diameter ($m_p r_p = m\lambda$). When increasing the wavelength, the travel speed decreases to the value $\alpha$ c. The fine structure constant relates the constant gravitational field G with constant magnetic field $\mu_0$.

$$G=\alpha^2 \mu_0$$

Gribbin was correct when he stated that the only alternative to the quantum probabilities, superposition of states, wave function collapse, and the spooky action at a distance, is that everything is predetermined [19].

### References

1. Smolin L (2003) Atoms of Space and Time. Sci Am 15.
2. Meessen A (2011) Space-Time Quantization, Elementary Particles and Dark Matter.
3. Meessen A (1999) Spacetime quantization, elementary particles and cosmology. Found Phys 29: 281-316.
4. Maggiore M (1994) Quantum Groups, Gravity, and the Generalized Uncertainty Principle. Phys Rev D 49: 5182.
5. Maggiore M (1993) The algebraic structure of the generalized uncertainty principle. Phys Lett B 319: 83-86.
6. Heisenberg W (1930) The physical principles of quantum theory.
7. Heisenberg W (1943) Die Beobachtbaren Grossen in der Theorie der Elemntarteilchen. Z Phys 120: 513.
8. Heisenberg W (1957) Quantum Theory of Fields and Elementary Particles. Rev Mod Phys 29: 269.
9. Martin S, Piero N, Marcus B (2012) Physics on Smallest - An Introduction to Minimal Length Phenomenology. Eur J Phys 33: 853-862.
10. Padmanabhan T (1985) Planck Length As The Lower Bound To All Physical Length Scale. Gen Rel Grav 17: 215-221.
11. Padmanabhan T (1985) Physical Significance of Planck Length. Ann Phys 165: 38-58.
12. Feynman RP, Hibbs AR (1965) Quantum Mechanics and Path Integrals. New York, McGraw-Hill, USA.
13. Bialynicki-Birula I (1994) Weyl, Dirac, and Maxwell equations on a lattice as unitary cellular automata. Phys Rev D 49: 6920-6927.
14. Farrelly TC and Short AJ (2014) Discrete Spacetime and Relativistic Quantum Particles. Phys Rev A 89: 062109.
15. Roberts MD (2002) Quantum Perturbative Approach to Discrete Redshift. Astrophys Space Sci 279: 305-342.
16. Cowan CL (1969) Periodic Clustering of Red-shifts in the Spectra of Quasistellar and other Unusual Objects. Nature 224: 655-656.
17. Karlsson KG (1971) Possible Discretization of Quasar Redshifts. Astron & Astrophys 13: 333-335.
18. Garrigues-Baixauli J (2015) Space, Time and Energy. Gs J.
19. Gribbin J (1990) The man who proved Einstein was wrong. New Scientist 128: 43-45.

# Nonholonomic Ricci Flows of Riemannian Metrics and Lagrange-Finsler Geometry

**Alexiou M[1], Stavrinos PC[2] and Vacaru SI[3]***

[1]*Department of Physics, National Technical University of Athens, Greece*
[2]*Department of Mathematics, University of Athens, Greece*
[3]*Rectors Department,University Al. I. Cuza, Greece*

**Abstract**

In this paper, the theory of the Ricci flows for manifolds is elaborated with nonintegrable (nonholonomic) distributions defining nonlinear connection structures. Such manifolds provide a unified geometrical arena for nonholonomic Riemannian spaces, Lagrange mechanics, Finsler geometry, and various models of gravity (the Einstein theory and string, or gauge, generalizations). Nonhlonomic frames are considered with associated nonlinear connection structure and certain defined classes of nonholonomic constraints on Riemann manifolds for which various types of generalized Finsler geometries can be modelled by Ricci flows. We speculate upon possible applications of the nonholonomic flows in modern geometrical mechanics and physics.

**Keywords:** Nonlinear connections; Nonholonomic Riemann manifolds; Lagrange and Finsler geometry; Geometric flows

## Introduction

A series of the most remarkable results in mathematics are related to Grisha Perelman's proof of the Poincare Conjecture [1-3] built on geometrization (Thurston) conjecture [4,5] for three dimensional Riemannian manifolds, and R. Hamilton's Ricci flow theory [6,7] see reviews and basic references explained by Kleiner [8-11]. Much of the works on Ricci flows has been performed and validated by experts in the area of geometrical analysis and Riemannian geometry. Recently, a number of applications in physics of the Ricci flow theory were proposed, by Vacaru [12-16].Some geometrical approaches in modern gravity and string theory are connected to the method of moving frames and distributions of geometric objects on (semi) Riemannian manifolds and their generalizations to spaces provided with nontrivial torsion, nonmetricity and/or nonlinear connection structures [17,18]. The geometry of nonholonomic manifolds and non–Riemannian spaces is largely applied in modern mechanics, gravity, cosmology and classical/quantum field theory expained by Stavrinos [19-35]. Such spaces are characterized by three fundamental geometric objects: nonlinear connection (N–connection), linear connection and metric. There is an important geometrical problem to prove the existence of the " best possible" metric and linear connection adapted to a N–connection structure. From the point of view of Riemannian geometry, the Thurston conjecture only asserts the existence of a best possible metric on an arbitrary closed three dimensional (3D) manifold. It is a very difficult task to define Ricci flows of mutually compatible fundamental geometric structures on non–Riemannian manifolds (for instance, on a Finsler manifold). For such purposes, we can also apply the Hamilton's approach but correspondingly generalized in order to describe nonholonomic (constrained) configurations. The first attempts to construct exact solutions of the Ricci flow equations on nonholonomic Einstein and Riemann–Cartan (with nontrivial torsion) manifolds, generalizing well known classes of exact solutions in Einstein and string gravity, were performed and explanied by Vacaru [13-16].

We take a unified point of view towards Riemannian and generalized Finsler–Lagrange spaces following the geometry of nonholonomic manifolds and exploit the similarities and emphasize differences between locally isotropic and anisotropic Ricci flows. In our works, it will be shown when the remarkable Perelman–Hamilton results hold true for more general non–Riemannian configurations. It should be noted that this is not only a straightforward technical extension of the Ricci flow theory to certain manifolds with additional geometric structures. The problem of constructing the Finsler–Ricci flow theory contains a number of new conceptual and fundamental issues on compatibility of geometrical and physical objects and their optimal configurations.There are at least three important arguments supporting the investigation of nonholonomic Ricci flows: 1) The Ricci flows of a Riemannian metric may result in a Finsler– like metric if the flows are subjected to certain nonintegrable constraints and modelled with respect to nonholonomic frames (we shall prove it in this work). 2) Generalized Finsler– like metrics appear naturally as exact solutions in Einstein, string, gauge and noncommutative gravity, parametrized by generic off–diagonal metrics, nonholonomic frames and generalized connections and methods explained by Vacaru S [33-35]. It is an important physical task to analyze Ricci flows of such solutions as well of other physically important solutions (for instance, black holes, solitonic and/pp–waves solutions, Taub NUT configurations [13-15] resulting in nonholonomic geometric configurations. 3) Finally, the fact that a 3D manifold establishes an appropriate Riemannian metric, which implies certain fundamental consequences (for instance) for our spacetime topology, allows us to consider other types of "also not bad" metrics with possible local anisotropy and nonholonomic gravitational interactions. What are the natural evolution equations for such configurations and how can we relate them to the topology of nonholonomic manifolds? We shall address such questions here (for regular Lagrange systems)

***Corresponding author:** Vacaru SI, Rectors Department, University Al. I. Cuza, Greece, E-mail: sergiu.vacaru@uaic.ro

and in further works. The notion of nonholonomic manifold was introduced independently by G. Vranceanu [36] and Horak [37] as there was a need for geometric interpretation of nonholonomic mechanical systems modern approaches, criticism and historical remarks explained by Vacaru [34,38,39]. A pair $(M,\mathcal{D})$, where $M$ is a manifold and $\mathcal{D}$ is a nonintegrable distribution on $M$, is called a nonholonomic manifold. Three well known classes of nonholonomic manifolds, where the nonholonomic distribution defines a nonlinear connection (N-connection) structure, are defined by the Finsler spaces [40-42] and their generalizations as Lagrange and Hamilton spaces [34,43] (usually such geometries are modelled on the tangent bundle $TM$) More recent examples, related to exact off-diagonal solutions and nonholonomic frames in Einstein/string/gauge/noncommutative gravity and nonholonomic Fedosov manifolds [33,34,44] also emphasize nonholonomic geometric structures.Let us now sketch the Ricci flow program for nonholonomic manifolds and Lagrange-Finsler geometries. Different models of "locally anisotropic" spaces can be elaborated for different types of fundamental geometric structures (metric, nonlinear and linear connections). In general, such spaces contain nontrivial torsion and nonmetricity fields. It would be a very difficult technical task to generalize and elaborate new proofs for all types of non-Riemannian geometries. Our strategy will be different: We shall formulate the criteria to determine when certain types of Finsler like geometries can be "extracted" (by imposing the corresponding nonholonomic constraints) from "well defined" Ricci flows of Riemannian metrics. This is possible because such geometries can be equivalently described in terms of the Levi Civita connections or by metric configurations with nontrivial torsion induced by nonholonomic frames. By nonholonomic transforms of geometric structures, we shall be able to generate certain classes of nonmetric geometries and/or generalized torsion configurations.The aim of this paper (the first one in a series of works) is to formulate the Ricci flow equations on nonholonomic manifolds and prove the conditions under which such configurations (of Finsler-Lagrange type and in modern gravity) can be extracted from well defined flows of Riemannian metrics and evolution of preferred frame structures. Further works will be dedicated to explicit generalizations of Perelman results [1-3] for nonholonomic manifolds and spaces provided with almost complex structure generated by nonlinear connections. We shall also construct new classes of exact solutions of nonholonomic Ricci flow equations, with noncommutative and/or Lie algebroid symmetry, defining locally anisotropic flows of black hole, wormhole and cosmological configurations and developing the results from work of Vacaru [13-15,33-35]. The paper is organized as follows: We start with preliminaries on geometry of nonholonomic manifolds provided with nonlinear connection (N-connection) structure in Section 2. We show how nonholonomic configurations can be naturally defined in modern gravity and the geometry of Riemann-Finsler and Lagrange spaces in Section 3. Section 4 is devoted to the theory of anholonomic Ricci flows: we analyze the evolution of distinguished geometric objects and speculate on nonholonomic Ricci flows of symmetric and nonsymmetric metrics. In Section 5, we prove that the Finsler-Ricci flows can be extracted from usual Ricci flows by imposing certain classes of nonholonomic constraints and deformations of connections. We also study regular Lagrange systems and consider generalized Lagrange-Ricci flows. The Appendix outlines some necessary results from the local geometry of N-anholonomic manifolds.

## Notation remarks

We shall use both the free coordinate and local coordinate formulas which are both convenient to introduce compact denotations and sketch some proofs. The left up/lower indices will be considered as labels of geometrical objects, for instance, on a nonholonomic Riemannian of Finsler space. The boldfaced letters will be used to denote that the objects (spaces) are adapted (provided) to (with) nonlinear connection structure.

## Preliminaries: Nonholonomic Manifolds

We recall some basic facts in the geometry of nonholonomic manifolds provided with nonlinear connection (N-connection) structure. The reader can refer to the concepts explained by Etayo [33,34,38,44] for details and proofs (for some important results we shall sketch the key points for such proofs). On nonholonomic vectors and (co-) tangent bundles and related Riemannian-Finsler and Lagrange-Hamilton geometries [34,41,42].

### N-connections

Consider a $(n+m)$-dimensional manifold $\mathbf{V}$, with $n \geq 2$ and $m \geq 1$ (for a number of physical applications, it is equivalently called to be a physical and/or geometric space). In a particular case, $\mathbf{V} = TM$, with $n=m$ (i.e. a tangent bundle), or $\mathbf{V} = \mathbf{E} = (E, M)$, $\dim M = n$, is a vector bundle on $M$, with total space $E$. In a general case, we can consider a manifold $\mathbf{V}$ provided with a local fibred structure into conventional "horizontal" and "vertical" directions. The local coordinates on $\mathbf{V}$ are denoted in the form $u = (x, y)$, or $u^{\alpha} = (x^i, y^a)$, where the "horizontal" indices run the values $i, j, k, \ldots = 1, 2, \ldots, n$ and the "vertical" indices run the values $a, b, c, \ldots = n+1, n+2, \ldots, n+m$. We denote by $\pi^{\top} : T\mathbf{V} \to TM$ the differential of a map $\pi : \mathbf{V} \to V$ defined by fiber preserving morphisms of the tangent bundles $TV$ and $TM$. The kernel of $\pi^{\top}$ is only the vertical subspace $vV$ with a related inclusion mapping $i : v\mathbf{V} \to T\mathbf{V}$.

**Definition 2.1:** *A nonlinear connection (N-connection)* $\mathbf{N}$ *on a manifold* $\mathbf{V}$ *is defined by the splitting on the left of an exact sequence*

$$0 \to v\mathbf{V} \xrightarrow{i} T\mathbf{V} \to T\mathbf{V} / v\mathbf{V} \to 0,$$

i. e. by a morphism of submanifolds $\mathbf{N} : T\mathbf{V} \to v\mathbf{V}$ such that $\mathbf{N} \circ \mathbf{i}$ is the unity in $v\mathbf{V}$

Locally, a N-connection is defined by its coefficients $N_i^a(u)$,

$$\mathbf{N} = N_i^a(u) dx^i \otimes \frac{\partial}{\partial y^a}. \tag{1}$$

Globalizing the local splitting, one proves:

**Proposition 2.1:** *Any N-connection is defined by a Whitney sum of conventional horizontal (h) subspace,* $(h\mathbf{V})$, *and vertical (v) subspace,* $(v\mathbf{V})$,

$$T\mathbf{V} = h\mathbf{V} \oplus v\mathbf{V}. \tag{2}$$

The sum (2) states on $TV$ a nonholonomic (equivalently, anholonomic, or nonintegrable) distribution of horizontal and vertical subspaces. The well known class of linear connections consists of a particular subclass with the coefficients being linear on $y^a$, i.e. $N_i^a(u) = \Gamma_{bj}^a(x) y^b$.

The geometric objects on $\mathbf{V}$ can be defined in a form adapted to a N-connection structure, following certain decompositions being invariant under parallel transports preserving the splitting (2). In this case, we call them to be distinguished (by the N-connection structure), i.e. d-objects. For instance, a vector field $\mathbf{X} \in T\mathbf{V}$ is expressed

$$\mathbf{X} = (hX, vX), \ or \ \mathbf{X} = X^{\alpha}\mathbf{e}_{\alpha} = X^i\mathbf{e}_i + X^a e_a,$$

where $hX = X^i\mathbf{e}_i$ and $vX = X^a e_a$ state, respectively, the adapted to the N-connection structure horizontal (h) and vertical (v) components of the vector. In brief, **X** is called a distinguished vector, in brief, d-vector). In a similar fashion, the geometric objects on TV like tensors, spinors, connections,... are called respectively d-tensors, d-spinors, d-connections if they are adapted to the N-connection splitting (2).

**Definition 2.2:** *The N-connection curvature is defined as the Neijenhuis tensor,*

$$\mathrm{W}(\mathbf{X},\mathbf{Y}) \doteqdot vX, vY] + v[\mathbf{X},\mathbf{Y}] - v[vX,\mathbf{Y}] - v[\mathbf{X}, vY]. \tag{3}$$

In local form, we have for (3)

$$\mathrm{W} = \frac{1}{2}\Omega^a_{ij}\, d^i \wedge d^j \otimes \partial_a,$$

with coefficients

$$\Omega^a_{ij} = \frac{\partial N^a_i}{\partial x^j} - \frac{\partial N^a_j}{\partial x^i} + N^b_i \frac{\partial N^a_j}{\partial y^b} - N^b_j \frac{\partial N^a_i}{\partial y^b}. \tag{4}$$

Any N-connection **N** may be characterized by an associated frame (vierbein) structure $\mathbf{e}_\nu = (\mathbf{e}_i, e_a)$, where

$$\mathbf{e}_i = \frac{\partial}{\partial x^i} - N^a_i(u)\frac{\partial}{\partial y^a} \text{ and } e_a = \frac{\partial}{\partial y^a}, \tag{5}$$

and the dual frame (coframe) structure $\mathbf{e}^\mu = (e^i, \mathbf{e}^a)$, where

$$e^i = dx^i \text{ and } \mathbf{e}^a = dy^a + N^a_i(u)dx^i. \tag{6}$$

These vielbeins are called respectively N-adapted frames and coframes. In order to preserve a relation with the previous denotations [33,34] we emphasize that $\mathbf{e}_\nu = (\mathbf{e}_i, e_a)$ and $\mathbf{e}^\mu = (e^i, \mathbf{e}^a)$ are correspondingly the former "N-elongated" partial derivatives $\delta_\nu = \delta / \partial u^\nu = (\delta_i, \partial_a)$ and N-elongated differentials $\delta^\mu = \delta u^\mu = (d^i, \delta^a)$. This emphasizes that the operators (5) and (6) define certain "N-elongated" partial derivatives and differentials which are more convenient for tensor and integral calculations on such nonholonomic manifolds.The vielbeins (6) satisfy the nonholonomy relations

$$[\mathbf{e}_\alpha, \mathbf{e}_\beta] = \mathbf{e}_\alpha \mathbf{e}_\beta - \mathbf{e}_\beta \mathbf{e}_\alpha = W^\gamma_{\alpha\beta}\mathbf{e}_\gamma \tag{7}$$

with (antisymmetric) nontrivial anholonomy coefficients $W^b_{ia} = \partial_a N^b_i$ and $W^a_{ji} = \Omega^a_{ij}$. The above presented formulas present the proof of

**Proposition 2.2:** *A N-connection on* **V** *defines a preferred nonholonomic N-adapted frame (vierbein) structure* $\mathbf{e} = (he, ve)$ *and its dual* $\tilde{\mathbf{e}} = (\tilde{he}, \tilde{ve})$ *with* **e** *and* $\tilde{\mathbf{e}}$ *linearly depending on N-connection coefficients.*

For simplicity, we shall work with a particular class of nonholonomic manifolds:

**Definition 2.3:** *A manifold* **V** *is N-anholonomic if its tangent space TV is enabled with a N-connection structure (2).*

There are two important examples of N-anholonomic manifolds, when *V=E*, or *TM*:

**Example 2.1:** *A vector bundle* $\mathbf{E} = (E, \pi, M, \mathbf{N})$, *defined by a surjective projection* $\pi : E \to M$, *with M being the base manifold,* $\dim M = n$, *and E being the total space,* $\dim E = n + m$, *and provided with a N-connection splitting (2) is called N-anholonomic vector bundle. A particular case is that of N-anholonomic tangent bundle* $\mathbf{TM} = (TM, \pi, M, \mathbf{N})$, *with dimensions n=m*

In a similar manner, we can consider different types of (super) spaces, Riemann or Riemann-Cartan manifolds, noncommutative bundles, or superbundles, provided with nonholonomc distributions (2) and preferred systems [33,34].

## Torsions and curvatures of d-connections and d-metrics

One can be defined N-adapted linear connection and metric structures:

**Definition 2.4:** *A distinguished connection (d-connection)* **D** *on a N-anholonomic manifold* **V** *is a linear connection conserving under parallelism the Whitney sum (2).*

For any d-vector **X**, there is a decomposition of **D** into h- and v-covariant derivatives,

$$\mathbf{D}_\mathbf{X} \doteqdot \mathbf{X} \rfloor \mathbf{D} = hX \rfloor \mathbf{D} + vX \rfloor \mathbf{D} = Dh_X + D_{vX} = hD_X + vD_X. \tag{8}$$

The symbol "$\rfloor$" in (8) denotes the interior product. We shall write conventionally that $\mathbf{D} = (hD, vD)$, or $\mathbf{D}_\alpha = (D_i, D_a)$. For convenience, in the Appendix, we present some local formulas for d-connections $\mathbf{D} = \{\Gamma^\gamma_{\alpha\beta} = (L^i_{jk}, L^a_{bk}, C^i_{jc}, C^a_{bc})\}$, with $hD = (L^i_{jk}, L^a_{bk})$ and $vD = (C^i_{jc}, C^a_{bc})$, see (6).

**Definition 2.5:** *The torsion of a d-connection* $\mathbf{D} = (hD, vD)$, *for any d-vectors* **X,Y** *is defined by d-tensor field*

$$\mathbf{T}(\mathbf{X},\mathbf{Y}) \doteqdot \mathbf{D}_\mathbf{X}\mathbf{Y} - \mathbf{D}_\mathbf{Y}\mathbf{X} - [\mathbf{X},\mathbf{Y}]. \tag{9}$$

One has a N-adapted decomposition

$$\mathbf{T}(\mathbf{X},\mathbf{Y}) = \mathbf{T}(hX, hY) + \mathbf{T}(hX, vY) + \mathbf{T}(vX, hY) + \mathbf{T}(vX, vY). \tag{10}$$

Considering h- and v-projections of (10) and taking into account that $h[vX, vY] = 0$, one proves

**Theorem 2.1:** *The torsion* **T** *of a d-connection* **D** *is defined by five nontrivial d-tensor fields adapted to the h- and v-splitting by the N-connection structure*

$$hT(hX, hY) \doteqdot D_{hX}\, hY - D_{hY}\, hX - h[\mathbf{X},\mathbf{Y}],$$

$$vT(hX, hY) \doteqdot v[hY, hX],$$

$$vT(hX, vY) \doteqdot -vD_{vY}\, hX - h[hX, vY],$$

$$vT(hX, vY) \doteqdot vD_{hX}\, vY - v[hX, vY],$$

$$vT(vX, vY) \doteqdot vD_X\, vY - vD_Y\, vX - v[vX, vY].$$

The d-torsions $hT(hX, hY), vT(vX, vY), \ldots$ are called respectively the *h (hh)*-torsion, *v(vv)*-torsion and so on. The local formulas (9) for torsion **T** are given in the Appendix.

**Definition 2.6:** *The curvature of a d-connection* **D** *is defined*

$$\mathbf{R}(\mathbf{X},\mathbf{Y}) \doteqdot \mathbf{D}_\mathbf{X}\mathbf{D}_\mathbf{Y} - \mathbf{D}_\mathbf{Y}\mathbf{D}_\mathbf{X} - \mathbf{D}_{[\mathbf{X},\mathbf{Y}]} \tag{11}$$

for any d-vectors **X,Y**

By straightforward calculations, one check the properties

$$hR(\mathbf{X},\mathbf{Y})\, vZ = 0,\ vR(\mathbf{X},\mathbf{Y})hZ = 0,$$

$$\mathbf{R}(\mathbf{X},\mathbf{Y})\mathbf{Z} = hR(\mathbf{X},\mathbf{Y})hZ + vR(\mathbf{X},\mathbf{Y})\, vZ,$$

for any for any d-vectors **X, Y, Z.**

**Theorem 2.2:** *The curvature* **R** *of a d-connection* **D** *is completely defined by six d-curvatures*

$$\mathbf{R}(hX, hY)hZ = \left(D_{hX}D_{hY} - D_{hY}D_{hX} - D_{[hX,hY]} - vD_{[hX,hY]}\right)hZ,$$

$$\mathbf{R}(hX, hY)\, vZ = \left(D_{hX}D_{hY} - D_{hY}D_{hX} - D_{[hX,hY]} - vD_{[hX,hY]}\right)vZ,$$

$$\mathbf{R}(vX, hY)hZ = \left(D_{hX}D_{hY} - D_{hY}\, D_{vX} - D_{[vX,hY]} - vD_{[vX,hY]}\right)hZ,$$

$$\mathbf{R}(vX, hY)\, vZ = \left(D_{vX}\, D_{hY} - D_{hY}\, D_{vX} - D_{h[vX,hY]} - D_{v[vX,hY]}\right)vZ,$$

$$\mathbf{R}(vX,vY)hZ = \left(D_{vX}D_{vY} - D_{vY}D_{vX} - D_{v[vX,vY]}\right)hZ,$$

$$\mathbf{R}(vX,vY)vZ = \left(D_{vX}D_{vY} - D_{vY}\ D_{vX} - D_{v[vX,vY]}\right)vZ.$$

The formulas for local coefficients of d-curvatures $\mathbf{R} = \{\mathbf{R}^{\alpha}_{\ \beta\gamma\delta}\}$ are given in the Appendix, see (11).

**Definition 2.7:** *A metric structure $\breve{g}$ on a N-anholonomic manifold* **V** *is a symmetric covariant second rank tensor field which is non degenerated and of constant signature in any point* $\mathbf{u} \in \mathbf{V}$.

In general, a metric structure is not adapted to a N-connection structure.

**Definition 2.8:** *A d-metric $\mathbf{g} = hg \oplus_N vg$ is a usual metric tensor which contracted to a d-vector results in a dual d-vector, d-covector (the duality being defined by the inverse of this metric tensor).*

The relation between arbitrary metric structures and d-metrics is established by

**Theorem 2.3:** *Any metric $\breve{g}$ can be equivalently transformed into a d-metric*

$$\mathbf{g} = hg(hX,hY) + vg(vX,vY) \tag{12}$$

adapted to a given N-connection structure.

**Proof:** e introduce $hg(hX,hY) = hg(hX,hY)$ and $v\breve{g}(vX,vY) = vg(vX,vY)$ and try to find a N-connection when

$$\breve{g}(hX,vY) = 0 \tag{13}$$

for any d-vectors **X,Y**. In local form, the equation (13) is an algebraic equation for the N-connection coefficients $N_i^a$, see formulas (1) and (2) in the Appendix.

A distinguished metric (in brief, d-metric) on a N-anholonomic manifold **V** is a usual second rank metric tensor **g** which with respect to a N-adapted basis (6) can be written in the form

$$\mathbf{g} = g_{ij}(x,y)\, e^i \otimes e^j + h_{ab}(x,y)\, \mathbf{e}^a \otimes \mathbf{e}^b \tag{14}$$

defining a N-adapted decomposition $\mathbf{g} = hg \oplus_{\mathbf{N}} vg = [hg, vg]$.

From the class of arbitrary d-connections **D** on **V**, one distinguishes those which are metric compatible (metrical d-connections) satisfying the condition

$$\mathbf{Dg} = \mathbf{0} \tag{15}$$

including all h- and v-projections

$$D_j g_{kl} = 0, D_a g_{kl} = 0, D_j h_{ab} = 0, D_a h_{bc} = 0.$$

Different approaches to Finsler-Lagrange geometry modelled on **TM** (or on the dual tangent bundle $\mathbf{T^*M}$, in the case of Cartan-Hamilton geometry) were elaborated for different d-metric structures which are metric compatible [34,40] or not metric compatible [34,42].

## (Non) adapted linear connections

For any metric structure **g** on a manifold **V**, there is the unique metric compatible and torsionless Levi Civita connection $\nabla$ for which ${}^{\nabla}\mathcal{T}^{\alpha} = 0$ and $\nabla\mathbf{g} = \mathbf{0}$. This is not a d-connection because it does not preserve under parallelism the N-connection splitting (2) (it is not adapted to the N-connection structure).

**Theorem 2.4** *For any d-metric $\mathbf{g} = [hg, vg]$ on a N-anholonomic manifold* **V**, *there is a unique metric canonical d-connection $\widehat{\mathbf{D}}$ satisfying the conditions $\widehat{\mathbf{D}}\mathbf{g} = 0$ and with vanishing $h(hh)$-torsion, $v(vv)$-torsion, i. e. $h\widehat{T}(hX,hY) = 0$ and $v\widehat{T}(vX,vY) = 0$.*

**Proof:** y straightforward calculations, we can verify that the d-connection with coefficients $\widehat{\Gamma}^{\gamma}_{\ \alpha\beta} = \left(\widehat{L}^i_{jk}, \widehat{L}^a_{bk}, \widehat{C}^i_{jc}, \widehat{C}^a_{bc}\right)$, see (15) in the Appendix, satisfies the condition of Theorem.

**Definition 2.9:** *A N-anholonomic Riemann-Cartan manifold ${}^{RC}\mathbf{V}$ is defined by a d-metric* **g** *and a metric d-connection* **D** *structures. For a particular case, we can consider that a space ${}^{R}\widehat{\mathbf{V}}$ is a N-anholonomic Riemann manifold if its d-connection structure is canonical, i.e., $\mathbf{D} = \widehat{\mathbf{D}}$.*

The d-metric structure **g** on ${}^{RC}\mathbf{V}$ is of type (14) and satisfies the metricity conditions (15). With respect to a local coordinate basis, the metric **g** is parametrized by a generic off-diagonal metric ansatz (2). For a particular case, we can take $\mathbf{D} = \widehat{\mathbf{D}}$ and treat the torsion $\widehat{\mathbf{T}}$ as a nonholonomic frame effect induced by a nonintegrable N-splitting. We conclude that a N-anholonomic Riemann manifold is with nontrivial torsion structure (9) (defined by the coefficients of N-connection (1), and d-metric (14) and canonical d-connection (15)). Nevertheless, such manifolds can be described alternatively, equivalently, as a usual (holonomic) Riemann manifold with the usual Levi Civita for the metric (1) with coefficients (2). We do not distinguish the existing nonholonomic structure for such geometric constructions.For more general applications, we have to consider additional torsion components, for instance, by the so-called *H*-field in string gravity [45].

**Theorem 2.5:** *The geometry of a (semi) Riemannian manifold V with prescribed (n+m)-splitting (nonholonomic h- and v-decomposition) is equivalent to the geometry of a canonical ${}^{R}\widehat{\mathbf{V}}$.*

**Proof:** et $g_{\alpha\beta}$ be the metric coefficients, with respect to a local coordinate frame, on *V*. The (*n*+*m*)-splitting states for a parametrization of type (2) which allows us to define the N-connection coefficients $N_i^a$ by solving the algebraic equations (3) (roughly speaking, the N-connection coefficients are defined by the "off-diagonal" N-coefficients, considered with respect to those from the blocks $n \times n$ and $m \times m$). Having defined $\mathbf{N} = \{N_i^a\}$, we can compute the N-adapted frames $\mathbf{e}_{\alpha}$ (5) and $\mathbf{e}^{\beta}$ (6) by using frame transforms (4) and (5) for any fixed values $e_i^{\underline{i}}(u)$ and $e_a^{\underline{a}}(u)$; for instance, for coordinate frames $e_i^{\underline{i}} = \delta_i^{\underline{i}}$ and $e_a^{\underline{a}} = \delta_a^{\underline{a}}$. As a result, the metric structure is transformed into a d-metric of type (14). We can say that *V* is equivalently re-defined as a N-anholonomic manifold **V**.

It is also possible to compute the coefficients of canonical d-connection $\widehat{\mathbf{D}}$ following formulas (15). We conclude that the geometry of a (semi) Riemannian manifold *V* with prescribed (*n*+*m*)-splitting can be described equivalently by geometric objects on a canonical N-anholonomic manifold ${}^{R}\widehat{\mathbf{V}}$ with induced torsion $\widehat{\mathbf{T}}$ with the coefficients computed by introducing (15) into (9). The inverse construction also holds true: A d-metric (14) on ${}^{R}\widehat{\mathbf{V}}$ is also a metric on *V* but with respect to certain N-elongated basis (6). It can be also rewritten with respect to a coordinate basis having the parametrization (2). From this Theorem, by straightforward computations with respect to N-adapted bases (6) and (5), one follows

**Corollary 2.1:** *The metric of a (semi) Riemannian manifold provided with a preferred N-adapted frame structure defines canonically two equivalent linear connection structures: the Levi Civita connection and the canonical d-connection.*

**Proof.** n a manifold ${}^{R}\widehat{\mathbf{V}}$, we can work with two equivalent linear connections. If we follow only the methods of Riemannian geometry, we have to choose the Levi Civita connection. In some cases, it may be optimal to elaborate a N-adapted tensor and differential calculus for nonholnomic structures, i.e. to choose the canonical d-connection. With respect to N-adapted frames, the coefficients of one connection can be expressed via coefficients of the second one, see formulas (16) and (15). Both such linear connections are defined by the same off-diagonal metric structure. For diagonal metrics with respect to local coordinate frames, the constructions are trivial.

Having prescribed a nonholonomic $n$+$m$ splitting on a manifold $V$, we can define two canonical linear connections $\nabla$ and $\widehat{\mathbf{D}}$. Correspondingly, these connections are characterized by two curvature tensors, ${}_{\shortmid}R^{\alpha}_{\beta\gamma\delta}(\nabla)$ (computed by introducing ${}_{\shortmid}\Gamma^{\alpha}_{\beta\gamma}$ into (7) and (10)) and $\mathbf{R}^{\alpha}_{\beta\gamma\delta}(\widehat{\mathbf{D}})$ (with the N-adapted coefficients computed following formulas (11)). Contracting indices, we can compute the Ricci tensor $Ric(\nabla)$ and the Ricci d-tensor $\mathbf{Ric}(\widehat{\mathbf{D}})$ following formulas (12), correspondingly written for $\nabla$ and $\widehat{\mathbf{D}}$. Finally, using the inverse d-tensor $\mathbf{g}^{\alpha\beta}$ for both cases, we compute the corresponding scalar curvatures ${}^{s}R(\nabla)$ and ${}^{s}\mathbf{R}(\widehat{\mathbf{D}})$, see formulas (13) by contracting, respectively, with the Ricci tensor and Ricci d-tensor.

## Metrization procedure and preferred linear connections

On a N-anholonomic manifold **V**, with prescribed fundamental geometric structures **g** and **N**, we can consider various classes of d-connections **D**, which, in general, are not metric compatible, i.e. $\mathbf{Dg} \neq \mathbf{0}$. The canonical d-connection $\widehat{\mathbf{D}}$ is the "simplest" metrical one, with respect to which other classes of d-connections $\mathbf{D} = \widehat{\mathbf{D}} + \mathbf{Z}$ can be distinguished by their deformation (equivalently, distorsion, or deflection) d-tensors **Z**. Every geometric construction performed for a d-connection **D** can be redefined for $\widehat{\mathbf{D}}$, and inversely, if **Z** is well defined.

Let us consider the set of all possible nonmetrical and metrical d-connections constructed only from the coefficients of a d-metric and N-connection structure, $g_{ij}, h_{ab}$ and $N^a_i$, and their partial derivatives. Such d-connections can be generated by two procedures of deformation,

$$\widehat{\mathsf{G}}^{\gamma}{}_{\alpha\beta} \rightarrow {}^{[K]}\mathsf{G}^{\gamma}{}_{\alpha\beta} = \mathsf{G}^{\gamma}{}_{\alpha\beta} + {}^{[K]}\mathbf{Z}^{\gamma}{}_{\alpha\beta} \ (\textit{Kawaguchi's metrization kaw1, kaw2}),$$

$$or \rightarrow {}^{[M]}\mathsf{G}^{\gamma}{}_{\alpha\beta} = \widehat{\mathsf{G}}^{\gamma}{}_{\alpha\beta} + {}^{[M]}\mathbf{Z}^{\gamma}{}_{\alpha\beta},$$

where ${}^{[K]}\mathbf{Z}^{\gamma}{}_{\alpha\beta}$ and ${}^{[M]}\mathbf{Z}^{\gamma}{}_{\alpha\beta}$ are deformation d-tensors.

**Theorem 2.6:** *For given d-metric $\mathbf{g}_{\alpha\beta} = [g_{ij}, h_{ab}]$ and N-connection $\mathbf{N} = \{N^a_i\}$ structures, the deformation d-tensor*

$${}^{[K]}\mathbf{Z}^{\gamma}{}_{\alpha\beta} = \{ {}^{[K]}Z^{i}{}_{jk} = \frac{1}{2} g^{im} D_j g_{mk}, {}^{[K]}Z^{a}{}_{bk} = \frac{1}{2} h^{ac} D_k h_{cb},$$

$${}^{[K]}Z^{i}{}_{ja} = \frac{1}{2} g^{im} D_a g_{mj}, {}^{[K]}Z^{a}{}_{bc} = \frac{1}{2} h^{ad} D_c h_{db} \}$$

transforms a d-connection $\mathsf{G}^{\gamma}{}_{\alpha\beta} = \left(L^i_{jk}, L^a_{bk}, C^i_{jc}, C^a_{bc}\right)$ into a metric d-connection

$${}^{[K]}\mathsf{G}^{\gamma}{}_{\alpha\beta} = \left(L^i_{jk} + {}^{[K]}Z^{i}{}_{jk}, L^a_{bk} + {}^{[K]}Z^{a}{}_{bk}, C^i_{jc} + {}^{[K]}Z^{i}{}_{ja}, C^a_{bc} + {}^{[K]}Z^{a}{}_{bc}\right).$$

**Proof:** t comes from a straightforward verification that the metricity conditions ${}^{[K]}\mathbf{Dg} = \mathbf{0}$ are satisfied (similarly to Chapter 1 in for generalized Finsler-affine spaces).

**Theorem 2.7:** *For fixed d-metric, $\mathbf{g}_{\alpha\beta} = [g_{ij}, h_{ab}]$, and N-connection, $\mathbf{N} = \{N^a_i\}$, structures the set of metric d-connections ${}^{[M]}\mathsf{G}^{\gamma}{}_{\alpha\beta} = \widehat{\mathsf{G}}^{\gamma}{}_{\alpha\beta} + {}^{[M]}\mathbf{Z}^{\gamma}{}_{\alpha\beta}$ is defined by the deformation d-tensors*

$${}^{[M]}\mathbf{Z}^{\gamma}{}_{\alpha\beta} = \{ {}^{[M]}Z^{i}{}_{jk} = {}^{[-]}O^{li}_{km} Y^m_{lj}, {}^{[M]}Z^{a}{}_{bk} = {}^{[-]}O^{ea}_{bd} Y^m_{ej},$$

$${}^{[M]}Z^{i}{}_{ja} = {}^{[+]}O^{mi}_{jk} Y^k_{mc}, {}^{[M]}Z^{a}{}_{bc} = {}^{[+]}O^{ea}_{bd} Y^d_{ec} \}$$

where the so-called Obata operators are defined

$${}^{[\pm]}O^{li}_{km} = \frac{1}{2}\left(\delta^l_k \delta^i_m \pm g_{km} g^{li}\right) \ and \ {}^{[\pm]}O^{ea}_{bd} = \frac{1}{2}\left(\delta^e_b \delta^a_d \pm h_{bd} h^{ea}\right)$$

and $Y^m_{lj}$, $Y^m_{ej}, Y^k_{mc}$, $Y^d_{ec}$ are arbitrary d-tensor fields.

**Proof:** t also comes from a straightforward verification. Here we note, that ${}^{[M]}\mathsf{G}^{\gamma}{}_{\alpha\beta}$ are generated with prescribed nontrivial torsion coefficients. If ${}^{[M]}\mathbf{Z}^{\gamma}{}_{\alpha\beta} = 0$, the canonical d-connection $\widehat{\mathsf{G}}^{\gamma}{}_{\alpha\beta}$ contains a nonholonomically induced torsion.

We can generalize the concept of N-anholonomic Riemann-Cartan manifold ${}^{RC}\mathbf{V}$ (see Definition 2.9):

**Definition 2.10:** *A N-anholonomic metric-affine manifold ${}^{ma}\mathbf{V}$ is defined by three fundamental geometric objects: 1) a d-metric $\mathbf{g}_{\alpha\beta} = [g_{ij}, h_{ab}]$, 2) a N-connection $\mathbf{N} = \{N^a_i\}$ and 3) a general d-connection **D**, with nontrivial nonmetricity d-tensor field $\mathbf{Q} = \mathbf{Dg}$.*

The geometry and classification of metric-affine manifolds and related generalized Finsler-affine spaces is considered in Part I of monograph explained by Vacaru [34]. From Theorems 2.6, 2.7 and 2.5, follows

**Conclusion 2.1:** The geometry of any manifold ${}^{ma}V$ can be equivalently modelled by deformation tensors on Riemann manifolds provided with preferred frame structure. The constructions are elaborated in N-adapted form if we work with the canonical d-connection, or not adapted to the N-connection structure if we apply the Levi Civita connection.

Finally, in this section, we note that if the torsion and nonmetricity fields of ${}^{ma}\mathbf{V}$ are defined by the d-metric and N-connection coefficients (for instance, in Finsler geometry with Chern or Berwald connection, see below section 5.1) we can equivalently (nonholonomically) transform ${}^{ma}\mathbf{V}$ into a Riemann manifold with metric structure of type (1) and (2).

## Gravity and Lagrange-Finsler Geometry

We study N-anholonomic structures in Riemmann-Finsler and Lagrange geometry modelled on nonholonomic Riemann-Cartan manifolds.

### Generalized lagrange spaces

If a N-anholonomic manifold is stated to be a tangent bundle, **V=TM** the dimension of the base and fiber space coincide, $n = m$, and we obtain a special case of N-connection geometry. For such geometric models, a N-connection is defined by Whithney sum

$$T\mathbf{TM} = h\mathbf{TM} \oplus v\mathbf{TM}, \tag{16}$$

with local coefficients $\mathbf{N} = \{N^a_i(x^i, y^a)\}$, where it is convenient to distinguish h-indices $i, j, k...$ from v-indices $a, b, c, ...$ On **TM**, there is an almost complex structure $\mathbf{F} = \{\mathbf{F}^{\beta}_{\alpha}\}$ associated to **N** defined by

$$\mathbf{F}(\mathbf{e}_i) = -e_i \ and \ \mathbf{F}(e_i) = \mathbf{e}_i, \tag{17}$$

where $\mathbf{e}_i = \partial / \partial x^i - N^k_i \partial / \partial y^k$ and $e_i = \partial / \partial y^i$ and $\mathbf{F}^{\beta}_{\alpha}\mathbf{F}^{\gamma}_{\beta} = -\delta^{\beta}_{\alpha}$. Similar constructions can be performed on N-anholonomic manifolds $\mathbf{V}^{n+n}$ where fibred structures of dimension $n$+$n$ are modelled.

A general d-metric structure (14) on $\mathbf{V}^{n+n}$, together with a prescribed N-connection **N**, defines a N-anholonomic Riemann-Cartan manifold of even dimension.

**Definition 3.1:** *A generalized Lagrange space is modelled on* $\mathbf{V}^{n+n}$ (by a d-metric with $g_{ij} = \delta_i^a \delta_j^b h_{ab}$, i.e.

$$^{c}\mathbf{g} = h_{ij}(x,y)\left(e^i \otimes e^j + \mathbf{e}^i \otimes \mathbf{e}^j\right). \quad (18)$$

One calls $\varepsilon = h_{ab}(x,y)\, y^a y^b$ to be the absolute energy associated to a $h_{ab}$ of constant signature.

**Theorem 3.1:** *For nondegenerated Hessians*

$$\tilde{h}_{ab} = \frac{1}{2}\frac{\partial^2 \varepsilon}{\partial y^a \partial y^b}, \quad (19)$$

when $\det|\tilde{h}| \neq 0$, there is a canonical N-connection completely defined by $h_{ij}$,

$$^{c}N_i^a(x,y) = \frac{\partial G^a}{\partial y^i} \quad (20)$$

where

$$G^a = \frac{1}{2}\tilde{h}^{ab}\left(y^k \frac{\partial^2 \varepsilon}{\partial y^b \partial x^k} - \delta_b^k \frac{\partial \varepsilon}{\partial x^k}\right).$$

**Proof:** ne has to consider local coordinate transformation laws for some coefficients $N_i^a$ preserving splitting (16). We can verify that $^{c}N_i^a$ satisfy such conditions. The sketch of proof is given and expained by Vacaru [34] for **TM**. We can consider any nondegenerated quadratic form $h_{a'b'}(x,y) = e^a_{a'} e^b_{b'} h_{ab}(x,y)$ on $\mathbf{V}^{n+n}$ if we redefine the v-coordinates in the form $y^{a'} = y^{a'}(x^i, y^a)$ and $x^{i'} = x^{i'}.x$

Finally, in this section, we state:

**Theorem 3.2:** *For any generalized Lagrange space, there are canonical N-connection* $^{c}\mathbf{N}$, *almost complex* $^{c}\mathbf{F}$, *d-metric* $^{c}\mathbf{g}$ *and d-connection* $^{c}\widehat{\mathbf{D}}$ *structures defined by an effective regular Lagrangian* $^{\varepsilon}L(x,y) = \sqrt{|\varepsilon|}$ *and its Hessian* $\tilde{h}_{ab}(x,y)$ *(19).*

**Proof:** t follows from formulas (19), (20), (17) and (19) and adapted d-connection (21) and d-metric structures (20) all induced by a $^{\varepsilon}L = \sqrt{|\varepsilon|}$.

## Lagrange-finsler spaces

The class of Lagrange-Finsler geometries is usually defined on tangent bundles but it is possible to model such structures on general N-anholonomic manifolds, for instance, in (pseudo) Riemannian and Riemann-Cartan geometry, if nonholonomic frames are introduced into consideration [33,34]. Let us consider two such important examples when the N-anholonomic structures are modelled on **TM**. One denotes by $\widetilde{TM} = TM \setminus \{0\}$ where {0} means the set of null sections of surjective map $\pi : TM \to M$.

**Example 3.1:** *A Lagrange space is a pair* $L^n = [M, L(x,y)]$ *with a differentiable fundamental Lagrange function* $L(x,y)$ *defined by a map* $L : (x,y) \in TM \to L(x,y) \in \mathbb{R}$ *of class* $C^\infty$ *on* $\widetilde{TM}$ *and continuous on the null section* $0 : M \to TM$ *of* $\pi$. *The Hessian (19) is defined*

$$^{L}g_{ij}(x,y) = \frac{1}{2}\frac{\partial^2 L(x,y)}{\partial y^i \partial y^j} \quad (21)$$

when $rank|g_{ij}| = n$ on $\widetilde{TM}$ and the left up "L" is an abstract label pointing that certain values are defined by the Lagrangian *L*.

The notion of Lagrange space was introduced by Kern [43] and elaborated as a natural extension of Finsler geometry. In a more particular case, we have

**Example 3.2:** *A Finsler space defined by a fundamental Finsler function* $F(x,y)$, *being homogeneous of type* $F(x,\lambda y) = \lambda F(x,y)$, *for nonzero* $\lambda \in \mathbb{R}$, *may be considered as a particular case of Lagrange geometry when* $L = F^2$.

Our approach to the geometry of N-anholonomic spaces (in particular, to that of Lagrange, or Finsler, spaces) is based on canonical d-connections. It is more related to the existing standard models of gravity and field theory allowing to define Finsler generalizations of spinor fields, noncommutative and supersymmetric models, discussed in by Vacaru [33,34]. Nevertheless, a number of schools and authors on Finsler geometry prefer linear connections which are not metric compatible (for instance, the Berwald and Chern connections, see below Definition 5.1) which define new classes of geometric models and alternative physical theories with nonmetricity field, see details in [34,40-42]. From a geometrical point of view [46,47], all such approaches are equivalent. It can be considered as a particular realization, for nonholonomic manifolds, of the Poincare's idea on duality of geometry and physical models stating that physical theories can be defined equivalently on different geometric spaces [48].

From the Theorem 3.2, one follows:

**Conclusion 3.1:** Any mechanical system with regular Lagrangian $L(x,y)$ (or any Finsler geometry with fundamental function $F(x,y)$) can be modelled as a nonhlonomic Riemann geometry with canonical structures $^{L}\mathbf{N}$, $^{L}\mathbf{g}$ and $^{L}\widehat{\mathbf{D}}$ (or $^{F}\mathbf{N}$, $^{F}\mathbf{g}$ and $^{F}\widehat{\mathbf{D}}$, for $L = F^2$) defined on a N-anholonomic manifold $\mathbf{V}^{n+n}$. In equivalent form, such Lagrange-Finsler geometries can be described by the same metric and N-anholonomic distributions but with the corresponding not adapted Levi Civita connections

Let us denote by $\mathbf{Ric}(\mathbf{D}) = C(1,4)\mathbf{R}(\mathbf{D})$, where $C(1,4)$ means the contraction on the first and fourth indices of the curvature $\mathbf{R}(\mathbf{D})$, and $\mathbf{Sc}(\mathbf{D}) = C(1,2)\mathbf{Ric}(\mathbf{D}) = {}^{s}\mathbf{R}$, where $C(1,2)$ is defined by contracting $\mathbf{Ric}(\mathbf{D})$ with the inverse d-metric, respectively, the Ricci tensor and the curvature scalar defined by any metric d-connection **D** and d-metric **g** on $^{RC}\mathbf{V}$, see also the component formulas (12), (13) and (14) in Appendix. The Einstein equations are

$$\mathbf{En}(\mathbf{D}) \doteqdot \mathbf{Ric}(\mathbf{D}) - \frac{1}{2}\mathbf{g}\,\mathbf{Sc}(\mathbf{D}) = ¡, \quad (22)$$

where the source ¡ reflects any contributions of matter fields and corrections from, for instance, string/brane theories of gravity. In a physical model, the equations (22) have to be completed with equations for the matter fields and torsion (for instance, in the Einstein-Cartan theory one considers algebraic equations [49] for the torsion and its source). It should be noted here that because of nonholonomic structure of $^{RC}\mathbf{V}$, the tensor $\mathbf{Ric}(\mathbf{D})$ is not symmetric and $\mathbf{D}[\mathbf{En}(\mathbf{D})] \neq 0$. This imposes a more sophisticated form of conservation laws on such spaces with generic "local anisotropy" [34], (a similar situation arises in Lagrange mechanics when nonholonomic constraints modify the definition of conservation laws). For $\mathbf{D} = \widehat{\mathbf{D}}$, all constructions can be equivalently redefined for the Levi Civita connection $\nabla$, when $\nabla[En(\nabla)] = 0$. A very important class of models can be elaborated when $¡ = diag[\lambda^h(\mathbf{u})\, hg, \lambda^v(\mathbf{u})\, vg]$, which defines the so-called N-anholonomic Einstein spaces with "nonhomogeneous" cosmological constant (various classes of exact solutions in gravity and nonholonomic Ricci flow theory were constructed and analyzed in [13-15,33,34].

## Anholonomic Ricci Flows

The Ricci flow theory was elaborated by Hamilton [6,7] and applied as a method approaching the Poincaré Conjecture and Thurston

Geometrization Conjecture [4,5] Perelman's works [1-3] and reviews of results [8,10].

## Holonomic Ricci flows

For a one parameter $\tau$ family of Riemannian metrics $\underline{g}(\tau)=\{\underline{g}_{\alpha\beta}(\tau,u^{\gamma})\}$ on a N-anholonomic manifold **V**, one introduces the Ricci flow equation

$$\frac{\partial \underline{g}_{\alpha\beta}}{\partial \tau} = -2\ {}_{\shortmid}\underline{R}_{\alpha\beta}, \tag{23}$$

where ${}_{\shortmid}\underline{R}_{\alpha\beta}$ is the Ricci tensor for the Levi Civita connection $\nabla=\{{}_{\shortmid}\Gamma^{\alpha}_{\beta\gamma}\}$ with the coefficients defined with respect to a coordinate basis $\partial_{\underline{\alpha}}=\partial/\partial u^{\underline{\alpha}}$. The equation (23) is a tensor nonlinear generalization of the scalar heat equation $\partial\phi/\partial\tau=\Delta\phi$, where $\Delta$ is the Laplace operator defined by $\underline{g}$. Usually, one considers normalized Ricci flows defined by

$$\frac{\partial}{\partial \tau} g_{\underline{\alpha\beta}} = -2\ {}_{\shortmid}R_{\underline{\alpha\beta}} + \frac{2r}{5} g_{\underline{\alpha\beta}}, \tag{24}$$

$$g_{\underline{\alpha\beta}}|_{\tau=0} = g^{[0]}_{\underline{\alpha\beta}}(u), \tag{25}$$

where the normalizing factor $r=\int R dV/dV$ is introduced in order to preserve the volume $V$, the boundary conditions are stated for $\tau$=0 and the solutions are searched for $\tau_0>\tau\geq 0$. For simplicity, we shall work with equations (23) if the constructions do not result in ambiguities.It is important to study the evolution of tensors in orthonormal frames and coframes on nonholonomic manifolds. Let $(\mathbf{V}, g_{\underline{\alpha\beta}}(\tau)), 0\leq\tau<\tau_0$, be a Ricci flow with ${}_{\shortmid}\underline{R}_{\alpha\beta} = {}_{\shortmid}R_{\underline{\alpha\beta}}$ and consider the evolution of basis vector fields

$$e_{\alpha}(\tau)=e^{\underline{\alpha}}_{\alpha}(\tau)\partial_{\underline{\alpha}} \ and \ e^{\beta}(\tau)=e^{\beta}_{\underline{\beta}}(\tau)du^{\underline{\beta}}$$

which are $\underline{g}(0)$-orthonormal on an open subset $\mathbf{U}\subset\mathbf{V}$. We evolve this local frame flows according to the formula

$$\frac{\partial}{\partial \tau} e^{\underline{\alpha}}_{\alpha} = g^{\underline{\alpha\beta}}\ {}_{\shortmid}R_{\underline{\beta\gamma}}\, e^{\underline{\gamma}}_{\alpha}. \tag{26}$$

There are unique solutions for such linear ordinary differential equations for all time $\tau\in 0,\tau_0)$.

Using the equations (24), (25) and (26), one can define the evolution equations under Ricci flow, for instance, for the Riemann tensor, Ricci tensor, Ricci scalar and volume form stated in coordinate frames (see, for example, the Theorem 3.13 in [10]. In this section, we shall consider such nonholnomic constraints on the evolution equation where the geometrical object will evolve in N-adapted form; we shall also model sets of N-anholnomic geometries, in particular, flows of geometric objects on nonholonomic Riemann manifolds and Finsler and Lagrange spaces.

## Ricci flows and N-anholonomic distributions

On manifold **V**, the equations (24) and (25) describe flows not adapted to the N-connections $N^a_i(\tau,u)$. For a prescribed family of such N-connections, we can construct from $\underline{g}_{\alpha\beta}(\tau,u^{\gamma})$ the corresponding set of d-metrics $\mathbf{g}_{\alpha\beta}(\tau,u)=[g_{ij}(\tau,u),h_{ab}(\tau,u)]$ and the set of N-adapted frames on $(\mathbf{V},\mathbf{g}_{\alpha\beta}(\tau)), 0\leq\tau<\tau_0$. The evolution of such N-adapted frames is not defined by the equations (26) but satisfies the

**Proposition 4.1:** *For a prescribed n+m splitting, the solutions of the system (24) and (25) define a natural flow of preferred N-adapted frame structures.*

**Proof:** Following formulas (1), (2) and (3), the boundary conditions (25) state the values $N^a_i(\tau=0,u)$ and $\mathbf{g}_{\alpha\beta}(\tau=0,u)=[g_{ij}(\tau=0,u),h_{ab}(\tau=0,u)]$. Having a well defined solution $\underline{g}_{\alpha\beta}(\tau,u)$, we can construct the coefficients of N-connection $N^a_i(\tau,u)$ and d-metric $\mathbf{g}(\tau,u)=[g(\tau,u),h(\tau,u)]$ for any $\tau\in 0,\tau_0)$: the associated set of frame (vielbein) structures $\mathbf{e}_{\nu}(\tau)=(\mathbf{e}_i(\tau),e_a)$, where

$$\mathbf{e}_i(\tau)=\frac{\partial}{\partial x^i}-N^a_i(\tau,u)\frac{\partial}{\partial y^a} \ and \ e_a=\frac{\partial}{\partial y^a}, \tag{27}$$

and the set of dual frame (coframe) structures $\mathbf{e}^{\mu}(\tau)=(e^i,\mathbf{e}^a(\tau))$, where

$$e^i=dx^i \ and \ \mathbf{e}^a(\tau)=dy^a+N^a_i(\tau,u)dx^i. \tag{28}$$

We conclude that prescribing the existence of a nonintegrable $(n+m)$-decomposition on a manifold for any $\tau\in 0,\tau_0)$, from any solution of the Ricci flow equations (26), we can extract a set of preferred frame structures with associated N-connections, with respect to which we can perform the geometric constructions in N-adapted form.

We shall need a formula relating the connection Laplacian on contravariant one-tensors with Ricci curvature and the corresponding deformations under N-anholonomic maps. Let **A** be a d-tensor of rank $k$. Then we define $\nabla^2\mathbf{A}$, for $\nabla$ being the Levi Civita connection, to be a contravariant tensor of rank $k$+2 given by

$$\nabla^2\mathbf{A}(\cdot,\mathbf{X},\mathbf{Y})=(\nabla_{\mathbf{X}}\nabla_{\mathbf{Y}}\mathbf{A})(\cdot)-\nabla_{\nabla_{\mathbf{X}}\mathbf{Y}}\mathbf{A}(\cdot). \tag{29}$$

This defines the (Levi Civita) Laplacian connection

$$\Delta\mathbf{A}\doteqdot\mathbf{g}^{\alpha\beta}\left(\nabla^2\mathbf{A}\right)(\mathbf{e}_{\alpha},\mathbf{e}_{\beta}), \tag{30}$$

for tensors, and

$$\Delta f \doteqdot tr\,\nabla^2 f=\mathbf{g}^{\alpha\beta}\left(\nabla^2 f\right)_{\alpha\beta},$$

for a scalar function on **V**. In a similar manner, by substituting $\nabla$ with $\widehat{\mathbf{D}}$, we can introduce the canonical d-connection Laplacian, for instance,

$$\widehat{\Delta}\mathbf{A}\doteqdot\mathbf{g}^{\alpha\beta}\left(\widehat{\mathbf{D}}^2\mathbf{A}\right)(\mathbf{e}_{\alpha},\mathbf{e}_{\beta}). \tag{31}$$

**Proposition 4.2** *The Laplacians $\widehat{\Delta}$ and $\Delta$ are related by formula*

$$\Delta\mathbf{A}=\widehat{\Delta}\mathbf{A}+{}_{\shortmid}\Delta\mathbf{A} \tag{32}$$

where the deformation d-tensor of the Laplacian, ${}_{\shortmid}\Delta$, is defined canonically by the N-connection and d-metric coefficients.

**Proof:** e sketch the method of computation ${}_{\shortmid}\Delta$. Using the formula (17), we have

$$\nabla_{\mathbf{X}}=\widehat{\mathbf{D}}_{\mathbf{X}}+{}_{\shortmid}Z_{\mathbf{X}} \tag{33}$$

where ${}_{\shortmid}Z_{\mathbf{X}}=\mathbf{X}^{\alpha}\ {}_{\shortmid}Z^{\gamma}_{\alpha\beta}$ is defined for any $\mathbf{X}^{\alpha}$ with ${}_{\shortmid}\mathbf{Z}^{\gamma}_{\alpha\beta}$ computed following formulas (17); all such coefficients depend on N-connection and d-metric coefficients and their derivatives, i.e. on generic off-diagonal metric coefficients (2) and their derivatives. Introducing (33) into (29) and (30), and separating the terms depending only on $\widehat{\mathbf{D}}_{\mathbf{X}}$ we get $\widehat{\Delta}\mathbf{A}$ (31). The rest of terms with linear or quadratic dependence on ${}_{\shortmid}Z^{\gamma}_{\alpha\beta}$ and their derivatives define

$${}_{\shortmid}\Delta\mathbf{A}\doteqdot\mathbf{g}^{\alpha\beta}(\ {}_{\Delta}\mathbf{Z}\mathbf{A}),$$

where

$${}_{\Delta}\mathbf{Z}\mathbf{A}=\widehat{\mathbf{D}}_{\mathbf{X}}({}_{\shortmid}Z_{\mathbf{Y}}\mathbf{A})+{}_{\shortmid}Z_{\mathbf{X}}\left(\widehat{\mathbf{D}}_{\mathbf{Y}}\mathbf{A}\right)+{}_{\shortmid}Z_{\mathbf{X}}(\ {}_{\shortmid}Z_{\mathbf{Y}}\mathbf{A})$$

$$-\widehat{\mathbf{D}}_{\ {}_{\shortmid}Z_{\mathbf{Y}}}\mathbf{A}-{}_{\shortmid}Z_{\widehat{\mathbf{D}}_{\mathbf{X}}\mathbf{Y}}\mathbf{A}-{}_{\shortmid}Z_{\ {}_{\shortmid}Z_{\mathbf{X}}\mathbf{Y}}\mathbf{A}.$$

In a similar form as for Proposition 4.2, we prove

**Proposition 4.3:** *The curvature, Ricci and scalar tensors of the Levi Civita connection* $\nabla$ *and the canonical d-connection* $\widehat{\mathbf{D}}$ *are defined by formulas*

$${}_{|}R(\mathbf{X},\mathbf{Y}) = \widehat{\mathbf{R}}(\mathbf{X},\mathbf{Y}) + {}_{|}\widehat{Z}(\mathbf{X},\mathbf{Y}),$$

$$Ric(\nabla) = \mathbf{Ric}(\widehat{\mathbf{D}}) + Ric({}_{|}\widehat{Z}),$$

$$Sc(\nabla) = \mathbf{Sc}(\widehat{\mathbf{D}}) + Sc({}_{|}\widehat{Z}),$$

where

$${}_{|}\widehat{Z}(\mathbf{X},\mathbf{Y}) = \mathbf{D}_{\mathbf{X}\,|}Z_{\mathbf{Y}} - {}_{|}Z_{\mathbf{Y}}\mathbf{D}_{\mathbf{X}} - {}_{|}Z_{[\mathbf{X},\mathbf{Y}]}$$

$$Ric({}_{|}\widehat{\mathbf{Z}}) = C(1,4)\,{}_{|}\widehat{Z},\ Sc({}_{|}\widehat{\mathbf{Z}}) = C(1,2)Ric({}_{|}\widehat{Z})$$

for $\widehat{\mathbf{R}}$ computed following formula (11) and $\mathbf{Sc}(\widehat{\mathbf{D}}) = {}^{s}\widehat{\mathbf{R}}$.

In the theory of Ricci flows, one considers tensors quadratic in the curvature tensors, for instance, for any given $\mathbf{g}^{\beta\beta'}$ and $\mathbf{D}$

$$\mathbf{B}_{\alpha\gamma\alpha'\gamma'} = \mathbf{g}^{\beta\beta'}\mathbf{g}^{\delta\delta'}\mathbf{R}_{\alpha\beta\gamma\delta}\mathbf{R}_{\alpha'\beta'\gamma'\delta'}, \quad (34)$$

$$\underline{\mathbf{B}}_{\alpha\gamma\alpha'\gamma'} \doteqdot \mathbf{B}_{\alpha\gamma\alpha'\gamma'} - \mathbf{B}_{\alpha\gamma\alpha'\gamma'} - \mathbf{B}_{\alpha\gamma'\gamma\alpha'} + \mathbf{B}_{\alpha\gamma\alpha'\gamma'},$$

$$\underline{\mathbf{B}}_{\alpha\gamma'} \doteqdot \mathbf{D}_{\alpha}\mathbf{D}_{\gamma'}\,{}^{s}\mathbf{R} - \mathbf{g}^{\beta\beta'}\left(\mathbf{D}_{\beta'}\mathbf{D}_{\alpha}\mathbf{R}_{\gamma'\beta} + \mathbf{D}_{\beta'}\mathbf{D}_{\gamma'}\mathbf{R}_{\alpha\beta}\right).$$

Using the connections $\nabla$, or $\widehat{\mathbf{D}}$, we similarly define and compute the values ${}_{|}B_{\alpha\gamma\alpha'\gamma'}$, ${}_{|}\underline{B}_{\alpha\gamma\alpha'\gamma'}$ and ${}_{|}\underline{B}_{\alpha\gamma'}$, or $\widehat{\mathbf{B}}_{\alpha\gamma\alpha'\gamma'}, \underline{\widehat{\mathbf{B}}}_{\alpha\gamma\alpha'\gamma'}$ and $\underline{\widehat{\mathbf{B}}}_{\alpha\gamma'}$.

## Evolution of distinguished geometrical objects

There are d-objects (d-tensors, d-connections) with N-adapted evolution completely defined by solutions of the Ricci flow equations (26).

**Definition 4.1:** *A geometric structure/object is extracted from a (Riemannian) Ricci flow (for the Levi Civita connection) if the corresponding structure/ object can be redefined equivalently, prescribing a* $(n+m)$ *-splitting, as a N-adapted structure/ d-object subjected to N-anholonomic flows.*

Following the Propositions 4.2 and 4.3 and formulas (34), we prove

**Theorem 4.1:** *The evolution equations for the Riemann and Ricci tensors and scalar curvature defined by the canonical d-connection are extracted respectively:*

$$\frac{\partial}{\partial\tau}\widehat{\mathbf{R}}_{\alpha\beta\gamma\delta} = \widehat{\Delta}\widehat{\mathbf{R}}_{\alpha\beta\gamma\delta} + 2\widehat{\underline{\mathbf{B}}}_{\alpha\beta\gamma\delta} + \widehat{\mathbf{Q}}_{\alpha\beta\gamma\delta},$$

$$\frac{\partial}{\partial\tau}\widehat{\mathbf{R}}_{\alpha\beta} = \widehat{\Delta}\widehat{\mathbf{R}}_{\alpha\beta} + \widehat{\mathbf{Q}}_{\alpha\beta},$$

$$\frac{\partial}{\partial\tau}\,{}^{s}\widehat{\mathbf{R}} = \widehat{\Delta}\,{}^{s}\widehat{\mathbf{R}} + 2\widehat{\mathbf{R}}_{\alpha\beta}\widehat{\mathbf{R}}^{\alpha\beta} + \widehat{\mathbf{Q}}$$

where, for

$${}_{|}R_{\alpha\beta\gamma\delta} = \widehat{\mathbf{R}}_{\alpha\beta\gamma\delta} + {}_{|}Z_{\alpha\beta\gamma\delta},\ {}_{|}\underline{B}_{\alpha\beta\gamma\delta} = \widehat{\underline{\mathbf{B}}}_{\alpha\beta\gamma\delta} + {}_{|}\widehat{\underline{Z}}_{\alpha\beta\gamma\delta},\ Z = \mathbf{g}^{\alpha\beta}\,{}_{|}Z_{\alpha\beta},$$

$${}_{|}R_{\alpha\beta} = \widehat{\mathbf{R}}_{\alpha\beta} + {}_{|}Z_{\alpha\beta},\ {}_{|}\underline{B}_{\alpha\gamma'} = \widehat{\underline{\mathbf{B}}}_{\alpha\gamma'} + {}_{|}\widehat{\underline{Z}}_{\alpha\gamma'},\ {}_{|}^{s}R = {}^{s}\widehat{\mathbf{R}} + Z,$$

the $Q$-terms (defined by the coefficients of canonical d-connection, $N_i^a$ and $\mathbf{g}_{\alpha\beta} = [g_{ij}, h_{ab}]$ and their derivatives) are

$$\widehat{\mathbf{Q}}_{\alpha\beta\gamma\delta} = -\frac{\partial}{\partial\tau}\,{}_{|}Z_{\alpha\beta\gamma\delta} + {}_{|}\Delta\widehat{\mathbf{R}}_{\alpha\beta\gamma\delta} + 2\,{}_{|}\widehat{\underline{Z}}_{\alpha\beta\gamma\delta},$$

$$\widehat{\mathbf{Q}}_{\alpha\beta} = -\frac{\partial}{\partial\tau}\,{}_{|}Z_{\alpha\beta} + {}_{|}\Delta\widehat{\mathbf{R}}_{\alpha\beta} + {}_{|}\widehat{\underline{Z}}_{\alpha\beta},$$

$$\widehat{\mathbf{Q}} = -\frac{\partial}{\partial\tau}Z + \widehat{\Delta}Z + {}_{|}\Delta\,{}^{s}\widehat{\mathbf{R}} + 2\widehat{\mathbf{R}}_{\alpha\beta}\,{}_{|}Z^{\alpha\beta} + 2\,{}_{|}Z_{\alpha\beta}\widehat{\mathbf{R}}^{\alpha\beta} + 2\,{}_{|}Z_{\alpha\beta}\,{}_{|}Z^{\alpha\beta}.$$

In Ricci flow theory, it is important to have the formula for the evolution of the volume form:

**Remark 4.1:** *The deformation of the volume form is stated by equation*

$$\frac{\partial}{\partial\tau}dvol(\tau,u^{\alpha}) = -\left({}^{s}\widehat{\mathbf{R}} + Z\right)dvol(\tau,u^{\alpha})$$

which is just that for the Levi Civita connection and $dvol(\tau,u^{\alpha}) \doteqdot \sqrt{|\det \underline{g}_{\alpha\beta}(\tau,u^{\gamma})|}$, where $\underline{g}_{\alpha\beta}(\tau)$ are metrics of type (1).

The evolution equations from Theorem 4.1 and Remark 4.1 transform into similar ones from Theorem 3.13 [10].

For any solution of equations (24) and (25), on $\mathbf{U} \subset \mathbf{V}$, we can construct for any $\tau \in 0,\tau_0)$ a parametrized set of canonical d-connections $\widehat{\mathbf{D}}(\tau) = \{\widehat{\mathbf{\Gamma}}^{\gamma}{}_{\alpha\beta}(\tau)\}$ (15) defining the corresponding canonical Riemann d-tensor (11), nonsymmetric Ricci d-tensor $\widehat{\mathbf{R}}_{\alpha\beta}$ (12) and scalar (13). The coefficients of d-objects are defined with respect to evolving N-adapted frames (27) and (28). One holds

**Conclusion 4.1:** *The evolution of corresponding d-objects on N-anholonomic Riemann manifolds can be canonically extracted from the evolution under Ricci flows of geometric objects on Riemann manifolds.*

In the sections 5.3 and 5.1, we shall consider how Finsler and Lagrange configurations can be extracted by more special parametrizations of metric and nonholonomic constraints.

## Nonholonomic ricci flows of (non) symmetric metrics

The Ricci flow equations were introduced by Hamilton [6] in a heuristic form similarly to how A. Einstein proposed his equations by considering possible physically grounded equalities between the metric and its first and second derivatives and the second rank Ricci tensor. On (pseudo) Riemannian spaces the metric and Ricci tensors are both symmetric and it is possible to consider the parameter derivative of metric and/or correspondingly symmetrized energy-momentum of matter fields as sources for the Ricci tensor.On N-anholonomic manifolds there are two alternative possibilities: The first one is to postulate the Ricci flow equations in symmetric form, for the Levi Civita connection, and then to extract various N-anholonomic configurations by imposing corresponding nonholonomic constraints. The bulk of our former and present work is related to symmetric metric configurations.

In the second case, we can start from the very beginning with a nonsymmetric Ricci tensor for a non-Riemannian space. In this section, we briefly speculate on such geometric constructions: The nonholonomic Ricci flows even beginning with a symmetric metric tensor may result naturally in nonsymmetric metric tensors $\hat{g}_{\alpha\beta} = \underline{g}_{\alpha\beta} + \overleftrightarrow{g}_{\alpha\beta}$, where $\overleftrightarrow{g}_{\alpha\beta} = -\overleftrightarrow{g}_{\beta\alpha}$. Nonsymmetric metrics in gravity were originally considered by Einstein [50] and Eisenhart [51], see modern approaches [52].

**Theorem 4.2:** *With respect to N-adapted frames, the canonical nonholonomic Ricci flows with nonsymmetric metrics defined by equations*

$$\frac{\partial}{\partial\tau}g_{ij} = -2\widehat{R}_{ij} + 2\lambda g_{ij} - h_{cd}\frac{\partial}{\partial\tau}(N_i^c N_j^d), \quad (35)$$

$$\frac{\partial}{\partial\tau}h_{ab} = -2\widehat{R}_{ab} + 2\lambda h_{ab}, \quad (36)$$

$$\frac{\partial}{\partial\tau}\overleftrightarrow{g}_{ia}=\widehat{R}_{ia},\frac{\partial}{\partial\tau}\overleftrightarrow{g}_{ai}=\widehat{R}_{ai} \quad (37)$$

where $\underline{g}_{\alpha\beta}=[g_{ij},h_{ab}]$ with respect to N-adapted basis (6), $\lambda=r/5$, $y^3=v$ and $\tau$ can be, for instance, the time like coordinate, $\tau=t$, or any parameter or extra dimension coordinate.

**Proof.** t follows from a redefinition of equations (24) with respect to N-adapted frames (by using the frame transform (4) and (5)), and considering respectively the canonical Ricci d-tensor (12) constructed from $[g_{ij},h_{ab}]$. Here we note that normalizing factor *r* is considered for the symmetric part of metric.

One follows:

**Conclusion 4.2:** *Nonholonomic Ricci flows (for the canonical d-connection) resulting in symmetric d-metrics are parametrized by the constraints*

$$\overleftrightarrow{g}_{\alpha\beta}=0 \text{ and } \widehat{R}_{ia}=\widehat{R}_{ai}=0. \quad (38)$$

The system of equations (35), (36) and (38), for "symmetric" nonholonomic Ricci flows, was introduced and analyzed in [13,14].

**Example 4.1:** *The conditions (38) are satisfied by any ansatz of type (14) in 3D, 4D, or 5D, with coefficients of type*

$$g_i=g_i(x^k),h_a=h_a(x^k,v),N_i^3=w_i(x^k,v),N_i^4=n_i(x^k,v), \quad (39)$$

for $i,j,...=1,2,3$ and $a,b,...=4,5$ (the 3D and 4D being parametrized by eliminating the cases $i=1$ and, respectively, $i=1,2$); $y^4=v$ being the so-called "anisotropic" coordinate. Such metrics are off-diagonal with the coefficients depending on 2 and 3 coordinates but positively not depending on the coordinate $y^5$.

We constructed and investigated various types of exact solutions of the nonholonomc Einstein equations and Ricci flow equations [33-35] and [13-15]. They are parametrized by ansatz of type (39) which positively constrains the Ricci flows to be with symmetric metrics. Such solutions can be used as backgrounds for investigating flows of Eisenhart (generalized Finsler-Eisenhart geometries) if the constraints (38) are not completely imposed. We shall not analyze this type of N-anholonomic Ricci flows in this series of works.

## Generalized Finsler-Ricci Flows

The aim of this section is to provide some examples illustrating how different types of nonholonomic constraints on Ricci flows of Riemannian metrics model different classes of N-anholonomic spaces (defined by Finsler metrics and connections, geometric models of Lagrange mechanics and generalized Lagrange geometries).

## Finsler-Ricci flows

Let us consider a τ-parametrized family (set) of fundamental Finsler functions $F(\tau)=F(\tau,x^i,y^a)$, see Example 3.2. For a family of nondegenerated Hessians

$${}^{F}h_{ij}(\tau,x,y)=\frac{1}{2}\frac{\partial^2F^2(\tau,x,y)}{\partial y^i\partial y^j}, \quad (40)$$

see formula (21) for effective $\varepsilon(\tau)=L(\tau)=F^2(\tau)$, we can model Finsler metrics on $\mathbf{V}^{n+n}$ (or on **TM**) and the corresponding family of canonical N-connections, see (20),

$${}^{c}N_i^a(\tau)=\frac{\partial G^a(\tau)}{\partial y^i}, \quad (41)$$

where $G^a(\tau)=\frac{1}{2}{}^{F}h^{ab}(\tau)\left(y^k\frac{\partial^2F^2(\tau)}{\partial y^b\partial x^k}-\delta_b^k\frac{\partial F^2(\tau)}{\partial x^k}\right)$ and ${}^{F}h^{ab}(\tau)$ are inverse to ${}^{F}h_{ij}(\tau)$.

**Proposition 5.1:** *Any family of fundamental Finsler functions $F(\tau)$ with nondegenerated ${}^{F}h_{ij}(\tau)$ defines a corresponding family of Sasaki type metrics*

$${}^{c}\mathbf{g}(\tau)={}^{F}h_{ij}(\tau,x,y)\left(e^i\otimes e^j+{}^{c}\mathbf{e}^i(\tau)\otimes{}^{c}\mathbf{e}^j(\tau)\right), \quad (42)$$

with ${}^{F}g_{ij}(\tau)={}^{F}h_{ij}(\tau,x,y)$, where ${}^{c}\mathbf{e}^a(\tau)=dy^a+{}^{c}N_i^a(\tau,u)dx^i$ are defined by the N-connection (41).

**Proof.** t follows from the explicit construction (42).

For $\mathbf{V}^{n+n}=\mathbf{TM}=(TM,\pi,M,{}^{c}N_i^a)$ with injective $\pi:TM\to M$, we can model by $F(\tau)$ various classes of Finsler geometries. In explicit form, we work on $\widetilde{TM}\doteqdot TM\setminus\{0\}$ and consider the pull-buck bundle $\pi^*TM$. One generates sets of geometric objects on pull-back cotangent bundle $\pi^*T^*M$ and its tensor products:

on $\pi^*T^*M\otimes\pi^*T^*M\otimes\pi^*T^*M$, a corresponding family of Cartan tensors

$$A(\tau)=A_{ijk}(\tau)dx^i\otimes dx^j\otimes dx^k,$$

$$A_{ijk}(\tau)\doteqdot\frac{F(\tau)}{2}\frac{\partial g_{ij}(\tau)}{\partial y^k};$$

on $\pi^*T^*M$, a family of Hilbert forms $\omega(\tau)\doteqdot\frac{\partial F(\tau)}{\partial y^i}dx^i$ and the d-connection 1-form

$$\omega_j^i(\tau)=L^i{}_{jk}(\tau)dx^k \quad (43)$$

$$L^i{}_{jk}(\tau)=\frac{1}{2}{}^{F}g^{ih}({}^{c}\mathbf{e}_k{}^{F}g_{jh}+{}^{c}\mathbf{e}_j{}^{F}g_{kh}-{}^{c}\mathbf{e}_h{}^{F}g_{jk}).$$

**Theorem 5.1:** *The set of fundamental Finsler functions $F(\tau)$ defines on $\pi^*TM$ a unique set of linear connections, called the Chern connections, characterized by the structure equations:*

$$d(dx^i)-dx^i\wedge\omega_i^i(\tau)=-dx^i\wedge\omega_i^i(\tau)=0,$$

i.e. the torsion free condition;

$$dg_{ij}(\tau)-{}^{F}g_{kj}(\tau)\omega_i^k(\tau)-{}^{F}g_{ik}(\tau)\omega_j^k(\tau)=2\frac{A_{ija}(\tau)}{F(\tau)}{}^{c}\mathbf{e}^a(\tau),$$

i.e. the almost metric compatibility condition.

**Proof:** t follows from straightforward computations. For any fixed value $\tau=\tau_0$, it is just the Chern's Theorem 2.4.1. from, In order to elaborate a complete geometric model on *TM*, which also allows us to perform the constructions for N-anholonomic manifolds, we have to extend the above considered forms with nontrivial coefficients with respect to ${}^{c}\mathbf{e}^a(\tau)$.

**Definition 5.1:** *A family of fundamental Finsler metrics $F(\tau)$ defines models of Finsler geometry (equivalently, space) with d-connections $\Gamma^\alpha_{\beta\gamma}(\tau)=(L^i{}_{jk}(\tau),\ C^i_{jk}(\tau))$ on a corresponding N-anholonomic manifold* **V**:

- of Cartan type if $L^i{}_{jk}(\tau)$ is that from (43) and

$$C^i{}_{jk}(\tau)=\frac{1}{2}{}^{F}g^{ih}(\tau)(\frac{\partial}{\partial x^k}{}^{F}g_{jh}(\tau)+\frac{\partial}{\partial x^j}{}^{F}g_{kh}(\tau)-\frac{\partial}{\partial x^h}{}^{F}g_{jk}(\tau)), \quad (44)$$

which is similar to formulas (21) but for $L=F^2(\tau)$;

- of Chern type if $L^i{}_{jk}(\tau)$ is given by (43) and $C^i{}_{jk}(\tau)=0$;
- of Berwald type if $L^i{}_{jk}(\tau)=\partial\,{}^{c}N^i_j/\partial y^k$ and $C^i{}_{jk}(\tau)=0$;
- of Hashiguchi type if $L^i{}_{jk}(\tau)=\partial\,{}^{c}N^i_j/\partial y^k$ and $C^i{}_{jk}(\tau)$ is given by (44).

Various classes of remarkable Finsler connections have been investigated by Bejancu [41,42]. On modelling Finsler like structures in

Einstein and string gravity and in noncommutative gravity. It should be emphasized that the models of Finsler geometry with Chern, Berwald or Hashiguchi type d-connections are with nontrivial nonmetricity field [33,34]. So, in general, a family of Finsler fundamental metric functions $F(\tau)$ may generate various types of N-anholonomic metric-affine geometric configurations, see Definition 2.10, but all components of such induced nonmetricity and/or torsion fields are defined by the coefficients of corresponding families of generic off-diagonal metrics of type (1), when the ansatz (2) is parametrized for $g_{ij} = h_{ij} = {}^F h_{ij}(\tau)$ and $N_i^a = {}^c N_i^a(\tau)$. Applying the results of Theorem 2.7, we can transform the families of "nonmetric" Finsler geometries into corresponding metric ones and model the Finsler configurations on N-anholonomic Riemannian spaces, see Conclusion 2.1. In the "simplest" geometric and physical manner (convenient both for applying the former Hamilton-Perelman results on Ricci flows for Riemannian metrics, as well for further generalizations to noncommutative Finsler geometry, supersymmetric models and so on...), we restrict our analysis to Finsler-Ricci flows with canonical d-connection of Cartan type when ${}^F\hat{\mathrm{G}}^{\alpha}_{\beta\gamma}(\tau) = (L^i_{\ jk}(\tau),\ C^i_{\ jk}(\tau))$ is with $L^i_{\ jk}(\tau)$ from (43) and $C^i_{\ jk}(\tau)$ from (44). This provides a proof for

**Lemma 5.1:** *A family of Finsler geometries defined by $F(\tau)$ can be characterized equivalently by the corresponding canonical d-connections (in N-adapted form) and Levi Civita connections (in not N-adapted form) related by formulas*

$$ {}^F_{\ \shortmid}\Gamma^{\gamma}_{\alpha\beta} = {}^F\hat{\mathrm{G}}^{\gamma}_{\alpha\beta} + {}_{\shortmid}Z^{\gamma}_{\alpha\beta} \tag{45}$$

where ${}_{\shortmid}Z^{\gamma}_{\alpha\beta}$ is computed following formulas (18) for $g_{ij} = h_{ij} = {}^F h_{ij}(\tau)$ and $N_i^a = {}^c N_i^a(\tau)$.

Following the Lemma 5.1 and section 4.1, we obtain the proof of

**Theorem 5.2:** *The Finsler-Ricci flows for fundamental metric functions*

$F(\tau)$ can be extracted from usual Ricci flows of Riemannian metrics parametrized in the form

$$ {}^F g_{\underline{\alpha\beta}}(\tau) = \begin{bmatrix} {}^F g_{ij} + {}^c N_i^a\, {}^c N_j^b\, {}^F g_{ab} & {}^c N_j^{e F} g_{ae} \\ {}^c N_i^e\, {}^F g_{be} & {}^F g_{ab} \end{bmatrix} \tag{46}$$

nd satisfying the equations (for instance, for normalized flows)

$$ \frac{\partial}{\partial\tau}\, {}^F g_{\underline{\alpha\beta}} = -2\, {}^F_{\ \shortmid}R_{\underline{\alpha\beta}} + \frac{2r}{5}\, {}^F g_{\underline{\alpha\beta}}, $$

$$ {}^F g_{\underline{\alpha\beta}|\tau=0} = {}^F g^{[0]}_{\underline{\alpha\beta}}(u). $$

The Finsler-Ricci flows are distinguished from the usual (unconstrained) flows of Riemannian metrics by existence of additional evolutions of preferred N-adapted frames (see Proposition 2.2):

**Corollary 5.1** *The evolution, for all "time" $\tau \in 0,\tau_0)$, of preferred frames on a Finsler space*

$$ {}^F\mathbf{e}_{\alpha}(\tau) = {}^F\mathbf{e}^{\underline{\alpha}}_{\alpha}(\tau,u)\partial_{\underline{\alpha}} $$

is defined by the coefficients

$$ {}^F\mathbf{e}^{\underline{\alpha}}_{\alpha}(\tau,u) = \begin{bmatrix} {}^F e_i^{\ \underline{i}}(\tau,u) & {}^c N_i^b(\tau,u)^F e_b^{\ \underline{a}}(\tau,u) \\ 0 & {}^F e_a^{\ \underline{a}}(\tau,u) \end{bmatrix}, \tag{47}$$

with ${}^F g_{ij}(\tau) = {}^F e_i^{\ \underline{i}}(\tau,u)^F e_j^{\ \underline{j}}(\tau,u)\eta_{\underline{ij}}$, where $\eta_{\underline{ij}} = diag[\pm1,\ldots\pm1]$ establish the signature of ${}^F g^{[0]}_{\underline{\alpha\beta}}(u)$, is given by equations

$$ \frac{\partial}{\partial\tau}\, {}^F e^{\underline{\alpha}}_{\alpha} = {}^F g^{\underline{\alpha\beta}}\, {}^F_{\ \shortmid}R_{\underline{\beta\gamma}}\, {}^F e^{\underline{\gamma}}_{\alpha}, \tag{48}$$

where ${}^F g^{\underline{\alpha\beta}}$ is inverse to (46) and ${}^F_{\ \shortmid}R_{\underline{\beta\gamma}}$ is the Ricci tensor constructed from the Levi Civita coefficients of (46).

**Proof.** e have to introduce the metric and N-connection coefficients (42) and (41), defined by $F(\tau)$, into (4). The equations (48) are similar to (26), but in our case for the N-adapted frames (47). We note that the evolution of the Riemann and Ricci tensors and scalar curvature defined by the Cartan d-connection, i.e. the canonical d-connection, ${}^F\hat{\mathrm{G}}^{\gamma}_{\alpha\beta}$, can be extracted as in Theorem 4.1 when the values are redefined for the metric (46) and (45).Finally, in this section, we conclude that the Ricci flows of Finsler metrics can be extracted from Ricci flows of Riemannian metrics by corresponding metric ansatz, nonholonomic constraints and deformations of linear connections, all derived canonically from fundamental Finsler functions.

## Ricci flows of regular lagrange systems

There were elaborated different approaches to geometric mechanics. We follow those related to formulations in terms of almost symplectic geometry [27] and generalized Finsler and Lagrange geometry [43]. We note that Lagrange-Finsler spaces can be equivalently modelled as almost Kähler geometries (see formulas (17) defining the almost complex structure) and, which is important for applications of the theory of anholonomic Ricci flows, modelled as nonholonomic Riemann manifolds, see Conclusion 3.1.

For regular mechanical systems, we can formulate the problem: Which fundamental Lagrange function $L(\tau) = L(\tau, x^i, y^j)$ from a class of Lagrangians parametrized by $\tau \in 0,\tau_0)$ will define the evolution of Lagrange geometry, from a theory of Ricci flows? The aim of this section is to present the key results solving this problem.

Following the formulas from Result 6.1 and the methods elaborated in previous section 5.1, when $F^2(\tau) \to L(\tau)$; ${}^F h_{ij}(\tau) \to {}^L g_{ij}(\tau)$, see (40) and (21); ${}^c N_i^a(\tau) \to {}^L N_j^i(\tau)$, see (41) and (19); ${}^c\mathbf{g}(\tau) \to {}^L\mathbf{g}(\tau)$, see (42) and (20); ${}^F\hat{\mathrm{G}}^{\alpha}_{\beta\gamma}(\tau) \to {}^L\hat{\mathrm{G}}^{\alpha}_{\beta\gamma}(\tau)$, see (45) and (21), where all values labeled by up-left "$L$" are canonically defined by $L(\tau)$, we prove (generalizations of Theorem 5.2 and Corollary 5.1):

**Theorem 5.3:** *The Lagrange-Ricci flows for regular Lagrangians $L(\tau)$ can be extracted from usual Ricci flows of Riemannian metrics parametrized as*

$$ {}^L g_{\underline{\alpha\beta}}(\tau) = \begin{bmatrix} {}^L g_{ij} + {}^L N_i^a\, {}^L N_j^b\, {}^L g_{ab} & {}^L N_j^{eL} g_{ae} \\ {}^L N_i^e\, {}^L g_{be} & {}^L g_{ab} \end{bmatrix} $$

and satisfying the equations (for instance, normalized)

$$ \frac{\partial}{\partial\tau}\, {}^L g_{\underline{\alpha\beta}} = -2\, {}^L_{\ \shortmid}R_{\underline{\alpha\beta}} + \frac{2r}{5}\, {}^L g_{\underline{\alpha\beta}}, $$

$$ {}^L g_{\underline{\alpha\beta}|\tau=0} = {}^L g^{[0]}_{\underline{\alpha\beta}}(u), $$

where ${}^L_{\ \shortmid}R_{\underline{\alpha\beta}}(\tau)$ are the Ricci tensors constructed from the Levi Civita connections of metrics ${}^L g_{\underline{\alpha\beta}}(\tau)$.

The Lagrange-Ricci flows are are characterized by the evolutions of preferred N-adapted frames (see Proposition 2.2):

**Corollary 5.2:** *The evolution, for all time $\tau \in 0,\tau_0)$, of preferred frames on a Lagrange space*

$$ {}^L\mathbf{e}_{\alpha}(\tau) = {}^L\mathbf{e}^{\underline{\alpha}}_{\alpha}(\tau,u)\partial_{\underline{\alpha}} $$

is defined by the coefficients

$$ {}^{L}\mathbf{e}_{\alpha}^{\underline{\alpha}}(\tau,u)=\begin{bmatrix} {}^{L}e_{i}^{\underline{i}}(\tau,u) & {}^{L}N_{i}^{b}(\tau,u)^{L}e_{b}^{\underline{a}}(\tau,u) \\ 0 & {}^{L}e_{a}^{\underline{a}}(\tau,u) \end{bmatrix}, $$

with ${}^{L}g_{ij}(\tau)={}^{L}e_{i}^{\underline{i}}(\tau,u)^{L}e_{j}^{\underline{j}}(\tau,u)\eta_{\underline{ij}}$, where $\eta_{\underline{ij}}=diag[\pm1,\ldots\pm1]$ establish the signature of ${}^{L}g_{\underline{\alpha\beta}}^{[0]}(u)$, is given by equations

$$ \frac{\partial}{\partial\tau}{}^{L}e_{\alpha}^{\underline{\alpha}}={}^{L}g^{\underline{\alpha\beta}}\ {}^{L}_{\ |}R_{\underline{\beta\gamma}}\ {}^{L}e_{\alpha}^{\underline{\gamma}}. $$

We conclude that the Ricci flows of Lagrange metrics can be extracted from Ricci flows of Riemannian metrics by corresponding metric ansatz, nonholonomic constraints and deformations of linear connections, all derived canonically for regular Lagrange functions.

## Generalized Lagrange-Ricci flows

We have the result that any mechanical system with a regular Lagrangian $L(x,y)$ can be geometrized canonically in terms of nonholonomic Riemann geometry, see Conclusion 3.1, and for certain conditions such configurations generate exact solutions of the gravitational field equations in the Einstein gravity and/or its string/gauge generalizations, see Result 6.2 and Theorem 6.1. In other words, for any symmetric tensor $g_{ij}=\delta_{i}^{a}\ \ \delta_{j}^{b}h_{ab}(x,y)$ on a manifold $\mathbf{V}^{n+n}$ we can generate a Lagrange space model, see section 3.1. The aim of this section is to show how we can construct nonholonomic Ricci flows with effective Lagrangians starting from an arbitrary family $g_{ij}(\tau)=\delta_{i}^{a}$ $\delta_{j}^{b}h_{ab}(\tau,x,y)$.

The values $h_{ab}(\tau)$ of constant signature defines a family of absolute energies $\varepsilon(\tau)=h_{ab}(\tau,x,y)\,y^{a}y^{b}$ and d-metrics of type (18),

$$ {}^{\varepsilon}\mathbf{g}(\tau)=h_{ij}(\tau,x,y)\left(e^{i}\otimes e^{j}+{}^{\varepsilon}\mathbf{e}^{i}(\tau)\otimes{}^{\varepsilon}\mathbf{e}^{j}(\tau)\right), $$

$$ {}^{\varepsilon}\mathbf{e}^{i}(\tau)=dy^{i}+{}^{\varepsilon}N_{i}^{a}(\tau,x,y)dx^{i}, \qquad (49) $$

where the τ-parametrized N-connection coefficients

$$ {}^{\varepsilon}N_{i}^{a}(\tau,x,y)=\frac{\partial\ {}^{\varepsilon}G^{a}(\tau)}{\partial y^{i}} \qquad (50) $$

with ${}^{\varepsilon}G^{a}(\tau)=\frac{1}{2}\,{}^{\varepsilon}\tilde{h}^{ab}(\tau)\left(y^{k}\frac{\partial^{2}\varepsilon(\tau)}{\partial y^{b}\partial x^{k}}-\delta_{b}^{k}\frac{\partial\varepsilon(\tau)}{\partial x^{k}}\right)$ are defined for nondegenerated Hessians

$$ {}^{\varepsilon}\tilde{h}_{ab}(\tau)=\frac{1}{2}\frac{\partial^{2}\varepsilon(\tau)}{\partial y^{a}\partial y^{b}}, \qquad (51) $$

when $\det|\tilde{h}|\neq0$.

For any fixed value of τ, the existence of fundamental geometric objects (49), (50) and (51) follows from Theorem 3.1. Similarly, the Theorem 3.2 states a modelling by $h_{ab}(\tau)$ of families of Lagrange spaces enabled with canonical N-connections ${}^{\varepsilon}\mathbf{N}(\tau)$, almost complex structure ${}^{\varepsilon}\mathbf{F}(\tau)$, d-metrics ${}^{\varepsilon}\mathbf{g}(\tau)$ and d-connections ${}^{\varepsilon}\widehat{\mathbf{D}}(\tau)$ structures defined respectively by effective regular Lagrangians ${}^{\varepsilon}L(\tau,x,y)=\sqrt{|\varepsilon(\tau,x,y)|}$ and theirHessians ${}^{\varepsilon}\tilde{h}_{ab}(\tau,x,y)$ (51). The results of previous section 5.3 can be reformulated in the form (with proofs being similar for those for Theorem 5.2 and Corollary 5.1, but with ${}^{\varepsilon}L$ instead of $F^{2}$ and ${}^{\varepsilon}N_{i}^{a}$ instead of ${}^{c}N_{i}^{a},\ldots$):

**Theorem 5.4:** *The generalized Lagrange-Ricci flows for regular effective Lagrangians ${}^{\varepsilon}L(\tau)$ derived from a family of symmetric tensors $h_{ab}(\tau,x,y)$ can be extracted from usual Ricci flows of Riemannian metrics parametrized in the form*

$$ {}^{\varepsilon}g_{\underline{\alpha\beta}}(\tau)=\begin{bmatrix} {}^{\varepsilon}\tilde{h}_{ij}+{}^{\varepsilon}N_{i}^{a}\ {}^{\varepsilon}N_{j}^{b}\ {}^{\varepsilon}\tilde{h}_{ab} & {}^{\varepsilon}N_{j}^{e}\ {}^{\varepsilon}\tilde{h}_{ae} \\ {}^{\varepsilon}N_{i}^{e}\ {}^{\varepsilon}\tilde{h}_{be} & {}^{\varepsilon}\tilde{h}_{ab} \end{bmatrix} $$

and satisfying the equations (for instance, normalized)

$$ \frac{\partial}{\partial\tau}{}^{\varepsilon}g_{\underline{\alpha\beta}}=-2\ {}^{\varepsilon}_{\ |}R_{\underline{\alpha\beta}}+\frac{2r}{5}\ {}^{\varepsilon}g_{\underline{\alpha\beta}}, $$

$$ {}^{\varepsilon}g_{\underline{\alpha\beta}|\tau=0}={}^{\varepsilon}g_{\underline{\alpha\beta}}^{[0]}(u), $$

where ${}^{\varepsilon}_{\ |}R_{\underline{\alpha\beta}}(\tau)$ are the Ricci tensors constructed from the Levi Civita connections of metrics ${}^{\varepsilon}g_{\underline{\alpha\beta}}(\tau)$.

The evolutions of preferred N-adapted frames (see Proposition 2.2) defined by generalized Lagrange-Ricci flows is stated by

**Corollary 5.3:** *The evolution, for all time $\tau\in0,\tau_{0})$, of preferred frames on an effective Lagrange space*

$$ {}^{\varepsilon}\mathbf{e}_{\alpha}(\tau)={}^{\varepsilon}\mathbf{e}_{\alpha}^{\underline{\alpha}}(\tau,u)\partial_{\underline{\alpha}} $$

is defined by the coefficients

$$ {}^{\varepsilon}\mathbf{e}_{\alpha}^{\underline{\alpha}}(\tau,u)=\begin{bmatrix} {}^{\varepsilon}e_{i}^{\underline{i}}(\tau,u) & {}^{\varepsilon}N_{i}^{b}(\tau,u)\ {}^{\varepsilon}e_{b}^{\underline{a}}(\tau,u) \\ 0 & {}^{\varepsilon}e_{a}^{\underline{a}}(\tau,u) \end{bmatrix}, $$

with ${}^{\varepsilon}\tilde{h}_{ij}(\tau)={}^{\varepsilon}e_{i}^{\underline{i}}(\tau,u)\ {}^{\varepsilon}e_{j}^{\underline{j}}(\tau,u)\eta_{\underline{ij}}$, where $\eta_{\underline{ij}}=diag[\pm1,\ldots\pm1]$ establish the signature of ${}^{\varepsilon}g_{\underline{\alpha\beta}}^{[0]}(u)$, is given by equations

$$ \frac{\partial}{\partial\tau}{}^{\varepsilon}e_{\alpha}^{\underline{\alpha}}={}^{\varepsilon}g^{\underline{\alpha\beta}}\ {}^{\varepsilon}_{\ |}R_{\underline{\beta\gamma}}\ {}^{\varepsilon}e_{\alpha}^{\underline{\gamma}}. $$

In Introduction and Part I of the monograph [34], it was proven that certain types of gravitational interactions can be modelled as generalized Lagrange-Finsler geometries and inversely, certain classes of generalized Finsler geometries can be modelled on N-anholonomic manifolds, even as exact solutions of gravitational field equations. The approach elaborated by Romanian geometers and physicists [33-35] originates from Vranceanu G and Horac Z works [36,37] on nonholonomic manifolds and mechanical systems, see a review of results and recent developments explained by Bejancu [38]. Recently, there were proposed various models of " analogous gravity", a review [53], which do not apply the methods of Finsler geometry and the formalism of nonlinear connections.

## Local Geometry Of N-Anholonomic Manifolds

Let us consider a metric structure on N-anholonomic manifold **V**,

$$ \breve{g}=\underline{g}_{\alpha\beta}(u)du^{\alpha}\otimes du^{\beta} \qquad (1) $$

defined with respect to a local coordinate basis $du^{\alpha}=\left(dx^{i},dy^{a}\right)$ by coefficients

$$ \underline{g}_{\alpha\beta}=\begin{bmatrix} g_{ij}+N_{i}^{a}N_{j}^{b}h_{ab} & N_{j}^{e}h_{ae} \\ N_{i}^{e}h_{be} & h_{ab} \end{bmatrix}. \qquad (2) $$

Such a metric (2) is generic off-diagonal, i.e. it can not be diagonalized by coordinate transforms if $N_{i}^{a}(u)$ are any general functions. The condition (13), for $hX\to e_{i}$ and $vY\to e_{a}$, transforms into

$$ \breve{g}(e_{i},e_{a})=0,\ \textit{equivalently}\ \underline{g}_{ia}-N_{i}^{b}h_{ab}=0, \qquad (3) $$

where $\underline{g}_{ia}\ \div g(\partial/\partial x^{i},\partial/\partial y^{a})$, which allows us to define in a unique form the coefficients $N_{i}^{b}=h^{ab}\underline{g}_{ia}$ where $h^{ab}$ is inverse to $h_{ab}$. We can write the metric $\breve{g}$ with ansatz (2) in equivalent form, as a d-metric (14) adapted to a N-connection structure, see Definition 2.8, if we define $g_{ij} \doteqdot \mathbf{g}(e_i, e_j)$ and $h_{ab} \doteqdot \mathbf{g}(e_a, e_b)$ and consider the vielbeins $\mathbf{e}_\alpha$ and $\mathbf{e}^\alpha$ to be respectively of type (5) and (6).

We can say that the metric $\breve{g}$ (1) is equivalently transformed into (14) by performing a frame (vielbein) transform

$$\mathbf{e}_\alpha = \mathbf{e}_\alpha^{\underline{\alpha}} \partial_{\underline{\alpha}} \text{ and } \mathbf{e}^\beta = \mathbf{e}^\beta_{\underline{\beta}} du^{\underline{\beta}}.$$

with coefficients

$$\mathbf{e}_\alpha^{\underline{\alpha}}(u) = \begin{bmatrix} e_i^{\,\underline{i}}(u) & N_i^b(u) e_b^{\,\underline{a}}(u) \\ 0 & e_a^{\,\underline{a}}(u) \end{bmatrix}, \quad (4)$$

$$\mathbf{e}^\beta_{\underline{\beta}}(u) = \begin{bmatrix} e^i_{\,\underline{i}}(u) & -N_k^b(u) e^k_{\,\underline{i}}(u) \\ 0 & e^a_{\,\underline{a}}(u) \end{bmatrix}, \quad (5)$$

being linear on $N_i^a$. We can consider that a N-anholonomic manifold $\mathbf{V}$ provided with metric structure $\breve{g}$ (1) (equivalently, with d-metric (14)) is a special type of a manifold provided with a global splitting into conventional "horizontal" and "vertical" subspaces (2) induced by the "off-diagonal" terms $N_i^b(u)$ and a prescribed type of nonholonomic frame structure (7).

The N-adapted components $\Gamma^\alpha_{\beta\gamma}$ of a d-connection $\mathbf{D}_\alpha = (\mathbf{e}_\alpha \rfloor \mathbf{D})$, where "$\rfloor$" denotes the interior product, are defined by the equations

$$\mathbf{D}_\alpha \mathbf{e}_\beta = \Gamma^\gamma_{\alpha\beta} \mathbf{e}_\gamma, \text{ or } \Gamma^\gamma_{\alpha\beta}(u) = (\mathbf{D}_\alpha \mathbf{e}_\beta) \rfloor \mathbf{e}^\gamma. \quad (6)$$

The N-adapted splitting into h- and v-covariant derivatives is stated by

$$h\mathbf{D} = \{\mathbf{D}_k = (L^i_{jk}, L^a_{bk})\}, \text{ and } v\mathbf{D} = \{\mathbf{D}_c = (C^i_{jc}, C^a_{bc})\},$$

where, by definition,

$$L^i_{jk} = (\mathbf{D}_k \mathbf{e}_j) \rfloor e^i, \quad L^a_{bk} = (\mathbf{D}_k e_b) \rfloor \mathbf{e}^a, C^i_{jc} = (\mathbf{D}_c \mathbf{e}_j) \rfloor e^i, \quad C^a_{bc} = (\mathbf{D}_c e_b) \rfloor \mathbf{e}^a.$$

The components $\Gamma^\gamma_{\alpha\beta} = (L^i_{jk}, L^a_{bk}, C^i_{jc}, C^a_{bc})$ completely define a d-connection $\mathbf{D}$ on a N-anholonomic manifold $\mathbf{V}$.

The simplest way to perform computations with d-connections is to use N-adapted differential forms like

$$\Gamma^\alpha_\beta = \Gamma^\alpha_{\beta\gamma} \mathbf{e}^\gamma \quad (7)$$

with the coefficients defined with respect to (6) and (5). For instance, torsion can be computed in the form

$$\mathcal{T}^\alpha \doteqdot \mathbf{D}\mathbf{e}^\alpha = d\mathbf{e}^\alpha + \Gamma^\alpha_\beta \wedge \mathbf{e}^\beta. \quad (8)$$

Locally it is characterized by (N-adapted) d-torsion coefficients

$$T^i_{jk} = L^i_{jk} - L^i_{kj}, T^i_{ja} = -T^i_{aj} = C^i_{ja}, T^a_{ji} = \Omega^a_{ji},$$

$$T^a_{bi} = -T^a_{ib} = \frac{\partial N^a_i}{\partial y^b} - L^a_{bi}, T^a_{bc} = C^a_{bc} - C^a_{cb}. \quad (9)$$

By a straightforward d-form calculus, we can find the N-adapted components of the curvature

$$\mathcal{R}^\alpha_\beta \doteqdot \mathbf{D}\Gamma^\alpha_\beta = \Gamma^\alpha_\beta - \Gamma^\gamma_\beta \wedge \Gamma^\alpha_\gamma = \mathbf{R}^\alpha_{\beta\gamma\delta} \mathbf{e}^\gamma \wedge \mathbf{e}^\delta, \quad (10)$$

of a d-connection $\mathbf{D}$, i.e. the d-curvatures from Theorem 2.2:

$$R^i_{hjk} = e_k L^i_{hj} - e_j L^i_{hk} + L^m_{hj} L^i_{mk} - L^m_{hk} L^i_{mj} - C^i_{ha} \Omega^a_{kj},$$

$$R^a_{bjk} = e_k L^a_{bj} - e_j L^a_{bk} + L^c_{bj} L^a_{ck} - L^c_{bk} L^a_{cj} - C^a_{bc} \Omega^c_{kj},$$

$$R^i_{jka} = e_a L^i_{jk} - D_k C^i_{ja} + C^i_{jb} T^b_{ka}, \quad (11)$$

$$R^c_{bka} = e_a L^c_{bk} - D_k C^c_{ba} + C^c_{bd} T^c_{ka},$$

$$R^i_{jbc} = e_c C^i_{jb} - e_b C^i_{jc} + C^h_{jb} C^i_{hc} - C^h_{jc} C^i_{hb},$$

$$R^a_{bcd} = e_d C^a_{bc} - e_c C^a_{bd} + C^e_{bc} C^a_{ed} - C^e_{bd} C^a_{ec}.$$

Contracting respectively the components of (11), one proves that the Ricci tensor $\mathbf{R}_{\alpha\beta} \doteqdot \mathbf{R}^\tau_{\ \alpha\beta\tau}$ is characterized by h- v-components, i.e. d-tensors,

$$R_{ij} \doteqdot R^k_{\ ijk}, \ R_{ia} \doteqdot -R^k_{\ ika}, R_{ai} \doteqdot R^b_{\ aib}, R_{ab} \doteqdot R^c_{\ abc}. \quad (12)$$

It should be noted that this tensor is not symmetric for arbitrary d-connections $\mathbf{D}$. The scalar curvature of a d-connection is

$${}^s\mathbf{R} \doteqdot \mathbf{g}^{\alpha\beta} \mathbf{R}_{\alpha\beta} = g^{ij} R_{ij} + h^{ab} R_{ab}, \quad (13)$$

defined by a sum the h- and v-components of (12) and d-metric (14).

The Einstein tensor is defined and computed in standard form

$$\mathbf{G}_{\alpha\beta} = \mathbf{R}_{\alpha\beta} - \frac{1}{2} \mathbf{g}_{\alpha\beta} {}^s\mathbf{R} \quad (14)$$

There is a minimal extension of the Levi Civita connection $\nabla$ to a canonical d-connection $\widehat{\mathbf{D}}$ which is defined only by a metric $\breve{g}$ is metric compatible, with $\widehat{T}^i_{\ jk} = 0$ and $\widehat{T}^a_{\ bc} = 0$ but $\widehat{T}^i_{\ ja}, \widehat{T}^a_{\ ji}$ and $\widehat{T}^a_{\ bi}$ are not zero, see (9). The coefficient $\widehat{\Gamma}^\gamma_{\ \alpha\beta} = (\widehat{L}^i_{jk}, \widehat{L}^a_{bk}, \widehat{C}^i_{jc}, \widehat{C}^a_{bc})$ of this connection, with respect to the N-adapted frames, are defined :

$$\widehat{L}^i_{jk} = \frac{1}{2} g^{ir} (e_k g_{jr} + e_j g_{kr} - e_r g_{jk}), \quad (15)$$

$$\widehat{L}^a_{bk} = e_b(N^a_k) + \frac{1}{2} h^{ac} (e_k h_{bc} - h_{dc}\, e_b N^d_k - h_{db}\, e_c N^d_k),$$

$$\widehat{C}^i_{jc} = \frac{1}{2} g^{ik} e_c g_{jk}, \widehat{C}^a_{bc} = \frac{1}{2} h^{ad} (e_c h_{bd} + e_c h_{cd} - e_d h_{bc}).$$

The Levi Civita linear connection $\nabla = \{{}_{\shortmid}\Gamma^\alpha_{\beta\gamma}\}$, uniquely defined by the conditions ${}_{\shortmid}\mathcal{T} = 0$ and $\nabla g = 0$, is not adapted to the distribution (2). Let us parametrize the coefficients in the form

$${}_{\shortmid}\Gamma^\alpha_{\beta\gamma} = ({}_{\shortmid}L^i_{jk}, {}_{\shortmid}L^a_{jk}, {}_{\shortmid}L^i_{bk}, {}_{\shortmid}L^a_{bk}, {}_{\shortmid}C^i_{jb}, {}_{\shortmid}C^a_{jb}, {}_{\shortmid}C^i_{bc}, {}_{\shortmid}C^a_{bc}),$$

where

$$\nabla_{\mathbf{e}_k}(\mathbf{e}_j) = {}_{\shortmid}L^i_{jk} \mathbf{e}_i + {}_{\shortmid}L^a_{jk} e_a, \nabla_{\mathbf{e}_k}(e_b) = {}_{\shortmid}L^i_{bk} \mathbf{e}_i + {}_{\shortmid}L^a_{bk} e_a,$$

$$\nabla_{e_b}(\mathbf{e}_j) = {}_{\shortmid}C^i_{jb} \mathbf{e}_i + {}_{\shortmid}C^a_{jb} e_a, \nabla_{e_c}(e_b) = {}_{\shortmid}C^i_{bc} \mathbf{e}_i + {}_{\shortmid}C^a_{bc} e_a.$$

A straightforward calculus[1] shows that the coefficients of the Levi-Civita connection can be expressed in the form

$${}_{\shortmid}L^i_{jk} = L^i_{jk}, \ {}_{\shortmid}L^a_{jk} = -C^i_{jb} g_{ik} h^{ab} - \frac{1}{2} \Omega^a_{jk}, \quad (16)$$

$${}_{\shortmid}L^i_{bk} = \frac{1}{2} \Omega^c_{jk} h_{cb} g^{ji} - \frac{1}{2} (\delta^i_j \delta^h_k - g_{jk} g^{ih}) C^j_{hb},$$

$${}_{\shortmid}L^a_{bk} = L^a_{bk} + \frac{1}{2} (\delta^a_c \delta^b_d + h_{cd} h^{ab}) [L^c_{bk} - e_b(N^c_k)],$$

$${}_{\shortmid}C^i_{kb} = C^i_{kb} + \frac{1}{2} \Omega^a_{jk} h_{cb} g^{ji} + \frac{1}{2} (\delta^i_j \delta^h_k - g_{jk} g^{ih}) C^j_{hb},$$

$${}_{\shortmid}C^a_{jb} = -\frac{1}{2} (\delta^a_c \delta^d_b - h_{cb} h^{ad}) [L^c_{dj} - e_d(N^c_j)], {}_{\shortmid}C^a_{bc} = C^a_{bc},$$

$${}_{\shortmid}C^i_{ab} = -\frac{g^{ij}}{2} \{ [L^c_{aj} - e_a(N^c_j)] h_{cb} + [L^c_{bj} - e_b(N^c_j)] h_{ca} \},$$

where $\Omega^a_{jk}$ are computed as in formula (4). For certain considerations, it is convenient to express

[1]Such results were originally considered by R. Miron and M. Anastasiei for vector bundles provided with N–connection and metric structures, see Ref. [?]. Similar proofs hold true for any nonholonomic manifold provided with a prescribed N–connection structures.

$$
{}_{|}\Gamma^{\gamma}_{\ \alpha\beta} = \widehat{\mathrm{G}}^{\gamma}_{\ \alpha\beta} + {}_{|}Z^{\gamma}_{\ \alpha\beta} \tag{17}
$$

where the explicit components of distorsion tensor ${}_{|}Z^{\gamma}_{\alpha\beta}$ can be defined by comparing the formulas (16) and (15):

$$
\begin{aligned}
&{}_{|}Z^{i}_{jk} = 0,\ {}_{|}Z^{a}_{jk} = -C^{i}_{jb} g_{ik} h^{ab} - \frac{1}{2}\Omega^{a}_{jk},\\
&{}_{|}Z^{i}_{bk} = \frac{1}{2}\Omega^{c}_{jk} h_{cb} g^{ji} - \frac{1}{2}(\delta^{i}_{j}\delta^{h}_{k} - g_{jk} g^{ih}) C^{j}_{hb},\\
&{}_{|}Z^{a}_{bk} = \frac{1}{2}(\delta^{a}_{c}\delta^{b}_{d} + h_{cd} h^{ab})\left[L^{c}_{bk} - e_{b}(N^{c}_{k})\right],\\
&{}_{|}Z^{i}_{kb} = \frac{1}{2}\Omega^{a}_{jk} h_{cb} g^{ji} + \frac{1}{2}(\delta^{i}_{j}\delta^{h}_{k} - g_{jk} g^{ih}) C^{j}_{hb},\\
&{}_{|}Z^{a}_{jb} = -\frac{1}{2}(\delta^{a}_{c}\delta^{d}_{b} - h_{cb} h^{ad})\left[L^{c}_{dj} - e_{d}(N^{c}_{j})\right],\ {}_{|}Z^{a}_{bc} = 0,\\
&{}_{|}Z^{i}_{ab} = -\frac{g^{ij}}{2}\left\{\left[L^{c}_{aj} - e_{a}(N^{c}_{j})\right] h_{cb} + \left[L^{c}_{bj} - e_{b}(N^{c}_{j})\right] h_{ca}\right\}.
\end{aligned} \tag{18}
$$

It should be emphasized that all components of ${}_{|}\Gamma^{\gamma}_{\alpha\beta}, \widehat{\mathrm{G}}^{\gamma}_{\ \alpha\beta}$ and ${}_{|}Z^{\gamma}_{\ \alpha\beta}$ are defined by the coefficients of d–metric (14) and N–connection (1), or equivalently by the coefficients of the corresponding generic off–diagonal metric (2).

For a differentiable Lagrangian $L(x,y)$, i.e. a fundamental Lagrange function, is defined by a map $L:(x,y) \in TM \to L(x,y) \in \mathbb{R}$ of class $C^{\infty}$ on $\widetilde{TM} = TM \backslash \{0\}$ and continuous on the null section $0: M \to TM$ of $\pi$ [34] the following results are derived:

## Result

1. The Euler–Lagrange equations

$$
\frac{d}{d\tau}\left(\frac{\partial L}{\partial y^{i}}\right) - \frac{\partial L}{\partial x^{i}} = 0
$$

where $y^{i} = \frac{dx^{i}}{d\varsigma}$ for ( ) depending on parameter $\varsigma$, are equivalent to the "nonlinear" geodesic equations

$$
\frac{d^{2}x^{i}}{d\tau^{2}} + 2G^{i}(x^{k}, \frac{dx^{j}}{d\varsigma}) = 0
$$

defining paths of a canonical semispray

$$
S = y^{i}\frac{\partial}{\partial x^{i}} - 2G^{i}(x,y)\frac{\partial}{\partial y^{i}}
$$

where

$$
2G^{i}(x,y) = \frac{1}{2}\,{}^{L}g^{ij}\left(\frac{\partial^{2}L}{\partial y^{i}\partial x^{k}}y^{k} - \frac{\partial L}{\partial x^{i}}\right)
$$

with ${}^{L}g^{ij}$ being inverse to (21).

2. There exists on $\mathbf{V} \simeq \widetilde{TM}$ a canonical N-connection

$$
{}^{L}N^{i}_{j} = \frac{\partial G^{i}(x,y)}{\partial y^{i}} \tag{19}
$$

defined by the fundamental Lagrange function $L(x,y)$, which prescribes nonholonomic frame structures of type (5) and (6), ${}^{L}\mathbf{e}_{\nu} = (\mathbf{e}_{i}, e_{a})$ and ${}^{L}\mathbf{e}^{\mu} = (e^{i}, \mathbf{e}^{a})$.

3. There is a canonical metric structure

$$
{}^{L}\mathbf{g} = g_{ij}(x,y)\, e^{i} \otimes e^{j} + g_{ij}(x,y)\, \mathbf{e}^{i} \otimes \mathbf{e}^{j} \tag{20}
$$

constructed as a Sasaki type lift from $M$ for $g_{ij}(x,y) = {}^{L}g_{ij}(x,y)$, see (21).

4. There is a unique metrical and, in this case, torsionless canonical d-connection ${}^{L}\mathbf{D} = (hD, vD)$ with the nontrivial coefficients with respect to ${}^{L}\mathbf{e}_{\nu}$ and ${}^{L}\mathbf{e}^{\mu}$ parametrized respectively ${}^{L}\widehat{\mathrm{G}}^{\alpha}_{\ \beta\gamma} = (\widehat{L}^{i}_{\ jk}, \widehat{C}^{a}_{\ bc})$, for

$$
\widehat{L}^{i}_{\ jk} = \frac{1}{2}g^{ih}(\mathbf{e}_{k}g_{jh} + \mathbf{e}_{j}g_{kh} - \mathbf{e}_{h}g_{jk}), \widehat{C}^{i}_{\ jk} = \frac{1}{2}g^{ih}(e_{k}g_{jh} + e_{j}g_{kh} - e_{h}g_{jk}) \tag{21}
$$

defining the generalized Christoffel symbols, where (for simplicity, we omitted the left up labels $(L)$ for N–adapted bases).

We conclude that any regular Lagrange mechanics can be geometrized as a nonholonomic Riemann manifold $\mathbf{V}$ equipped with canonical N-connection (19) and adapted d-connection (21) and d-metric structures (20) all induced by a $L(x,y)$.

Let us show how N-anholonomic configurations can defined in gravity theories explained by Vacaru [33,34]. In this case, it is convenient to work on a general manifold $\mathbf{V}, \dim \mathbf{V} = n + m$ enabled with a global N-connection structure, instead of the tangent bundle $\widetilde{TM}$.

**Result 6.2:** *Various classes of vacuum and nonvacuum exact solutions of (22) parametrized by generic off–diagonal metrics, nonholonomic vielbeins and Levi Civita or non–Riemannian connections in Einstein and extra dimension gravity models define explicit examples of N–anholonomic Einstein–Cartan (in particular, Einstein) spaces.*

It should be noted that a subclass of N–anholonomic Einstein spaces was related to generic off–diagonal solutions in general relativity by such nonholonomic constraints when $\mathbf{Ric}(\widehat{\mathbf{D}}) = Ric(\nabla)$ even $\widehat{} \neq \nabla$ where $\widehat{\mathbf{D}}$ is the canonical d–connection and $\nabla$ is the Levi–Civita connection.

A direction in modern gravity is connected to analogous gravity models when certain gravitational effects and, for instance, black hole configurations are modelled by optical and acoustic media. Following our approach on geometric unification of gravity and Lagrange regular mechanics in terms of N–anholonomic spaces, one holds

**Theorem 6.1:** A Lagrange (Finsler) space can be canonically modelled as an exact solution of the Einstein equations (22) on a N–anholonomic Riemann–Cartan space if and only if the canonical N–connection ${}^{L}\mathbf{N}$ (${}^{F}\mathbf{N}$)' d–metric ${}^{L}\mathbf{g}$ (${}^{F}\mathbf{g}$) and d–connection ${}^{L}\widehat{\mathbf{D}}$ (${}^{F}\widehat{\mathbf{D}}$) structures defined by the corresponding fundamental Lagrange function $L(\mathbf{x},\mathbf{y})$ (Finsler function $F(\mathbf{x},\mathbf{y})$) satisfy the gravitational field equations for certain physically reasonable sources

**Proof.** t can be performed in local form by considering the Einstein tensor (14) defined by the ${}^{L}\mathbf{N}$ (${}^{F}\mathbf{N}$) in the form (19) and ${}^{L}\mathbf{g}$ (${}^{F}\mathbf{g}$) in the form (20) inducing the canonical d–connection ${}^{L}\widehat{\mathbf{D}}$ (${}^{F}\widehat{\mathbf{D}}$). For certain zero or nonzero ;, such N–anholonomic configurations may be defined by exact solutions of the Einstein equations for a d–connection structure [53].

### Acknowledgement

A series of results were obtained during visits at IMAFF CSIC Madrid (2005), Fields Institute (2008), CERN (2013) and with partial support from the project IDEI: PN-II-ID-PCE-2011-3-0256. Recent papers contain review of results on metric compatible and noncompatible Finsler-Ricci flows and possible applications of such geometric methods in modern gravity, cosmology and particle physics.

### References

1. Perelman G (2002) The Entropy Formula for the Ricci Flow and its Geometric Applications.
2. Perelman G (2003) Ricci Flow with Surgery on Three–Manifolds.
3. Perelman G (2003) Finite Extinction Time for the Solutions to the Ricci Flow on Certain Three–Manifolds.

4. Thurston W (1982) Three Dimensional Manifolds, Kleinian Groups and Hyperbolic Geometry. Bull Amer Math Soc 6: 357-381.

5. Thurston W (1997) The Geometry and Topology of Three–Manifolds.Princeton University Press.

6. Hamilton RS (1982) Three Manifolds of Positive Ricci Curvature. J Diff Geom 17: 255-306.

7. Hamilton RS (1995) Formation of Singularities in the Ricci flow, in: Surveys in Differential Geometry. Matematica Contemporanea 34: 103-133.

8. Cao HD, Chow B, Chu SC, Yau ST (2003) Ricci Flow in Series in Geometry and Topology. International Press Somerville.

9. Cao HD, Zhu HP (2006) Hamilton–Perelnam's proof of the Poincaré conjecture and the geometrization conjecture, Asian J Math 10: 165-495.

10. Morgan JW, Tian G (2007) Ricci flow and the Poincaré conjecture.

11. Kleiner B, Lott J (2008) Notes on Perelman's papers. Geometry and Topology 12: 2587-2855.

12. Vacaru S (2008) Nonholonomic Ricci Flows: II. Evolution Equations and Dynamics. J Math Phys 49: 043504

13. Vacaru S (2006) Ricci Flows and Solitonic pp–Waves. Int J Mod Phys A 21: 4899-4912.

14. Vacaru S, Visinescu M (2007) Nonholonomic Ricci Flows and Running Cosmological Constant: I. 4D Taub–NUT Metrics. Int J Mod Phys A 22: 1135-1159.

15. Vacaru S, Visinescu M (2008) Romanian Reports in Physics 60: 218-238.

16. Carstea SA, Visinescu M (2005) Special solutions for Ricci flow equation in 2D using the linearization approach. Mod Phys Lett A 20: 2993-2998.

17. Cartan E (1963) La Methode du Repere Mobile, la Theorie des Groupes Continus et les Espaces Generalises.

18. Deligne P (2014) Quantum Fields and Strings: A Course for Mathematicians. American Mathematical Society.

19. Stavrinos P, Vacaru O, Vacaru S (2014) Modified Einstein and Finsler Like Theories on Tangent Lorentz Bundles, Int J Mod Phys D 23: 145-194.

20. Basilakos S, Kouretsis A, Saridakis E, Stavrinos PC (2013) Resembling Dark Energy and Modified Gravity with Finsler-Randers Cosmology. Phys Rev D 88: 123510.

21. Kouretsis AP (2014) Cosmic Magnetization in Curved and Lorentz Violating Space-Times. Eur Phys Journal C 74: 2879.

22. Kouretsis AP, Stahakopoulos M, Stavrinos PC (2013) Relativistic Finsler geometry. Mathematical Methods in the Applied Sciences 37: 223-229.

23. Mavromatos N, Sarkar S, Vergou A (2010) Stringy Space-Time Foam, Finsle-like Metrics and Dark Matter Relics. Phys Lett B 696: 300-304.

24. Mavromatos N, Mitsou V, Sarkar S, Vergou A (2012) Implications of a Stochastic Microscopic Finsler Cosmology 72: 1956.

25. Arcus C, Peyghan E, Sharahi E (2014) Weyl's Theory in the Generalized Lie Algebroids Framework. J Math Phys 55: 123505.

26. Peyghan E, Tyebi A (2014) Finsler Manifolds with a Special Class of g–natural Metrics. J Contemporary Math. Analysis - Armenian Academy of Sciences 49: 260-269.

27. Marsden JE, Ratiu T (1999) Introduction to Mechanics and Symmetry, texts in Applied Mathematics. Springer.

28. Castro C (2012) Quaternionic-valued Gravitation in 8D, Grand Univfication and Finsler Geometry. Int J Theor Phys 51: 3318-3329.

29. Castro C (2012) Gravity in Curved Phase-Spaces, Finsler Geometry and Two-Times Physics. Int J Mod Phys A.

30. Gallego R (2006) A Finslerian Version of 't Hooft Deterministic Quantum Models. J Math Phys.

31. Esrafilian E, Azizpour E (2004) Nonlinear Connections and Supersprays in Finsler Superspaces. Rep Math Phys 54: 365-372.

32. Rezalii MM, Azizpour E (2005) On a Superspray in Lagrange Superspaces. Rep Math Phys 56: 257-269.

33. Vacaru S (2005) Exact Solutions with Noncommutative Symmetries in Einstein and gauge Gravity. J Mat Phys 46: 042503.

34. www.mathem.pub.ro/dgds/mono/va-t.pdf and gr-qc/0508023

35. Vacaru S (2006) Clifford Algebroids and Nonholonomic Einstein–Dirac Structures. J Math Phys 47: 093504.

36. Vrnceanum G (1936) Sur les espaces non holonomes. C R Acad Sci Paris 183: 1083-1085.

37. Horak Z (1927) Sur les systèmes non holonomes, Bull. Internat. Acad. Sci. Bohème.

38. Bejancu A, Farran HR (2005) Foliations and Geometric Structures.Springer.

39. Grozman P, Leites D (2007) The Nonholonomic Riemann and Weyl Tensors for Flag Manifolds. Theory Math Physics 153: 1511-1538.

40. Cartan E (1935) Les Espaces de Finsler (Paris, Hermann).

41. Bejancu A (1990) Finsler Geometry and Applications.

42. Bao D, Chern SS, Chen Z (2000) An Introduction to Riemann–Finsler Geometry. Springer.

43. Kern J (1974) Lagrange Geomety. Arch Math 25: 438-443.

44. Etayo F, Santamara R, Vacaru S (2005) Lagrange-Fedosov Nonholonomic Manifolds. J Math Phys 46: 032901.

45. Deligne P (1994) Quanum Fields and Strings: A Course for Mathematicians. American Mathematical Society.

46. Kawaguchi A (1937) Bezienhung zwischen einer metrischen linearen Ubertragung unde iener micht–metrischen in einem allemeinen metrischen Raume akad Wetensch Amsterdam Proc 40: 596-601.

47. Kawaguchi A (1952) On the theory of nonlinear connections. I. Introduction to the theory of general nonlinear connections.

48. Poincare H (1952) Science and Hypeothesis

49. Barcelo C, Liberati S, Visser M (2005) Analogue Gravity. Living Rev Rel 8: 12.

50. Einstein A (1945) A Generaliziation of the Relativitistic Theory of Gravitation. Ann Math 46: 578-584.

51. Eisenhart LP (1952) Generalized Riemann Spaces and General Relativity II. Proc Natl Acad sci USA 40: 463-466.

52. Moffat JW (1995) Nonsymmetric Gravitational Theory. Phys Lett B 355: 447-452.

53. Vacaru V (2012) Metric Compatible or Noncompatible Finsler–Ricci Flows. Int J Geom Meth Mod Phy 09 1250041.

# Puiseux Series Expansions for the Eigenvalues of Transfer Matrices and Partition Functions from the Newton Polygon Method for Nanotubes and Ribbons

**Jeffrey R Schmidt*** **and Dileep Karanth**

*Department of Physics, University of Wisconsin-Parkside, USA*

**Abstract**

For certain classes of lattice models of nanosystems the eigenvalues of the row-to-row transfer matrix and the components of the corner transfer matrix truncations are algebraic functions of the fugacity and of Boltzmann weights. Such functions can be expanded in Puiseux series using techniques from algebraic geometry. Each successive term in the expansions in powers a Boltzmann weight is obtained exactly without modifying previous terms. We are able to obtain useful analytical expressions for any thermodynamic function for these systems from the series in circumstances in which no exact solutions can be found.

## Introduction

### Transfer matrices and partition functions

The partition function $Z$ for vertex and face models of lattice systems in statistical mechanics (such as those with nearest neighbor interactions) can be computed by application of matrix mathematics to the problem in at least two ways. For models with nearest neighbor interactions only, a row-to-row transfer matrix [1] can be constructed whose eigenvalues determine $Z$. For example let the states of a lattice site $i$ in a row of length $m$ of the lattice be enumerated by a state-variable $t_i$, $1 \le i \le m$. The state-variables of a row in the lattice form a vector $\mathbf{t} = (t_1, t_2, \cdots, t_m)$, and the Hamiltonian for the system decomposes as

$$H = \sum_{(\mathbf{t},\mathbf{t}')} h(\mathbf{t},\mathbf{t}') + \sum_{\mathbf{t}} k(\mathbf{t}) \qquad (1)$$

in which we sum over all pairs $(\mathbf{t},\mathbf{t}')$ of adjacent rows. We will use $t$ and $s$ for state-variables appearing in row-to-row transfer matrices, and $a,b$ and $c$ for state-variables appearing in corner transfer matrices.

If the lattice has a periodic boundary condition on its $N$ rows, the partition function is writable in terms of eigenvalues $x_i$ of the row-to-row transfer matrix $T$,

$$Z = \sum_{\{\mathbf{t}\}} \sum_{\{\mathbf{t}'\}} \cdots \sum_{\{\mathbf{t}''\}} T_{\mathbf{t},\mathbf{t}'} \cdots T_{\mathbf{t}'',\mathbf{t}'''} = trT^N = \sum_i x_i^N, \qquad T_{\mathbf{t},\mathbf{t}'} = e^{-\beta(h(\mathbf{t},\mathbf{t}')+k(\mathbf{t})+k(\mathbf{t}')2)} \qquad (2)$$

in which the sums are over all of the possible states of each row. If $N$ is very large, the free energy per site will be determined by a single eigenvalue, the largest, which we will label as $x_1$, $x_1 > x_i$, for $i > 1$

$$f = -\lim_{N\to\infty} kT \ln ZN = -kT \ln x_1 - \lim_{N\to\infty} kTN \ln\left(1 + \left(\frac{x_2}{x_1}\right)^N + \cdots\right) = -kT \ln x_1 \qquad (3)$$

Computation of $N$-point functions ($n \ge 2$) require additional eigenvalues [1]. The characteristic equation $0 = \det(T - X \cdot 1)$ for the row-to-row transfer matrix $T = T(z)$ will be a multi-variable polynomial in $X$ and Boltzmann weights $z$ (or fugacity $z$ if $Z$ is the grand canonical partition function), so the solutions $X = x_i$ will be algebraic functions of these weights. It is quite a challenge to compute these eigenvalues exactly in dimensions higher than one.

The most extensively studied two-dimensional lattice models are the $m \times N$ lattice vertex and face models in the thermodynamic limit of both $m, N \to \infty$, the row-to-row transfer matrix eigenvalues are obtained by the Bethe Ansatz or methods descended from it. A number of interesting cases have not yet been solved exactly (hard squares, the Ising model with external fields, to name two). In such cases series expansions can provide some answers to questions about th physical properties. There is a vast body of literature on series expansions for the row-to row $Z$ and the free energy per particle for interacting lattice models [2], and the subject continues to receive attention [3].

Corner transfer matrices can be used to define a variety of partition functions, on a triangular lattice three partition functions $Z_A, Z_2$, (Figure 2) and $Z_w$ (Supplementary Figure 1 in the appendix), can be constructed [1,4,5] from matrices $A(a), F(a,b)$ which are functions of the state-variables $a,b$ of central sites (we discuss a few details in the appendix);

$$Z_A = \sum_a Tr\, A^6(a)$$

$$Z_2 = \sum_{a,b} Tr\, A^2(a) F(a,b) A^2(b) F(b,a)$$

(the third partition function $Z_w = \sum_{a,b,c} w(a,b,c) F(a,b) A(b) F(b,c) A(c) F(c,a) A(a)$ is given in the appendix, *w(a,b,c)* is the weight of the face of the lattice whose vertices are *a,b,c*) and from these one can express the partition function per lattice site as $\kappa = \frac{Z_A Z_w^2}{Z_2^3}$ as the lattice size becomes infinite. By taking variations of Boltzmann parameters that leave the partition function per site $\kappa$ unchanged, two corner transfer matrix equations (on the triangular lattice) can be obtained [4,5].

$$\sum_c w(a,b,c) F(b,c) A(c) F(c,a) = \left(\frac{Z_w}{Z_2}\right) A(b) F(b,a) A(a) \equiv \eta\, A(b) F(b,a) A(a) \qquad (4)$$

$$\sum_b F(a,b) A^2(b) F(b,a) = \left(\frac{Z_2}{Z_A}\right) A^4(a) \equiv \xi\, A^4(a), \qquad \kappa = \frac{\eta^2}{\xi}$$

***Corresponding author:** Dr. Jeffrey R Schmidt, Associate Professor–Physics, Mathematics and Physics, University of Wisconsin-Parkside, 900 Wood Road, Kenosha, Wisconsin 53141, USA, E-mail: schmidt@uwp.edu

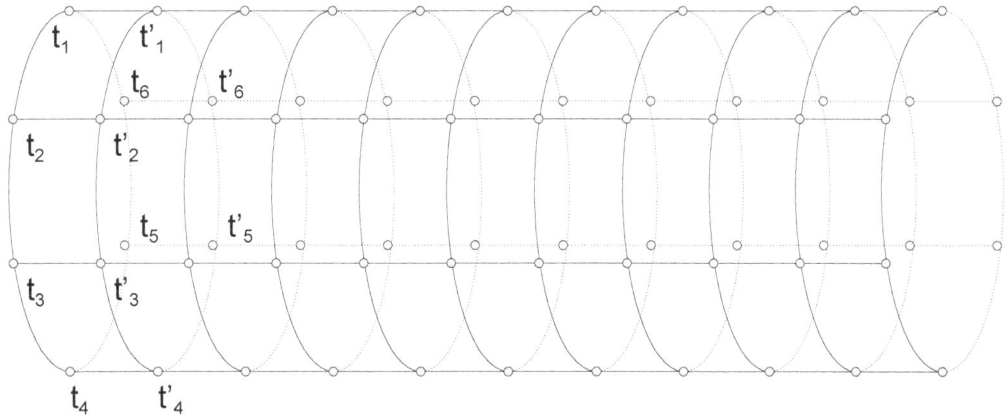


**Figure 1:** Adjacent rows of sites in a $6 \times N$ Ising model tube.

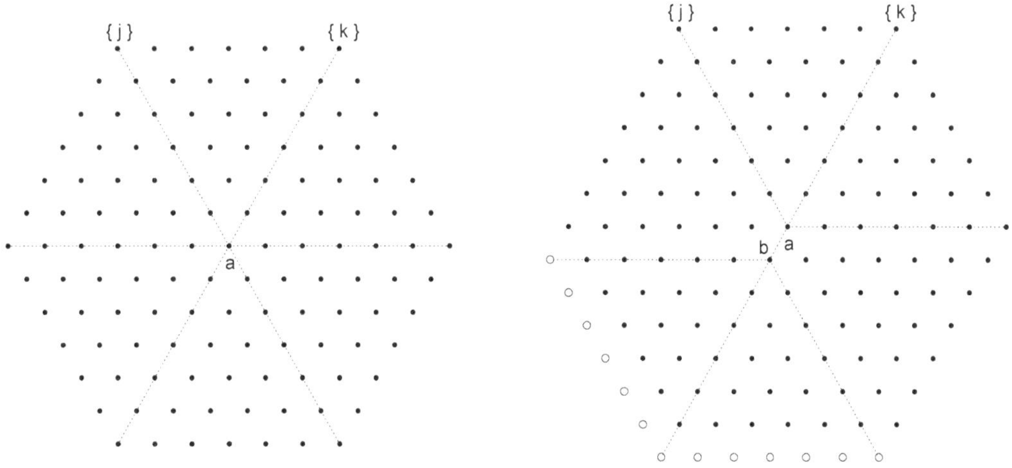


$$Z_A = \sum_a Tr A^6(a)$$

$$Z_2 = \sum_{a,b} Tr A^2(a) F(a,b) A^2(b) F(b,a)$$

**Figure 2:** Two partition functions using corner transfer matrices for a triangular lattice.

The square lattice case is a bit more complex, with eight partition functions. Computation of the partition function per site $\kappa$ from the corner transfer matrix equations is even more difficult than the row-to-row matrices because of their nonlinearity. Approximation methods truncate the infinite-dimensional matrices $A(a)$ and $F(a,b)$ to some finite size in representations with $A(a)$ diagonal, and even the most extreme approximation (truncation to one by one matrices, the Kramers-Wannier approximation) gives surprisingly good results for $\kappa$, for a non-interacting lattice gas this approximation is in fact exact.

We have found that for a number of important models on $m \times N$ lattices with $m$ finite and $N \to \infty$, the eigenvalues of the row-to-row transfer matrix $T$, and the corner transfer-matrix "eigenvalues" $\eta$ and $\xi$ can be expanded in Puiseux series in the Boltzmann weights, by a powerful method from the theory of algebraic geometry, the Newton polygon. This is a type of series expansion that does not appear to have been considered before in statistical physics (but has been used recently in cosmological calculations [6]). These ideas allow us to obtain analytical formulas for thermodynamical quantities for nanosystems such as whiskers, filaments, and tubes, in which one dimension is small and one is large.

Many symbolic-numerical methods for finding eigenvalues of matrices or roots of polynomials can fail on degenerate roots [7,8] but this does not appear to happen for the problems that we investigated by the techniques described in this article. We found the largest root is always non-degenerate (a consequence of the Frobenius-Perron theorem), and we are able to find expansions of degenerate "less significant" roots without any difficulties, in fact there were tell-tale signs at stages in the calculations that revealed what the degeneracies were. The way in which these series expansions are constructed is fundamentally different from traditional developments of power series expansions of functions such as $Z$ and $f$ in that it is not a perturbation or cluster expansion or a counting of graphs. The expansions gotten by successive iterations of the Newton polygon method are not progressive refinements; rather they are extensions, each of which gives one new term in the series exactly, and does not alter preceding terms.

## The Geometry of Algebraic Functions

We will first introduce the Puiseux series of an algebraic function, and the procedure for obtaining them using the method of the Newton polygon. Next we explain what the series represent, and why the method works the way it does.

### Puiseux series

If $K$ is a field then we can define the closed non-Archimedean field $K[[z]]$ of Puiseux series over $K$ (for our purposes $K$=C or R) with coefficients $f_k \in K$ as the set of formal expressions of the form [9-11].

$$\bar{f}(z) = \sum_{k=k_0}^{\infty} f_k z^{\frac{k}{n}} \quad (5)$$

where $n$ and $k_0$ are a nonzero natural number and an integer respectively (which are part of the datum of $f$): in other words, Puiseux series differ from formal Laurent series in that they allow for fractional exponents of the indeterminate as long as these fractional exponents have bounded denominator (here $n$), and like Laurent series, Puiseux series allow for negative exponents of the indeterminate as long as these negative exponents are bounded (here by $k_0$).

This type of series will arise in statistical physics when we seek expansions of the roots (eigenvalues) $X = \bar{x}_i$, $i = 1, 2, \cdots, p$ of characteristic polynomials of the row-to-row or corner transfer matrix

$$\det(T(z) - X \cdot 1) = P(X, z) = \bar{a}_0(z) + \bar{a}_1(z)X + \bar{a}_2(z)X^2 + \cdots \bar{a}_p(z)X^p = 0 \quad (6)$$

which defines $X = \bar{x}_i$ as an algebraic function of $Z$ (the fugacity or a Boltzmann weight)

$$\bar{x}_i = x_{i,1} z^{\xi_{i,1}} + x_{i,2} z^{\xi_{i,1}+\xi_{i,2}} + x_{i,3} z^{\xi_{i,1}+\xi_{i,2}+\xi_{i,3}} + \cdots \quad (7)$$

We introduce some terminology used in algebraic geometry: if $\bar{x}_i \in K[[z]]$ with $x_{i,1} \neq 0$ and with the exponents $\xi_i = \sum_j \xi_{i,j}$ ordered $\xi_1 < \xi_2 < \cdots$, we call the **order of** $\bar{x}_i$ the function $O(\bar{x}_i) = \xi_{i,1}$ (exponent of the lowest power of the indeterminate), and the **initial coefficient** of $\bar{x}_i$ is $In(\bar{x}_i) = x_{i,1}$. These would correspond to the exponents and coefficient of the leading term with respect to a local ordering in the case of polynomials or power series.

The polynomial (in $X$) Eq. 6 defines an algebraic curve in the $(X, z)$ plane. The problem of parameterizing such a curve can be summarized in a few steps [10]. First a theorem; in some suitable coordinate system any parameterization of the $i^{th}$ branch is equivalent to one of the form

$$z = z(t) = t^n, \quad X = \bar{x}_i = x_{i,1}t^{n_1} + x_{i,2}t^{n_2} + x_{i,3}t^{n_3} + \cdots, \quad 0 < n, \quad 0 < n_1 < n_2 < \cdots \quad (8)$$

(introducing parameter $t$). Fractional power series occur when we eliminate $t$ and put $X \in K[z^{1/n}] \subset K[[z]]$. The second step is the development of the expansion with respect to some monomial ordering (which is a local ordering in $K[z]$). Let $P$ be a polynomial in $X$ (integer powers of $X$) with coefficients in $K[[z]]^* = \bigcup_{n=1}^{\infty} K\left[z^{\frac{1}{n}}\right]$. Write it as

$$P(X, z) = \sum_{j=0}^{p} \bar{a}_j(z)X^j = \bar{a}_p(z)\prod_{j=1}^{p}(X - \bar{x}_j(z)), \qquad P(\bar{x}_i, z) = 0, \quad i = 1, 2, \cdots, p \quad (9)$$

Then solutions are of the form Eq. 7 (suppressing the subscript $i$ labeling the branches)

$$X = cz^{\gamma} + c'z^{\gamma+\gamma_1} + c''z^{\gamma+\gamma_1+\gamma_2} + \cdots, \qquad \gamma = O(X), \qquad \gamma_i > 0 \quad (10)$$

and the series is built term by term. The idea of the proof of this statement is as follows; write $X = z^{\gamma}(c + X_1(z))$ with $O(X_1) = \gamma_1 > 0$ substitute and expand Eq. 9

$$P = (\bar{a}_0(z) + \bar{a}_1(z)cz^{\gamma} + \bar{a}_2(z)c^2z^{2\gamma} + \cdots + \bar{a}_p(z)c^pz^{p\gamma}) + g(X_1(z)) \quad (11)$$

This can be zero if the terms with like powers of $z$ cancel; begin with the leading term according to our monomial ordering and let

$$f = min(O(\bar{a}_j(z)c^jz^{j\gamma})) = min(O(\bar{a}_j) + j\gamma) \quad (12)$$

be the lowest power of $z$ appearing on the right side, then if $In(\bar{a}_\ell)$ is the coefficient of the leading term of $\bar{a}_\ell(x)$

$$\sum_{\ell}^{O(\bar{a}_\ell(z))+\ell\gamma=f} In(\bar{a}_\ell)c^\ell = 0 \quad (13)$$

which we solve for $c$. We now iterate the process on what is left over, using the next term in the series solution to kill the next lowest powers of $z$. The lowest powers of $z$ will come from those terms of $P(X,z)$ corresponding to the very lowest points of the Newton diagram of $P(X,z)$.

### The Newton diagram of $P$ $(X, z)$

The algorithm for carrying out this construction of the series is called Newton's method. The first term $cz^{\gamma} = x_{i,1}z^{\xi_{i,1}}$ of the series can be found by purely geometric considerations, the next term will be obtained by applying Newton's method to the remainder after the leading term elimination; $P(X + x_{i,1}z^{\xi_{i,1}}, z) = 0$, and higher terms computed by further iterations [11,12].

For small $z$, candidates for the dominant term are found geometrically. For every term $cX^iz^r$ in $P(X,z)$ draw a dot $(i,r)$ in the $i \in \mathbb{Z}$, $r \in \mathbb{Q}$ plane. These constitute the Newton diagram of $P(X,z)$, which will be columns of dots at integer intervals horizontally. The **Newton polygon** is the convex hull of the dots at the bottom of the columns. The Newton polygon will give us the numbers $x_1$ and $\xi_1$ in the Puiseux expansion of the roots above. For example [11] the polynomial

$$P(X,z) = z - 2z^2X^2 - X^3 + zX^4 + zX^5 \quad (14)$$

has Newton diagram consisting of the points $\{(0,1),(2,2),(3,0),(4,1),(5,1)\}$ (Figure 3). To create the convex lower hull, find the set of points at the bottom of each column, which in this case is all of the points. Order them by increasing first component to get points $\{M_0, M_1, \cdots\}$, and draw a vertical line downwards from $M_0$ (in this case $(0,1)$) to serve as an

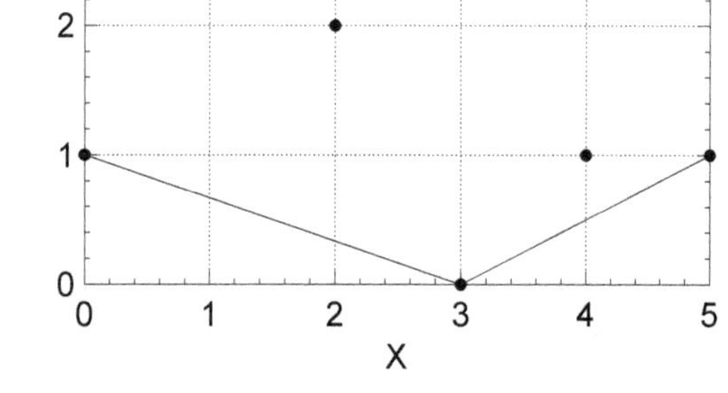


**Figure 3**: Newton diagram of Eq. 14

indicator as to how the next line is drawn. **You must always turn left.** Draw a connecting line from $M_0$ to $M_i$ if the vertical line from $M_0$ downwards hits $M_i$ as it is slowly rotated counterclockwise. Then repeat the process **but now the indicator line that you rotate lies along** $\overline{M_0 M_i}$. There are systematic algorithms, we state the Jarvis Gift Wrap algorithm [13]: create the vectors $(0,0)-M_0$, and $M_i - M_0$, in this case these are $(0,-1)$ and $\{(2,1),(3,-1),(4,0),(5,0)\}$. Now compute the angle between the line connecting the previous two points of the polygon with the last point added to the polygon, and each of the remaining lowest points in each column of points, and select as the next point the one that gives the **smallest positive value of this angle**. This will ensure that every point of the Newton diagram lies on or above the Newton polygon. The algorithm may not be the most efficient, but it is simple to program.

Once the Newton polygon is found, the algorithm for construction of the roots of Eq. 6 is as follows [11]; each segment $E=[M,\cdots,M']$ of the Newton polygon of *P* corresponds to a new homogeneous sub-polynomial (a slight departure from the notation of [11], we want to display the *z*-dependence) $Q$ of $P$ of the form

$$Q(P,E,X,z)=\sum_h In(\bar{a}_h)z^{O(\bar{a}_h)}X^h, \quad M_h \in E, \quad M_h=\left(h,\frac{m_h}{m}\right), \quad O(\bar{a}_h)=\frac{m_h}{m} \tag{15}$$

with *h* running over the exponents of *X* for points of the Newton diagram on the corresponding segment. The substitution $X=cz^\gamma=x_1 z^{\xi_1}$ for a unique $\gamma$ lets one factor out *z* from $Q$, transforming Eq. 15 into Eq. 13, to be solved for $c$. [10,11], we have only tried to summarize the basic ideas that distinguish the series from the more familiar series such as Taylor-Maclaurin. The statement in theorem form Basu et al. [11] is; **let $\bar{x}$ be a nonzero root of one of these characteristic polynomials $Q$ of a segment $E$ with slope $-\xi_1$, of multiplicity *r*, then a Puiseux series for a root of *P* starting with $x_1 z^{\xi_1}$ is**

$$\bar{x}=x_1 z^{\xi_1}+x_2 z^{\xi_1+\xi_2}+\cdots, \qquad P(\bar{x},z)=0 \tag{16}$$

We find that this first term alone is a very good approximation to the roots. The two Newton polynomials for the polynomial $P(X,z)$ of Eq. 14, [11]

$$Q=z-X^3, \qquad Q=X^3(zX^2-1) \tag{17}$$

(the two segments in Figure 1) which have roots $X=\bar{x}=1\cdot z^{\frac{1}{3}}, e^{\frac{2\pi i}{3}}\cdot z^{\frac{1}{3}}, e^{\frac{4\pi i}{3}}\cdot z^{\frac{1}{3}}$ and $X=\pm 1\cdot z^{-\frac{1}{2}}$, and slopes $\xi=-\frac{1}{3}$ and $\frac{1}{2}$, so that the roots of the polynomial have Puiseux series beginning with

$$\bar{x}=z^{\frac{1}{3}}, \qquad e^{\frac{2\pi i}{3}}z^{\frac{1}{3}}, \qquad e^{\frac{4\pi i}{3}}z^{\frac{1}{3}}, \qquad \pm z^{-\frac{1}{2}} \tag{18}$$

The first term in the Puiseux series expansions for the eigenvalues of the transfer matrix can be computed from a small fraction of the total number of terms in the secular equation, all we need is the lower convex hull. This means that the first terms can be computed by hand even for fairly large transfer matrices. Obtaining the secular (characteristic) equation computationally is more intensive than computing the first terms for the Puiseux series of its roots. A few words are said about this step in the appendix.

### Iteration

Suppose that we start with a transfer matrix secular equation Eq. 9 with *z* a Boltzmann weight, and obtain the Newton polygon solutions (the first terms of the series expansions) $\bar{x}$ for $X$

$$\bar{x}=\{x_{1,1}z^{\xi_{1,1}}, x_{2,1}z^{\xi_{2,1}}, \cdots, x_{p,1}z^{\xi_{p,1}}\} \tag{19}$$

If one performs the substitution

$$X \to X + x_{i,1}z^{\xi_{i,1}}, \qquad P'=P(X+x_{i,1}z^{\xi_{i,1}},z) \tag{20}$$

and applies the polygon procedure to $P'$, one obtains the next term in the Puiseux series for this solution, of the form seen in Eq. 16. This process is now repeated until the series attains the desired precision, in other words for the $i^{th}$ solution

$$X \to X + x_{i,2}z^{\xi_{i,1}+\xi_{i,2}}, \qquad P''=P'(X+x_{i,2}z^{\xi_{i,1}+\xi_{i,2}},z) \tag{21}$$

results in the next correction $x_{i,3}z^{\xi_{i,1}+\xi_{i,2}+\xi_{i,3}}$.

The calculations are straightforward, but there can be vast numbers of terms in these polynomials. We use the REDUCE computer algebra system [14] to perform the calculations. Singular [15] also has a Puiseux expansion (Hamburger-Noether expansion) library "hnoether.lib", but we prefer our own in REDUCE because we have designed it to make it easier to follow a specific branch. Our REDUCE program *NPplus.red* returns a list of Newton polygon lines $\{slope, intercept, Q(x,z)\}$. We illustrate its use by expanding the roots of the multivariate polynomial

$$P=X^3-4Xz+z^2 \tag{22}$$

(in each iteration we follow the top of first branch $=Q(X,z)$ selection)

$$\textit{First iteration} \quad P(X,z)=X^3-4Xz+z^2 \to \begin{cases} Q(X,z)=X^3-4Xz \\ Q(X,z)=z^2-4Xz \end{cases}$$

$$x_{1,1}z^{\xi_{1,1}}=2z^{\frac{1}{2}}$$

$$\textit{Second iteration} \quad P(X,z)=P(X+2\sqrt{z},z) \to \begin{cases} Q(X,z)=z^2+8Xz \\ Q(X,z)=X^3+8Xz+6X^2\sqrt{z} \end{cases}$$

$$x_{1,2}z^{\xi_{1,1}+\xi_{1,2}}=-\frac{1}{8}z$$

$$\textit{Third iteration} \quad P(X,z)=P(X-z/8,z) \to \begin{cases} Q(X,z)=\frac{3}{32}z^{\frac{5}{2}}+8Xz \\ Q(X,z)=X^3+8Xz+6X^2\sqrt{z} \end{cases}$$

$$x_{1,3}z^{\xi_{1,1}+\xi_{1,2}+\xi_{1,3}}=-\frac{3}{256}z^{\frac{3}{2}}$$

$$\textit{Fourth iteration} \quad P(X,z)=P(X-\frac{3}{256}z^{\frac{3}{2}},z) \to \begin{cases} Q(X,z)=8Xz+\frac{1}{64}z^3 \\ Q(X,z)=X^3+8Xz+6X^2\sqrt{z} \end{cases}$$

from which we deduce that a Puiseux expansion of one of the roots of this polynomial is

$$\bar{x}_1=2z^{\frac{1}{2}}-\frac{1}{8}z-\frac{3}{256}z^{\frac{3}{2}}-\frac{1}{512}z^2+\cdots \tag{23}$$

and a simple test for numerical values $0\le z\le 0.5$ shows that these few terms give a precision in the value of a root of five decimal places

$$\bar{x}_1^3-4z\bar{x}_1+z^2=0.003204345703125z^{\frac{7}{2}}+O(z^4) \tag{24}$$

The series that are being obtained are not expansions about a point, like a Taylor-Maclaurin or Laurent series, but are instead series that are the solutions to an algebraic equation correct to a particular order in *z*. These series may actually truncate or terminate, we will see this occur in our study of $m\times N$ Ising models. It should be pointed out that these series solutions are formal, and so are very useful as generating functions for combinatorial data, but convergence under substitution of explicit *z*-values is not guaranteed. As we see here and will see again shortly, convergence for sufficiently large/small *z*-values is actually quite spectacular.

## Application to *m*×*N* Lattice Gases with Exclusion

Now we turn our attention to the application of these ideas to statistical physics. The procedure described in the previous section is perfectly suited to computation of high and low temperature expansions for such systems as planar lattice gas ribbons and tubes; the former being a lattice gas on an $m \times N$ lattice with periodic boundary conditions in the "long dimension" $N$ only, and the latter case with periodic boundary conditions on both "long" and "short" dimensions.

## Hard squares at low fugacity z<<1

Consider the case of hard squares, a square lattice with site occupancy $t_i = 0,1$ such that no two adjacent sites can be simultaneously occupied. This model has not yet been solved exactly, but in the thermodynamic limit ($N \times N$ lattice, $N \to \infty$) corner transfer matrices provide us with some information [5,16]. On a square lattice $\kappa = \eta\xi$ is the partition function per particle

$$\sum_{b=0,1} F(a,b)A^2(b)F(b,a) = \xi A^2(a) \tag{25}$$

$$\sum_{c,d} w(a,c,d,b)F(a,c)A(c)F(c,d)A(d)F(d,b) = \eta A(a)F(a,b)A(b)$$

For hard squares the face weight function is

$$w(a,b,c,d) = z^{(a+b+c+d)/4}\varepsilon(a,b)\varepsilon(b,c)\varepsilon(c,d)\varepsilon(d,a), \qquad \varepsilon(a,b) = 1-ab \tag{26}$$

In the Kramers-Wannier approximation one treats the matrices as numbers (one by one matrices), the equations Eq. 25 simplify to $A(0)=1$, $A(1)=A$, $F(1,1)=0$, $F(0,1)\cdot F(1,0)=\xi A^2$, $F^2(0,0)=\xi(1-A^4)$, $\eta = \kappa\xi$, elimination of $\eta$ and $\xi$ results in

$$\kappa = 2z^{\frac{1}{4}}A^3 + \left(1-A^4\right) \tag{27}$$

$$\kappa A = z^{\frac{1}{4}}(1-A^4) + z^{\frac{1}{2}}A^3$$

Eliminate $\kappa$ and apply the Newton polygon method to expand $A$ in a Puiseux series in $Z$

$$A = z^{\frac{1}{4}}(1 - z + 4z^2 - 21z^3 + 125z^4 - 800z^5 + 5368z^6 - 37240z^7 + 264828z^8 - 1919690z^9 + \cdots) \tag{28}$$

and use the first of Eqs. 27 to get the partition function per particle [16]

$$\kappa = \lim_{N\to\infty} Z^{\frac{1}{N^2}} = 1 + z - 2z^2 + 8z^3 - 40z^4 + 225z^5 - 1362z^6 + 8670z^7 - 57254z^8 + 388830z^9 + \cdots \tag{29}$$

where agreement of this result with computer calculations that perform graph counting is discussed. Since the hard squares model has not been solved analytically, series provide the only information available for the thermodynamic limit case.

For a $m \times N$ hard-square ribbon lattice gas with $m$ finite and $N \to \infty$ we order the (row to row) transfer matrix rows and columns as

$$T_{1+t_1+2t_2+\cdots+2^{m-1}t_m,\,1+t'_1+2t'_2+\cdots+2^{m-1}t'_m} = \left(\prod_{j=1}^{m}(1-t_jt'_j)(1-t_jt_{j+1})(1-t'_jt'_{j+1})\right)z^{\sum_{i=1}^{m}(t_i+t'_i)/2} \tag{30}$$

and solve for the largest eigenvalue $X = x_{max}$ of polynomial $P(X,z) = \det(T(z) - X\cdot 1)$ using our simple Newton polygon library. Our results for the $N \to \infty$ partition function per lattice site for the $4 \times N$ ribbon and tube, respectively

$$Z^{1/N}_{4\times N,rib} = x_{max} = 1 + 4z - z^2 + 8z^3 - 39z^4 + 206z^5 - 1152z^6 + 6710z^7 - 40277z^8 + \cdots \tag{31}$$

$$\kappa = Z^{\frac{1}{4N}} = 1 + z - \frac{7}{4}z^2 + \frac{25}{4}z^3 - \frac{899}{32}z^4 + \frac{4557}{32}z^5 - \frac{99441}{128}z^6 + \cdots$$

$$Z^{1/N}_{4\times N,tube} = x_{max} = 1 + 4z - 2z^2 + 12z^3 - 62z^4 + 352z^5 - 2122z^6 + 13344z^7 - 86566z^8 + \cdots$$

$$\kappa = Z^{\frac{1}{4N}} = 1 + z - 2z^2 + 8z^3 - \frac{159}{4}z^4 + \frac{885}{4}z^5 + \frac{2639}{2}z^6 + \cdots$$

It is reassuring to see that this is a good match for the $N \times N$ case in thermodynamic limit, at least for the lowest order terms. For the $5 \times N$ ribbon and tube respectively we obtain

$$Z^{1/N}_{5\times N,rib} = x_{max} = 1 + 5z + z^2 + 7z^3 - 37z^4 + 208z^5 - 1225z^6 - 7485z^7;\cdots \tag{32}$$

$$\kappa = Z^{\frac{1}{5N}} = 1 + z - \frac{9}{5}z^2 + \frac{33}{5}z^3 - \frac{762}{25}z^4 + 159z^5 - \frac{111814}{125}z^6 + \cdots$$

$$Z^{1/N}_{5\times N,tube} = x_{max} = 1 + 5z + 10z^3 - 55z^4 + 325z^5 - 2025z^6 + 13120z^7 - 87545z^8 + \cdots$$

$$\kappa = Z^{\frac{1}{5N}} = 1 + z - 2z^2 + 8z^3 - 40z^4 + \frac{1124}{5}z^5 - \frac{6791}{5}z^6 + \cdots$$

For the $6 \times N$ ribbon and tube, respectively

$$Z^{1/N}_{6\times N,rib} = x_{max} = 1 + 6z + 4z^2 + 6z^3 - 32z^4 + 190z^5 - 1164z^6 + 7352z^7 - 47630z^8;\cdots \tag{33}$$

$$\kappa = Z^{\frac{1}{16N}} = 1 + z - \frac{11}{6}z^2 + \frac{41}{6}z^3 - \frac{2309}{72}z^4 + \frac{4081}{24}z^5 - \frac{1260845}{1296}z^6 + \cdots$$

$$Z^{1/N}_{6\times N,tube} = x_{max} = 1 + 6z + 3z^2 + 8z^3 - 45z^4 + 276z^5 - 1760z^6 + 11610z^7 - 78681z^8 + \cdots$$

$$\kappa = Z^{\frac{1}{16N}} = 1 + z - 2z^2 + 8z^3 - 40z^4 + 225z^5 - \frac{8171}{6}z^6 + \frac{51997}{6}z^7 - \frac{171577}{3}z^8 + \cdots$$

(note that six must be a good approximation to infinity for eighth order; the $z^8$ term agrees with the corner transfer matrix result for $N \times N$, $N \to \infty$ to within 11 parts in 10,000). We can go on

$$Z^{1/N}_{7\times N,tube} = 1 + 7z + 7z^2 + 7z^3 - 35z^4 + 224z^5 - 1463z^6 + 9807z^7 - 67270z^8 + 470442z^9 + \cdots \tag{34}$$

$$\kappa = Z^{\frac{1}{17N}} = 1 + z - 2z^2 + 8z^3 - 40z^4 + 225z^5 - 1362z^6 + \frac{60689}{7}z^7 - \frac{400744}{7}7z^8 + \cdots$$

$$Z^{1/N}_{8\times N,tube} = 1 + 8z + 12z^2 + 8z^3 - 26z^4 + 176z^5 - 1180z^6 + 8048z^7 - 55886z^8 + 394488z^9 + \cdots$$

$$\kappa = Z^{\frac{1}{18N}} = 1 + z - 2z^2 + 8z^3 - 40z^4 + 225z^5 - 1362z^6 + 8670z^7 - \frac{458023}{8}z^8 + \cdots$$

REDUCE is built for comfort, not for speed, nevertheless each iteration takes only a second or two even for the $64 \times 64$ transfer matrix of the $6 \times N$ case.

## Hard squares at high fugacity z>>1

The high fugacity expansions for the lattice gases do in general result in true Puiseux series, with both positive and negative fractional exponents. For the simplest case, which can be worked out by hand, the characteristic polynomial for the $1 \times N$ hard-squares tube is

$$T_{1+t_1,1+t'_1} = (1-t_1t'_1)z^{(t_1+t'_1)/2}, \qquad P(X,z) = \det(T - X\cdot 1) = X^2 - X - z \tag{35}$$

To obtain the high-fugacity expansion we transform $z \to 1u$, and multiply the entire polynomial by a sufficient number $q$ of $u$ factors (the order of $P(x,z)$ as a function of $z$) in order to make all $u$-exponents positive;

$$P(x,z) \to u^q P(x,u^{-1}) = uP(X,u^{-1}) = uX^2 - uX - 1 \tag{36}$$

and apply the Newton construction, the largest eigenvalue will be seeded by the lowest power of $u$, and finally perform $u \to z^{-1}$ to obtain

$$Z^{1/N}_{1\times N} = 1 + z - z^2 + 2z^3 - 5z^4 + 14z^5 - 42z^6 + 132z^7 - 429z^8 + \cdots \qquad z \ll 1 \tag{37}$$

$$Z^{1/N}_{1\times N} = z + \frac{1}{2} + \frac{1}{8}z^{-\frac{1}{2}} - \frac{1}{128}z^{-\frac{3}{2}} + \frac{1}{1024}z^{-\frac{5}{2}} + \cdots \qquad z \gg 1$$

We get a better picture of what transformation Eq. 36 does for the $3 \times N$ hard squares tube, compare the Newton diagrams (Figure 4) in each regime

For $z \ll 1$ the characteristic polynomial is

$$P(X,z) = X^4(X^4 - X^3 - 3X^2(z^2+z) - X(2z^3+3z^2) - z^3) \tag{38}$$

and for $u = 1z \ll 1$

$$u^3P(X,u) = X^4(X^4u^3 - X^3u^3 - 3X(u^2+u) - X(3u+2) - 1) \tag{39}$$

The lowest level construction results in $\bar{x} = 1$ and a doubly-degenerate $-z$ for the former case and $\bar{x} = -12, -u^{-1}, -u^{-1}, 2u^{-1}$ for the latter. Since the Puiseux series solutions Eq. 7 have ascending positive

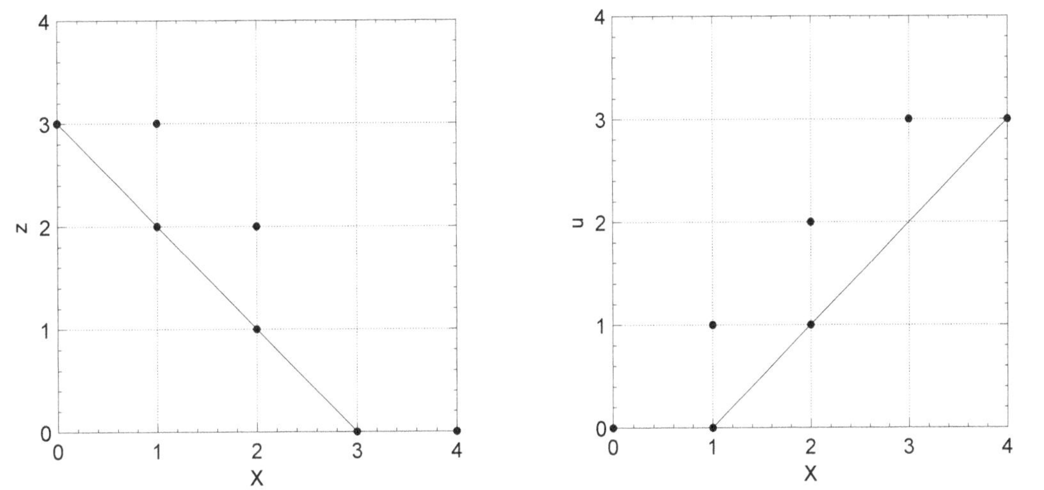


**Figure 4:** $P(x,z) \to u^q P(x,u^{-1})$ is a symmetry transforming upper convex hull of one diagram into lower convex hull.

exponents, when $z << 1$ the solution starting with $\bar{x} = 1$ will be largest, and when $u^{-1} = z >> 1$ or $u << 1$ the series starting with $2u^{-1}$ will be largest;

$$Z^{1/N}_{3\times N,tube} = \frac{2}{u} + \frac{3}{2} - \frac{3}{8}u + \frac{3}{8}u^2 - \frac{57}{128}u^3 + \frac{75}{128}u^4 - \frac{843}{1024}u^5 + \cdots \tag{40}$$

$$= 2z + \frac{3}{2} - \frac{3}{8}z^{-1} + \frac{3}{8}z^{-2} - \frac{57}{128}z^{-3} + \frac{75}{128}z^{-4} - \frac{843}{1024}z^{-5} + \cdots \qquad z >> 1$$

$$Z^{1/N}_{3\times N,tube} = 1 + 3z - 3z^2 + 12z^3 - 57z^4 + 300z^5 + 1686z^6 + \cdots \qquad z << 1$$

$$Z^{1/N}_{4\times N,tube} = z^2 + 2z + 4 - 2z^{-1} - 8z^{-2} + 22z^{-3} - 190z^{-5} + 120z^{-6} + 1442z^{-7} - 3268z^{-8}, \qquad z >> 1$$

$$Z^{1/N}_{4\times N,tube} = 1 + 4z - 2z^2 + 12z^3 - 62z^4 + 352z^5 - 2122z^6 + 13344z^7 - 86566z^8 + \cdots \qquad z << 1$$

$$Z^{1/N}_{5\times N,tube} = 1 + 5z + 10z^3 - 55z^4 + 325z^5 - 20225z^6 + 13120z^7 - 87545z^8 + \cdots, z << 1$$

$$Z^{1/N}_{5\times N,tube} = 2z^2 + \frac{9}{2}z + \frac{11}{8} - \frac{15}{16}z^{-1} + \frac{339}{128}z^{-2} - \frac{1765}{256}z^{-3} + \frac{16831}{1024}z^{-4} - \frac{72975}{2048}z^{-5} + \cdots z >> 1$$

$$Z^{1/N}_{6\times N,tube} = 1 + 6z + 3z^2 + 8z^3 - 45z^4 + 276z^5 - 1760z^6 + 11610z^7 - 78681z^8 + \cdots, \qquad z << 1$$

$$Z^{1/N}_{6\times N,tube} = z^3 + 3z^2 + 3z + 14 - 9z^{-1} + 45z^{-2} - 371z^{-3} + 1134z^{-4} - 4959z^{-5} + 27798z^{-6}$$

$$-121002z^{-7} + 570303z^{-8} + \cdots \qquad z >> 1$$

$$Z^{1/N}_{8\times N,tube} = 1 + 8z + 12z^2 + 8z^3 - 26z^4 + 176z^5 - 1180z^6 + 8048z^7 - 55886z^8 + 39448z^9 + \cdots \quad z << 1$$

$$Z^{1/N}_{8\times N,tube} = z^4 + 4z^3 + 6z^2 + 8z + 44 - 12z^{-1} + 166z^{-2} - 612z^{-3} - 1444z^{-4} + \cdots \quad z >> 1$$

We take a moment to address convergence. These formulas are in excellent agreement with high-precision numerical computation of the eigenvalues (obtained using Gnu octave).

| | HS $5\times N$ tube | | HS $6\times N$ tube | |
|---|---|---|---|---|
| $z$ | $\bar{x}_{max}$ (octave) | $\bar{x}_{max}$ (Eq. 40) | $\bar{x}_{max}$ (octave) | $\bar{x}_{max}$ (Eq. 40) |
| 0.01 | 1.05000948 | 1.0500094 6 | 1.06030757595 | 1.06030757595 |
| 0.025 | 1.12513751 | 1.12513 307 | 1.15198475 | 1.15198475 |
| 0.05 | 1.25098387 | 1.250 69862 | 1.30828428 | 1.30828 350 |
| 5 | 7.37564404e+01 | 7.375 75383 e+01 | 2.27193463e+02 | 2.27 949742 e+02 |
| 10 | 2.46302186e+02 | 2.463021 95 e+02 | 1.34326223e+03 | 1.34326 421 e+03 |
| 20 | 8.91333977e+02 | 8.91333977+02 | 9.27362202e+03 | 9.27362202e+03 |
| 40 | 3.38135312e+03 | 3.38135312e+03 | 6.89337977288e+04 | 6.89337977288e+04 |

The last of Eqs. 40 are approaching the infinite lattice results for $z >> 1$ reported in Appendix B of [4] computed by numerical computations of corner transfer matrix truncations, since for the $m \times n$ lattice $Z^{1/N}_{m\times N} = \kappa^m$

$$z^{-1}\kappa^2(z^{-1}) = 1 + z^{-1} + z^{-3} - z^{-4} + 5z^{-5} - 10z^{-6} + 39z^{-7} - 95z^{-8} \tag{41}$$

$$\kappa^6 = z^3 + 3z^2 + 3z + 4 + 3z^{-1} + 12z^{-2} + 69z^{-4} + \cdots$$

$$\kappa^8 = z^4 + 4z^3 + 6z^2 + 8z + 9 + 20z^{-1} + 18z^{-2} + 92z^{-3} + 36z^{-4} + \cdots$$

## Hard hexagons

The hard hexagon model can be realized as a lattice gas on a square lattice with an additional interaction along one diagonal [1,16]

$$T_{1+t_1+2t_2+\cdots+2^{m-1}t_m,1+t'_1+2t'_2+\cdots+2^{m-1}t'_m} = \left(\prod_{j=1}^{m}(1-t_j t_{j+1})(1-t'_j t'_{j+1})(1-t_j t'_j)(1-t_j t'_{j+1})\right) z^{\sum_{j=1}^{m}(t_j+t'_j)2} \tag{42}$$

This matrix is considerably sparser than the hard squares transfer matrix, with many more null vectors. We find that for low fugacity $z << 1$, $m \times N$ periodic boundary conditions (tubes) result in

$$Z^{1/N}_{4\times N,tube} = 1 + 4z - 6z^2 + 32z^3 - 212z^4 + 1568z^5 - 12400z^6 + 102592z^7 - 876944z^8 + 7683456z^9 + \cdots \tag{43}$$

$$\kappa = Z^{\frac{1}{4}} = 1 + z - 3z^2 + 16z^3 - \frac{423}{4}z^4 + \frac{3125}{4}z^5 - \frac{24707}{4}z^6 + 51115z^7 - \frac{13987571}{32}z^8 + \frac{122615141}{32}z^9 + \cdots$$

$$Z^{1/N}_{6\times N,tube} = 1 + 6z - 3z^2 + 26z^3 - 186z^4 + 1440z^5 - 11836z^6 + 101586z^7 - 900342z^8 + 8178268z^9 + \cdots$$

$$\kappa = Z^{\frac{1}{6}} = 1 + z - 3z^2 + 16z^3 - 106z^4 + 789z^5 - \frac{37907}{6}z^6 + \frac{319129}{6}z^7 - 464388z^9 + \frac{25008499}{6}z^9 + \cdots$$

$$Z^{1/N}_{8\times N,tube} = 1 + 8z + 4z^2 + 16z^3 - 134z^4 + 1104z^5 - 9384z^6 + 82312z^7 - 741276z^8 + 6821040z^9 + \cdots$$

$$\kappa = Z^{\frac{1}{8}} = 1 + z - 3z^2 + 16z^3 - 106z^4 + 789z^5 - 6318z^6 + 53198z^7 - \frac{7317383}{8}z^8 + \frac{33392609}{8}z^9 + \cdots$$

These agree exactly up to and including the $z^{m-1}$ term with the known low-fugacity series expansions for the $N \times N$ model with $N \to \infty$, we begin to see finite size deviations beyond that point. We compare with the corner transfer matrix computation in the Kramers-Wannier approximation [16]:

$$\sum_b F(a,b)A^2(b)F(b,a) = \xi A^4(a) \tag{44}$$

$$\sum_c w(a,b,c)F(a,c)A(c)F(c,b) = \eta A(a)F(a,b)A(b)$$

$$w(a,b,c) = z^{\frac{a+b+c}{6}}\varepsilon(a,b)\varepsilon(b,c)\varepsilon(c,a), \qquad \kappa = \frac{\eta^2}{\xi}$$

in which $\varepsilon(a,b) = 1 - ab$, $a,b \in \{0,1\}$. Truncating to $1\times 1$ matrices we get

$$\xi A^4(0) = A^2(1)F(0,1)F(1,0) + A^2(0)F^2(0,0) \tag{45}$$

$$\xi A^4(1) = A^2(0)F(0,1)F(1,0) + A^2(1)F^2(1,1)$$
$$\eta A(0)F(0,0)A(0) = z^{\frac{1}{6}} A(1)F(0,1)F(1,0) + A(0)F^2(0,0)$$
$$\eta A(0)F(0,1)A(1) = z^{\frac{1}{6}} A(0)F(0,1)F(0,0)$$
$$\eta A(1)F(1,0)A(0) = z^{\frac{1}{6}} A(0)F(1,0)F(0,0)$$
$$\eta A(1)F(1,1)A(1) = 0$$

from which we deduce

$$F(1,1)=0, \qquad F(0,1)=F(1,0) \tag{46}$$

Set $A(0)=1$, $A(1)=A$, $F(0,1)=f$, $F(0,0)=g$, the corner transfer matrix equations define a one-dimensional affine variety in $(A, f, g, \xi, \kappa, z)$ space [17]

$$\xi = A^2 f^2 + g^2 \tag{47}$$
$$\xi A^4 = f^2$$
$$\eta A = z^{\frac{1}{6}} g$$
$$\eta g = z^{\frac{1}{6}} A f^2 + g^2$$

Eliminating $f$, $g$ and $\xi$ we obtain the elimination ideal generated by

$$z^{\frac{1}{6}} A^6 - (1-A^6)(z^{\frac{1}{6}} - A) = 0, \qquad \kappa = \frac{(1-A^6) z^{\frac{1}{3}}}{A^2} \tag{48}$$

Expand the first of these in a Puiseux series by the Newton polygon method, and use the second to obtain $\kappa$:

$$A = z^{\frac{1}{6}}(1 - z + 5z^2 - 34z^3 + 265z^4 - 2232z^5 + 19766z^6 - 181300z^7 + 1706737z^8 + \cdots) \tag{49}$$

$$\kappa = 1 + z - 3z^2 + 16z^3 - 106z^4 + 789z^5 - 6319z^6 + 53228z^7 - 465258z^8 + \cdots$$

a result in excellent agreement [16] with series computations produced in the 60's and 70's. The Newton polygon method is very well-suited to the study of corner transfer matrices if one can establish some of the elimination ideals of the variety defined by the corner transfer matrix equations. We include this example because the first of Eq. 49 is a true Puiseux series, and are an approximation for the $N \to \infty$ limit with which to compare Eq. 43.

Returning to the row-to-row transfer matrix, we can produce high-fugacity expansions valid for $z >> 1$:

$$Z_{4\times N}^{1/N} = z + 4 - 20z^{-1} + 236z^{-2} - 3468z^{-3} + 57044z^{-4} - 1005124z^{-5} + 18552156z^{-6} - 354084764z^{-7} + \cdots \tag{50}$$

$$Z_{6\times N}^{1/N} = z^2 + 2z + 15 + 62z^{-1} - 92z^{-2} - 3392z^{-3} - 17462z^{-4} + 95678z^{-5} + 1919540z^{-6} + 7151412z^{-7} + \cdots$$

$$Z_{8\times N}^{1/N} = z^{\frac{5}{2}} + z^2 + 4z^{\frac{3}{2}} + 7z + 21z^{\frac{1}{2}} - 5 - \frac{65}{2} z^{-\frac{1}{2}} - 325z^{-1} - \frac{593}{2} z^{-32} + 919z^{-2} + 6221z^{-\frac{5}{2}} + \cdots$$

We also examined a variation (perhaps unstudied) on the hard hexagon model with Boltzmann weights for a face $w(a,b,c) = z^{(a+b+c)/6}(1-abc)$, and found excellent agreement between the $6\times N$ tube and truncated corner transfer matricies for the infinite lattice

$$\kappa_{6\times N} = (1 + z - 2z^3 + 7z^4 - 12z^5 - 15z^6 + 194z^7 + \cdots) \tag{51}$$
$$\kappa_{TCTM} = (1 + z - 2z^3 + 7z^4 - 12z^5 - 15z^6 + 193z^7 + \cdots)$$

## Application to *m*×*N* Lattice Ising Models

Ising $m\times N$ ribbons and tubes have transfer matrices with all matrix elements non-zero, and one finds that eigenvalues occur in approximate $\pm$-pairs, a spectral feature revealed by the first level Newton polygon segments, with very large gaps between largest and second largest eigenvalues of the same sign.

Ising ribbons on $m\times N$ lattices and tubes on $2m\times N$ lattices have transfer matrices with the property

$$P\cdot T(z)\cdot P' = T(z^{-1}), \qquad z = e^{-\beta J} \tag{52}$$

in which $P, P'$ are permutation (orthogonal) matrices with additional properties

$$[P, P'] = 0, \qquad P\cdot P' = S, \qquad [T(z), S] = 0 \tag{53}$$

Consider a $3\times N$ ribbon (Figure 5) of magnetic dipoles $s_i$ each of which can be in two states $s_i = \pm 1$, with magnetic dipole-dipole interactions between neighboring spins only.

The transfer matrix for this ribbon is $8\times 8$ since each row of three spins has $2^3$ states, and the states of two adjacent rows label the rows and columns of the transfer matrix;

$$T(z)_{(s_1,s_2,s_3),(s_1',s_2',s_3')} = e^{-\beta J((s_1s_2+s_2s_3+s_1's_2'+s_2's_3')/2+s_1s_1'+s_2s_2'+s_3s_3')}, \quad \beta = \frac{1}{kT}, \quad T \geq 0 \tag{54}$$
$$= z^{((s_1s_2+s_2s_3+s_1's_2'+s_2's_3')/2+s_1s_1'+s_2s_2'+s_3s_3')}, \qquad z = e^{-\beta J}$$

If $J < 0$ $(0 < z < 1)$ the model is ferromagnetic, otherwise anti-ferromagnetic. Note that the matrix is real and symmetric, and has symmetry under flipping all spins

$$T(z)_{(s_1,s_2,s_3),(s_1',s_2',s_3')} = T(z)_{(-s_1,-s_2,-s_3),(-s_1',-s_2',-s_3')} \tag{55}$$
$$T(z)_{(s_1,s_2,s_3),(s_1',s_2',s_3')} = T(z)_{(s_1',s_2',s_3'),(s_1,s_2,s_3)}$$

The inversion property Eq. 52 is due to the exponent being negated by flipping alternating spins in adjacent rows according to the pattern (for example again the $3\times N$ case)

$$T(z)_{(s_1,s_2,s_3),(s_1',s_2',s_3')} = T(z^{-1})_{(s_1,-s_2,s_3),(-s_1',s_2',-s_3')} \tag{56}$$

## The matrices *P*, *P'*, and *S*

Switch to a "bit-ordered" presentation of the states of each row of

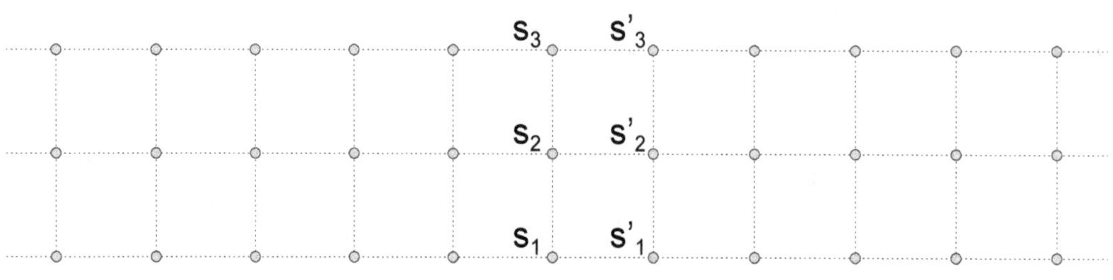

**Figure 5:** A section of the 3×*N* Ising ribbon.

the lattice, let

$$t_i = \begin{cases} 0 & if \quad s_i = -1 \\ 1 & if \quad s_i = 1 \end{cases}, \qquad s_i = 2t_i - 1 \tag{57}$$

and label the rows and columns of the transfer matrix accordingly in binary, for the $3\times N$ example

$$T(z)_{(s_1,s_2,s_3),(s_1',s_2',s_3')} = T(z)_{1+t_1+2t_2+4t_3,1+t_1'+2t_2'+4t_3'} = T(z)_{i,j}, \qquad i,j = 1,2,\cdots,8 \tag{58}$$

Define

$$\bar{t}_i = \begin{cases} 1 & if \quad t_i = 0 \\ 0 & if \quad t_i = 1 \end{cases} \tag{59}$$

then symmetry Eq. 52 can be expressed as

$$T(z)_{1+t_1+2t_2+4t_3,1+t_1'+2t_2'+4t_3'} = T(z^{-1})_{1+t_1+2\bar{t}_2+4t_3,1+\bar{t}_1+2t_2'+4\bar{t}_3} \tag{60}$$

Let *P* and *P'* be stochastic (permutation) matrices that are zeroes everywhere except

$$P_{1+t_1+2t_2+4t_3,1+t_1+2\bar{t}_2+4t_3} = 1, \qquad P'_{1+t_1+2t_2+4t_3,1+\bar{t}_1+2t_2'+4\bar{t}_3} = 1 \tag{61}$$

Note that [*P, P'*]=0, and *P. P'*=*S* is the matrix with all ones down the upper right to lower left diagonal. The significance is that the transformation $z \to z^{-1}$ is a linear transformation of the transfer matrix

$$T(z^{-1}) = P\cdot T(z)\cdot P' \tag{62}$$

and from the form of *P* and *P'* one can see that

$$[P,P'] = 0, \qquad (P\cdot P')^2 = 1 = S^2 \tag{63}$$

and the spin-negation symmetry Eq. 55 is simply

$$S\cdot T(z)\cdot S = T(z) \tag{64}$$

## Proof of the inversion property of largest eigenvalue

Let $T(z)$ be a symmetric matrix ($2^n \times 2^n$) all of whose matrix elements are of the form $z^q$ where *q* is a finite rational number, and *z* is a **non-negative** real variable.

The matrix $T(z)$ has the property that there exists two stochastic matrices *P, P'* (each row/column are unit vectors with only a single component non-zero, they are permutation matrices)

$$P\cdot T(z)\cdot P' = T(z^{-1}) \tag{65}$$

$$[P,P'] = 0, \qquad P\cdot P' = S, \qquad S^2 = 1, \qquad [T(z),S] = 0 \tag{66}$$

however $T(z)$ and $T(z^{-1})$ are **not** similar. Then it is easy to show that

$$(T(z^{-1}))^N = P\cdot(T(z))^N\cdot P' \tag{67}$$

for any **odd** integer $N$.

$$(T(z^{-1}))^N = (P\cdot T(z)\cdot P')(P\cdot T(z)\cdot P')\cdots(P\cdot T(z)\cdot P')(P\cdot T(z)\cdot P') \tag{68}$$

$$= P\cdot T(z)\cdot(P'\cdot P)\cdot T(z)\cdot(P'\cdot P)\cdots(P'\cdot P)\cdot T(z)\cdot(P'\cdot P)\cdot T(z)\cdot P'$$

$$= P\cdot T(z)\cdot(S)\cdot T(z)\cdot(S)\cdots(S)\cdot T(z)\cdot(S)\cdot T(z)\cdot P'$$

$$= P\cdot(S)^{N-1}\cdot(T(z))^N\cdot P'$$

$$= P\cdot(T(z))^N\cdot P'$$

For any vector $\mathbf{v}$ we have

$$|P\cdot\mathbf{v}| = |P'\cdot\mathbf{v}| = |\mathbf{v}| \tag{69}$$

Let $\mathbf{V}$ be a vector of length one, all of whose components are positive. There are numbers $\{a_i | i = 1,2,\cdots,2^N\}$ and $\{b_i | i = 1,2,\cdots,2^N\}$ ($T(z)$ and $T(z^{-1})$ are Hermitian) such that

$$P'\cdot\mathbf{v} = \sum_i a_i\mathbf{u}_i, \qquad \mathbf{v} = \sum_i b_i\mathbf{w}_i \tag{70}$$

where $\{\bar{x}_i(z),\mathbf{u}_i\}$ are the ordered eigenvalue/eigenvector pairs of $T(z)$ and $\{\bar{y}_i(z),\mathbf{w}_i\}$ are the ordered eigenvalue/eigenvector pairs of $T(z^{-1})$

$$\bar{x}_i \ge \bar{x}_{i+1}, \qquad \bar{y}_i \ge \bar{y}_{i+1} \tag{71}$$

By the Frobenius-Perron theorem the largest eigenvalues $\bar{x}_1$ and $\bar{y}_1$ are non-degenerate, positive and correspond to eigenvectors with non-negative components.

$$|(T(z^{-1}))^N\cdot\mathbf{v}| = |P\cdot(T(z))^N\cdot P'\cdot\mathbf{v}|, \qquad |\sum_i b_i\bar{y}_i^N\mathbf{w}_i| = |P\cdot\sum_i a_i\bar{x}_i^N\mathbf{u}_i| \tag{72}$$

As *N* becomes very large, the two sides become

$$\bar{y}_1^N|b_1\mathbf{w}_1| + \mathcal{O}(\frac{\bar{y}_2}{\bar{y}_1})^N = \bar{y}_1^N b_1 + \mathcal{O}(\frac{\bar{y}_2}{\bar{y}_1})^N = \bar{x}_1^N|a_1 P\cdot\mathbf{u}_1| + \mathcal{O}(\frac{\bar{x}_2}{\bar{x}_1})^N = \bar{x}_1^N a_1 + \mathcal{O}(\frac{\bar{x}_2}{\bar{x}_1})^N \tag{73}$$

and therefore for any $z$, by dividing these equations for consecutive odd integers $N$ and $N+2$ as $N\to\infty$

$$\bar{x}_1^2 = \bar{y}_1^2, \qquad \bar{x}_1 = \bar{y}_1 \tag{74}$$

the phases being set to unity by the Frobenius-Perron theorem.

Even more can be said about the spectra of $T(z)$ and $T(z^{-1})$ (ferromagnetic/anti-ferromagnetic Ising models). It is easy to show that

$$\bar{y}_i = \pm\bar{x}_i, \qquad i > 1 \tag{75}$$

Since $[T(z^{-1}),S] = 0$ there is a unitary matrix *U* such that

$$U\cdot T(z^{-1})\cdot U^{-1} = \Delta(z), \qquad U\cdot S\cdot U^{-1} = \Sigma \tag{76}$$

in which $\Delta(z) = diag(\bar{y}_1,\bar{y}_2,\cdots)$ and $\Sigma$ is diagonal with half of its diagonal entries being 1, the other half being $-1$. Then since *P* is orthogonal, $P\cdot T(z)\cdot P^{-1}$ and $T(z)$ are similar

$$P\cdot T(z)\cdot P' = T(z^{-1}) \tag{77}$$

$$P\cdot T(z)\cdot(P^{-1}\cdot P)\cdot P' = P\cdot T(z)\cdot P^{-1}\cdot S$$

$$P\cdot T(z)\cdot P^{-1} = T(z^{-1})\cdot S$$

$$U\cdot(P\cdot T(z)\cdot P^{-1})\cdot U^{-1} = U\cdot T(z^{-1})\cdot S\cdot U^{-1}$$

$$U\cdot(P\cdot T(z)\cdot P^{-1})\cdot U^{-1} = (U\cdot T(z^{-1})\cdot U^{-1})\cdot(U\cdot S\cdot U^{-1})$$

$$(U\cdot P)\cdot T(z)\cdot(U\cdot P)^{-1} = \Delta(z)\cdot\Sigma$$

The right side of this expression is a diagonal matrix, therefore the left side must be as well, call it $\Lambda(z) = diag(\bar{x}_{i_1},\bar{x}_{i_2},\cdots)$

$$\Lambda(z) = \Delta(z)\cdot\Sigma \tag{78}$$

An interesting bit of algebraic geometry is that models whose transfer matrix $T(z)$ has property Eq. 52 have characteristic polynomials $P(x,z) = \det(T(z) - x)$ which have Newton diagrams with convex hulls possessing a reflection symmetry about a horizontal line Figure 6.

## Results for the Ising ribbons and tubes

The ribbons possess the inversion property for all $m$, but the tubes only for $m$ even (and recall that $z = e^{-\beta J}$).

$$Z_{3\times N,rib}^{1/N} = z^{-5} + 3z + 7z^3 + 12z^5 + 18z^7 + 13z^9 - 41z^{11} - 222z^{13} - 648z^{15} - 1295z^{17} + \cdots \quad z \ll 1 \tag{79}$$

$$Z_{3\times N,rib}^{1/N} = z^5 + 3z^{-1} + 7z^{-3} + 12z^{-5} + 18z^{-7} + 13z^{-9} - 41z^{-11} - 222z^{-13} - 648z^{-15} - 1295z^{-17}\cdots \quad z \gg 1$$

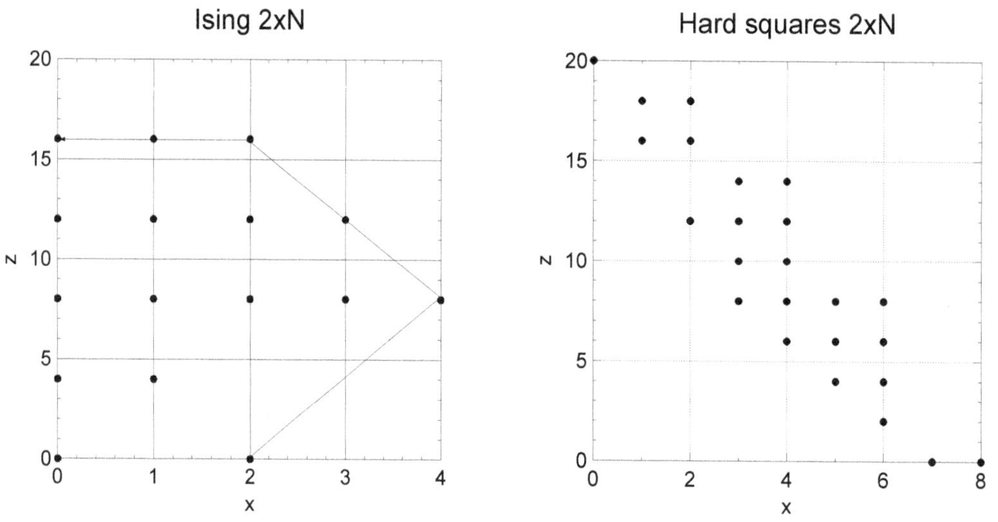


**Figure 6:** A comparison of models with Newton diagram hulls with and without the reflection symmetry.

$$Z^{1/N}_{3\times N,tube} = z^{-4} + 2z^{-2} + 2 + z^{2} + 3z^{4} - 3z^{10} - 3z^{12} + 12z^{14} - 18z^{16} + 45z^{18} - 69z^{20} \cdots \quad z << 1$$

$$Z^{1/N}_{3\times N,tube} = z^{6} + 1 + 3z^{-2} + 6z^{-4} + 6z^{-6} + 9z^{-8} + 9z^{-10} - 6z^{-12} + 36z^{-14} - 99z^{-16} - 213z^{-18} + \cdots \quad z >> 1$$

| | Ising $3\times N$ ribbon | | Ising $3\times N$ tube | |
|---|---|---|---|---|
| $z$ | $\overline{x}_{max}$ (octave) | $\overline{x}_{max}$ (Eq. 79) | $\overline{x}_{max}$ (octave) | $\overline{x}_{max}$ (Eq. 79) |
| 0.05 | 3.20000015087876 E+06 | 3.20000015087876 E+06 | 1.60802002518750 E+05 | 1.60802002518750 E+05 |
| 0.1 | 1.0000030712181 3 E+05 | 1.0000030712181 2 E+05 | 1.02020102999997 E+04 | 1.02020102999997 E+04 |
| 0.2 | 3.125660076011 46 E+03 | 3.125660076011 53 E+03 | 6.77044799682371 E+02 | 6.77044799682371 E+02 |
| 0.25 | 1.0248722282 0783 E+03 | 1.0248722282 1178 E+03 | 2.900742157512 70 E+02 | 2.900742157512 68 E+02 |
| 0.3 | 4.12644867 125248 E+02 | 4.12644867 229159 E+02 | 1.47793293548 279 E+02 | 1.47793293548 156 E+02 |
| 4 | 1.0248722282 0783 E+03 | 1.0248722282 1178 E+03 | 4.09721254 773772 E+03 | 4.09721254 800625 E+03 |
| 10 | 1.00000307121813 E+05 | 1.00000307121813 E+05 | 1.00000103060609 E+06 | 1.00000103060609 E+06 |
| 20 | 3.20000015087876 E+06 | 3.20000015087876 E+06 | 6.40000010075376 E+07 | 6.40000010075376 E+07 |

$$Z^{1/N}_{4\times N,rib} = z^{-7} + 2z^{-1} + 5z + 10z^{3} + 28z^{5} + 54z^{7} + 84z^{9} + 86z^{11} - 56z^{13} - 698z^{15} + \cdots \quad z << 1 \quad (80)$$

$$Z^{1/N}_{4\times N,rib} = z^{7} + 2z + 5z^{-1} + 10z^{-3} + 28z^{-5} + 54z^{-7} + 84z^{-9} + 86z^{-11} - 56z^{-13} - 698z^{-15} + \cdots \quad z >> 1$$

$$Z^{1/N}_{4\times N,tube} = z^{-8} + 5 + 20z^{4} + 64z^{8} + 116z^{12} - 140z^{16} + 2556z^{20} + \cdots \quad z << 1$$

$$Z^{1/N}_{4\times N,tube} = z^{8} + 5 + 20z^{-4} + 64z^{-8} + 116z^{-12} - 140z^{-16} + 2556z^{-20} + \cdots \quad z >> 1$$

| | Ising $4\times N$ ribbon | | Ising $4\times N$ tube | |
|---|---|---|---|---|
| $z$ | $\overline{x}_{max}$ (octave) | $\overline{x}_{max}$ (Eq. 80) | $\overline{x}_{max}$ (octave) | $\overline{x}_{max}$ (Eq. 80) |
| 0.05 | 1.28000004025 e+09 | 1.28000004025 e+09 | 2.56000000050001 e+10 | 2.56000000050001 e+10 |
| 0.1 | 1.00000205103 e+07 | 1.00000205103 e+07 | 1.00000005002001 e+08 | 1.00000005002001 e+08 |
| 0.2 | 7.81360896959 e+04 | 7.81360896959 e+04 | 3.90630032164314 e+05 | 3.90630032164314 e+05 |
| 0.25 | 1.63934372289 e+04 | 1.639343722 91 e+04 | 6.55410791084417 e+04 | 6.554107910844 64 e+04 |
| 0.3 | 4.58099200670 e+03 | 4.5809920 1152 e+03 | 1.52467452875790 e+04 | 1.5246745287 7609 e+04 |
| 4 | 1.63934372289 e+04 | 1.639343722 91 e+03 | 6.55410791084417 e+04 | 6.554107910844 64 e+04 |
| 10 | 1.00000205102855 e+07 | 1.00000205103 e+07 | 1.00000005002001 e+08 | 1.00000005002001 e+08 |
| 20 | 1.28000004025126 e+09 | 1.28000004025 e+09 | 2.56000000050001 e+10 | 2.56000000050001 e+10 |

$$Z^{1/N}_{5\times N,rib} = z^{-9} + 2z^{-3} + 5z^{-1} + 5z + 20z^{3} + 52z^{5} + 129z^{7} + 245z^{9} + \cdots \quad z << 1 \quad (81)$$

$$Z^{1/N}_{5\times N,rib} = z^{9} + 2z^{3} + 5z^{1} + 5z^{-1} + 20z^{-3} + 52z^{-5} + 129z^{-7} + 245z^{-9} + \cdots \quad z >> 1$$

$$Z^{1/N}_{5\times N,tube} = z^{-8} + 2z^{-6} + 2z^{-4} + 2z^{-2} + 7 + 11z^{2} + 20z^{4} + 10z^{6} - 25z^{8} + \cdots \quad z << 1$$

$$Z^{1/N}_{5\times N,tube} = z^{10} + 5z^{2} + 1 + 10z^{-2} + 20z^{-4} + 35z^{-6} + 105z^{-8} + 110z^{-10} + 280z^{-12} + \cdots \quad z >> 1$$

| | Ising $5\times N$ ribbon | | Ising $5\times N$ tube | |
|---|---|---|---|---|
| $z$ | $\overline{x}_{max}$ (octave) | $\overline{x}_{max}$ (Eq. 81) | $\overline{x}_{max}$ (octave) | $\overline{x}_{max}$ (Eq. 81) |
| 0.05 | 5.12000016100253 e+11 | 5.12000016100253 e+11 | 2.5728320807 e+10 | 2.5728320807 e+10 |
| 0.1 | 1.00000205052053 e+09 | 1.00000205052053 e+09 | 1.0202020711 e+08 | 1.0202020711 e+08 |
| 0.2 | 1.9534011784 2540 e+06 | 1.9534011784 1664 e+06 | 4.2318247279 e+05 | 4.23182472 58 e+05 |
| 0.25 | 2.62293622 193879 e+05 | 2.62293622 089386 e+05 | 7.4279768402 e+04 | 7.427976 7685 e+04 |
| 0.3 | 5.089820435 87032 e+04 | 5.089820356 06666 e+04 | 1.8262359688 e+04 | 1.826235 67048 e+04 |
| 4 | 2.62293622 193880 e+05 | 2.62293622 089386 e+05 | 1.0486577134 e+06 | 1.048657713 3 e+06 |
| 10 | 1.00000205052053 e+09 | 1.00000205052053 e+09 | 1.0000000501 e+10 | 1.0000000501 e+10 |
| 20 | 5.12000016100253 e+11 | 5.12000016100253 e+11 | 1.0240000002 e+13 | 1.0240000002 e+13 |

The precision of the series expansions become astounding at $m = 6$;

$$Z^{1/N}_{6\times N,tube} = z^{-12} + 6z^{-4} + 13 + 75z^{4} + 378z^{8} + 1420z^{12} + 3933z^{16} + \cdots \quad z << 1 \quad (82)$$

$$Z^{1/N}_{6\times N,tube} = z^{12} + 6z^{4} + 13 + 75z^{-4} + 378z^{-8} + 1420z^{-12} + 3933z^{-16} + \cdots \quad z >> 1$$

| Ising $6 \times N$ tube | | |
|---|---|---|
| $z$ | $\overline{x}_{max}$ (octave) | $\overline{x}_{max}$ (Eq. 82) |
| 0.05 | 4.09600000096002e+15 | 4.09600000096001e+15 |
| 0.1 | 1.00000006001301e+12 | 1.00000006001301e+12 |
| 0.2 | 2.44144388120973e+08 | 2.44144388120973e+08 |
| 0.25 | 1.67787652988221e+07 | 1.67785092988221e+07 |
| 0.3 | 1.88243079697202e+06 | 1.88230734018170e+06 |
| 4 | 1.67787652988222e+07 | 1.67787652988221e+07 |
| 10 | 1.00000006001301e+12 | 1.00000006001301e+12 |
| 20 | 4.09600000096001e+15 | 4.09600000096001e+15 |

We report the partition functions per particle $\kappa_{m\times N} = Z_{m\times N}^{1mN}$

$$\kappa_{3\times N} = z^2 + \frac{1}{3}z^{-4} + z^{-6} + 2z^{-8} + \frac{17}{9}z^{-10} + \frac{7}{3}z^{-12} + \frac{2}{3}z^{-14} + \cdots \quad (83)$$

$$\kappa_{4\times N} = z^2 + \frac{5}{4}z^{-6} + 5z^{-10} + \frac{437}{32}z^{-14} + \frac{41}{4}z^{-18} + \cdots$$

$$\kappa_{5\times N} = z^2 + z^{-6} + 15z^{-8} + 2z^{-10} + 4z^{-12} + 5z^{-14} + 1015z^{-16} + \cdots$$

$$\kappa_{6\times N} = z^2 + z^{-6} + \frac{13}{6}z^{-10} + 10z^{-14} + \frac{313}{6}z^{-18} + \cdots$$

## Polynomial and degenerate eigenvalues

We discover some interesting surprises scattered among the eigenvalues computed by these methods. Some Ising systems possess eigenvalues with truncated series expansion, polynomials. For the $3\times N$ Ising tube we find that there are two pairs of degenerate eigenvalues whose series truncate after a few terms. The signal for this termination is that the Newton polygon construction gives no new segments.

$$\overline{x}_{max} = \overline{x}_1 = z^6 + 1 + 3z^{-2} + 6z^{-4} + 6z^{-6} + 9z^{-8} + 9z^{-10} - 6z^{-12} - 36z^{-14} - 99z^{-16} + \cdots \quad (84)$$

$$\overline{x}_2 = z^6 - 1 + 3z^{-2} - 6z^{-4} + 6z^{-6} - 9z^{-8} + 9z^{-10} + 6z^{-12} - 36z^{-14} + 99z^{-16} + \cdots$$

$$\overline{x}_3 = z^2 + 2 - z^{-2} - 5z^{-4} - 6z^{-6} - 9z^{-8} - 9z^{-10} + 6z^{-12} + 36z^{-14} + 99u^{-16} + \cdots$$

$$\overline{x}_{4,5} = z^2 + 1 - z^{-2} - z^{-4}$$

$$\overline{x}_{6,7} = z^2 - 1 - z^{-2} + z^{-4}$$

$$\overline{x}_8 = z^2 - 2 - z^{-2} + 5z^{-4} - 6z^{-6} + 9z^{-8} - 9z^{-10} - 6z^{-12} + 36z^{-14} - 99u^{-16} + \cdots$$

For the $4\times N$ Ising tube, $z \to u^{-1}$ we obtain three branches at the first level

$$Q = (xu^8 - 1)^2, \qquad x^2(xu^4 - 1)^{12}, \qquad (x-1)^2 \quad (85)$$

indicating two (large) roots starting with $u^{-8} = z^8$, twelve starting with $u^{-4} = z^4$ and two starting with 1.

Take the twelve-fold branch and substitute $x \to x + u^{-4}$, the new Newton polygon exhibits branches

$$Q = (u^2x-2)^2(u^2x+2)^2(u^4x^2-8), \qquad x - 3u^4, \qquad Q = x(x+4)(x+2)^4 \quad (86)$$

Follow the branch $x \to x + u^{-4}$, it does not fork, there is only one new (degenerate) Newton polygon segment

$$Q = (x + 2u^2)^2 \quad (87)$$

which we follow with $x \to x - 2u^2$, again it does not fork, there is only one new (degenerate) segment

$$Q = (x + u^4)^2 \quad (88)$$

which terminates at the next level (no new segments are produced), giving the (double) polynomial root (far left branch set of Figure 7).

$$\overline{x} = u^{-4} + 2u^{-2} - 2u^2 - u^4 = z^4 + 2z^2 - 2z^{-2} - z^{-4} \quad (89)$$

Following another branch of the twelve-fold $x \to x + u^{-4} \to x - 2u^{-2}$ segment, namely

$$Q = (x - 2u^2)^2 \quad (90)$$

another step to

$$Q = (x + u^4)^2 \quad (91)$$

we come to another terminating line, giving the double root

$$\overline{x} = u^{-4} - 2u^{-2} + 2u^2 - u^4 = z^4 - 2z^2 + 2z^{-2} - z^{-4} \quad (92)$$

Return to second node, and follow the branch $x \to x + u^{-4} \to x - 2$ to yet another fork

$$Q = x^2 - 4, \qquad Q = (x - u^4)^4 \quad (93)$$

the second branch of this fork promptly terminates on the quadruple root

$$\overline{x} = u^{-4} - 2 + u^4 = z^4 - 2 + z^{-4} \quad (94)$$

The significance of polynomial roots, and especially degenerate polynomial roots, is that they are rather easy to find by the Newton polygon method since they require few iterations. They appear to follow a clear pattern. Once they are found, they can be factored out of the characteristic polynomial exactly, reducing its order substantially. This not only simplifies calculation of the remaining, more significant roots, but makes it possible to tackle even larger systems.

This example contains the first case that we have illustrated in which a root has non-rational coefficients: look at the branch beginning with the factor $(u^4x^2 - 8)$. The best way to deal with such a case is by treating it symbolically [12]

let !C^2-8=0;

and develop the root in a series in the two variables $C$ and $u$, obtaining in fact two roots (the third largest and third from smallest) since there are two solutions to $8 = 0$

$$x = u^{-4} + 8C^{-1}u^{-2} + 2 - 8C^{-1}u^2 - 7u^4 - 48C^{-1}u^6 - 32u^8 - 144C^{-1}u^{10} - 56u^{12} - 112C^{-1}u^{14} + \cdots \quad (95)$$

$$\overline{x}_3 = z^4 + \sqrt{8}z^2 + 2 - \sqrt{8}z^{-2} - 7z^{-4} - 6\sqrt{8}z^{-6} - 32z^{-8} - 18\sqrt{8}z^{-10} - 56z^{-12} - 14\sqrt{8}z^{-14} + \cdots$$

$$\overline{x}_{14} = z^4 - \sqrt{8}z^2 + 2 + \sqrt{8}z^{-2} - 7z^{-4} + 6\sqrt{8}z^{-6} - 32z^{-8} + 18\sqrt{8}z^{-10} - 56z^{-12} + 14\sqrt{8}z^{-14} + \cdots$$

the third largest eigenvalue of the transfer matrix.

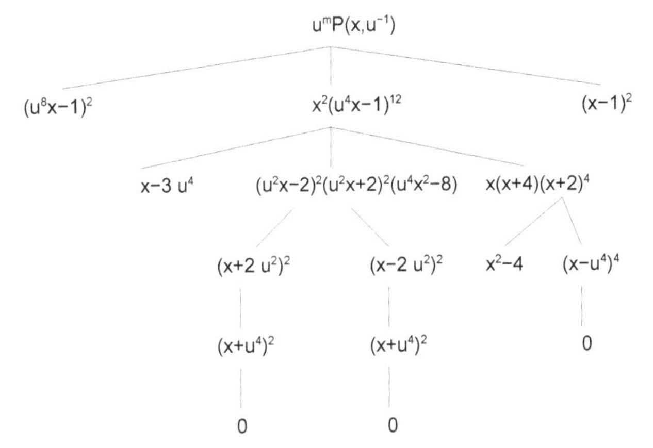


**Figure 7**: Illustration of the lineage ($Q$-polynomials) of degenerate polynomial eigenvalues for the $4\times N$ Ising tube.

The full set of eigenvalue expansions for the $4\times N$ ferromagnetic Ising tube are as follows:

$$\overline{x}_1 = z^8 + 5 + 20z^{-4} + 64z^{-8} + 116z^{-12} - 140z^{-16} + 2556z^{-20} + \cdots \quad (96)$$

$$\overline{x}_2 = z^8 + 3 - 4z^{-4} - 12z^{-8} + 12z^{-12} + 84z^{-16} - 84z^{-20} + \cdots$$

$$\overline{x}_3 = z^4 + \sqrt{8}z^2 + 2 - \sqrt{8}z^{-2} - 7z^{-4} - 6\sqrt{8}z^{-6} - 32z^{-8} - 18\sqrt{8}z^{-10} - 56z^{-12} - 14\sqrt{8}z^{-14} + 64z^{-16} + \cdots$$

$$\overline{x}_{4,5} = z^4 + 2z^2 - 2z^{-2} - z^{-4}$$

$$\overline{x}_6 = z^4 + 3z^{-4} - 4z^{-8} - 12z^{-12} + 12z^{-16} + 84z^{-20} + \cdots$$

$$\overline{x}_{7,8,9,10} = z^4 - 2 + z^{-4}$$

$$\overline{x}_{11} = z^4 - 4 + 3z^{-4} + 12z^{-8} - 12z^{-12} - 84z^{-16} + 84z^{-20} + \cdots$$

$$\overline{x}_{12,13} = z^4 - 2z^2 + 2z^{-2} - z^{-4}$$

$$\overline{x}_{14} = z^4 - \sqrt{8}z^2 + 2 + \sqrt{8}z^{-2} - 7z^{-4} + 6\sqrt{8}z^{-6} - 32z^{-8} + 18\sqrt{8}z^{-10} - 56z^{-12} + 14\sqrt{8}z^{-14} + 64z^{-16} + \cdots$$

$$\overline{x}_{15} = 1 - 4z^{-4} + 3z^{-8} + 12z^{-12} - 12z^{-16} - 84z^{-20} + \cdots$$

$$\overline{x}_{16} = 1 - 4z^{-4} + z^{-8} - 4z^{-12} + 12z^{-16} + 140z^{-20} + 560z^{-24} + \cdots$$

The $4\times N$ hard-squares model on a tube (low fugacity) also has polynomial (in fact monomial) eigenvalues, as well as a large null-space

$$\overline{x}_1 = 1 + 4z - 2z^2 + 12z^3 - 62z^4 + 352z^5 - 2122z^6 + 13344z^7 - 86566z^8 + \cdots \qquad z \ll 1 \quad (97)$$

$$\overline{x}_2 = z^2 - 2z^3 + 4z^4 - 6z^5 + 50z^7 - 268z^8 + 990z^9 + \cdots$$

$$\overline{x}_3 = z^2 - 2z^3 + 6z^4 - 22z^5 + 90z^6 - 394z^7 + 1806z^8 - 8558z^9 + \cdots$$

$$\overline{x}_{4-12} = 0$$

$$\overline{x}_{13} = -z + 2z^2 - 10z^3 + 58z^4 - 346z^5 + 2122z^6 - 13394z^7 + 86834z^8 + \cdots$$

$$\overline{x}_{14,15} = -z$$

$$\overline{x}_{16} = -z - 2z^2 + 2z^3 - 6z^4 + 22z^5 - 90z^6 + 394z^7 - 1806z^8 + \cdots$$

and for high fugacity we also find polynomial (monomial) roots

$$\overline{x}_1 = z^2 + 2z + 4 - 2z^{-1} - 8z^{-2} + 22z^{-3} - 190z^{-5} + 120z^{-6} + 1442z^{-7} - 3268z^{-8}, \qquad z \gg 1 \quad (98)$$

$$\overline{x}_2 = z - 2 + 6z^{-1} - 22z^{-2} + 90z^{-3} - 394z^{-4} + 1806z^{-5} - 8558z^{-6} + 41586z^{-7} - 206098z^{-8} + \cdots$$

$$\overline{x}_3 = z - 2 + 2z^{-1} + 6z^{-2} - 22z^{-3} - 10z^{-4} + 186z^{-5} - 178z^{-6} - 1390z^{-7} + 3630z^{-8} + \cdots$$

$$\overline{x}_{4\to 12} = 0$$

$$\overline{x}_{13} = -1 + 2z^{-2} - 10z^{-4} + 4z^{-5} + 58z^{-6} - 52z^{-7} - 362z^{-8} + 544z^{-9} + \cdots$$

$$\overline{x}_{14,15} = -z$$

$$\overline{x}_{16} = -z^2 - 2z + 2 - 6z^{-1} + 22z^{-2} - 90z^{-3} + 394z^{-4} - 1806z^{-5} + 8558z^{-6} - 41586z^{-7} + 206098z^{-8} + \cdots$$

Correlation functions and other interesting properties can be computed if one has all of the eigenvalues of the transfer matrix. Let $U$ be a unitary matrix that diagonalizes $T(z)$, in other words let $U\cdot T(z)\cdot U^{-1} = \Lambda(z) = diag(\overline{x}_1, \overline{x}_2, \cdots)$, then ($\overline{x}_1$ is the largest eigenvalue) with $\sigma = U\cdot s\cdot U^{-1}$

$$\langle s_j s_k\rangle = \frac{Tr\left(\sigma\cdot\Lambda^{N-(j-k)}(z)\cdot\sigma\cdot\Lambda^{j-k}\right)}{Tr\left(\Lambda^N(z)\right)} \to \sum_{\ell\neq 1} a_\ell \left(\frac{\overline{x}_\ell}{\overline{x}_1}\right)^{j-k}, \quad N\to\infty \quad (99)$$

the coefficients $a_\ell$ being given by the various components of $U$. Hensel constructions can be used to construct the corresponding eigenvector components in Puiseux series for non-degenerate eigenvalues [18].

## The Baxter-Wu model

The Baxter-Wu model, with Hamiltonian

$$H = -J\sum_{faces\, i} s_{i,1}s_{i,2}s_{i,3}, \qquad s_{i,n}\in\{-1,1\} \quad (100)$$

in which we sum over all faces of the (triangular) lattice, each face having three vertices (with states $s_{i,n}$, $n=1,2,3$) provides us with an example in which second-largest magnitude eigenvalues are complex. For the $4\times N$ tube we find

$$\overline{x}_{max} = z^{-8} + 4z^4 + 13z^8 + 64z^{12} + 356z^{16} + 1980z^{20} + 10792z^{24} + \cdots, \qquad z = e^{\beta J} \quad (101)$$

| **Baxter-Wu $4\times N$ tube** | | |
|---|---|---|
| $z$ | Octave | Series |
| 0.1 | 1.00000000000400e+08 | 1.00000000000e+0.8 |
| 0.2 | 3.90625006433545e+05 | 3.90625006434e+5 |
| 0.3 | 1.52416123161342e+04 | 15241.6123161 |
| 0.4 | 1.52599107786097e+03 | 1525.99107738 |
| 0.5 | 256.324688942023101 | 256.324369907 |

with eight complex second-largest roots. The secular equation has a branch $x = \omega z^{-6}$ with $\omega^8 = 1$, $\omega = e^{2\pi ip/8}$. The primitives ($p = 1,3,5,7$) give four roots

$$\overline{x}(p) = wz^{-6} - \frac{1}{2}(1-w^6)z^{-4} + \frac{1}{2}wz^{-2} - \frac{17}{32}wz^2 + \frac{3}{2}(1-w^6)z^4 + \frac{337}{128}wz^6 + 4(1-w^6)z^8 + \frac{21915}{2048}wz^{10} + \cdots \quad (102)$$

all of which have equal magnitude real and imaginary parts. The roots for which $p = 0,4$ ($\omega = \pm 1$) give eigenvalues

$$\overline{x}(p) = \omega z^{-6} + z^{-4} + \frac{3}{2}\omega z^{-2} + 4 + \frac{67}{8}\omega z^2 + 15z^4 + \frac{359}{16}\omega z^6 - 10z^8 - \frac{15261}{128}\omega z^{10} + \cdots \quad (103)$$

and $p = 2,6$ ($\omega = \pm i$) result in polynomials

$$\overline{x}(p) = \omega(z^{-6} + z^{-2} - z^2 - z^6) \quad (104)$$

The $4\times N$ Baxter-Wu ribbon is interesting for having its four largest-magnitude eigenvalues all of the same leading order in z, the first level has four branches $Q = (xz^6)^4 - (xz^6)^3 - (xz^6)^2 + 1$, or $x = z^{-6}$, $x = \omega z^{-6}$ with $\omega = e^{2\pi ip/3}$, $p = 0,1,2$;

$$\overline{x}_{max} = \overline{x}_1 = z^{-6} + 4 + 10z^2 + 30z^4 + 78z^6 + 174z^8 + 307z^{10} + \cdots$$

$$\overline{x}_{2,3,4} = \omega\left(z^{-6} + \frac{4}{3} - \frac{2}{3}z^2 - 2z^4 - \frac{34}{9}z^6 - \frac{26}{9}z^8 + \frac{59}{9}z^{10} + \cdots\right) \quad (105)$$

In all of the examples in this subsection, there is agreement with purely numerical determination of the eigenvalues for $z = e^{\beta J} = 0.1, 0.2$ to ten decimal places or more, for both real and imaginary parts. Our purpose here is not to tout the precision (but it is a welcome feature of the results), but rather to point out that our methods do not suffer from any of the problems associated with eigenvalue degeneracy or near-degeneracy that plague many purely numerical approaches.

## Conclusions and Conjectures

The Newton polygon algorithm offers some distinct advantages for the study of nanotubes and ribbons. In the first place series expansions for all eigenvalues of the transfer matrix can be found

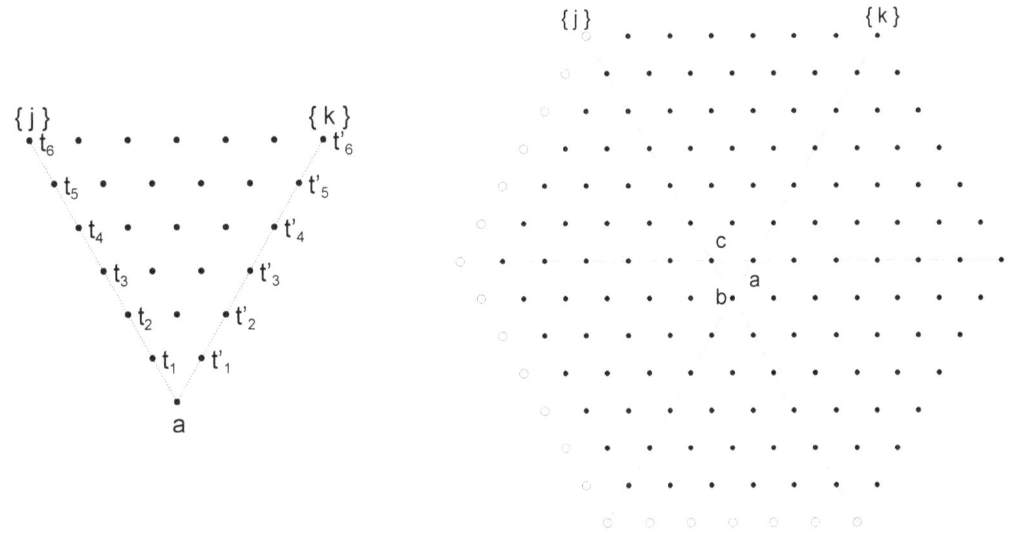


**Figure 8:** Graphical representation of $A(a) = A_{\mathbf{k},\mathbf{j}}(a)$ (left) and $Z_w$ (right).

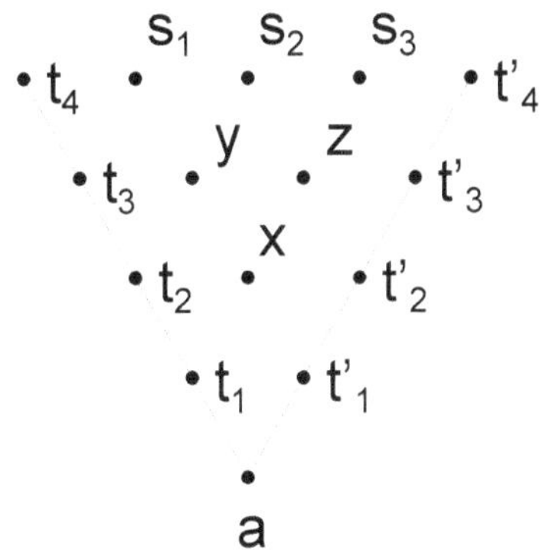


**Figure 9**: An explicit case of *A(a)* for *N = 4*.

symbolic-numerically: as functions of some quantity such as fugacity or a Boltzmann weight. Both very large and very low fugacity regimes, and the cases of ferromagnetic or anti-ferromagnetic systems, are equally approachable. Secondly the technique is not an approximation. Each iteration gives the next order in the series expansion exactly. Furthermore if a root truncates to a finite sum of terms, the Newton polygon process itself simply halts, no new segments are produced once the last term in the series is reached. A third point is that degeneracy of a root presents none of the problems normally encountered by numerical root extraction methods in this case. The level of degeneracy of a root has tell-tale signs in the calculation.

The technique is not limited to a single expansion variable *z*, Inaba [19] has used an extended Hensel construction based upon the Newton polygon algorithm to factor multivariate polynomials.

## Leading exponents

There **appears** to be a correlation between the leading terms of eigenvalues, the number of eigenvalues with this leading term, and the trace of the transfer matrix for the ferromagnetic Ising tubes.

| **Ferromagnetic Ising *m*×*N* tube** | | |
|---|---|---|
| *m* | Level 1 leading term (degeneracy) | *Tr*(*T*) |
| 3 | $z^6$ (2),<br>$z^2$ (6) | $2Z^2(3+z^4)$ |
| 4 | $z^8$ (2),<br>$z^4$ (12),<br>$z^0$ (2) | $2(1+6z^4+z^8)$ |
| 5 | $z^{10}$ (2),<br>$z^6$ (20),<br>$z^2$ (10) | $2z^2(z^8+10z^4+5)$ |
| 6 | $z^{12}$ (2),<br>$z^8$ (30),<br>$z^4$ (30),<br>$z^0$ (2) | $2(z^{12}+ 15z^8+15z^4+1)$ |

The trace is rather easy to calculate in the general case ( $m \times N$ Ising tube):

$$Tr(T) = \sum_{s_1=\pm1} \cdots \sum_{s_m=\pm1} z^{\sum_{i=1}^{m} s_i^2 + \sum_i^{PBC} s_i s_{i+1}} \tag{106}$$

$$= z^m \sum_{s_1=\pm1} \cdots \sum_{s_m=\pm1} \prod_i^{PBC} z^{s_i s_{i+1}}$$

$$= z^m \sum_{s_1=\pm1} \cdots \sum_{s_m=\pm1} \prod_i^{PBC} (\cosh(\beta J) + s_i s_{i+1} \sinh(\beta J)), \quad z = e^{\beta J}$$

$$= z^m 2^m (\cosh^m(\beta J) + \sin^m(\beta J)) = (z^2+1)^m + (z^2-1)^m$$

For hard squares we can make no such conjecture, since the trace of the transfer matrix (the sum of Boltzmann weights for all allowed configurations with adjacent "rungs" of the ladder in identical states) is simply one for all *m*.

| **Hard squares *m*×*N* tube** | | |
|---|---|---|
| *m* | Level 1 leading term (degeneracy) | *Tr*(*T*) |
| 4 | 0(9), $z^0$(1), $z^1$(4), $z^2$(2) | 1 |
| 5 | 0(21), $z^0$(1), $z^1$(5), $z^2$(5) | 1 |
| 6 | 0(46), $z^0$(1), $z^1$(6), $z^2$(9), $z^3$(2) | 1 |
| 7 | $z^0$(1), $z^1$(7), $z^2$(14), $z^3$(7) | 1 |
| 8 | 0(209), $z^0$(1), $z^1$(8), $z^2$(20), $z^3$(16), $z^4$(2) | 1 |

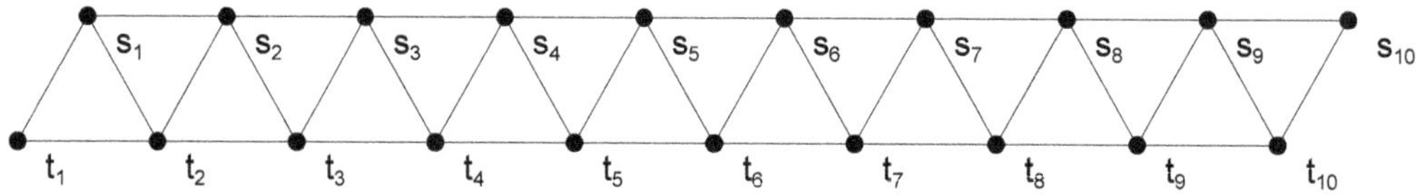


**Figure 10:** Two rows of the lattice, with sites labeled appropriately for constructing the row-to-row transfer matrix.

## References


1. RJ Baxter (1982) Exactly Solved Models in Statistical Mechanics. Dover, New York.
2. Domb C, Green MS (1974) Phase Transitions and Critical Phenomena (Volume 3). Series Expansions for Lattice Models. Academic Press, London.
3. Jaan Oitmaa, Chris Hamer, Weihong Zheng (2006) Series Expansion Methods For Strongly Interacting Lattice Models. Cambridge University Press, Cambridge.
4. Baxter RJ (1981) Corner transfer matrices. Physica 106A: 18-27.
5. Baxter RJ (2007) Corner transfer matrices in statistical mechanics. J Phys A: Math Theor40: 12577-12588.
6. Celine Cattoen, Matt Visser (2006) Generalized Puisieux series expansion for cosmological milestones. arXiv:gr-qc/0609073.
7. Takua Kitamoto (1994) Approximate eigenvalues, eigenvectors and inverse of a matrix with polynomial entries. J Indust Appl Math 11: 73-85.
8. Takeaki Sasaki, Fujio Kako (1999) Solving multivariate algebraic equation by Hensel construction.J Indust Appl Math16: 257-285.
9. David A Cox, John Little, Donal O'Shea (2005) Using Algebraic Geometry (2nd edn). Springer.
10. Robert JW (1950) Algebraic Curves. Dover, New York.
11. S Basu, R Pollack, Marie-Françoise, C Roy (2006) Algorithms in Real Algebraic Geometry (2nd edn), Springer.
12. Kung HT, Traub JF (1978)All Algebraic Functions Can Be Computed Fast. JACM 25: 245-260.
13. Thomas H Cormen, Charles E Leisterson, Ronald L Rivest, Clifford Stein (2009) Introduction to Algorithms (3rd edn). The MIT Press, Cambridge.
14. http://reduce-algebra.sourceforge.net/.
15. http://www.singular.uni-kl.de
16. Baxter RJ (1999) Planar lattice gases with nearest-neighbor exclusion. Annals of Combinatorics 3: 191-203.
17. David Cox, John Littel, Donal O'Shea (1997) Ideals, Varieties and Algorithms. Springer, New York, USA.
18. Takua Kitamoto (1994) Approximate eigenvalues, eigenvectors and inverse of a matrix with polynomial entries. J Indust Appl Math 11: 73-85.
19. Daiju Inaba (2005) Factorization of multivariate polynomials by extended Hensel construction. ACM SIGSAM Bulletin. 39: 2-14.

# The Generalized Triple Difference Lacunary Statistical on $\Gamma^3$ Over P-Metric Spaces Defined by Musielak Orlicz Function

**Mishra LN[1*], Deepmala[2] and Subramanian N[3]**

[1]*Department of Mathematics, National Institute of Technology, Silchar – 788 010, India*
[2]*SQC & OR Unit, Indian Statistical Institute, Kolkata-700 108, India*
[3]*Department of Mathematics, Sastra University, Thanjavur-613 401, India*

**Abstract**

We introduce the generalized triple sequence spaces of entire difference lacunary statistical convergence and discuss general topological properties also inclusion theorems are with respect to a sequence of Musielak-Orlicz function.

**Keywords:** Analytic sequence; Triple sequences; Difference sequence; $\Gamma^3$ space; Musielak-Orlicz function; Lacunary sequence; Statistical convergence

## Introduction

A triple sequence (real or complex) can be defined as a function $x:\mathbb{N}\times \mathbb{N}\times\mathbb{N}\to\mathbb{R}\ (\mathbb{C})$ where $\mathbb{N},\mathbb{R}$ and $\mathbb{C}$ denote the set of natural numbers, real numbers and complex numbers respectively. The different types of notions of triple sequence was introduced and investigated at the initial by Sahiner [1,2], Esi [3-5], Datta [6], Subramanian [7], Debnath [8] and many others.

A triple sequence $x=(x_{mnk})$ is said to be triple analytic if

$$sup_{m,n,k}\left|x_{mnk}\right|^{\frac{1}{m+n+k}} < \infty.$$

The space of all triple analytic sequences are usually denoted by $\Lambda^3$. A triple sequence $x=(x_{mnk})$ is called triple entire sequence if

$$\left|x_{mnk}\right|^{\frac{1}{m+n+k}} \to 0 \text{ as } m,n,k\to\infty.$$

The notion of difference sequence spaces (for single sequences) was introduced by Kizmaz [9] as follows

$$Z(\Delta)=\{x=(x_k)\in w:(\Delta x_k)\in Z\}$$

For $Z=c,c_0$ and $\ell_\infty$ where $\Delta x_k = x_k - x_{k+1}$ for all $k\in\mathbb{N}$

The difference triple sequence space was introduced by Debnath et al. (see [8]) and is defined as

$$\Delta x_k= \begin{array}{l}\Delta x_{mnk} = x_{mnk} - x_{m,n+1,k} - x_{m,n,k+1} + x_{m,n+1,k+1} \\ -x_{m+1,n,k} + x_{m+1,n+1,k} + x_{m+1,n,k+1} - x_{m+1,n+1,k+1}\end{array}$$

and $\Delta^0 x_{mnk} = \langle x_{mnk}\rangle$.

## Definitions and Preliminaries

Throughout the article $w^3,\Gamma^3(\Delta)$, $\Lambda^3(\Delta)$ denote the spaces of all, triple entire difference sequence spaces and triple analytic difference sequence spaces respectively.

Subramanian introduced by a triple entire sequence spaces, triple analytic sequences spaces and triple gai sequence spaces [7]. The triple sequence spaces of $\Gamma^3(\Delta)$, $\Lambda^3(\Delta)$ are defined as follows:

$$\Gamma^3(\Delta)=\left\{x\in w^3 : \left|\Delta x_{mnk}\right|^{1/m+n+k} \to 0 \quad as \quad m,n,k\to\infty\right\},$$

$$\Lambda^3(\Delta)=\left\{x\in w^3 : sup_{m,n,k}\left|\Delta x_{mnk}\right|^{1/m+n+k} < \infty\right\}.$$

### 1. Definition

An Orlicz function is a function $M:[0,\infty)\to[0,\infty)$ which is continuous, non-decreasing and convex with $M(0)=M(x)>0$ for$M(x)>0$ $x>0$ and $M(x)\to\infty$ as $x\to\infty$ [10]. If convexity of Orlicz function $M$ is replaced by $M(x+y)\leq M(x)+M(y)$ then this function is called modulus function. $M:[0,\infty)\to[0,\infty)$

Lindenstrauss and Tzafriri [11] used the idea of Orlicz function to construct Orlicz sequence space. A sequence $g=(g_{mn})$ defined by

$$g_{mn}(v)=sup\{|v|u-(f_{mnk})(u):u\geq 0\}, m,n,k=1,2,\cdots$$

is called the complementary function of a Musielak-Orlicz function $f$. For a given Musielak-Orlicz function $f$ the Musielak-Orlicz sequence space $t_f$ is defined as follows [12]

$$t_f=\left\{x\in w^3 : I_f\left(\left|x_{mnk}\right|\right)^{1/m+n+k} \to 0 \quad as \quad m,n,k\to\infty\right\},$$

Where $I_f$ is a convex modular defined by

$$I_f(x)=\sum_{m=1}^{\infty}\sum_{n=1}^{\infty}\sum_{k=1}^{\infty} f_{mnk}\left(\left|x_{mnk}\right|\right)^{1/m+n+k}, x=(x_{mnk})\in t_f.$$

We consider $t_f$ equipped with the Luxemburg metric

$$d(x,y)=\sum_{m=1}^{\infty}\sum_{n=1}^{\infty}\sum_{k=1}^{\infty} f_{mnk}\left(\frac{\left|x_{mnk}\right|^{1/m+n+k}}{mnk}\right)$$

is an exteneded real number.

### 2. Definition

Let $n\in\mathbb{N}$ and $X$ be a real vector space of dimension $w$ where $nm$. A real valued function $d_p(x_1,\ldots,x_n)=\|(d_1(x_1,0),\ldots,d_n(x_n,0))\|_p$ on $X$ satisfying the following four conditions:

***Corresponding author:** Mishra LN, Department of Mathematics, National Institute of Technology, India, E-mail: lakshminarayanmishra04@gmail.com

(i) $\|(d_1(x_1,0),\ldots,d_n(x_n,0))\|_p = 0$ if and and only if $d_1(x_1,0),\ldots d_n(x_n,0)$ are linearly dependent,

(ii) $\|(d_1(x_1,0),\ldots,d_n(x_n,0))\|_p$ is invariant under permutation,

(iii) $\|(\alpha d_1(x_1,0),\ldots,d_n(x_n,0))\|_p = |\alpha| \|(d_1(x_1,0),\ldots,d_n(x_n,0))\|_p, \alpha \in \mathbb{R}$

(iv) $d_p((x_1,y_1),(x_2,y_2)\cdots(x_n,y_n)) = (d_X(x_1,x_2,\cdots x_n)^p + d_Y(y_1,y_2,\cdots y_n)^p)^{1/p}$ $for 1 \le p < \infty;$ (or)

(v) $d((x_1,y_1),(x_2,y_2),\cdots(x_n,y_n)) := \sup\{d_X(x_1,x_2,\cdots x_n), d_Y(y_1,y_2,\cdots y_n)\}$,

For $x_1, x_2, \ldots x_n \in X, y_1, y_2, \ldots y_n \in Y$ is called the $p$ product metric of the Cartesian product of $n$ metric spaces [13].

## 3. Definition

Let $X$ be a linear metric space. A function $\rho: X \to \mathbb{R}$ is called paranorm, if

(1) $\rho(x) \ge 0$ for all $x \in X$;

(2) $\rho(-x) = \rho(x)$ for all $x \in X$;

(3) $\rho(x+y+z) \le \rho(x) + \rho(y) + \rho(z)$, for all $x,y,z \in X$,

(4) If $(\sigma_{mnk})$ is a sequence of scalars with $\sigma_{mnk} \to \sigma$ as $m,n,k \to \infty$ and $x=(x_{mk})$ is a sequence of vectors with

$\rho(x_{mk} - x) \to 0$ as $m,n,k \to \infty$ then $\rho(\sigma_{mnk} x_{mk} - \sigma x) \to 0$ as $m,n,k \to \infty$

## 4. Definition

The triple sequence $\theta_{i,\ell,j} = \{(m_i, n_\ell, k_j)\}$ is called triple lacunary if there exist three increasing sequences of integers such that

$m_0 = 0, h_i = m_i - m_{r-1} \to \infty$ as $i \to \infty$ and

$n_0 = 0, \overline{h_\ell} = n_\ell - n_{\ell-1} \to \infty$ as $\ell \to \infty$

$k_0 = 0, \overline{h_j} = k_j - k_{j-1} \to \infty$ as $j \to \infty$

Let $m_{i,\ell,j} = m_i n_\ell k_j, h_{i,\ell,j} = h_i \overline{h_\ell h_j}$, and $\theta_{i,\ell,j}$ is determine by

$$I_{i,\ell,j} = \{(m,n,k): m_{i-1} < m < m_i \text{ and } n_{\ell-1} < n \le n_\ell \text{ and } k_{j-1} < k \le k_j\},$$

$$q_k = \frac{m_k}{m_{k-1}}, \overline{q_\ell} = \frac{n_\ell}{n_{\ell-1}}, \overline{q_j} = \frac{k_j}{k_{j-1}}.$$

## Main Results

The notion of λ-triple entire and triple analytic sequences as follows: Let $\lambda = (\lambda_{mnk})_{m,n,k=0}^{\infty}$ be a strictly increasing sequences of positive real numbers tending to infinity, that is

$0 < \lambda_{000} < \lambda_{111} < \ldots$ and $\lambda_{mnk} \to \infty$ as $m,n,k \to \infty$

and said that a sequence $x=(x_{mnk}) \in w^3$

is λ-convergent to 0, called a the λ-limit of $x$ if $\mu_{mnk}(x) \to 0$ as $m,n,k \to \infty$ where

$$\mu_{mnk}(x) = \frac{1}{\varphi_{rst}} \sum_{m \in I_{rst}} \sum_{n \in I_{rst}} \sum_{k \in I_{rst}} (\Delta^{m-1}\lambda_{m,n,k}) - (\Delta^{m-1}\lambda_{m,n+1,k}) - (\Delta^{m-1}\lambda_{m,n,k+1}) + (\Delta^{m-1}\lambda_{m,n+1,k+1}) - (\Delta^{m-1}\lambda_{m+1,n,k}) + (\Delta^{m-1}\lambda_{m+1,n+1,k}) + (\Delta^{m-1}\lambda_{m+1,n,k+1}) - (\Delta^{m-1}\lambda_{m+1,n+1,k+1}) \left|\Delta^{m+1} x_{mnk}\right|^{1/m+n+k}.$$

The sequence $x=(x_{mnk}) \in w^3$ is λ-triple difference analytic if $sup_{uvs} |\mu_{mnk}(x)| < \infty$. If $lim_{xnk} x_{mnk}(x) = 0$ in the ordinary sense of convergence, then

$$lim_{mnk} \frac{1}{\varphi_{rst}} \sum_{m \in I_{rst}} \sum_{n \in I_{rst}} \sum_{k \in I_{rst}} (\Delta^{m-1}\lambda_{m,n,k}) - (\Delta^{m-1}\lambda_{m,n+1,k}) - (\Delta^{m-1}\lambda_{m,n,k+1}) + (\Delta^{m-1}\lambda_{m,n+1,k+1}) - (\Delta^{m-1}\lambda_{m+1,n,k}) + (\Delta^{m-1}\lambda_{m+1,n+1,k}) + (\Delta^{m-1}\lambda_{m+1,n,k+1}) - (\Delta^{m-1}\lambda_{m+1,n+1,k+1}) \left|\Delta^{m+1} x_{mnk}\right|^{1/m+n+k} = 0$$

This implies that

$$lim_{mnk} |\mu_{mnk}(x) - 0| = lim_{mnk} \frac{1}{\varphi_{rst}} \sum_{m \in I_{rst}} \sum_{n \in I_{rst}} \sum_{k \in I_{rst}} (\Delta^{m-1}\lambda_{m,n,k}) - (\Delta^{m-1}\lambda_{m,n+1,k}) - (\Delta^{m-1}\lambda_{m,n,k+1}) + (\Delta^{m-1}\lambda_{m,n+1,k+1}) - (\Delta^{m-1}\lambda_{m+1,n,k}) + (\Delta^{m-1}\lambda_{m+1,n+1,k}) + (\Delta^{m-1}\lambda_{m+1,n,k+1}) - (\Delta^{m-1}\lambda_{m+1,n+1,k+1}) \left|\Delta^{m+1} x_{mnk} - 0\right|^{1/m+n+k} = 0$$

which yields that $lim_{uvs} \mu_{mnk}(x) = 0$ and hence $x=(x_{mnk}) \in w^3$ is λ-convergent to 0. Let $I^3$- be an admissible ideal of $3^{N \times N \times N}$, $\theta_{rst}$ be a triple difference lacunary sequence, $f=f_{mnk}$ be a Musielak-Orlicz function and $\left(X, \|(d(x_1,0), d(x_2,0), \cdots, d(x_{n-1},0))\|_p\right)$ be a $p$-metric space, $q=(q_{mnk})$ be triple difference analytic sequence of strictly positive real numbers. By $w^3(p-X)$ we denote the space of all sequences defined over

$$\left(X, \|(d(x_1,0), d(x_2,0), \cdots, d(x_{n-1},0))\|_p\right).$$

In the present paper we define the following sequence spaces:

$$\left[\Gamma_{f\mu}^{3\Delta^m q}, \|(d(x_1,0), d(x_2,0), \cdots, d(x_{n-1},0))\|_p^{\varphi}\right]_{\theta_{rst}}^{I^3} = \left\{r,s,t \in I_{rst} : \left[f_{mnk}\left(\|\mu_{mnk}(x), (d(x_1,0), d(x_2,0), \cdots, d(x_{n-1},0))\|_p\right)\right]^{q_{mnk}} \ge \varepsilon\right\} \in I^3,$$

$$\left[\Lambda_{f\mu}^{3\Delta^m q}, \|(d(x_1,0), d(x_2,0), \cdots, d(x_{n-1},0))\|_p^{\varphi}\right]_{\theta_{rst}}^{I^3} = \left\{r,s,t \in I_{rst} : \left[f_{mnk}\left(\|\mu_{mnk}(x), (d(x_1,0), d(x_2,0), \cdots, d(x_{n-1},0))\|_p\right)\right]^{q_{mnk}} \ge K\right\} \in I^3,$$

If we take $f_{mnk}(x) = x$ we get

$$\left[\Gamma_{f\mu}^{3\Delta^m q}, \|(d(x_1,0), d(x_2,0), \cdots, d(x_{n-1},0))\|_p^{\varphi}\right]_{\theta_{rst}}^{I^3} = \left\{r,s,t \in I_{rst} : \left[\left(\|\mu_{mnk}(x), (d(x_1,0), d(x_2,0), \cdots, d(x_{n-1},0))\|_p\right)\right]^{q_{mnk}} \ge \varepsilon\right\} \in I^3,$$

$$\left[\Lambda_{f\mu}^{3\Delta^m q}, \|(d(x_1,0), d(x_2,0), \cdots, d(x_{n-1},0))\|_p^{\varphi}\right]_{\theta_{rst}}^{I^3} = \left\{r,s,t \in I_{rst} : \left[\left(\|\mu_{mnk}(x), (d(x_1,0), d(x_2,0), \cdots, d(x_{n-1},0))\|_p\right)\right]^{q_{mnk}} \ge K\right\} \in I^3,$$

If we take $q=(q_{mnk})=1$ we get

$$\left[\Gamma_{f\mu}^{3\Delta^m}, \|(d(x_1,0), d(x_2,0), \cdots, d(x_{n-1},0))\|_p^{\varphi}\right]_{\theta_{rst}}^{I} = \left\{r,s,t \in I_{rst} : \left[f_{mnk}\left(\|\mu_{mnk}(x), (d(x_1,0), d(x_2,0), \cdots, d(x_{n-1},0))\|_p\right)\right] \ge \varepsilon\right\} \in I^3,$$

$$\left[\Lambda_{f\mu}^{3\Delta^m}, \|(d(x_1,0), d(x_2,0), \cdots, d(x_{n-1},0))\|_p^{\varphi}\right]_{\theta_{rst}}^{I^3} = \left\{r,s,t \in I_{rst} : \left[f_{mnk}\left(\|\mu_{mnk}(x), (d(x_1,0), d(x_2,0), \cdots, d(x_{n-1},0))\|_p\right)\right] \ge K\right\} \in I^3$$

In the present paper we plan to study some topological properties and inclusion relation between the above defined sequence spaces. $\left[\Gamma_{f\mu}^{3\Delta^m q}, \|(d(x_1,0), d(x_2,0), \cdots, d(x_{n-1}))\|_p^{\varphi}\right]_{\theta_{rst}}^{I^2}$ and $\left[\Lambda_{f\mu}^{3\Delta^m q}, \|(d(x_1,0), d(x_2,0), \cdots, d(x_{n-1},0))\|_p^{\varphi}\right]_{\theta_{rst}}^{I^3}$ which we shall discuss in this paper.

## 1. Theorem

Let $f=f_{mnk}$ be a Musielak-Orlicz function, $q=(q_{mnk})$ be a triple analytic difference sequence of strictly positive real numbers, the sequence spaces

$$\left[\Gamma^{3\Delta^m q}_{f\mu}, \|(d(x_1,0),d(x_2,0),\cdots,d(x_{n-1},0))\|_p^{\varphi}\right]^{I^3}_{\theta_{rst}} \text{ and}$$

$$\left[\Lambda^{3\Delta^m q}_{f\mu}, \|(d(x_1,0),d(x_2,0),\cdots,d(x_{n-1},0))\|_p^{\varphi}\right]^{I^3}_{\theta_{rst}} \text{ are linear spaces.}$$

**Proof:** It is routine verification. Therefore the proof is omitted.

## 2. Theorem

Let $f=f_{mnk}$ be a Musielak-Orlicz function, $q=(q_{mnk})$ be a triple analytic difference sequence of strictly positive real numbers, the sequence space

$$\left[\Gamma^{3\Delta^m q}_{f\mu}, \|(d(x_1,0),d(x_2,0),\cdots,d(x_{n-1},0))\|_p^{\varphi}\right]^{I^3}_{\theta_{rst}}$$ is a paranormed space with respect to the paranorm defined by

$$g(x)=\inf\left\{\left[f_{mnk}\left(\|\mu_{mnk}(x),(d(x_1,0),d(x_2,0),\cdots,d(x_{n-1},0))\|_p\right)\right]^{q_{mnk}}\le 1\right\}.$$

**Proof:** Clearly $g(x)\ge 0$ for

$$x=(x_{mnk})\in\left[\Gamma^{3\Delta^m q}_{f\mu}, \|(d(x_1,0),d(x_2,0),\cdots,d(x_{n-1},0))\|_p^{\varphi}\right]^{I^3}_{\theta_{rst}}$$

Since $f_{mnk}(0)=0$ we get $g(0)=0$

Conversely, suppose that $g(x)$ then

$$\inf\left\{\left[f_{mnk}\left(\|\mu_{mnk}(x),(d(x_1,0),d(x_2,0),\cdots,d(x_{n-1},0))\|_p\right)\right]^{q_{mnk}}\le 1\right\}=0.$$

Suppose that $\mu_{mnk}(x)\neq 0$. for each $m,n,k\in\mathbb{N}$ Then

$\|\mu_{mnk}(x),(d(x_1,0),d(x_2,0),\cdots,d(x_{n-1},0))\|_p^{\varphi}\to\infty$. It follows that

$$\left(\left[f_{mnk}\left(\|\mu_{mnk}(x),(d(x_1,0),d(x_2,0),\cdots,d(x_{n-1},0))\|_p\right)\right]^{q_{mnk}}\right)^{1/H}\to\infty$$

which is a contradiction. Therefore $\mu_{mnk}(x)=0$. Let

$$\left(\left[f_{mnk}\left(\|\mu_{mnk}(x),(d(x_1,0),d(x_2,0),\cdots,d(x_{n-1},0))\|_p\right)\right]^{q_{mnk}}\right)^{1/H}\le 1$$

and

$$\left(\left[f_{mnk}\left(\|\mu_{mnk}(y),(d(x_1,0),d(x_2,0),\cdots,d(x_{n-1},0))\|_p\right)\right]^{q_{mnk}}\right)^{1/H}\le 1$$

Then by using Minkowski's inequality, we have

$$\left(\left[f_{mnk}\left(\|\mu_{mnk}(x+y),(d(x_1,0),d(x_2,0),\cdots,d(x_{n-1},0))\|_p\right)\right]^{q_{mnk}}\right)^{1/H}$$
$$\le\left(\left[f_{mnk}\left(\|\mu_{mnk}(x),(d(x_1,0),d(x_2,0),\cdots,d(x_{n-1},0))\|_p\right)\right]^{q_{mnk}}\right)^{1/H}$$
$$+\left(\left[f_{mnk}\left(\|\mu_{mnk}(y),(d(x_1,0),d(x_2,0),\cdots,d(x_{n-1},0))\|_p\right)\right]^{q_{mnk}}\right)^{1/H}.$$

So we have

$$g(x+y)=\inf\left\{\left[f_{mnk}\left(\|\mu_{mnk}(x+y),(d(x_1,0),d(x_2,0),\cdots,d(x_{n-1},0))\|_p\right)\right]^{q_{mnk}}\le 1\right\}$$
$$\le\inf\left\{\left[f_{mnk}\left(\|\mu_{mnk}(x),(d(x_1,0),d(x_2,0),\cdots,d(x_{n-1},0))\|_p\right)\right]^{q_{mnk}}\le 1\right\}+\inf$$
$$\left\{\left[f_{mnk}\left(\|\mu_{mnk}(y),(d(x_1,0),d(x_2,0),\cdots,d(x_{n-1},0))\|_p\right)\right]^{q_{mnk}}\le 1\right\}$$

Therefore,

$$g(x+y)\le g(x)+g(y).$$

Finally, to prove that the scalar multiplication is continuous. Let λ be any complex number. By definition,

$$g(\lambda x)=\inf\left\{\left[f_{mnk}\left(\|\mu_{mnk}(\lambda x),(d(x_1,0),d(x_2,0),\cdots,d(x_{n-1},0))\|_p\right)\right]^{q_{mnk}}\le 1\right\}.$$

Then

$$g(\lambda\ \ x)=\inf\left\{((|\lambda|t)^{q_{mnk}/H}:\left[f_{mnk}\left(\|\mu_{mnk}(\lambda x),(d(x_1,0),d(x_2,0),\cdots,d(x_{n-1},0))\|_p\right)\right]^{q_{mnk}}\le 1\right\}$$

where $t=\frac{1}{|\lambda|}$. Since $|\lambda|^{q_{mnk}}\le\max\left(1,|\lambda|^{\sup\ q_{mnk}}\right)$, we have

$$g(\lambda\ \ x)\le\max\left(1,|\lambda|^{\sup q_{mnk}}\right)\inf$$
$$\left\{t^{q_{mnk}/H}:\left[f_{mnk}\left(\|\mu_{mnk}(\lambda x),(d(x_1,0),d(x_2,0),\cdots,d(x_{n-1},0))\|_p\right)\right]^{q_{mnk}}\le 1\right\}$$

This completes the proof.

## 3. Theorem

(i) If the Musielak Orlicz function $(f_{mnk})$ satisfies $\Delta_2$- condition, then

$$\left[\Gamma^{3\Delta^m q}_{f\mu}, \|\mu_{mnk}(x),(d(x_1,0),d(x_2,0),\cdots,d(x_{n-1},0))\|_p^{\varphi}\right]^{I^{3\alpha}}_{\theta_{rst}}=$$
$$\left[\Gamma^{3\Delta^m q\mu}_{g}, \|\mu_{uvs}(x),(d(x_1,0),d(x_2,0),\cdots,d(x_{n-1},0))\|_p^{\varphi}\right]^{I^3}_{\theta_{rst}}.$$

(ii) If the Musielak Orlicz function $(g_{mnk})$ satisfies $\Delta_2$- condition, then

$$\left[\Gamma^{3\Delta^m q\mu}_{g}, \|\mu_{mnk}(x),(d(x_1,0),d(x_2,0),\cdots,d(x_{n-1},0))\|_p^{\varphi}\right]^{I^{3\alpha}}_{\theta_{rst}}=$$
$$\left[\Gamma^{3\Delta^m q}_{f\mu}, \|\mu_{mnk}(x),(d(x_1,0),d(x_2,0),\cdots,d(x_{n-1},0))\|_p^{\varphi}\right]^{I^3}_{\theta_{rst}}$$

**Proof:** Let the Musielak Orlicz function $(f_{mk})$ satisfies $\Delta_2$-condition, we get

$$\left[\Gamma^{3\Delta^m q\mu}_{g}, \|\mu_{mnk}(x),(d(x_1,0),d(x_2,0),\cdots,d(x_{n-1},0))\|_p^{\varphi}\right]^{I^3}_{\theta_{rst}}.$$
$$\subset\left[\Gamma^{3\Delta^m q}_{f\mu}, \|\mu_{mnk}(x),(d(x_1,0),d(x_2,0),\cdots,d(x_{n-1},0))\|_p^{\varphi}\right]^{I^{3\alpha}}_{\theta_{rst}}\cdots\cdots \quad (1)$$

To prove the inclusion

$$\left[\Gamma^{3\Delta^m q}_{f\mu}, \|\mu_{mnk}(x),(d(x_1,0),d(x_2,0),\cdots,d(x_{n-1},0))\|_p^{\varphi}\right]^{I^{3\alpha}}_{\theta_{rst}}$$
$$\subset\left[\Gamma^{3\Delta^m q\mu}_{g}, \|\mu_{mnk}(x),(d(x_1,0),d(x_2,0),\cdots,d(x_{n-1},0))\|_p^{\varphi}\right]^{I^3}_{\theta_{rst}},$$

let $a\in\left[\Gamma^{3\Delta^m q}_{f\mu}, \|\mu_{mnk}(x),(d(x_1,0),d(x_2,0),\cdots,d(x_{n-1},0))\|_p^{\varphi}\right]^{I^{3\alpha}}_{\theta rst}$.

Then for all $\{x_{mnk}\}$ with

$$(x_{mnk})\in\left[\Gamma^{3\Delta^m q}_{f\mu}, \|\mu_{mnk}(x),(d(x_1,0),d(x_2,0),\cdots,d(x_{n-1},0))\|_p^{\varphi}\right]^{I^3}_{\theta_{rst}}$$

we have

$$\sum_{m=1}^{\infty}\sum_{n=1}^{\infty}\sum_{k=1}^{\infty}\left|\Delta^m x_{mnk}a_{mnk}\right|<\infty. \quad (1)$$

Since the Musielak Orlicz function $(f_{mk})$ satisfies condition, then

$(y_{mnk})\in\left[\Gamma^{3\Delta^m q}_{f\mu}, \|\mu_{mnk}(x),(d(x_1,0),d(x_2,0),\cdots,d(x_{n-1},0))\|_p^{\varphi}\right]^{I^3}_{\theta_{rst}}$, we get

$\sum_{m=1}^{\infty}\sum_{n=1}^{\infty}\sum_{k=1}^{\infty}\left|\frac{\varphi_{rst}y_{mnk}a_{mnk}}{\Delta^m\lambda_{mnk}}\right|<\infty$. by (1). Thus

$(\varphi_{rst}a_{mnk})\in\left[\Gamma_{f\mu}^{3\Delta^m q},\|\mu_{mnk}(x),(d(x_1,0),d(x_2,0),\cdots,d(x_{n-1},0))\|_p^{\varphi}\right]_{\theta_{rst}}^{I^3}=$ and hence $\left[\Gamma_{g}^{3\Delta^m q\mu},\|\mu_{mnk}(x),(d(x_1,0),d(x_2,0),\cdots,d(x_{n-1},0))\|_p^{\varphi}\right]_{\theta_{rst}}^{I^3}$

$(a_{mnk})\in\left[\Gamma_{g}^{3\Delta^m q\mu},\|\mu_{mnk}(x),(d(x_1,0),d(x_2,0),\cdots,d(x_{n-1},0))\|_p^{\varphi}\right]_{\theta_{rst}}^{I^3}$. This gives that

$$\left[\Gamma_{f\mu}^{3\Delta^m q},\|\mu_{mnk}(x),(d(x_1,0),d(x_2,0),\cdots,d(x_{n-1},0))\|_p^{\varphi}\right]_{\theta_{rst}}^{I^{3\alpha}}$$

$$\subset\left[\Gamma_{g}^{3\Delta^m q\mu},\|\mu_{mnk}(x),(d(x_1,0),d(x_2,0),\cdots,d(x_{n-1},0))\|_p^{\varphi}\right]_{\theta_{rs}}^{I^3}\cdots\cdots\cdots(2)$$

we are granted with (1) and (2)

$$\left[\Gamma_{f\mu}^{3\Delta^m q},\|\mu_{mnk}(x),(d(x_1,0),d(x_2,0),\cdots,d(x_{n-1},0))\|_p^{\varphi}\right]_{\theta_{rst}}^{I^{3\alpha}}=$$

$$\left[\Gamma_{g}^{3\Delta^m q\mu},\|\mu_{mnk}(x),(d(x_1,0),d(x_2,0),\cdots,d(x_{n-1},0))\|_p^{\varphi}\right]_{\theta_{rst}}^{I^3}$$

(ii)Similarly,onecanprovethat

$$\left[\Gamma_{g}^{3\Delta^m q\mu},\|\mu_{mnk}(x),(d(x_1,0),d(x_2,0),\cdots,d(x_{n-1},0))\|_p^{\varphi}\right]_{\theta_{rst}}^{I^{3\alpha}}$$

$$\subset\left[\Gamma_{f\mu}^{3\Delta^m q},\|\mu_{mnk}(x),(d(x_1,0),d(x_2,0),\cdots,d(x_{n-1},0))\|_p^{\varphi}\right]_{\theta_{rst}}^{I^3}$$

if the Musielak Orlicz function $(g_{mk})$ satisfies $\Delta_2$-condition.

## 1. Proposition

The sequence space

$\left[\Gamma_{f\mu}^{3\Delta^m q},\|\mu_{mnk}(x),(d(x_1,0),d(x_2,0),\cdots,d(x_{n-1},0))\|_p^{\varphi}\right]_{\theta_{rst}}^{I^3}$ is not solid

**Proof:** The result follows from the following example.

**Example:** Consider

$$\Delta^m x=(\Delta^m x_{mnk})=\begin{bmatrix}1 & 1 & \cdots & 1\\ 1 & 1 & \cdots & 1\\ \cdot & & & \\ \cdot & & & \\ \cdot & & & \end{bmatrix}\in\left[\Gamma_{f\mu}^{3\Delta^m q},\|\mu_{mnk}(x),(d(x_1,0),d(x_2,0),\cdots,d(x_{n-1},0))\|_p^{\varphi}\right]_{\theta_{rst}}^{I^3}.$$ Let

$$\Delta^m\alpha_{mnk}=\begin{bmatrix}-1^{m+n+k} & -1^{m+n+k} & \cdots & -1^{m+n+k}\\ -1^{m+n+k} & -1^{m+n+k} & \cdots & -1^{m+n+k}\\ \cdot & & & \\ \cdot & & & \\ \cdot & & & \end{bmatrix},$$ for all $m,n,k\in\mathbb{N}$

Then $\Delta^m\alpha_{mnk}x_{mnk}\notin\left[\Gamma_{f\mu}^{3q},\|\mu_{mnk}(x),(d(x_1,0),d(x_2,0),\cdots,d(x_{n-1},0))\|_p^{\varphi}\right]_{\theta_{rst}}^{I^3}$. Hence

$\left[\Gamma_{f\mu}^{3\Delta^m q},\|\mu_{mnk}(x),(d(x_1,0),d(x_2,0),\cdots,d(x_{n-1},0))\|_p^{\varphi}\right]_{\theta_{rst}}^{I^3}$ is not solid.

## 2. Proposition

The sequence space

$$\left[\Gamma_{f\mu}^{3\Delta^m q},\|\mu_{mnk}(x),(d(x_1,0),d(x_2,0),\cdots,d(x_{n-1},0))\|_p^{\varphi}\right]_{\theta_{rst}}^{I^3}$$

is not monotone.

**Proof:** The proof follows from Proposition 3.4.

## 3. Proposition

The sequence space

$\left[\Lambda_{f\mu}^{3\Delta^m q},\|\mu_{mnk}(x),(d(x_1,0),d(x_2,0),\cdots,d(x_{n-1},0))\|_p^{\varphi}\right]_{\theta_{rst}}^{I^3}$ is not solid.

## 4. Proposition

The sequence space

$$\left[\Lambda_{f\mu}^{3\Delta^m q},\|\mu_{mnk}(x),(d(x_1,0),d(x_2,0),\cdots,d(x_{n-1},0))\|_p^{\varphi}\right]_{\theta_{rst}}^{I^3}$$

is not monotone.

# Conclusion

Through this paper we studied some topological properties and inclusion relation with respect to a sequence of Musielak-Orlicz function.

### References

1. Sahiner A, Gurdal M, Duden FK (2007) Triple sequences and their statistical convergence. Selcuk J Appl Math 8: 49-55.
2. Sahiner A, Tripathy BC (2008) Some *I* related properties of triple sequences. Selcuk J Appl Math 9: 9-18.
3. Esi A (2014) On some triple almost lacunary sequence spaces defined by Orlicz functions. Research and Reviews:Discrete Mathematical Structures 1: 16-25.
4. Esi A, Catalbas MN (2014) Almost convergence of triple sequences. Global Journal of Mathematical Analysis 2: 6-10.
5. Esi A, Savas E (2015) On lacunary statistically convergent triple sequences in probabilistic normed space. Appl Mathand Inf Sci 9: 2529-2534
6. Datta AJ, Esi A, Tripathy BC (2013) Statistically convergent triple sequence spaces defined by Orlicz function. Journal of Mathematical Analysis 4: 16-22.
7. Subramanian N, Esi A (2015) The generalized tripled difference of $\chi^3$ sequence spaces. Global Journal of Mathematical Analysis 3: 54-60.
8. Debnath S, Sarma Das BC (2015) Some generalized triple sequence spaces of real numbers. Journal of nonlinear analysis and optimization. 6: 71-79.
9. Kizmaz H (1981) On certain sequence spaces. Canadian Mathematical Bulletin 24: 169-176.
10. Kamthan PK, Gupta M (1981) Sequence spaces and series. Pure and Applied Mathematics, USA.
11. Lindenstrauss J, Tzafriri L (1971) On Orlicz sequence spaces. Israel J Math 10: 379-390.
12. Musielak J (1983) Orlicz Spaces and Modular Spaces.Springer
13. Subramanian N, Murugesan C(2016) The entire sequence over Musielak *p*-metric space. Journal of the Egyptian Mathematical Society 24: 233-238.

# Permissions

All chapters in this book were first published in JPM, by OMICS International; hereby published with permission under the Creative Commons Attribution License or equivalent. Every chapter published in this book has been scrutinized by our experts. Their significance has been extensively debated. The topics covered herein carry significant findings which will fuel the growth of the discipline. They may even be implemented as practical applications or may be referred to as a beginning point for another development.

The contributors of this book come from diverse backgrounds, making this book a truly international effort. This book will bring forth new frontiers with its revolutionizing research information and detailed analysis of the nascent developments around the world.

We would like to thank all the contributing authors for lending their expertise to make the book truly unique. They have played a crucial role in the development of this book. Without their invaluable contributions this book wouldn't have been possible. They have made vital efforts to compile up to date information on the varied aspects of this subject to make this book a valuable addition to the collection of many professionals and students.

This book was conceptualized with the vision of imparting up-to-date information and advanced data in this field. To ensure the same, a matchless editorial board was set up. Every individual on the board went through rigorous rounds of assessment to prove their worth. After which they invested a large part of their time researching and compiling the most relevant data for our readers.

The editorial board has been involved in producing this book since its inception. They have spent rigorous hours researching and exploring the diverse topics which have resulted in the successful publishing of this book. They have passed on their knowledge of decades through this book. To expedite this challenging task, the publisher supported the team at every step. A small team of assistant editors was also appointed to further simplify the editing procedure and attain best results for the readers.

Apart from the editorial board, the designing team has also invested a significant amount of their time in understanding the subject and creating the most relevant covers. They scrutinized every image to scout for the most suitable representation of the subject and create an appropriate cover for the book.

The publishing team has been an ardent support to the editorial, designing and production team. Their endless efforts to recruit the best for this project, has resulted in the accomplishment of this book. They are a veteran in the field of academics and their pool of knowledge is as vast as their experience in printing. Their expertise and guidance has proved useful at every step. Their uncompromising quality standards have made this book an exceptional effort. Their encouragement from time to time has been an inspiration for everyone.

The publisher and the editorial board hope that this book will prove to be a valuable piece of knowledge for researchers, students, practitioners and scholars across the globe.

# List of Contributors

**Djeghloul N and Tahiri M**
Laboratory of Theoretical Physics of Oran (LPTO), University of Oran, Algeria

**Rodrigues FG and Capelas de Oliveira E**
Department of Applied Mathematics, IMECC - UNICAMP, Brazil

**Abbas MI**
Physics Department, Faculty of Science, Alexandria University, Alexandria, Egypt

**Ibrahim OA**
Physics Department, Faculty of Science, Beirut Arab University, Beirut, Lebanon

**Anguraj A and Banupriya K**
P.S.G. College of Arts and Science, Coimbatore, Tamil Nadu, India

**Palraj Jothiappan**
Department of Mathematics, PSG College of Arts and Science, Coimbatore, Tamil Nadu, India

**Fethi Soltani**
Department of Mathematics, Faculty of Science, Jazan University, Saudi Arabia

**Deyssenroth H**
Senior Researcher, Germany

**Eugene Kreymer**
Institute for Physics and Engineering, Donetsk, 83114, Ukraine

**Traino AC and Piccinno M**
Unit of Medical Physics, University Hospital Pisana, Italy

**Boni G**
Unit of Nuclear Medicine, University Hospital Pisana, Italy

**Bargellini I and Bozzi E**
S.D Radiologia Vascular and Interventional, University Hospital Pisana, Italy

**Pestov IB**
Bogoliubov Laboratory of Theoretical Physics, Joint Institute for Nuclear Research, 141980, Dubna, Moscow Region, Russia

**Akhmet'ev PM**
Professor in IZMIRAN, Troitsk, Moscow region, Russia

**Kumar VR and Raju MC**
Department of Mathematics, Annamacharya Institute of Technology and Sciences, India

**Raju GSS**
Department of Mathematics, JNTUA College of Engineering, India

**Varma SVK**
Department of Mathematics, S.V. University, India

**Su LD**
North-Eastern Federal University, Belinskogo, Yakutsk, Russia
Department of Mathematics, Linyi University, Linyi, P.R. China

**Jiang ZW and Jiang TS**
Department of Mathematics, Linyi University, Linyi, P.R. China

**Ming Bao Yu**
407 Oak Tree Square, Athens, Georgia 30606, USA

**Hassan HK and Stepanyants YA**
University of Southern Queensland, Toowoomba, Australia

**Zied Driss, Tarek Chelbi, Walid Barhoumi and Mohamed Salah Abid**
Laboratory of Electro-Mechanic Systems (LASEM), National School of Engineers of Sfax (ENIS), University of Sfax, B.P. 1173, Road Soukra 3038, Sfax, Tunisia, Africa

**Ahmed Kaffel**
University of Maryland College Park, MD 20742, USA

**Hajji MA**
Department of Mathematical Sciences, United Arab Emirates University, United Arab Emirates

**Lawrence M**
Maldwyn Centre for Theoretical Physics, Cranfield Park, Burstall, Suffolk, UK

**Ferdows M**
Department of Applied Mathematics, University of Dhaka, Bangladesh

**Liu D**
Department of Engineering and Science, Louisiana Tech University, USA

**Sladkov P**
Independent Researcher, Russia

**Makanae M**
Researcher, Representative Free Web College, Japan

**Abbas MI**
Department of Physics, Faculty of Science, Alexandria University, Egypt

**Ibrahim OA**
Department of Physics, Faculty of Science, Beirut Arab University, Lebanon

**Ibrahim T and Sakr M**
Department of Physics, Faculty of Science, Alexandria University, Egypt
Department of Physics, Faculty of Science, Beirut Arab University, Lebanon

**Mazurkin PM**
Doctor of Engineering, Academician of Russian Academy of Natural History and Russian Academy of Natural Sciences, Volga State University of Technology, Russia

**Kthiri H**
Department of Mathematics, University of Sfax, Tunisa

**Mazurkin PM**
Doctor of Engineering, Academician of Russian Academy of Natural History and Russian Academy of Natural Sciences, Volga State University of Technology, Russia

**Baixauli JG**
Independent Researcher, Spain

**Alexiou M**
Department of Physics, National Technical University of Athens, Greece

**Stavrinos PC**
Department of Mathematics, University of Athens, Greece

**Vacaru SI**
Rectors Department,University Al. I. Cuza, Greece

**Jeffrey R Schmidt and Dileep Karanth**
Department of Physics, University of Wisconsin-Parkside, USA

**Mishra LN**
Department of Mathematics, National Institute of Technology, Silchar - 788 010, India

**Deepmala**
SQC & OR Unit, Indian Statistical Institute, Kolkata-700 108, India

**Subramanian N**
Department of Mathematics, Sastra University, Thanjavur-613 401, India

# Index

www.ingramcontent.com/pod-product-compliance
Lightning Source LLC
Chambersburg PA
CBHW080638200326
41458CB00013B/4668
* 9 7 8 1 6 3 2 3 8 6 4 6 5 *